Nano- and Biocatalysts for Biodiesel Production

Nano- and Biocatalysts for Biodiesel Production

Edited by

Avinash P. Ingle
Biotechnology Centre
Department of Agricultural Botany
Dr. Panjabrao Deshmukh Krishi Vidyapeeth
Akola, Maharashtra
India

WILEY

This edition first published 2021

Registered Offices
John Wiley & Sons, Inc., 111 River Street, Hoboken, NJ 07030, USA
John Wiley & Sons Ltd, The Atrium, Southern Gate, Chichester, West Sussex, PO19 8SQ, UK

Editorial Office
The Atrium, Southern Gate, Chichester, West Sussex, PO19 8SQ, UK

For details of our global editorial offices, customer services, and more information about Wiley products visit us at www.wiley.com.

Wiley also publishes its books in a variety of electronic formats and by print-on-demand. Some content that appears in standard print versions of this book may not be available in other formats.

For general information on our other products and services or for technical support, please contact our Customer Care Department within the United States at (800) 762-2974, outside the United States at (317) 572-3993 or fax (317) 572-4002.
Wiley also publishes its books in a variety of electronic formats. Some content that appears in print may not be available in electronic formats. For more information about Wiley products, visit our web site at ww.wiley.com.

Library of Congress Cataloging-in-Publication Data is applied for
ISBN: 978-1-119-73000-2

Cover Design: Wiley
Cover Image: Gas pump nozzle © Corona Borealis Studio/Shutterstock, Olive oil © Valentyn Volkov/Shutterstock, Buckyball © maggio07/Getty Images, NpmA methyltransferase © LAGUNA DESIGN/Science Photo Library RF/Getty Images

Set in 9.5/12.5pt STIXTwoText by Straive, Chennai, India
Printed and bound by CPI Group (UK) Ltd, Croydon, CR0 4YY

C9781119730002_080621

Contents

Preface

The ever increasing population and global industrialization considerably increase the energy demand. The petroleum-based fuels (fossil fuels), coal, natural gas, nuclear, and hydropower are the major energy sources currently available for use. However, the utilization of these energies exerts several negative impacts on the environment including issues like global warming, greenhouse gas emission, depletion of fossil fuel reserves, etc. Environmental pollution due to release of various particulate matters and contaminants exerts hazardous effects on human health. Therefore, it is an urgent need to search renewable and sustainable alternative energy sources having novel features like biodegradability, ecofriendly nature, low toxicity, and economical viability. These features are usually possessed by biofuels like biodiesel. Since last few decades, biodiesel has attracted a great deal of attention from scientific as well as a political community due to their many advantages over petroleum diesel like a significant reduction in greenhouse gas emissions, non-sulfur emissions, non-particulate matter pollutants, possess very low toxicity, biodegradable nature, and renewability.

Although technologies for industrial production of biodiesel are already developed, these conventional technologies reported to have some drawbacks because those are energy and labor-intensive, expensive, time consuming, required high amount of water, etc. In this context, scientists are looking towards nanotechnology as a new hope because it has potential to revolutionize different areas of research as well as life. Recent studies, already confirmed that nanomaterials having a size in the range of 1-100 nm play a crucial role in different processes (i.e. esterification and transesterification) used for biodiesel production. Nanomaterials exhibit novel and outstanding properties including strong catalytic activity due to its minute size. To date, various nanomaterials have been investigated and employed as nano and nanobiocatalysts in enhanced biodiesel production from various renewable sources like vegetable oils, microbial oils, waste cooking oils, animal fats, waste materials, etc. Considering these facts, the editor attempted to discuss the recent advances and role of different nanocatalysts, biocatalysts and nanobiocatalysts in biodiesel production through this book.

In this book, there are total 13 chapters, which are broadly focused on the recent advances and the role of different nano and biocatalysts for the production of biodiesel through esterification and transesterification. Chapter 1 is an introductory chapter, it mainly focused on important feedstocks used for the production of biodiesel. In addition, it also emphasized on various conventional methods commonly employed and different factors affecting the industrial production of biodiesel. Chapter 2 is based on role of different nano(bio)catalysts in the production of biodiesel using waste cooking oil. It also highlighted on the issues of waste cooking oil disposal and its management through their use as important feedstock for biodiesel production. Authors also described the processes for the development of nano(bio)catalysts via different enzyme immobilization techniques. In Chapter 3 authors reviewed the role of different bio/nano structured heterogeneous catalysts in biodiesel production. Chapter 4 specifically focused on the role of calcium oxide nanocatalysts in the production of biodiesel because it is revealed that among different nanocatalysts, metal oxide-based nanocatalysts showed better catalytic performances as far as transesterification approach is concerned. In the similar line, Chapter 5 discusses the role of titanium di-oxide nanocatalysts in biodiesel production. In addition, authors also described the different precursors commonly employed in synthesis of titanium di-oxide nanoparticles and different synthesis methods. Chapter 6 deals with recent advances in the synthesis of zinc-based nanocatalysts and their effective applications in biodiesel production. Moreover, this chapter briefly highlighted about the different common feedstocks used for the production of biodiesel and conventional approaches routinely in practice for industrial production of biodiesel. Chapter 7 examines the applications of another important kind of nanocatalysts i.e. carbon-based heterogeneous nanocatalysts for the production of biodiesel. Chapter 8 emphasizes on one of the most important category of nanocatalysts (functionalized magnetic nanocatalysts) in biodiesel production. In recent past, scientific community working in this fields focusing on the use of magnetic nanocatalysts for rapid and economically viable production of biodiesel. The main advantage of using functionalized magnetic nanocatalysts is they can be easily recovered due to their strong magnetic nature and reuse for multiple times making the down-stream processing easy and cost-effective. Chapter 9 is about the application of bio-based nanocatalysts for biodiesel production. Chapter 10 is dedicated to the use of various heterogeneous nanocatalysts in the conversion of waste to biodiesel. It discussed about the effectiveness of heterogeneous nanocatalysts over other conventional catalysts. Chapter 11 focused on application of rare-earth cation exchanged nanozeolite as a support for the immobilization of fungal lipase and their use in biodiesel production. Nanozeolites are considered to be one of the most suitable support for the immobilization of enzymes commonly employed in biodiesel production because these kind of supports provides high stability to enzymes and protects it from inactivations. Chapter 12 emphasizes on application of lipase immobilized magnetic nanoparticles as promising nanobiocatalysts in biodiesel production. This chapters briefly discussed about various techniques used for enzyme immobilization of lipase on magnetic nanoparticles and their advantages over other catalysts as far as biodiesel production in concerned. Final chapter 13 is focused on the most important and relevant aspects i.e. techno-economical analysis of biodiesel production using different feedstocks. The techno-economical analysis of any products is of utmost important from industrial point of view. This is the only analysis which help to run the industry smoothly.

Overall, this book covers very informative chapters written by one or more specialists, experts in the concerned topic. In this way, I would like to offer a very rich guide for researchers in this field, undergraduate or graduate students of various disciplines like biotechnology, nanotechnology, biofuel sectors, biorefining fields, etc. and allied subjects.

In addition, this book is useful for people working in various biorefining industries, regulatory bodies, and energy related organizations.

I would like to thank all the contributors for their outstanding efforts to provide state-of-the-art information on the subject matter of their respective chapters. Their efforts will certainly enhance and update the knowledge of the readers about the role of nanotechnology in general and nano(bio)catalysts in particular for biodiesel production. I also thank everyone in the Wiley team for their constant help and constructive suggestions particularly to Higginbotham Sarah (Senior Editor), Stefanie, Nivetha, and other team members. I am also thankful to Science and Engineering Research Board (SERB), Department of Science and Technology, Government of India, New Delhi for providing financial assistance in the form of "Ramanujan Fellowship".

I hope that the book will be useful for all the readers to find the required information on the latest research and advances in the field of biorefinery and biofuel industries.

Avinash P. Ingle

List of Contributors

Mary Adejoke Ajala
Department of Chemical Engineering
University of Ilorin
Ilorin, Kwara State
Nigeria

Elijah Olawale Ajala
Department of Chemical Engineering
University of Ilorin
Ilorin, Kwara State
Nigeria

Shehu-Ibrahim Akinfalabi
Institute of Advanced Technology
University Putra Malaysia
Serdang
Malaysia

Rahul Bhagat
Department of Biotechnology
Government Institute of Science
Aurangabad, Maharashtra
India

Nilutpal Bhuyan
Department of Energy
Tezpur University
Assam
India

Neelam Bora
Department of Energy
Tezpur University
Assam
India

Manash J. Borah
Department of Energy
Tezpur University
Assam
India

Dhanapati Deka
Department of Energy
Tezpur University
Assam
India

Fábio Rogério de Moraes
Physics Department
São Paulo State University - UNESP
São Paulo
Brazil

Adriano de Vasconcellos
Physics Department
São Paulo State University - UNESP
São Paulo
Brazil

Hossein Esmaeili
Department of Chemical Engineering
Bushehr Branch
Islamic Azad University
Bushehr
Iran

Rushikesh Fopase
Bio-Interface & Environmental
Engineering Lab, Department of
Biosciences and Bioengineering
Indian Institute of Technology Guwahati
Guwahati
India

Shemelis Nigatu Gebremariam
Hawassa University
Wondo Genet College of Forestry and
Natural Resources
Shashemene
Ethiopia

João Gomes
CERENA – Center for Natural Resources
and Instituto Superior Técnico
Lisbon University
Lisbon
Portugal

Chemical Engineering Department
Instituto Superior de Engenharia de Lisboa
Lisbon Polytechnic, Lisbon
Portugal

Guilherme de Paula Guarnieri
Physics Department
São Paulo State University – UNESP
São Paulo
Brazil

Indarchand Gupta
Department of Biotechnology
Government Institute of Science
Aurangabad, Maharashtra
India

Naeemah A. Ibrahim
Institute of Advanced Technology
University Putra Malaysia
Serdang
Malaysia

Avinash P. Ingle
Biotechnology Centre
Department of Agricultural Botany
Dr. Panjabrao Deshmukh Krishi
Vidyapeeth
Akola
Maharashtra
India

Rupam Kataki
Department of Energy
Tezpur University
Assam
India

Jorge Marchetti
Faculty of Sciences and Technology
Norwegian University of Life Sciences
Ås, Norway

Archit Mohapatra
Gujarat Biotechnology Research Centre
Gandhinagar, Gujarat
India

Hamid Mukhtar
Institute of Industrial Biotechnology
Government College University
Lahore
Pakistan

Muhammad Waseem Mumtaz
Department of Chemistry
University of Gujrat
Gujrat
Pakistan

José Geraldo Nery
Physics Department
São Paulo State University – UNESP
São Paulo
Brazil

Chawalit Ngamcharussrivichai
Center of Excellence in Catalysis for Bioenergy and Renewable Chemicals (CBRC), Faculty of Science
Chulalongkorn University
Pathumwan
Thailand

Center of Excellence on Petrochemical and Materials Technology (PETROMAT)
Chulalongkorn University
Pathumwan
Thailand

Harris Panakkal
Department of Biotechnology
Government Institute of Science
Aurangabad, Maharashtra
India

Lalit M. Pandey
Bio-Interface & Environmental Engineering Lab, Department of Biosciences and Bioengineering
Indian Institute of Technology Guwahati
Guwahati
India

Priti R. Pandit
Gujarat Biotechnology Research Centre
Gandhinagar, Gujarat
India

Jaime Puna
CERENA – Center for Natural Resources and Instituto Superior Técnico
Lisbon University
Lisbon
Portugal

Chemical Engineering Department
Instituto Superior de Engenharia de Lisboa
Lisbon Polytechnic, Lisbon
Portugal

Kalyani Rajkumari
Department of Chemistry
National Institute of Technology Silchar
India

Department of Chemistry
C.V. Raman Global University
Bhubaneswar
India

Umer Rashid
Institute of Advanced Technology
University Putra Malaysia
Serdang
Malaysia

Lalthazuala Rokhum
Department of Chemistry
National Institute of Technology Silchar
India

Department of Chemistry
University of Cambridge
Cambridge, UK

Dipanka Saikia
Department of Energy
Tezpur University
Assam
India

Harvis Bamidele Saka
Department of Chemical Engineering
University of Ilorin
Ilorin, Kwara State
Nigeria

Samuel Santos
CERENA – Center for Natural Resources and
Instituto Superior Técnico, Lisbon University
Lisbon
Portugal

Swati Sharma
Bio-Interface & Environmental Engineering Lab, Department of Biosciences and Bioengineering
Indian Institute of Technology Guwahati
Guwahati
India

Sajad Tamjidi
Department of Chemical Engineering
Bushehr Branch
Islamic Azad University
Bushehr
Iran

Tooba Touqeer
Department of Chemistry
University of Gujrat
Gujrat
Pakistan

1

Biodiesel: Different Feedstocks, Conventional Methods, and Factors Affecting its Production

Hossein Esmaeili[1] *and Sajad Tamjidi*[2]

[1] *Department of Chemical Engineering, Bushehr Branch, Islamic Azad University, Bushehr, Iran*
[1] *Department of Chemical Engineering, Shiraz Branch, Islamic Azad University, Shiraz, Iran*

1.1 Introduction

Fossil fuels are a non-renewable source of energy whose reserves are limited, and they take millions of years to develop (Banković–Ilić et al. 2014). The widespread use of petroleum derivatives in recent decades has led to energy crisis, global climate change, environmental pollution, and many medical problems, such as cardiovascular diseases and cancers (Dhiraj and Mangesh 2012). Collectively, all these concerns, along with others like global warming and greenhouse gas emissions, have spurred the search for alternative fuels (e.g., biohydrogen, biodiesel, bioethanol, biomethanol, biogas, natural gas, and bioelectricity) that have relatively less adverse impacts on and greater compatibility with the environment (Demirbas 2004; Nascimento et al. 2011; Fahd et al. 2014). According to a report presented by the International Energy Agency (IEA), by 2035, world energy consumption will increase by 33% (International Energy Agency 2013), and it is anticipated that 40% of the growth will come from renewable sources. Among the renewable energies, biodiesel has recently experienced significant developments thanks to its outstanding advantages, including higher cetane number (CN), nontoxicity, and higher flashpoint compared with fossil fuels (Liu et al. 2010; Hasheminejad et al. 2011). Biodiesel is a biofuel with properties closely mimicking those of diesel, but without unfavorable contents such as nitrogen, sulfur, and polycyclic aromatics. This renewable biofuel is a mono-alkyl esters of long-chain fatty acids that is produced from vegetable oil, waste edible oil (WEO), waste cooking oil (WCO), waste non-edible oil, animal fats, and microorganisms such as algae, fungi, and bacteria (Kralova and Sjoblom 2010; Nabi and Hoque 2008).

There are four primary methods of biodiesel production: blending, microemulsion, thermal cracking (pyrolysis), and transesterification. Among these, the transesterification reaction is the most commonly used for the conversion of oils into biodiesel, because the fuel produced by this method has been found to be highly compatible with conventional diesel engines. Direct use of the vegetable oil-derived biodiesel damages such engines due to the high viscosity of the oil (Ramli et al. 2017). The most common short-chain alcohols used for this purpose include methanol and ethanol; thanks to its lower price, methanol

Nano- and Biocatalysts for Biodiesel Production, First Edition. Edited by Avinash P. Ingle.

is the economic alcohol of choice (Ramadhas et al. 2005). Also, removal and recovery of methanol from the final product (biodiesel) is easier than for other alcohols (Allah and Alexandru 2016).

There are three kinds of catalysts commonly used for this process: alkaline, acidic, and enzymatic. Acidic and alkaline catalysts come in both homogeneous and heterogeneous types. Homogeneous alkaline catalysts (HACs) like NaOH and KOH have some disadvantages, including corrosion problems, nonrecyclability, and the production of a large amount of waste, while heterogeneous catalysts have several advantages, including appropriate recyclability, no requirement for a washing step, and higher efficiency in biodiesel production compared to homogeneous ones. However, enzymatic catalysts also have several limitations, including a low reaction rate, high cost of the enzymes, and deactivation of the catalyst, especially when used in industry (Lin et al. 2011; Talha and Sulaiman 2016). Given these limitations, the use of nanocatalysts for biodiesel production has been extensively increased in recent years. Nanocatalysts have several outstanding advantages, including reusability, high catalytic activity, high surface area, and high efficiency (Ambat et al. 2019; Seffati et al. 2020). Apart from the catalyst, there are several other factors that affect the efficiency of biodiesel production. These mainly include reaction time, reaction temperature, alcohol to oil molar ratio, type and concentration of catalyst, stirring rate, and feedstock (oil) used (Verma and Sharma 2016).

Biodiesel is a long-chain fatty-acid methyl ester (FAME) derived by the reaction between alcohol, oil, and an appropriate catalyst (Pasupulety et al. 2013). In the conventional process, the oil reacts with the alcohol (methanol, ethanol, propanol, or butanol) in the presence of a catalyst (alkaline, acidic, or enzymatic catalyst) to produce a FAME (biodiesel) as the main product and glycerin as a byproduct (Sundus et al. 2017). According to the United States Environmental Protection Agency (US EPA), the use of biodiesel in a vehicle's engine decreases the emission of hydrocarbon (HC) (about 70%) and of carbon monoxide and particulate matter (50%) compared to diesel fuel, but increases that of NO_x (about 10%) (Geller and Goodrum 2004). Phosphorus is another hazardous gas present in diesel that can harm the catalytic part of the control system in the vehicle engine. Therefore, the concentration of phosphorus in the oil must be controlled in order to protect the system. Also, the presence of sulfur can damage the catalytic converter and emission control system. At present, the sulfur content in commercial biodiesel is nearly zero, which is one of its main advantages compared to petro-diesel (Chen et al. 2018). Moreover, the presence of water in the oil causes hydrolysis of triglyceride to free fatty acid (FFA) and therefore leads to soap formation. If water concentration in the oil is more than 0.05 wt%, water must be removed (Chen et al. 2018). Deposition of metals such as calcium, magnesium, sodium, and potassium can further cause a lot of problems in vehicle engines (Balasubramaniyan 2016).

The aim of this chapter is to discuss the various feedstocks (i.e. oil sources) available (e.g. vegetable oil, microalga oil, animal fat, and waste oil) and their conversion in biodiesel using different conventional heterogeneous catalysts. In addition, recent advances and the application and impact of heterogeneous nanocatalysts on biodiesel production are briefly discussed, and the impact of various reaction parameters such as temperature, reaction time, catalyst content, and alcohol to oil ratio on the transesterification reaction is described. Special focus is given to the physical properties of biodiesel, including pour point, flash point, kinematic viscosity, CN, density, acid number, cloud point, and K + Na + Mg + Ca concentration, and their comparison with international standards.

1.2 Different Feedstocks for Biodiesel Production

More than 70% of the cost of biodiesel production is related to the raw materials. Biodiesel and petro-diesel cost about 3.03 and 2.46 US$/gal, respectively (Chen et al. 2018). In the literature, a number of different feedstocks are commonly used for biodiesel production, including vegetable oils, WEOs, animal fats, and microalga oil (Channi et al. 2016).

1.2.1 Vegetable Sources

The vegetable oils used to produce biodiesel can be either edible or non-edible. More than 95% of the biodiesel produced in the world is produced from edible oil, which is easily obtained from the agricultural industries. However, large-scale production of edible oil-derived biodiesel may have a negative effect on human life, because it leads to reduction of food supply (Gui et al. 2008). The most common edible oil sources used for biodiesel production include sunflower oil (Visser et al. 2011), rapeseed oil, peanut oil, sesame oil, rice oil, coconut oil (Karmakar et al. 2010), soybean oil, corn oil (Alptekin and Canakci 2008), and hazelnut oil (Sanli and Canakci 2008).

In contrast to edible oil, non-edible oils can't be consumed by humans, because of the presence of toxic compounds in these sources (Gui et al. 2008). Common examples are palm oil (Salamatinia et al. 2013), jatropha oil (Pramanik 2003), cottonseed oil (Nabi et al. 2009), castor oil (Visser et al. 2011), *Moringa oleifera* seed oil (Ogbunugafor et al. 2011), neem oil, jojoba oil, and sea mango (Gui et al. 2008). Castor oil may be the best option for biodiesel production because it does not require heat and energy, which are necessary for other sources of vegetable oil (Manan 2013).

Currently, edible oils are in use in several countries, leading to increases in their cost, and hence the cost of the biodiesel produced. In this context, it is economically more efficient to use non-edible oils (Karmakar et al. 2010). Table 1.1 shows the biodiesel conversion yield (BCY) of different vegetable oils used in the presence of HACs, such as KOH and NaOH. The yield of biodiesel production is calculated through Eq. (1.1), proposed by Seffati et al. (2019):

$$\text{BCY} = \frac{\text{Weight of biodiesel produced}}{\text{Weight of initial oil}} \times 100 \tag{1.1}$$

1.2.2 Waste Oils

WEOs are oil-based substances containing animal or vegetable matter that can be used for the preparation of food or in cooking but are not suitable for consumption by human beings. The amount of WEO produced in any country around the world is large, varying depending on the quantity of edible oil consumed. More than 15 million tons of WEO are produced annually around the globe, mainly by countries like China (4.5 million tons), Malaysia (0.5 million tons), the United States (10 million tons), Taiwan (0.07 million tons), Canada (0.12 million tons), and Japan (0.45–0.57 million tons), as well as European nations (0.7–1 million tons) (Gui et al. 2008). This oil source can be converted to biodiesel via catalytic and non-catalytic reaction (supercritical transesterification process) (Gui et al. 2008). The disposal of WEOs and WCOs is a major problem as they can pollute the environment. Developed

Table 1.1 Biodiesel production from vegetable oils in the presence of HACs (NaOH and KOH).

Oil source	Catalyst type	BCY (%)	References
Jojoba oil	KOH	83.5	Bouaid et al. (2007)
Rapeseed oil	KOH	97	Jazie et al. (2012)
Rapeseed oil	NaOH	92	Jazie et al. (2012)
Peanut oil	KOH	95	Jazie et al. (2012)
Peanut oil	NaOH	88	Jazie et al. (2012)
Jatropha curcas oil	NaOH	98	Chitra et al. (2005)
Jatropha curcas oil	NaOH	90	Berchmans and Hirata (2008)
Jatropha curcas oil	KOH	95	Patil and Deng (2009)
Jatropha curcas oil	KOH	99	Syam et al. (2009)
Jatropha curcas oil	KOH	93	Sahoo and Das (2009)
Jatropha curcas oil	KOH	99	Tiwari et al. (2007)
Rapeseed oil	KOH	96	Rashid and Anwar (2008)
Sunflower oil	NaOH	97.1	Winayanuwattikun et al. (2008)
Peanut oil	NaOH	89	Winayanuwattikun et al. (2008)
Corn oil	KOH	96	Winayanuwattikun et al. (2008)
Camelina oil	KOH	97.9	Frohlich and Rice (2005)
Canola oil	KOH	95	Winayanuwattikun et al. (2008)
Cotton oil	NaOH	96.9	Winayanuwattikun et al. (2008)
Pumpkin oil	NaOH	97.5	Winayanuwattikun et al. (2008)
Jatropha curcas oil	NaOH	98	Winayanuwattikun et al. (2008)
Pongamia pinnata oil	KOH	98	Sahoo and Das (2009)

countries have adopted policies that penalize the disposal of WEOs or WCOs via water drainage. Biodiesel production may thus be the best approach to their disposal, being economically viable and efficient. Information on diesel demand and the availability of WCOs in different countries shows that WCO-derived biodiesel may not be sufficient to completely replace petro-diesel. However, a significant amount of biodiesel can be produced from WCOs, helping reduce dependency on oil-based fuel. The amount of WCO produced in any one country varies according to the utilization of vegetable oil (Kulkarni and Dalai 2006).

It is well investigated that WEO or WCO can be used as a low-cost feedstock for the production of biodiesel. However, due to the presence of particulate contaminants and impurities in WCOs, biodiesel produced from these sources shows relatively high values of pour point and cloud point. Therefore, pretreatment or modification of such oils is essential prior to their use for biodiesel production. This can be achieved using particular chemical processes (Ghanei et al. 2014). According to Allah and Alexandru (2016), the overall cost involved in the production of biodiesel using WCOs is comparatively less than that with vegetable oils and diesel fuel. Table 1.2 shows the BCY of biodiesel production from WCOs in the presence of different catalysts.

Table 1.2 Biodiesel production from WCOs in the presence of different catalysts.

Catalyst	Temp (°C)	Alcohol to oil ratio	Amount of catalyst (wt%)	Time (h)	BCY (%)	References
KOH	57.31	9.05	0.99	1.28	96.33	Dhingra et al. (2016)
CaO	50	8 : 1	1	1.5	96	Degfie et al. (2019)
Ba/CaO	65	6 : 1	3	3	88	Balakrishnan et al. (2013)
Copper/zinc oxide	55	8 : 1	12	0.833	97.71	Gurunathan and Ravi (2015)
4Mn–6Zr/CaO	80	15 : 1	3	3	92.1	Mansir et al. (2018)
MgO	65	24 : 1	2	1	93.3	Ashok et al. (2018)
Calcium diglyceroxide	60	9 : 1	1	0.5	93.5	Gupta et al. (2015)
KOH/clinoptilolite	65	2.25 : 1	8.1	0.223	97.45	Mohadesi et al. (2020)
SO_4/Fe-Al-TiO_2	90	10 : 1	3	2.5	96	Gardy et al. (2018)
Butyl-methyl imidazolium hydrogensulfate	160	15 : 1	5	1	95.65	Ullah et al. (2015)
Waste eggshell	65	9 : 1	5	2.75	87.8	Peng et al. (2018)
NaOH	69.37	16.7 : 1	4.571	7.08	94.6	Leung and Guo (2006)
KOH	94	1.5	1	6 : 1	60	Foroutan et al. (2018a)
NaOH	85	1.5	1	6 : 1	60	Foroutan et al. (2018b)
NaOH	88.8	0.33	1.1	7 : 1	60	Lam et al. (2010)
KOH	87	2	6	9 : 1	87	Lam et al. (2010)
H_2SO_4	99	4	41.8	245 : 1	70	Lam et al. (2010)

1.2.3 Animal Fats

The oils obtained from animal fats are another kind of non-edible oil used for biodiesel production. The important animal fats that are used as sources of these oils include chicken fat (Seffati et al. 2019), goat fat (Chakraborty and Sahu 2014), duck fat (Liu and Wang 2013), mutton fat (Mutreja et al. 2011), and lamb, cow, and pork fats (Banković–Ilić et al. 2014). Among these, goat and mutton fat are the most preferentially used feedstocks for biodiesel production. The total population of goats in the world is around 861.9 million, of which 514.4 million are in Asia (Hassan et al. 2016). The global sheep population is 1078.2 million and that in Asia is 452.3 million, which is about 42% of the world total (Aziz 2010). Therefore, these two sources of feedstocks can be used as an important oil source for biodiesel production in Asia. The final cost of biodiesel produced from raw oils (e.g. vegetable oils) is comparatively higher than that of petroleum-derived fuels because the cost of feedstock (i.e. the oil) represents about 70–90% of the total expense of biodiesel production (Zhang et al. 2003; Dorado et al. 2006). However, the cost of biodiesel production can be decreased by using non-edible oils such as waste oils and animal fats. Figure 1.1 shows the shares of

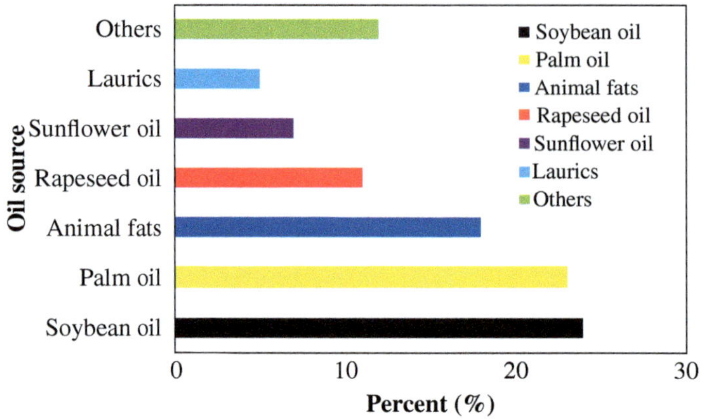

Figure 1.1 Percentages of various oil sources used in biodiesel production.

Table 1.3 Biodiesel production from animal fats in the presence of different catalysts.

Oil source	Catalyst type	BCY (%)	References
Tallow oil	H_2SO_4	98.28	Bhatti et al. (2008)
Nile tilapia oil	KOH	98.2	Santos et al. (2010)
Poultry oil	H_2SO_4	99.72	Bhatti et al. (2008)
Beef tallow	KOH	90.8	Banković-Ilić et al. (2014)
Pork lard	KOH	91.4	Banković-Ilić et al. (2014)
Mutton fat	KOH	78.3	Banković-Ilić et al. (2014)
Chicken fat	H_2SO_4	99	Banković-Ilić et al. (2014)
Goat tallow	NaOH	96	Esmaeili and Foroutan (2018)
Goat tallow	KOH	98	Esmaeili and Foroutan (2018)
Goat fat	MgO	93.12	Rasouli and Esmaeili (2019)
Chicken fat	CaO	94.4	Keihani et al. (2018)
Chicken fat	CaO	75.4	Awaluddin et al. (2010)

different sources of oil in biodiesel production. As can be seen, the most common sources of biodiesel production are vegetable oils, such as soybean and palm oil, which are used in about 50% of all biodiesel produced globally, while less than 20% of biodiesel is produced by means of animal fats (Gnanaprakasam et al. 2013). Moreover, Table 1.3 shows the BCY of different animal fats in the presence of acidic and base catalysts.

1.2.4 Microalga Oil

The high oil contents present in microalgae make them a promising source of oil for biodiesel production. The cost of microalga oil is less than or equal to that of animal and vegetable oils (Chen et al. 2018). Microalgae seem to be the only biodiesel source with

Table 1.4 Biodiesel production from microalga oils in the presence of different catalysts.

Oil source	Catalyst type	Temp. (°C)	Time (h)	Catalyst conc. (wt%)	Alcohol to oil ratio	BCY (%)	References
Chlorella protothecoides	KOH	68	1.33	0.75	6 : 1	98.6	Yaşar and Altun (2018)
Spirulina maxima	KOH	65	0.33	0.75	9 : 1	86.1	Rahman et al. (2017)
Nannochloropsis sp.	$Ca(OCH_3)_2$	80	3	3	30 : 1	99	Teo et al. (2016)
Microalga oil	CaO	55	—	2	9 : 1	96.3	Siva and Marimuthu (2015)
Chlorella pyrenoidosa biomass	Sulfuric acid	120	3	—	—	92.5	Cao et al. (2013)
Neochloris oleoabundans	ultrasonic-assisted H_2SO_4	—	—	—	—	98	Singh et al. (2017)
Neochloris oleoabundans	H_2SO_4 and NaOH two-step transesterification process	65	1	10	—	91	Singh et al. (2017)
Microalga oil	K-Pumice	60	2	10	18 : 1	77	Cercado et al. (2017)
Chlorella sp.	H_2SO_4	23	8	115	—	79.9	Ehimen et al. (2010)
Microalga oil	NaOH	60	0.2	2	12 : 1	85	Cercado et al. (2018)
Microalga oil	KOH	60	0.2	3	12 : 1	85	Cercado et al. (2018)
Microalga oil	LiOH	60	0.2	5	12 : 1	55	Cercado et al. (2018)

the potential to entirely replace fossil diesel. They contain high amounts of oil that can be converted to biodiesel and have a higher capability to produce it compared to other feedstocks such as soybean, corn, canola, coconut, jatropha, and palm oil (Chisti 2007) Moreover, microalgae generate many different types of HCs, lipids, and other compounds that are required to produce biodiesel. They have large productivity and an affordable cost (Chisti 2007; Chen et al. 2018).

The capability of some feedstocks for biodiesel production has been studied, and the results show that microalga oil can replace other oil sources (Chen et al. 2018). *Chlorella protothecoides* (Chen et al. 2012), *Nannochloropsis oculate*, *Phaeodactylum tricornutum*, *Scenedesmus dimorphus* (Islam et al. 2013), *Chlorella emersonii*, *Chlorella salina*, and *Chlorella vulgaris* (Talebi et al. 2013) are some microalgae that show high capability for biodiesel production. Table 1.4 provides a comparison between different microalga oils.

1.3 Conventional Methods of Biodiesel Production

Direct use of animal fats and vegetable oils (edible and non-edible), as well as other sources for biodiesel production, is not practical because of the high viscosity and reactiveness of unsaturated HCs. Several methods have been proposed and routinely used to reduce the kinematic viscosity of oils in order to reach the required quality for use in diesel engines, including direct use and mixing, pyrolysis, microemulsion, and transesterification. Among them, the transesterification method is most commonly applied, due to its high BCY (Kirubakaran and Arul Mozhi Selvan 2018; Wahlund et al. 2004).

1.3.1 Microemulsion

According to the International Union of Pure and Applied Chemistry (IUPAC), a microemulsion is a thermodynamically stable system developed upon dispersion of water in oil or oil in water in the presence of surfactant particles, in which one phase is continuous and the other is dispersed in the continuous phase with a size between 0.001 and 0.15 μm (Esmaeili et al. 2014, 2018). The biodiesel produced by this method has a proper CN. The alcohols used are methanol, ethanol, and 1-butanol. The use of alcohol can reduce the fuel viscosity and improve separation of oils and alkyl nitrate; in addition, alcohol can also increase the CN of the fuel (Kirubakaran and Arul Mozhi Selvan 2018; Gebremariam and Marchetti 2017). The microemulsion method is simple and environmentally friendly, as it produces very small amounts of pollutants. However, it requires high temperatures and expensive devices, and the purity of the biodiesel produced is low (Lin et al. 2011).

1.3.2 Pyrolysis or Thermal Cracking

In pyrolysis, or thermal cracking, chemical changes are applied using heat in the presence of air or nitrogen (Yaman 2004). Several studies have been performed on the thermal cracking of oils to obtained diesel fuel as end product. Thermal decomposition of oil results in the production of several groups of components, including the alkanes, alkenes, alkadienes, carboxylic acids, and aromatics. Different types of vegetable and plant oils undergo different structural changes after thermal decomposition. The biodiesel produced exhibits a low viscosity and a high CN compared to vegetable oil. The process has some disadvantages, including low volatility, high viscosity, and stability against its own simplicity (Lin et al. 2011). In general, the biodiesel produced through microemulsion and thermal cracking methods demonstrates comparatively low CN, leading to incomplete combustion, which makes the approach nonconvenient (Lin et al. 2011; Abbaszaadeh et al. 2012).

1.3.3 Transesterification

Among all proposed methods, the transesterification reaction is the most reliable and effective for biodiesel production on both experimental and industrial scales, because it requires low temperature and pressure and a comparatively short reaction time. It has a high conversion yield and is a simple conversion process (Gebremariam and Marchetti 2017; Lin et al. 2011).

The transesterification process is the reaction between an oil and an alcohol in the presence of an appropriate catalyst to produce methyl ester (biodiesel) and glycerol. The role of the catalyst in this process is to speed up the reaction. After the reaction, the viscosity of oil reduces, while maintaining its heating value. The alcohols used are mostly methanol, ethanol, propanol, and butanol, particularly the former two; methanol is most commonly used to produce biodiesel because of its low cost and comparatively high reactivity, and because it results in the production of FAMEs with higher volatility in comparison to the fatty-acid ethyl esters (FAEEs). The viscosity of FAEEs is slightly higher than that of FAMEs, but their pour point and cloud point are slightly lower (Seffati et al. 2020; Foroutan et al. 2020). Though the transesterification process is reversible, the process efficiency is affected by several factors, including the reactant ratio, catalyst content, and reaction conditions (Demirbas 2008). Furthermore, the use of more methanol results in more biodiesel production, but excess use leads to higher costs. Usually, catalysts are used to increase the yield and rate of the reaction; these may be alkaline, acidic, or enzymatic, with the alkaline catalysts leading to a faster reaction than the acidic catalysts (Bozbas 2008; Canakci 2007).

1.4 Catalysts Used in Biodiesel Production

The catalysts used for biodiesel production are classified into four groups: homogeneous catalysts, heterogeneous catalysts, enzymatic catalysts, and nanocatalysts (Narasimharao et al. 2007). These will be briefly discussed in this section.

1.4.1 Homogeneous Catalysts

1.4.1.1 Homogeneous Alkaline Catalysts

At industrial scale, biodiesel is usually produced using HACs such as potassium hydroxide (KOH) and sodium hydroxide (NaOH). These types of catalysts are used in industrial applications for many reasons. The reports available show that the reaction rate in the presence of an alkaline catalyst is about 4000-fold faster than that using an acidic catalyst, though this type of catalyst is limited to vegetable-derived oils with FFA concentration less than 6 wt%. If an oil (e.g. WEO) contains FFA higher than 6 wt%, an alkaline catalyst cannot be an appropriate option as it may react with the FFA to produce soap, which is undesirable, as it can deactivate the alkaline catalyst, affecting the biodiesel production yield (Esmaeili and Foroutan 2018; Kulkarni and Dalai 2006).

1.4.1.2 Homogeneous Acidic Catalysts

When the FFA content in oil is high, liquid acidic catalysts can be used. Two important acidic catalysts are commonly employed: sulfuric acid and chloric acid. However, while these catalysts have several advantages, they also have some disadvantages, including a comparatively low reaction rate, the need for high reaction temperatures, high alcohol to oil ratios, catalyst separation, and corrosion issues. Other acidic catalysts like HCl, H_3PO_4, and organic sulfonate acids can also be used to produce biodiesel (Jacobson et al. 2008). The key parameters for an acidic catalyst are the protonation (generation of positive charges) of the carbonyl group on the triglycerides and the alcohol, which attacks the positive-charged

Table 1.5 Biodiesel production from different oils using homogeneous alkali and acidic catalysts.

Oil source	Catalyst type	Temp. (°C)	Time (h)	Catalyst conc. (wt%)	Alcohol to oil ratio	BCY (%)	References
Jojoba oil	KOH	25	—	1.35	—	83.5	Bouaid et al. (2007)
Peanut oil	KOH	60	1.5	0.5	6 : 1	95	Jazie et al. (2012)
Peanut oil	NaOH	60	1.5	0.5	6 : 1	88	Jazie et al. (2012)
Rapeseed oil	KOH	60	1.5	1	6 : 1	97	Jazie et al. (2012)
Rapeseed oil	NaOH	60	1.5	1	6 : 1	92	Jazie et al. (2012)
Microalga oil	LiOH	60	0.2	5	12 : 1	55	Cercado et al. (2018)
Chicken fat	H_2SO_4	50	24	—	30 : 1	99.72	Bhatti et al. (2008)
Mutton fat	H_2SO_4	60	24	—	30 : 1	98.28	Bhatti et al. (2008)
Cotton seed oil	Phosphoric Acid	120	10	5	10 : 1	99.79	Dholakiya (2012)

carbon atom to form a tetrahedral intermediate (Demirbas 2008; Koberg and Gedanken 2013; Aransiola et al. 2014).

Table 1.5 shows the BCYs of various oil sources using homogeneous acidic and alkaline catalysts.

1.4.2 Heterogeneous Catalysts

1.4.2.1 Heterogeneous Alkaline Catalysts

Heterogeneous catalysts are superior to homogeneous catalysts because they allow easy separation of the biodiesel and byproducts like glycerol. They also entail lower production costs and are environmentally friendly (Seffati et al. 2019; Saoud 2018). Oxides and derivatives of alkaline metals (e.g. Ba, Mg, Ca, Be, Sr, and Ra) have been studied by several researchers. Among them, Mg and Sr oxides have been extensively applied as catalysts for biodiesel production due to their desirable heterogeneous nature (Rasouli and Esmaeili 2019; Yoo et al. 2010). In addition, waste materials can be used to produce heterogeneous catalysts. Several materials, such as eggshell, mollusk shell, and bone, can be used to produce heterogeneous alkaline catalysts like calcium oxide (CaO). They also help manage the problem of waste materials, which is an important concern. The CaO produced from waste materials offers great potential as a catalyst in the production of biodiesel and has been widely studied (Keihani et al. 2018; Foroutan et al. 2020; Chakraborty and Sahu 2014).

1.4.2.2 Heterogeneous Acid Catalysts

The transesterification reaction of fats and oils in the presence of HACs such as NaOH, NaOMe, KOH, and KOMe has several problems resulting from large amounts of FFA.

Though these catalysts can be applied to produce biodiesel, a large amount of methanol and catalyst is consumed. Also, homogeneous acid catalysts such as H_2SO_4, HCl, and H_3PO_4 have less BCY than the HACs, and they require higher temperatures, methanol to oil ratios, pressures, and catalyst contents to produce biodiesel with an adequate BCY. Heterogeneous acid catalysts can be easily used in a packed bed continuous flow reactor: one of their main advantages. Also, the use of heterogeneous acid catalysts enables easy separation and purification of the product and reduces waste production (Boey et al. 2013; Melero et al. 2009). Recent research looking for new catalysts has demonstrated that heterogeneous acidic catalysts such as zirconium oxide (ZrO_2) and cationic resins show great potential for replacing the liquid homogeneous acidic catalysts (Jacobson et al. 2008).

Table 1.6 shows the BCYs of various oil sources using heterogeneous acidic and alkali catalysts.

1.4.3 Enzymatic Catalysts

The transesterification reaction using enzymatic catalysts has some advantages. The reaction in the presence of enzymatic catalysts occurs without formation of soap and continues free of purification, washing, and neutralization problems. Also, the transesterification reaction can be carried out under normal conditions, and this catalyst has no problem with oils containing large amounts of FFA (Shahid and Jamal 2011). Other

Table 1.6 Biodiesel production from different oils using heterogeneous acidic and alkaline catalysts.

Oil source	Catalyst type	Temp. (°C)	Time (h)	Catalyst conc. (wt%)	Alcohol to oil ratio	BCY (%)	References
Nannochloropsis sp.	$Ca(OCH_3)_2$	80	3	3	30 : 1	99	Teo et al. (2016)
Microalga oil	K-Pumice	60	2	10	18 : 1	77	Cercado et al. (2017)
Waste cooking oil	KOH/clinoptilolite	65	0.223	8.1	2.25 : 1	97.45	Mohadesi et al. (2020)
Waste cooking oil	Ba/CaO	65	3	3	6 : 1	88	Balakrishnan et al. (2013)
Waste cooking oil	Copper/zinc oxide	55	0.833	12	8 : 1	97.71	Gurunathan and Ravi (2015)
Waste cooking oil	4Mn-6Zr/CaO	80	3	3	15 : 1	92.1	Mansir et al. (2018)
Waste cooking oil	Calcium diglyceroxide	60	0.5	1	9 : 1	93.5	Gupta et al. (2015)
Waste cooking oil	Waste eggshell	65	2.75	5	9 : 1	87.8	Peng et al. (2018)
Waste cooking oil	$SO_4/Fe\text{-}Al\text{-}TiO_2$	90	2.5	3	10 : 1	96	Gardy et al. (2018)
Waste cooking oil	Butyl-methyl imidazolium hydrogen sulfate	160	1	5	15 : 1	95.65	Ullah et al. (2015)

advantages include low energy usage, reusability of the catalyst, and zero wastage of water (Norjannah et al. 2016). However, these catalysts are expensive and require long reaction times (about 12–24 hours), which limits their application (Norjannah et al. 2016; Shahid and Jamal 2011). The lipase enzyme is mostly used for converting oil to biodiesel (triacylglycerol acylhydrolase) (Arumugam and Ponnusami 2017; Norjannah et al. 2016). Lipases can be derived from different sources, such as yeast (*Candida parapsilosis, Candida deformans, Candida quercitrusa, Candida antarctica, Geotrichum candidum, Pichia burtonii, Pichia xylosa, Saccharomyces lipolytica*, and *Pichia sivicola*), bacteria (e.g. *Pseudomonas aeruginosa, Chromobacterium viscosum, Bacillus subtilis, Aeromonas hydrophilia, Staphylococcus canosus, Staphylococcus aureus, Burkholderia glumae*, and *Achromobacter lipolyticum*), and fungi (e.g. *Alternaria brassicicola, Streptomyces exfoliates, Rhizopus chinensis, Mucor miehei, Aspergillus niger*, and *Rhizopus oryzae*) (Norjannah et al. 2016). The enzymatic transesterification process for biodiesel production occurs via the ping-pong bi-bi mechanism, whereby substrates react to generate products via formation of a substrate–enzyme intermediate. Mechanisms have three steps: (i) alcoholysis of monoglycerides, diglycerides, and triglycerides (glycerides) into fatty acid alkyl ester; (ii) hydrolysis (converting glycerides into FFA) followed by esterification (converting FFA into esters); and (iii) simultaneous reactions of hydrolysis and alcoholysis following esterification (Norjannah et al. 2016).

Table 1.7 shows the BCYs of various oil sources using enzymatic catalysts.

1.4.4 Nanocatalysts

Biodiesel production using the transesterification reaction in the presence of acidic or alkaline catalysts is associated with some problems, like long ester formation and reaction completion times. Moreover, these catalysts are not reusable and form stable emulsions that make ester separation more difficult (Seffati et al. 2019, 2020). In recent years, the use of nanocatalysts has significantly increased due to their improved characteristics compared with other conventional catalysts (Tamjidi et al. 2019; Tamjidi and Esmaeili 2019). A nanocatalyst is a material with a high catalytic activity that exists in a nanoscale dimension. It has a large surface/volume ratio, and its effectiveness can be increased by increasing its specific surface area for reacting with reactants. There are several types of nanocatalysts for the production of biodiesel, including metal oxide-based nanocatalysts (e.g. CaO, MgO, and ZnO), metal oxide nanocatalysts supported by metals (e.g. Cs–Al–Fe_3O_4 and Au/ZnO), and metal oxide supported by metal oxide (e.g. KF/CaO and KF/CaO/Fe_3O_4) and alloys (e.g. Cu–Co) (Saoud 2018). One of the most important catalysts in this field is CaO nanocatalyst. This is widely used to produce biodiesel because of its high activity and low cost. It can be easily synthesized from different sources, such as eggshell and snail shell, through calcination in a furnace at 800 °C for four hours (Seffati et al. 2019). In addition, the development and application of magnetic nanocatalysts is found to be most promising, and these catalysts considerably enhance biodiesel production compared with other conventional acidic and alkaline catalysts. Moreover, due to their magnetic properties, they can be recovered and reused in multiple cycles of transesterification

Table 1.7 Biodiesel production from different oils in the presence of enzymatic catalysts.

Oil source	Catalyst type	Temp. (°C)	Time (h)	Catalyst conc. (wt%)	Alcohol to oil ratio	BCY (%)	References
Jatropha oil	*Pseudomonas fluorescens*	40	48	3	4 : 1	72	Devanesan et al. (2007)
Palm oil	*Pseudomonas fluorescens*	40	24	20	18 : 1	98	Wang et al. (2014)
Soybean oil	Novozym 435	40	10	30	12 : 1	92	Wang et al. (2014)
Chlorella oil	*Candida* sp.	38	12	75	3 : 1	98.2	Wang et al. (2014)
Chlorella oil	Penicillium expansum lipase	50	48	20	4 : 1	90.7	Wang et al. (2014)
Jatropha oil	*Chromobacterium viscosum*	40	24	10	4 : 1	92	Wang et al. (2014)
Nannochloropsis oceanica IMET1 oil	Novozym 435	25	4	20	12 : 1	99.1	Wang et al. (2014)
Rapeseed oil	*Lipozyme TLIM*	35	12	3	4 : 1	95	Li et al. (2006)
Triolein	Immobilized *Pseudomonas cepacia*	40	24	1.46	3 : 1	90	Salis et al. (2005)
Tallow	Immobilized *Mucor miehei*	45	5	10	3 : 1	77.8	Nelson et al. (1996)
Cotton seed oil	Novozym 435	40	24	10	3 : 1	62.1	Su et al. (2007)
Waste cooking oil	Novozym 435	60	24	10	6 : 1	77.87	Gharat and Rathod (2013)
Jatropha oil	Immobilized *Rhizopus oryzae* whole-cell	30	60	4	1 : 1	80	Tamalampudi et al. (2008)

reactions. Among other advantages, nanocatalysts have a high recovery factor, high energy consumption recovery, and low reaction temperature requirement (Shahid and Jamal 2011; Ingle et al. 2020). Such outstanding properties of magnetic nanoparticles (e.g. low cost, low toxicity, environment friendliness, and easy and fast separation using an external magnetic field, which most commonly leads to the elimination of the centrifugal stage) have made them useful and popular heterogeneous catalysts for biodiesel production (Liu et al. 2018; Dos Santos-Durndell et al. 2018).

Table 1.8 shows a summary of biodiesel production yields from different oil sources in the presence of nanocatalysts, while Table 1.9 shows the advantages and disadvantages of various types of catalysts used in biodiesel production.

Table 1.8 Summary of biodiesel production yield from different oil sources in the presence of nanocatalysts.

Catalyst	Feedstock	Temp. (°C)	Alcohol to oil ratio	Catalyst content (wt%)	Time	BCY (%)	References
SO4/Fe-Al-TiO_2	Cooking oil	90	10 : 1	3	2.5 h	96	Gardy et al. (2018)
CaO@MgO	Waste edible oil	69.37	16.7 : 1	4.571	7.08 h	98.37	Foroutan et al. (2020)
KOH/AL_2O_3	Palm oil	70	15 : 1	25	3 h	91.1	Noiroj et al. (2009)
Lipozyme TLIM Nanocatalyst and magnetic catalyst	Rapeseed oil	35	4 : 1	3	12 h	95	Li et al. (2006)
KF/CaO	Tallow seed oil	65	12 : 1	4	2 h	96.8	Wen, L et al. (2010)
MgO	Goat fat	70	12 : 1	1	3 h	93.12	Rasouli and Esmaeili (2019)
Li/ZnO–Fe_3O_4	Rapeseed oil	35	12 : 1	0.8	0.58 h	99.8	Kelarijani et al. (2019)
Li/Fe_3O_4	Rapeseed oil	35	12 : 1	0.8	0.58 h	99.8	Kelarijani et al. (2019)
CaO/Al/Fe_3O_4	Rapeseed oil	—	15 : 1	6	3 h	98.71	Tang et al. (2012)
SO_4/Mg–Al–Fe_3O_4	Waste cooking oil	95	9 : 1	4	5 h	98.5	Gardy et al. (2019)
CaO–NiO	Waste cottonseed oil	65	15 : 1	5	4 h	>99	Kaur and Ali (2011)
CaO–CeO_2	Refined palm oil	85	20 : 1	5	3 h	95	Thitsartarn and Kawi (2011)
CaO–ZrO_2	Waste cooking oil	65	30 : 1	10	2 h	92.1	Dehkordi and Ghasemi (2012)
CaO–Al_2O_3	Palm oil	65	12 : 1	6	5 h	98.64	Zabeti et al. (2010)
CaO–La_2O_3	Jatropha curcas oil	65	24 : 1	4	6 h	86.51	Teo et al. (2014)
MgO	*Moringa oleifera* seeds oil	45	12 : 1	1	4 h	93.69	Esmaeili et al. (2019)
MgO/CaO	Waste cooking oil	75	7 : 1	3	6 h	98.95	Tahvildari et al. (2015)
CaO	Chicken oil	65	9 : 1	1	5 h	94.4	Keihani et al. (2018)
CaO/$CuFe_2O_4$	Chicken oil	70	15 : 1	3	4 h	94.52	Seffati et al. (2019)

Table 1.8 (Continued)

Catalyst	Feedstock	Temp. (°C)	Alcohol to oil ratio	Catalyst content (wt%)	Time	BCY (%)	References
$CaO/CuFe_2O_4$@AC	Chicken oil	65	12 : 1	3	4 h	95.6	Seffati et al. (2020)
CaO–MgO	Jatropha curcas oil	120	25 : 1	3	3 h	90	Taufiq-Yap et al. (2011)
CaO–ZnO	Sunflower oil	60	12 : 1	3	45 min	>90	Rubio-Caballero et al. (2009)
$CaO–La_2O_3$	Soybean oil	58	20 : 1	5	1 h	94.3	Yan et al. (2009)
CaO–ZnO	Sunflower oil	60	10 : 1	2	4 h	97.5	Kesić et al. (2012)
$KF–CaO–Fe_3O_4$	Stillingia oil	65	12 : 1	4	3 h	>95	Hu et al. (2011)
MgO–ZnO	Jatropha curcas oil	120	25 : 1	3	3 h	83	Lee et al. (2013)
$TiO_2–MgO$	Waste cooking oil	150	30 : 1	5	6 h	>85	Wen, Z et al. (2010)
LaMgO	Cottonseed oil	65	54 : 1	5	33 min	96	Mutreja et al. (2014)
MgAlCe	Soybean oil	67	9 : 1	5	3 h	>90	Dias et al. (2012)
Sr–MgO	Refined palm oil	60	6 : 1	3	1.25 h	96	Faungnawakij et al. (2012)

1.5 Effects of Different Factors on Biodiesel Production Yield

The most important operating parameters in biodiesel production include temperature, alcohol to oil ratio, contact time, and catalyst content (Hariharan et al. 2009; Marchetti et al. 2007), as briefly explained in this section.

1.5.1 Reaction Temperature

The reaction temperature is a critical factor that affects the biodiesel production yield significantly. The BCY increases with increasing reaction temperature. In other words, the viscosity of the oil reduces with increasing reaction temperature (Leung et al. 2010). Seffati et al. (2019) studied the impact of different temperature ranges (50–80 °C) on biodiesel production. The results obtained suggest that an increase from 50 to 70 °C increases the BCY from 41.28 to 94.25%. However, at temperatures above 70 °C, the methanol evaporates, decreasing the contact between it and the oil, which results in a decrease in the BCY. Therefore, the temperature must be less than the boiling point of the alcohol used in the reaction to prevent its evaporation (Seffati et al. 2019).

Table 1.9 Advantages and disadvantages of various types of catalysts.

Type of catalyst	Advantages	Disadvantages	References
Homogeneous alkaline catalysts	Reaction at low temperature and atmospheric pressure, requires less energy, easy accessibility, economical, high conversion rate, less corrosive	Sensitivity to FFA content in the oil, performance, trouble creating the product	Gebremariam and Marchetti (2017), Talha and Sulaiman (2016)
Homogeneous acidic catalysts	High biodiesel yield, insensitive to FFA content in the oil, no soap production	Harder separation of the catalyst from the final product, low reaction rate, high reaction temperature, corrosive	Jacobson et al. (2008), Talha and Sulaiman (2016)
Heterogeneous alkaline catalysts	Easy separation of catalyst from the final product, high reusability, less energy intensive, high selectivity	Allergy to FFA content in the oil, catalyst poisoned in the ambient air, higher alcohol to oil molar ratio	Gebremariam and Marchetti (2017) Talha and Sulaiman (2016)
Heterogeneous acidic catalysts	Insensitive to FFA content in the oil, esterification and transesterification occur simultaneously, easy reusability and recovery of the catalyst	High reaction temperature required, higher alcohol to oil molar ratio, high cost of synthesized catalysts	Gebremariam and Marchetti (2017), Talha and Sulaiman (2016)
Enzymatic catalysts	Insensitive to FFA content of the oil, reusable, low amount of alcohol needed	Slow process, sensitive to alcohol type, high cost of enzyme	Gebremariam and Marchetti (2017), Talha and Sulaiman (2016)
Nanocatalysts and magnetic catalysts	High specific surface area, high efficiency, environmentally friendly, low toxicity	Relatively high alcohol requirement, higher cost for suitable catalysts, sedimentation and agglomeration	Kulkarni, and Dalai (2006)

1.5.2 Alcohol to Oil Molar Ratio

The alcohol to oil molar ratio is another significant variable that impacts on biodiesel production. The higher the alcohol to oil molar ratio, the greater the yield of converting fat to esters. Therefore, the BCY will improve with increasing alcohol to oil molar ratio up to an optimum value (Mathiyazhagan and Ganapathi 2011). Increasing the amount of alcohol beyond the optimum value not only increases the cost of the process but also stops affecting the BCY positively (Leung et al. 2010). Recently, Seffati et al. (2020) studied the effect of various methanol to oil molar ratios (4 : 1, 6 : 1, 8 : 1, 12 : 1, 15 : 1, 18 : 1, and 24 : 1) on BCY. Their investigations showed that an increase from 4 : 1 to 12 : 1 resulted in an improvement in BCY from 56.37 to 94.56%, and the 12 : 1 molar ratio was determined as the optimum value, with further increases beyond this point having no significant effect on the BCY because of the greater difficulty of separation of the glycerol from the biodiesel at higher ethanol contents (Seffati et al. 2020).

1.5.3 Reaction Time

The reaction time is among the most critical parameters affecting the transesterification process (Seffati et al. 2019). The BCY decreases with increasing contact time between the oil and the methanol (beyond the optimum value), with excess time leading to the formation of soap (Mathiyazhagan and Ganapathi 2011). A study associated with the impact of reaction time (two to eight hours) on BCY indicated that the BCY was much less at the beginning of the reaction and increased with increasing reaction time from two to six hours, before decreasing again at a reaction time of eight hours (Salimi and Hosseini 2019).

1.5.4 Catalyst Dosage

The catalyst content (dosage) has an important impact on the transesterification process, as it affects the reaction rate while contributing to hydrolysis and soap formation (Seffati et al. 2019, 2020). The increase in the dose of catalyst increases the conversion rate of triglyceride, and hence also the biodiesel production efficiency. However, an inadequate amount of catalyst leads to incomplete conversion of triglyceride into the fatty-acid esters (Leung et al. 2010; Mathiyazhagan and Ganapathi 2011). In one study, it was found that BCY increases with increasing catalyst content (1–10 wt.%), because the available surface area on the catalyst increases with increasing catalyst dosage (Ali et al. 2017).

1.5.5 pH

pH does not have an effect on biodiesel production when the transesterification reaction is carried out out using acidic or alkaline catalysts. However, it will have a significant impact when enzymatic catalysts are applied to produce biodiesel, because at low or high values of pH, breakdown or degradation of the enzyme takes place (Allah and Alexandru 2016).

1.5.6 Mixing Rate

Mixing can increase the rate of reaction and the diffusion of oil, alcohol, and catalyst in a transesterification reaction. Thus, the mixing rate should be optimized to achieve the maximum BCY. Also, by determining the optimum mixing rate, the lowest energy requirement for biodiesel production can be obtained (Allah and Alexandru 2016).

1.5.7 Fatty Acids

In the transesterification reaction, the vegetable oil should have an acid value (AV) lower than 1 and all substances should be anhydrous. If the AV is more than 1, NaOH or KOH should be injected to neutralize the FFA (Demirbas 2009). WCO has a high FFA content compared to fresh cooking oil. Also, higher FFA contents will result in the formation of soap and water. Moreover, if the content of FFA is more than 3%, the transesterification process will not occur, even with homogenous alkaline catalysts. To solve this problem, it is suggested that before performing a reaction using heterogeneous catalysts, the oil should be pretreated with acidic homogenous catalysts or acidic heterogeneous catalysts to esterify the FFA to FFA ester (Gnanaprakasam et al. 2013).

1.5.8 Water Content

Water content is another critical parameter in the transesterification of vegetable oils. The presence of FFA and water has negative impacts, including formation of soap, consumption of catalyst, and reduction of catalyst effectiveness (Demirbas 2009). The resulting soaps can increase viscosity and gel and foam formation, and make it difficult to separate glycerol (Demirbas 2009).

1.6 Physical Properties of Biodiesel

The physical characteristics of biodiesel mainly include kinematic viscosity, density, flash point, pour point, cloud point, Ca + Mg + K + Na content, AV, and CN (Seffati et al. 2020). Each of these is important as far as the quality of the biodiesel is concerned.

The CN represents the diesel's combustion quality, similar to the octane number for the petrol, revealing the knocking feature of petro-diesel fuel (Nag 2008). The greater the CN, the more desirable the fuel. It also measures the value of fuel combustion delay (Sivaramakrishnan and Ravikumar 2012). The density has a significant impact on the fuel injection system and is a measure of the fuel's ability to flow, and hence affects the quality of fuel atomization, droplet size, and diffusion. The rate and time of fuel injection are also directly affected by this factor (Canakci and Sanli 2008). According to the international standards, the density of diesel must be in the range of 860–900 kg/m^3 (Seffati et al. 2020).

The viscosity plays an important role in fuel spraying, formation of mixture, and the combustion process (Canakci and Sanli 2008). Fuels with higher viscosity may form larger droplets during injection, which decreases atomization during spraying. It may also increase the sediment in the engine and increase the energy required for fuel pumping. These problems lead to poor combustion, increasing exhaust gases and greenhouse gas emissions (Ramírez-Verduzco et al. 2012; Fernando et al. 2007). Similarly, flash point is an important property as far as storage and transportation are concerned. It depends on the fuel's volatility and acts as a critical property in starting and warming up an engine (Gouveia et al. 2017). The flash point of biodiesel is greater than that of fossil diesel, which shows that biodiesel is much safer to use and store (Adebayo et al. 2011).

The cloud point is the temperature at which crystals form very rapidly, turning the color of the liquid opaque. The pour point is the lowest temperature at which the fuel can be

Table 1.10 Standard values of some important physical properties of diesel fuel.

Property	EN 14214	ASTM D 6751	EN 14538	References
Kinematic viscosity (mm^2/s)	3.5–5	1.9–6	—	Seffati et al. (2020)
Density (kg/m^3)	860–900	—	—	Seffati et al. (2020)
Flash point (°C)	>120	>130	—	Seffati et al. (2020)
Pour point (°C)	—	—	—	—
Cloud point (°C)	—	—	—	—
Oxidation stability (h)	Minimum 6	Minimum 3	—	Seffati et al. (2020)
Acid value (mgKOH/g)	Maximum 0.5	Maximum 0.5	—	Seffati et al. (2020)
Ca + Mg (mg/Kg)	—	—	Maximum 5	Balasubramaniyan (2016)
K + Na (mg/Kg)	—	—	Maximum 5	Balasubramaniyan (2016)

pumped; it is only a little higher than the freezing point (Canakci and Sanli 2008). Both pour point and cloud point determine the usability of a fuel at low temperatures. They must be below 0 °C; if they are higher, the biodiesel will freeze in cold regions.

AV is the amount of KOH (in mg) required to neutralize the free acids present in 1 g of oil. If other acidic components like amino acids are present in the oil, the AV can be estimated (Dagne 2015). According to the international standard ASTM D 6751, the AV for biodiesel should be less than 0.50 mg KOH/g oil. ASTM D 664 is a standard method for measuring the AV for biodiesel and diesel (Mahajan et al. 2006).

The concentrations of sodium (Na), calcium (Ca), magnesium (Mg), and potassium (K) are other important properties of a fuel, because the presence of these metals in a vehicle's engine system will lead to the deposition of solid, causing many problems. The maximum amount of such metals in biodiesel should be 5 mg/kg of oil, according to standard EN 14538 (Balasubramaniyan 2016). Oxidation stability is another important characteristic of biodiesel fuel, affecting its stability during long-term storage. This property refers to the tendency of a fuel to react with oxygen at ambient temperature and describes its relative sensitivity to oxidation (Pullen and Saeed 2012).

Table 1.10 shows the standard values for various properties of biodiesel.

1.7 Conclusions

Biodiesel is a biodegradable, nontoxic, and renewable fuel that is a good candidate for replacing diesel fuel. Among the several methods of producing biodiesel, transesterification is the most commonly used. In the transesterification process, an oil (e.g. vegetable oil, animal fat, alga oil, or WCO) reacts with methanol in the presence of a proper catalyst. Different types of catalysts are conventionally used for biodiesel production, mainly acidic catalysts, enzymatic catalysts, and alkaline catalysts. Alkaline catalysts include homogeneous and heterogeneous catalysts, between which heterogeneous catalysts have attracted much more attention due to their high potential to produce biodiesel. Moreover, heterogeneous nanocatalysts show novel properties like high activity and high surface area, and hence allow high biodiesel yield. In addition, the development of magnetic nanocatalysts is a path-breaking discovery as far as biodiesel production is concerned, because these catalysts possess many advantages such as high efficiency, easy recovery through application of an external magnetic field, and reusability for multiple cycles of transesterification, which ultimately helps reduce the total process cost.

A number of factors can greatly affect the BCY, principally reaction temperature, reaction time, alcohol to oil molar ratio, and type and content of different catalysts. Appropriate selection is crucial. Several nanocatalysts, such as MCM-41/ECH/Na_2SiO_3/Fe_3O_4, CaO@γ-Fe_2O_3, Li/ZnO-Fe_3O_4, $MgFe_2O_4$@CaO, K-Fe_3O_4-CeO_2, CaO/CuF_2O_4, AC/CaO@$CuFe_2O_4$, MgO@CaO, and Li/Fe_3O_4, have been found to be particularly effective for biodiesel production. Moreover, physical characteristics of biodiesel such as CN, density, kinematic viscosity, flash point, cloud point, pour point, AV, and Ca, Na, K, and Mg content are very important in deciding the product quality and other related standards.

References

Abbaszaadeh, A., Ghobadian, B., Omidkhah, M.R., and Najafi, G. (2012). Current biodiesel production technologies: a comparative review. *Energy Conversion and Management* 63: 138–148.

Adebayo, G.B., Ameen, O.M., and Abass, L.T. (2011). Physico-chemical properties of biodiesel produced from *Jatropha Curcas* oil and fossil diesel. *Journal of Microbiology and Biotechnology Research* 1 (1): 12–16.

Ali, M.A., Al-Hydary, I.A., and Al-Hattab, T.A. (2017). Nano-magnetic catalyst CaO-Fe_3O_4 for biodiesel production from date palm seed oil Bull. *Bulletin of Chemical Reaction Engineering & Catalysis* 12 (3): 460–468.

Allah, F.U.M. and Alexandru, G. (2016). Waste cooking oil as source for renewable fuel in Romania. *IOP Conference Series: Materials Science and Engineering* 147 (1): 012133.

Alptekin, E. and Canakci, M. (2008). Determination of the density and the viscosities of biodiesel – diesel fuel blends. *Renewable Energy* 33 (12): 2623–2630.

Ambat, I., Srivastava, V., Haapaniemi, E., and Sillanpää, M. (2019). Nano-magnetic potassium impregnated ceria as catalyst for the biodiesel production. *Renewable Energy* 139: 1428–1436.

Aransiola, E.F., Ojumu, T.V., Oyekola, O.O. et al. (2014). A review of current technology for biodiesel production: state of the art. *Biomass and Bioenergy* 61: 276–297.

Arumugam, A. and Ponnusami, V. (2017). Production of biodiesel by enzymatic transesterification of waste sardine oil and evaluation of its engine performance. *Heliyon* 3 (12): 00486.

Ashok, A., Kennedy, L.J., Vijaya, J.J., and Aruldoss, U. (2018). Optimization of biodiesel production from waste cooking oil by magnesium oxide nanocatalyst synthesized using co precipitation method. *Clean Technologies and Environmental Policy* 20: 1219–1231.

Awaluddin, A., Saryono, Prayitno, A., and Amri, T.A. (2010). Transesterification of waste chicken fats for synthesizing biodiesel by CaO as heterogeneous base catalyst. *Proceedings of the International Conference on Energy and Sustainable Development: Issues and Strategies* (ESD 2010): 1–4.

Aziz, M.A. (2010). Present status of the world goat populations and their productivity. *Lohmann Information* 45 (2): 42–52.

Balakrishnan, K., Olutoye, M.A., and Hameed, B.H. (2013). Synthesis of methyl esters from waste cooking oil using construction waste material as solid base catalyst. *Bioresource Technology* 128: 788–791.

Balasubramaniyan, D.D. (2016). Development of homogeneous and heterogeneous catalysts for the production of biodiesel and aviation fuel. Doctoral dissertation, University of Nevada, Reno, NV.

Banković-Ilić, I.B., Stojković, I.J., Stamenković, O.S. et al. (2014). Waste animal fats as feedstocks for biodiesel production. *Renewable and Sustainable Energy Reviews* 32: 238–254.

Berchmans, H.J. and Hirata, S. (2008). Biodiesel production from crude *Jatropha curcas* L. seed oil with a high content of free fatty acids. *Bioresource Technology* 99 (6): 1716–1721.

Bhatti, H.N., Hanif, M.A., Qasim, M., and Ata-ur, R. (2008). Biodiesel production from waste tallow. *Fuel* 87 (13–14): 2961–2966.

Boey, P.L., Ganesan, S., Maniam, G.P. et al. (2013). A new heterogeneous acid catalyst for esterification: optimization using response surface methodology. *Energy Conversion and Management* 65: 392–396.

Bouaid, A., Bajo, L., Martinez, M., and Aracil, J. (2007). Optimization of biodiesel production from jojoba oil. *Process Safety and Environmental Protection* 85 (5): 378–382.

Bozbas, K. (2008). Biodiesel as an alternative motor fuel: production and policies in the European Union. *Renewable and Sustainable Energy Reviews* 12 (2): 542–552.

Canakci, M. (2007). Combustion characteristics of a turbocharged DI compression ignition engine fueled with petroleum diesel fuels and biodiesel. *Bioresource Technology* 98 (6): 1167–1175.

Canakci, M. and Sanli, H. (2008). Biodiesel production from various feedstocks and their effects on the fuel properties. *Journal of Industrial Microbiology and Biotechnology* 35: 431–441.

Cao, H., Zhang, Z., Wu, X., and Miao, X. (2013). Direct biodiesel production from wet microalgae biomass of *Chlorella pyrenoidosa* through in situ transesterification. *BioMed Research International*: 1–6.

Cercado, A.P.I., Ballesteros, F.C., and Capareda, S.C. (2017). Biodiesel from microalgae using synthesized novel alkali catalyst. *Applied Ecology and Environmental Research* 15 (4): 1859–1870.

Cercado, A.P., Ballesteros, F.C., and Capareda, S.C. (2018). Biodiesel from three microalgae transesterification processes using different homogenous catalysts. *International Journal of Technology* 9 (4): 645–651.

Chakraborty, R. and Sahu, H. (2014). Intensification of biodiesel production from waste coal tallow using infrared radiation: process evaluation through response surface methodology and artificial neural network. *Applied Energy* 114: 827–836.

Channi, H.K., Gupta, S., Singh, H., and Gill, S.S. (2016). Different biodiesel production technologies: a review. *International Journal of Research in Engineering and Applied Sciences* 6 (9): 123–128.

Chen, Y.H., Huang, B.Y., Chiang, T.H., and Tang, T.C. (2012). Fuel properties of microalgae (*Chlorella protothecoides*) oil biodiesel and its blends with petroleum diesel. *Fuel* 94: 270–273.

Chen, J., Li, J., Dong, W. et al. (2018). The potential of microalgae in biodiesel production. *Renewable and Sustainable Energy Reviews* 90: 336–346.

Chisti, Y. (2007). Biodiesel from microalgae. *Biotechnology Advances* 25 (3): 294–306.

Chitra, P., Venkatachalam, P., and Sampathrajan, A. (2005). Optimisation of experimental conditions for biodiesel production from alkali-catalysed transesterification of *Jatropha curcas* oil. *Energy for Sustainable Development* 9 (3): 13–18.

Dagne, H. (2015). Optimization and characterization of biodiesel production from leather industry fleshing wastes. Doctoral dissertation, Addis Ababa University, Addis Ababa, Ethiopia.

Degfie, T.A., Mamo, T.T., and Mekonnen, Y.S. (2019). Optimized biodiesel production from waste cooking oil (WCO) using calcium oxide (CaO) nano-catalyst. *Scientific Reports* 9 (1): 1–8.

Dehkordi, A.M. and Ghasemi, M. (2012). Transesterification of waste cooking oil to biodiesel using ca and Zr mixed oxides as heterogeneous base catalysts. *Fuel Processing Technology* 97: 45–51.

Demirbas, A. (2004). Combustion characteristics of different biomass fuels. *Progress in Energy and Combustion Science* 30 (2): 219–230.

Demirbas, A. (2008). Comparison of transesterification methods for production of biodiesel from vegetable oils and fats. *Energy Conversion and Management* 49 (1): 125–130.

Demirbas, A. (2009). Progress and recent trends in biodiesel fuels. *Energy Conversion and Management* 50 (1): 14–34.

Devanesan, M., Viruthagiri, T., and Sugumar, N. (2007). Transesterification of Jatropha oil using immobilized *Pseudomonas fluorescens*. *African Journal of Biotechnology* 6 (21): 2497–2501.

Dhingra, S., Bhushan, G., and Dubey, K.K. (2016). Validation and enhancement of waste cooking sunflower oil based biodiesel production by the trans-esterification process. *Energy Sources, Part A: Recovery, Utilization, and Environmental Effects* 38 (10): 1448–1454.

Dhiraj, S.D. and Mangesh, M.D. (2012). Biodiesel production from animal fats and its impact on the diesel engine with ethanol-diesel blends: a review. *International Journal of Emerging Technology and Advanced Engineering* 2 (10): 179–185.

Dholakiya, B.Z. (2012). Super phosphoric acid catalyzed biodiesel production from low cost feed stock. *Archives of Applied Science Research* 4 (1): 551–561.

Dias, A.P.S., Bernardo, J., Felizardo, P., and Correia, M.J.N. (2012). Biodiesel production over thermal activated cerium modified Mg-Al hydrotalcites. *Energy* 41: 344–353.

Dorado, M.P., Cruz, F., Palomar, J.M., and Lopez, F.J. (2006). An approach to the economics of two vegetable oil-based biofuels in Spain. *Renewable Energy* 31 (8): 1231–1273.

Dos Santos-Durndell, V.C., Peruzzolo, T.M., Ucoski, G.M. et al. (2018). Magnetically recyclable nanocatalysts based on magnetite: an environmentally friendly and recyclable catalyst for esterification reactions. *Biofuel Research Journal* 18: 806–812.

Ehimen, E.A., Sun, Z.F., and Carrington, C.G. (2010). Variables affecting the in situ transesterification of microalgae lipids. *Fuel* 89 (3): 677–684.

Esmaeili, H. and Foroutan, R. (2018). Optimization of biodiesel production from goat tallow using alkaline catalysts and combining them with diesel. *Chemistry and Chemical Technology* 12 (1): 44–52.

Esmaeili, H., Esmaeilzadeh, F., and Mowla, D. (2014). Effect of surfactant on stability and size distribution of gas condensate droplets in water. *Journal of Chemical & Engineering Data* 59 (5): 1461–1467.

Esmaeili, H., Esmaeilzadeh, F., and Mowla, D. (2018). Destabilization and separation of gas condensate from wastewater using different surfactant demulsifiers. *Tenside Surfactants Detergents* 55 (2): 153–161.

Esmaeili, H., Yeganeh, G., and Esmaeilzadeh, F. (2019). Optimization of biodiesel production from *Moringa oleifera* seeds oil in the presence of nano-MgO using Taguchi method. *International Nano Letters* 9: 257–263.

Fahd, M., Lee, P., Chou, S.K. et al. (2014). Experimental study and empirical correlation development of fuel properties of waste cooking palm biodiesel and its diesel blends at elevated temperatures. *Renewable energy* 68: 282–288.

Faungnawakij, K., Yoosuk, B., Namuangruk, S. et al. (2012). Sr–Mg mixed oxides as biodiesel production catalysts. *Chem Cat Chem* 4 (2): 209–216.

Fernando, S., Karra, P., Hernandez, R., and Jha, S.K. (2007). Effect of incompletely converted soybean oil on biodiesel quality. *Energy* 32 (5): 844–851.

Foroutan, R., Esmaeili, H., and Ghyasi, F.F. (2018a). Optimization of biodiesel production from waste cooking oil by alkaline catalysts. *Journal of Food Technology and Food* 1 (1): 1–7.

Foroutan, R., Esmaeili, H., Mousavi, S.M. et al. (2018b). The physical properties of biodiesel-diesel fuel produced via transesterification process from different oil sources. *Physical Chemistry Research* 7 (2): 415–424.

Foroutan, R., Mohammadi, R., Esmaeili, H. et al. (2020). Transesterification of waste edible oils to biodiesel using calcium oxide@magnesium oxide nanocatalyst. *Waste Management* 105: 373–383.

Frohlich, A. and Rice, B. (2005). Evaluation of Camelina sativa oil as a feedstock for biodiesel production. *Industrial Crops and Products* 21 (1): 25–31.

Gardy, J., Osatiashtiani, A., Céspedes, O. et al. (2018). A magnetically separable SO_4/Fe-Al-TiO_2 solid acid catalyst for biodiesel production from waste cooking oil. *Applied Catalysis B: Environmental* 234: 268–278.

Gardy, J., Nourafkan, E., Osatiashtiani, A. et al. (2019). A core-shell SO4/Mg-Al-Fe_3O_4 catalyst for biodiesel production. *Applied Catalysis B: Environmental* 259: 118093.

Gebremariam, S.N. and Marchetti, J.M. (2017). Biodiesel production technologies. *AIMS Energy* 5 (3): 425–457.

Geller, D.P. and Goodrum, J.W. (2004). Effects of specific fatty acid methyl esters on diesel fuel lubricity. *Fuel* 83 (17–18): 2351–2356.

Ghanei, R., Heydarinasab, A., Moradi, G.R., and Falauti, V. (2014). Process optimization for biodiesel production from the waste frying oil with constructional lime. *International Journal of Biosciences* 4 (8): 36–45.

Gharat, N. and Rathod, V.K. (2013). Enzyme catalyzed transesterification of waste cooking oil with dimethyl carbonate. *Journal of Molecular Catalysis B: Enzymatic* 88: 36–40.

Gnanaprakasam, A., Sivakumar, V.M., Surendhar, A. et al. (2013). Recent strategy of biodiesel production from waste cooking oil and process influencing parameters: a review. *Journal of Energy* 2013: 926392.

Gouveia, L., Oliveira, A.C., Congestri, R. et al. (2017). Biodiesel from microalgae. In: *Microalgae-Based Biofuels and Bioproducts* (eds. C. Gonzalez-Fernandez and R. Muñoz), 235–258. Cambridge: Woodhead.

Gui, M.M., Lee, K.T., and Bhatia, S. (2008). Feasibility of edible oil vs. non-edible oil vs. waste edible oil as biodiesel feedstock. *Energy* 33 (11): 1646–1653.

Gupta, A.R., Yadav, S.V., and Rathod, V.K. (2015). Enhancement in biodiesel production using waste cooking oil and calcium diglyceroxide as a heterogeneous catalyst in presence of ultrasound. *Fuel* 158: 800–806.

Gurunathan, B. and Ravi, A. (2015). Biodiesel production from waste cooking oil using copper doped zinc oxide nanocomposite as heterogeneous catalyst. *Bioresource Technology* 188: 124–127.

Hariharan, V.S., Reddy, K.V., and Rajagopal, K. (2009). Study of the performance, emission and combustion characteristics of a diesel engine using sea lemon oil based fuels. *Indian Journal of Science and Technology* 2: 43–47.

Hasheminejad, M., Tabatabaei, M., Mansourpanah, Y. et al. (2011). Upstream and downstream strategies to economize biodiesel production. *Bioresource Technology* 102 (2): 461–468.

Hassan, M.M., Hossain, M.E., and Uddin, M.B. (2016). Perception of smallholding goat farmers on diseases condition of Bangladesh. *International Journal of Natural Sciences* 6 (1): 43–48.

Hu, S., Guan, Y., Wang, Y., and Han, H. (2011). Nano-magnetic catalyst KF/CaO–Fe_3O_4 for biodiesel production. *Applied Energy* 88 (8): 2685–2690.

Ingle, A.P., Chandel, A.K., Philippini, R. et al. (2020). Advances in nanocatalysts mediated biodiesel production: a critical appraisal. *Symmetry* 12 (2): 256.

International Energy Agency (2013). *World Energy Outlook*. Paris: International Energy Agency.

Islam, M., Magnusson, M., Brown, R. et al. (2013). Microalgal species selection for biodiesel production based on fuel properties derived from fatty acid profiles. *Energies* 6 (3): 5676–5702.

Jacobson, K., Gopinath, R., Meher, L.C., and Dalai, A.K. (2008). Solid acid catalyzed biodiesel production from waste cooking oil. *Applied Catalysis B: Environmental* 85 (1–2): 86–91.

Jazie, A.A., Sinha, A.S.K., and Pramanik, H. (2012). Optimization of biodiesel production from peanut and rapeseed oils using response surface methodology. *International Journal of Biomass & Renewables* 1 (2): 9–18.

Karmakar, A., Karamkar, S., and Mukherjee, S. (2010). Properties of various plants and animals feedstocks for biodiesel production. *Bioresource Technology* 101 (19): 7201–7210.

Kaur, M. and Ali, A. (2011). Lithium ion impregnated calcium oxide as nano catalyst for the biodiesel production from karanja and Jatropha oils. *Renewable Energy* 36 (11): 2866–2871.

Keihani, M., Esmaeili, H., and Rouhi, P. (2018). Biodiesel production from chicken fat using nano-calcium oxide catalyst and improving the fuel properties via blending with diesel. *Physical Chemistry Research* 6 (3): 521–529.

Kelarijani, A.F., Zanjani, N.G., and Pirzaman, A.K. (2019). Ultrasonic assisted transesterification of rapeseed oil to biodiesel using nano magnetic catalysts. *Waste and Biomass Valorization* 11: 2613–2621.

Kesić, Ž., Lukić, I., Brkić, D. et al. (2012). Mechanochemical preparation and characterization of CaOZnO used as catalyst for biodiesel synthesis. *Applied Catalysis A: General* 427–428: 58–65.

Kirubakaran, M. and Arul Mozhi Selvan, V. (2018). A comprehensive review of low cost biodiesel production from waste chicken fat. *Renewable and Sustainable Energy Reviews* 82 (1): 390–401.

Koberg, M. and Gedanken, A. (2013). Using microwave radiation and SrO as a catalyst for the complete conversion of oils, cooked oils, and microalgae to biodiesel. In: *New and Future Developments in Catalysis* (ed. S. Suib), 209–227. Amsterdam: Elsevier.

Kralova, I. and Sjoblom, J. (2010). Biofuels – renewable energy sources: a review. *Journal of Dispersion Science and Technology* 31 (3): 409–425.

Kulkarni, M.G. and Dalai, A.K. (2006). Waste cooking oil-an economical source for biodiesel production: a review. *Industrial & Engineering Chemistry* 45 (9): 2901–2913.

Lam, M.K., Lee, K.T., and Mohamed, A.R. (2010). Homogeneous, heterogeneous and enzymatic catalysis for transesterification of high free fatty acid oil (waste cooking oil) to biodiesel: a review. *Biotechnology Advances* 28 (4): 500–518.

Lee, H.V., Taufiq-Yap, Y., Hussein, M.Z., and Yunus, R. (2013). Transesterification of Jatropha oil with methanol over Mg–Zn mixed metal oxide catalysts. *Energy* 49: 12–18.

Leung, D.Y.C. and Guo, Y. (2006). Transesterification of neat and used frying oil: optimization for biodiesel production. *Fuel Processing Technology* 87 (10): 883–890.

Leung, D.Y.C., Wu, X., and Leung, M.K.H. (2010). A review on biodiesel production using catalyzed transesterification. *Applied Energy* 87 (4): 1083–1095.

Li, L., Du, W., Liu, D. et al. (2006). Lipase-catalyzed transesterification of rapeseed oils for biodiesel production with a novel organic solvent as the reaction medium. *Journal of Molecular Catalysis B: Enzymatic* 43 (1–4): 58–62.

Lin, L., Cunshan, Z., Vittayapadung, S. et al. (2011). Opportunities and challenges for biodiesel fuel. *Applied Energy* 88 (4): 1020–1031.

Liu, K. and Wang, R. (2013). Biodiesel production by transesterification of duck oil with methanol in the presence of alkali catalyst. *Petroleum & Coal* 55 (1): 68–72.

Liu, C., Lv, P.M., Yuan, Z.H. et al. (2010). The nanometer magnetic solid base catalyst for production of biodiesel. *Renewable Energy* 35 (7): 1531–1536.

Liu, K., Wang, R., and Yu, M. (2018). An efficient recoverable solid base catalyst of magnetic bamboo charcoal: preparation, characterization and performance in biodiesel production. *Renewable Energy* 127: 531–538.

Mahajan, S., Konar, S.K., and Boocock, D.G.B. (2006). Determining the acid number of biodiesel. *Journal of the American Oil Chemists Society* 83 (6): 567–570.

Manan, F.A.B.A. (2013). Exploring of the transesterification of castor oil to produce biodiesel using quicklime as a basic catalyst. Doctoral dissertation submitted to University Malaysia Pahang, Pahang, Malaysia.

Mansir, N., Teo, S.H., Rabiu, I., and Taufiq-Yap, Y.H. (2018). Effective biodiesel synthesis from waste cooking oil and biomass residue solid green catalyst. *Chemical Engineering Journal* 347: 137–144.

Marchetti, J.M., Miguel, V.U., and Errazu, A.F. (2007). Possible methods for biodiesel production. *Renewable and Sustainable Energy Reviews* 11 (6): 1300–1311.

Mathiyazhagan, M. and Ganapathi, A. (2011). Factors affecting biodiesel production. *Research in Plant Biology* 1 (2): 1–5.

Melero, J.A., Iglesias, J., and Morales, G. (2009). Heterogeneous acid catalysts for biodiesel roduction: current status and future challenges. *Green Chemistry* 11 (9): 1285–1308.

Mohadesi, M., Aghel, B., Maleki, M., and Ansari, A. (2020). The use of KOH/Clinoptilolite catalyst in pilot of microreactor for biodiesel production from waste cooking oil. *Fuel* 263: 116659.

Mutreja, V., Singh, S., and Ali, A. (2011). Biodiesel from mutton fat using KOH impregnated MgO as heterogeneous catalysts. *Renewable Energy* 36 (8): 2253–2258.

Mutreja, V., Singh, S., and Ali, A. (2014). Potassium impregnated nano crystalline mixed oxides of La and Mg as heterogeneous catalysts for transesterification. *Renewable Energy* 62: 226–233.

Nabi, M. and Hoque, N. (2008). Biodiesel production from linseed oil and performance study of a diesel engine with diesel bio-diesel fuels. *Journal of Mechanical Engineering* 39 (1): 40–44.

Nabi, M.N., Rahman, M.M., and Akhter, M.S. (2009). Biodiesel from cotton seed oil and its effect on engine performance and exhaust emissions. *Applied Thermal Engineering* 29 (11–12): 2265–2270.

Nag, A. (2008). Biofuels refining and performance. *NICE (News & Information for Chemical Engineers)* 26 (2): 165–165.

Narasimharao, K., Lee, A., and Wilson, K. (2007). Catalysts in production of biodiesel: a review. *Journal of Biobased Materials and Bioenergy* 1 (1): 19–30.

Nascimento, L.A.S., Angélica, R.S., da Costa, C.E.F. et al. (2011). Conversion of waste produced by the deodorization of palm oil as feedstock for the production of biodiesel using a catalyst prepared from waste material. *Bioresource Technology* 102 (17): 8314–8317.

Nelson, L.A., Foglia, T.A., and Marmer, W.N. (1996). Lipase-catalyzed production of biodiesel. *Journal of the American Oil Chemists' Society* 73 (9): 1191–1195.

Noiroj, K., Intarapong, P., Luengnaruemitchai, A., and Jai-In, S. (2009). A comparative study of KOH/Al_2O_3 and KOH/NaY catalysts for biodiesel production via transesterification from palm oil. *Renewable Energy* 34 (4): 1145–1150.

Norjannah, B., Ong, H.C., Masjuki, H.H. et al. (2016). Enzymatic transesterification for biodiesel production: a comprehensive review. *RSC Advances* 6 (65): 60034–60055.

Ogbunugafor, H.A., Eneh, F.U., and Ozumba, A.N. (2011). Physico-chemical and antioxidant properties of *Moringa oleifera* seed oil. *Pakistan Journal of Nutrition* 10 (5): 409–414.

Pasupulety, N., Gunda, K., Liu, Y. et al. (2013). Production of biodiesel from soybean oil on CaO/Al_2O_3 solid base catalysts. *Applied Catalysis A: General* 452: 189–202.

Patil, P.D. and Deng, S. (2009). Optimization of biodiesel production from edible and nonedible vegetable oils. *Fuel* 88 (7): 1302–1306.

Peng, Y.P., Amesho, K.T.T., Chen, C.E. et al. (2018). Optimization of biodiesel production from waste cooking oil using waste eggshell as a base catalyst under a microwave heating system. *Catalysts* 8 (2): 81.

Pramanik, K. (2003). Properties and use of *Jatropha curcas* oil and diesel fuel blends in compression ignition engine. *Renewable Energy* 28 (2): 239–248.

Pullen, J. and Saeed, K. (2012). An overview of biodiesel oxidation stability. *Renewable and Sustainable Energy Reviews* 16 (8): 5924–5950.

Rahman, M.A., Aziz, M.A., Al-khulaidi, R.A. et al. (2017). Biodiesel production from microalgae *Spirulina* maxima by two step process: optimization of process variable. *Journal of Radiation Research and Applied Sciences* 10 (2): 140–147.

Ramadhas, A.S., Jayaraj, S., and Muraleedharan, C. (2005). Biodiesel production from high FFA rubber seed oil. *Fuel* 84 (4): 335–340.

Ramírez-Verduzco, L.F., Rodríguez-Rodríguez, J.E., and Jaramillo-Jacob, A.R. (2012). Predicting cetane number, kinematic viscosity, density and higher heating value of biodiesel from its fatty acid methyl ester composition. *Fuel* 91 (1): 102–111.

Ramli, A., Farooq, M., Naeem, A. et al. (2017). Bifunctional heterogeneous catalysts for biodiesel production using low cost feedstocks: a future perspective. In: *Frontiers in Bioenergy and Biofuels* (ed. E. Jacob-Lopes). Rijeka: InTech, ch. 14.

Rashid, U. and Anwar, F. (2008). Production of biodiesel through optimized alkaline catalyzed transesterification of rapeseed oil. *Fuel* 87 (3): 265–273.

Rasouli, H. and Esmaeili, H. (2019). Characterization of MgO nanocatalyst to produce biodiesel from goat fat using transesterification process. *3 Biotech* 9 (11): 429.

Rubio-Caballero, J.M., Santamaría-González, J., Mérida-Robles, J. et al. (2009). Calcium zincate as precursor of active catalysts for biodiesel production under mild conditions. *Applied Catalysis B: Environmental* 91 (1–2): 339–346.

Sahoo, P.K. and Das, L.M. (2009). Process optimization for biodiesel production from Jatropha, Karanja and Polanga oils. *Fuel* 88 (9): 1588–1594.

Salamatinia, B., Hashemizadeh, I., and Ahmad Zuhairi, A. (2013). Alkaline earth metal oxide catalysts for biodiesel production from palm oil elucidation of process behaviors and modeling using response surface methodology. *Iranian Journal of Chemistry and Chemical Engineering* 32 (1): 113–126.

Salimi, Z. and Hosseini, S.A. (2019). Study and optimization of conditions of biodiesel production from edible oils using $ZnO/BiFeO_3$ nano magnetic catalyst. *Fuel* 239: 1204–1212.

Salis, A., Pinna, M., Monduzzi, M., and Solinas, V. (2005). Biodiesel production from triolein and short chain alcohols through biocatalysis. *Journal of Biotechnology* 119 (3): 291–299.

Sanli, H. and Canakci, M. (2008). Effects of different alcohol and catalyst usage on biodiesel production from different vegetable oils. *Energy & Fuels* 22 (4): 2713–2719.

Santos, F.F., Malveira, J.Q., Cruz, M.G., and Fernandes, F.A. (2010). Production of biodiesel by ultrasound assisted esterification of *Oreochromis niloticus* oil. *Fuel* 89 (2): 275–279.

Saoud, K. (2018). Nanocatalyst for biofuel production: a review. In: *Green Nanotechnology for Biofuel Production* (eds. N. Srivastava, M. Srivastava, H. Pandey, et al.), 36–62. Cham: Springer.

Seffati, K., Honarvar, B., Esmaeili, H., and Esfandiari, N. (2019). Enhanced biodiesel production from chicken fat using $CaO/CuFe_2O_4$ nanocatalyst and its combination with diesel to improve fuel properties. *Fuel* 235: 1238–1244.

Seffati, K., Esmaeili, H., Honarvar, B., and Esfandiari, N. (2020). $AC/CuFe_2O_4@CaO$ as a novel nanocatalyst to produce biodiesel from chicken fat. *Renewable Energy* 147 (1): 25–34.

Shahid, E.M. and Jamal, Y. (2011). Production of biodiesel: a technical review. *Renewable and Sustainable Energy Reviews* 15 (9): 4732–4745.

Singh, A., Pal, A., and Maji, S. (2017). Biodiesel production from microalgae oil through conventional and ultrasonic methods. *Energy Sources, Part A: Recovery, Utilization, and Environmental Effects* 39 (8): 806–810.

Siva, S. and Marimuthu, C. (2015). Production of biodiesel by transesterification of algae oil with an assistance of nano-CaO catalyst derived from egg shell. *International Journal of Chem Tech Research* 7 (4): 2112–2116.

Sivaramakrishnan, K. and Ravikumar, P. (2012). Determination of cetane number of biodiesel land it's influence on physical properties. *ARPN Journal of Engineering and Applied Sciences* 7 (2): 205–211.

Su, E.Z., Zhang, M.J., Zhang, J.G. et al. (2007). Lipase-catalyzed irreversible transesterification of vegetable oils for fatty acid methyl esters production with dimethyl carbonate as the acyl acceptor. *Biochemical Engineering Journal* 36 (2): 167–173.

Sundus, F., Fazal, M.A., and Masjuki, H.H. (2017). Tribology with biodiesel: a study on enhancing biodiesel stability and its fuel properties. *Renewable and Sustainable Energy Reviews* 70: 399–412.

Syam, A.M., Yunus, R., Ghazi, T.I.M., and Yaw, T.C.S. (2009). Methanolysis of Jatropha oil in the presence of potassium hydroxide catalyst. *Journal of Applied Science* 9 (17): 3161–3165.

Tahvildari, K., Anaraki, Y.N., Fazaeli, R. et al. (2015). The study of CaO and MgO heterogenic nano-catalyst coupling on transesterification reaction efficacy in the production of biodiesel from recycled cooking oil. *Journal of Environmental Health Science and Engineering* 13, Art. No.:73.

Talebi, A.F., Mohtashami, S.K., Tabatabaei, M. et al. (2013). Fatty acids profiling: a selective criterion for screening microalgae strains for biodiesel production. *Algal Research* 2 (3): 258–267.

Talha, N.S. and Sulaiman, S. (2016). Overview of catalysts in biodiesel production. *ARPN Journal of Engineering and Applied Sciences* 11 (1): 439–448.

Tamalampudi, S., Talukder, M.R., Hama, S. et al. (2008). Enzymatic production of biodiesel from Jatropha oil: a comparative study of immobilized-whole cell and commercial lipases as a biocatalyst. *Biochemical Engineering Journal* 39 (1): 185–189.

Tamjidi, S. and Esmaeili, H. (2019). Chemically modified CaO/Fe_3O_4 nanocomposite by sodium dodecyl sulfate for Cr (III) removal from water. *Chemical Engineering & Technology* 42 (3): 607–616.

Tamjidi, S., Esmaeili, H., and Kamyab Moghadas, B. (2019). Application of magnetic adsorbents for removal of heavy metals from wastewater: a review study. *Materials Research Express* 6 (10): 102004.

Tang, S., Wang, L., Zhang, Y. et al. (2012). Study on preparation of $Ca/Al/Fe_3O_4$ magnetic composite solid catalyst and its application in biodiesel transesterification. *Fuel Processing Technology* 95: 84–89.

Taufiq-Yap, Y., Lee, H.V., Yunus, R., and Juan, J.C. (2011). Transesterification of non-edible *Jatropha curcas* oil to biodiesel using binary Ca–Mg mixed oxide catalyst: effect of stoichiometric composition. *Chemical Engineering Journal* 178: 342–347.

Teo, S.H., Rashid, U., and Taufiq-Yap, Y.H. (2014). Biodiesel production from crude *Jatropha Curcas* oil using calcium based mixed oxide catalysts. *Fuel* 136: 244–252.

Teo, S.H., Islam, A., and Taufiq-Yap, Y.H. (2016). Algae derived biodiesel using nanocatalytic transesterification process. *Chemical Engineering Research and Design* 111: 362–370.

Thitsartarn, W. and Kawi, S. (2011). An active and stable $CaO–CeO_2$ catalyst for transesterification of oil to biodiesel. *Green Chemistry* 13 (12): 3423–3430.

Tiwari, A.K., Kumar, A., and Raheman, H. (2007). Biodiesel production from jatropha oil (*Jatropha curcas*) with high free fatty acids: an optimized process. *Biomass and Bioenergy* 31 (8): 569–575.

Ullah, Z., Bustam, M.A., and Man, Z. (2015). Biodiesel production from waste cooking oil by acidic ionic liquid as a catalyst. *Renewable Energy* 77: 521–526.

Verma, P. and Sharma, M.P. (2016). Review of process parameters for biodiesel production from different feedstocks. *Renewable and Sustainable Energy Reviews* 62: 1063–1071.

Visser, E.M., Oliveira, F.,.M., Martins, M.A., and Steward Brian, L. (2011). Bioethanol production potential from Brazilian biodiesel co-products. *Biomass and Bioenergy* 35 (1): 489–494.

Wahlund, B., Yan, J., and Westermark, M. (2004). Increasing biomass utilization in regional energy systems: a comparative study of CO_2 reduction and cost for different bioenergy processing options. *Biomass and Bioenergy* 26 (6): 531–544.

Wang, Y., Liu, J., Gerken, H. et al. (2014). Highly-efficient enzymatic conversion of crude algal oils into biodiesel. *Bioresource Technology* 172: 143–149.

Wen, L., Wang, Y., Lu, D. et al. (2010). Preparation of KF/CaO nanocatalyst and its application in biodiesel production from Chinese tallow seed oil. *Fuel* 89 (9): 2267–2271.

Wen, Z., Yu, X., Tu, S.-T. et al. (2010). Biodiesel production from waste cooking oil catalyzed by TiO_2–MgO mixed oxides. *Bioresource Technology* 101 (24): 9570–9576.

Winayanuwattikun, P., Kaewpiboon, C., Piriyakananon, K. et al. (2008). Potential plant oil feedstock for lipase-catalyzed biodiesel production in Thailand. *Biomass and Bioenergy* 32 (12): 1279–1286.

Yaman, S. (2004). Pyrolysis of biomass to produce fuels and chemical feedstocks. *Energy Conversion and Management* 45 (5): 651–671.

Yan, S., Kim, M., Salley, S.O., and Ng, K.S. (2009). Oil transesterification over calcium oxides modified with lanthanum. *Applied Catalysis A: General* 360 (2): 163–170.

Yaşar, F. and Altun, Ş. (2018). Biodiesel properties of microalgae (*Chlorella protothecoides*) oil for use in diesel engines. *International Journal of Green Energy* 15 (14–15): 941–946.

Yoo, S.J., Lee, H.S., Bambang, V. et al. (2010). Synthesis of biodiesel from rapeseed oil using supercritical methanol with metal oxide catalysts. *Bioresource Technology* 101 (22): 8686–8689.

Zabeti, M., Daud, W.M.A.W., and Aroua, M.K. (2010). Biodiesel production using alumina-supported calcium oxide: an optimization study. *Fuel Processing Technology* 91 (2): 243–248.

Zhang, Y., Dube, M.A., McLean, D.D., and Kates, M. (2003). Biodiesel production from waste cooking oil: 2. Economic assessment and sensitivity analysis. *Bioresource Technology* 90 (1): 229–240.

2

Nano(Bio)Catalysts: An Effective Tool to Utilize Waste Cooking Oil for the Biodiesel Production

Rushikesh Fopase, Swati Sharma and Lalit M. Pandey

Bio-Interface & Environmental Engineering Lab, Department of Biosciences and Bioengineering, Indian Institute of Technology Guwahati, Guwahati, India

2.1 Introduction

Crude oil is the primary source for the fuels used in automobiles and various machinery. However, oil reserves are predicted to become exhausted within the next five decades (Shafiee and Topal 2009; Stephens et al. 2010). Also, the burning of fossil fuels causes pollution, which is increasing above permitted levels, with the release of carbon dioxide and methane gas linked to global warming (Letcher 2019; Ahima 2020). Therefore, to fulfill the increasing global demand for energy and tackle the pollution problem, various alternative fuel sources are being explored by researchers.

Biofuels offer the advantages of being environmentally friendly, sustainable, and renewable (Surriya et al. 2015; Prasad & Ingle 2019). There are three categories of biofuel: solid (biochar), liquid (biodiesel, bioethanol, bio-oil, etc.), and gaseous (biohydrogen, bio syngas, and biogas) (Demirbas 2008; Datta et al. 2018). Biodiesel has very similar characteristics to the conventional fuels used in internal combustion engines (Gog et al. 2012; Arumugam and Ponnusami 2017). The production process of biodiesel is based on the available biomass, and thus the biodiesel produced does not contain any source-originated toxic chemicals. Also, combustion of biodiesel does not release harmful gases (CO, NOx, SOx), unlike that of petroleum fuels (Rodionova et al. 2017; Hosseinzadeh-Bandbafha et al. 2018). Further, researchers have reported that the application of biodiesel can improve the performance of combustion engines (Chincholkar et al. 2005; Van Gerpen 2005; Ashraful et al. 2014).

The Indian government has adopted several policies to meet fuel demands by blending biodiesels with petroleum fuels. Under these policies, the Ministry of New and Renewable Energy, Ministry of Petroleum and Natural Gas and Ministry of Agriculture are working toward the production and application of biofuels such as biodiesel and bioethanol (Chandel et al. 2017; Saravanan et al. 2018). Figure 2.1 shows data on the global energy supply from different types of fuel source. It can clearly be seen that the use of oil has been reduced and that of biofuels increased. This represents the outcome of various government initiatives aimed at the popularization of biofuels.

Nano- and Biocatalysts for Biodiesel Production, First Edition. Edited by Avinash P. Ingle.

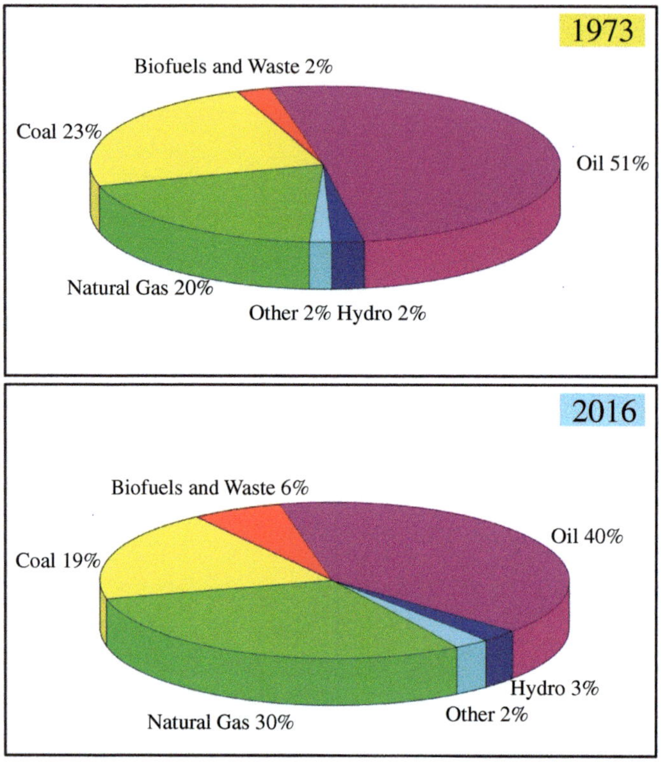

Figure 2.1 Global energy supply from different types of energy sources. Source: Adapted from Saravanan et al. (2018) with copyright permission from Elsevier.

Biodiesel is the product of transesterification of fatty acids present in sources such as vegetable and animal oils, which are converted to fatty acid alkyl esters. The conversion is carried out in the presence of short-chain alcohols (methanol or ethanol). Use of ethanol gives fatty acids ethyl ester (FAEE), while methanol produces fatty acids methyl ester (FAME) (Lam et al. 2010; Chesterfield et al. 2012). However, methanol is more widely used due to its low cost and availability. The transesterification reaction is catalyzed by chemical (acid, alkali, nanomaterials) or biochemical catalysts in order to lower the activation energy barrier (Gog et al. 2012; Sharma et al. 2018; Saxena et al. 2019). The chemical catalysts are low-cost materials but work at relatively high temperatures (60–250 °C) and cause secondary pollution (Sharma et al. 2018), while the biochemical catalysts (like enzymes) are specific, work at mild temperatures, and are eco-friendly. However, their production and purification is expensive (Leung et al. 2010), making it necessary to reuse them.

Reuse of enzyme catalysts is restricted by the difficulties in their separation from the reaction mixture. The easiest way to separate an enzyme catalyst is to attach or entrap it on to a distinct and separable support that can be easily separated out by centrifugation, filtration, or magnetic fields. Enzymes are easily immobilized on these nanosize particles and used in the transesterification process (Jegannathan et al. 2008; Meunier et al. 2017). The assembly of enzyme and nanomaterial is referred as a nanobiocatalyst and has been reported to be very efficient in terms of efficiency, reusability, and process economics (Fatima et al. 2020; Ingle et al. 2020a).

Apart from the catalyst, the overall production cost of a biodiesel also depends on the feedstock used (de Jong et al. 2017). Most biodiesel production processes use vegetable oils as raw material. However, use of edible oil in biofuel production leads to the food versus fuel controversy (Tomei and Helliwell 2016). Therefore, the use of low-cost oil feedstock like waste cooking oils (WCOs) is preferable. The focus of this chapter is on the utilization of WCOs for the production of biodiesels using nanobiocatalysts. The mechanism of transesterification is also briefly discussed, as are applications of microbial lipases in biodiesel production. Moreover, various methods for the immobilization of lipases and their utility in the transesterification process are examined, along with recent examples.

2.2 Waste Cooking Oils

WCOs are a superior alternative for the production of biodiesel as they are produced in huge volumes globally on a daily basis. They represent the remainder of oils used in the preparation of food items, which must be disposed of properly in order to avoid pollution. When deposited in the environment, WCOs form thin films over water bodies, disrupting diffusion of oxygen (Singh-Ackbarali et al. 2017). Some bacterial strains, such as *Bacillus subtilis* and *Pseudomonas aeruginosa*, are found to degrade WCOs and produce value-added products like biosurfactants (Sharma et al. 2019; Sharma and Pandey 2020). However, the rate of biodegradation is not sufficient for the amount of WCOs generated. Therefore, the utilization of WCOs for biodiesel production offers environmental benefits in addition to the economic ones (Lam et al. 2010).

WCOs are generally vegetable oils and consist of various components, as listed in Table 2.1. Their properties vary depending on the sources and the exposure temperature (Knothe and Steidley 2009). Elevated temperature and water cause accelerated hydrolysis of the virgin oil triglycerides, increasing the free fatty acid (FFA) content. The average FFA content of a refined oil is within 0.5 wt%, while that of WCOs ranges from 0.5 to 15 wt% (Wang et al. 2006). Higher amounts affect the efficiency of the transesterification process (Bautista et al. 2009). Bouaid et al. (2016) reported a decrease in the yield of methyl esters (i.e. biodiesel) from 97.2% to 95% with increasing FFA values up to 4% in the WCO substrate. Higher FFA contents also lead to saponification (Gnanaprakasam et al. 2013). Similarly, increasing oil viscosity significantly increases the number of heating cycles. Increases in peroxides values, acid values, saponification values, and polar compounds also occur following the heating of oils (Bandyopadhyay et al. 2017; Uslu and Özcan 2018; Yahya et al. 2019). Therefore, the removal of these impurities is recommended, by pretreating the collected WCO feedstocks.

2.3 Pretreatment of WCOs

The FFA content of WCOs can be removed by esterification using acids like sulfuric acid and methanol or ion-exchange resins (Marchetti and Errazu 2008). Özbay and coworkers (2008)

Table 2.1 Chemical compositions of a WCO (Asli et al. 2012).

Component	Amount (%)
Lauric acid (C12 : 0)	0.34
Myristic acid (C14 : 0)	1.03
Palmitic acid (C16 : 0)	38.35
Stearic acid (C18 : 0)	4.33
Oleic acid (C18 : 2n9c)	43.67
Linoleic acid (C18 : 2n6c)	11.39
γ-Linolenic acid (C18 : 3n6)	0.37
Linolenic acid (C18 : 3n3)	0.29
Cis-11-Eicosenoic acid (C20 : 1)	0.16
Heneicosanoic acid (C21 : 0)	0.08

reported the use of ion-exchange resin (A-15) to reduce FFA contents with a conversion of 45.7% at 60 °C. Soap formation can be avoided by neutralization with alkali followed by distillation. Researchers have reported the maintenance of the saponification value of WCOs at 298.97 mg/KOH g (GB/T5530-1998) while using lipase-catalyzed transesterification (Chen et al. 2009; Ali et al. 2017). The water content can be easily removed by heat treating the WCOs at above 100 °C; various industries also use vacuum distillation at 0.05 bar (Felizardo et al. 2006). Solid impurities can be separated by filtration or centrifugation of WCOs and washing with sterile water (Ali et al. 2017).

2.4 Transesterification Process

In the transesterification process, constituent fatty acids are separated from the glycerol chain to produce fatty acid esters and glycerol (Otera 1993). This process requires low-molecular-weight alcohols, and some catalysts to speed it up. The free fatty esters thus produced are used as biodiesel. The acyl group released from the reaction is accepted by the added short-chain alcohols such as methanol and ethanol (Zhao et al. 2014; Marx 2016). The alcohols may sometimes affect the enzyme activity. Therefore, the use of different acyl receptors like dimethyl carbonate is being investigated for better chemical activity, prevention of glycerol formation, and eco-friendly characteristics (Gharat and Rathod 2013).

The general mechanism of the transesterification suggests the formation of 3 mol of biodiesel or FAME and glycerol when 1 mol of triglycerides from WCOs reacts with 3 mol of methanol (Figure 2.2). However, the transesterification occurs in a stepwise manner (Figure 2.3) with the formation of intermediate products as (i) conversion of fatty acids into diglycerides, (ii) formation of monoglycerides from diglycerides, and (iii) final conversion of monoglycerides into biodiesel and glycerol (Mansir et al. 2018; Sharma et al. 2018).

The transesterification reaction is accelerated by different catalysts. The types of catalyst used classify the transesterification process as acid-, base-, heterogeneous-, or

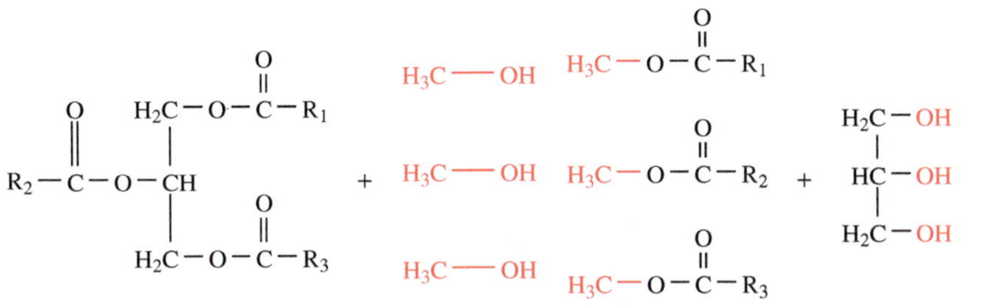

Figure 2.2 General transesterification reaction. Source: Adapted from Sharma et al. (2018) an open access article.

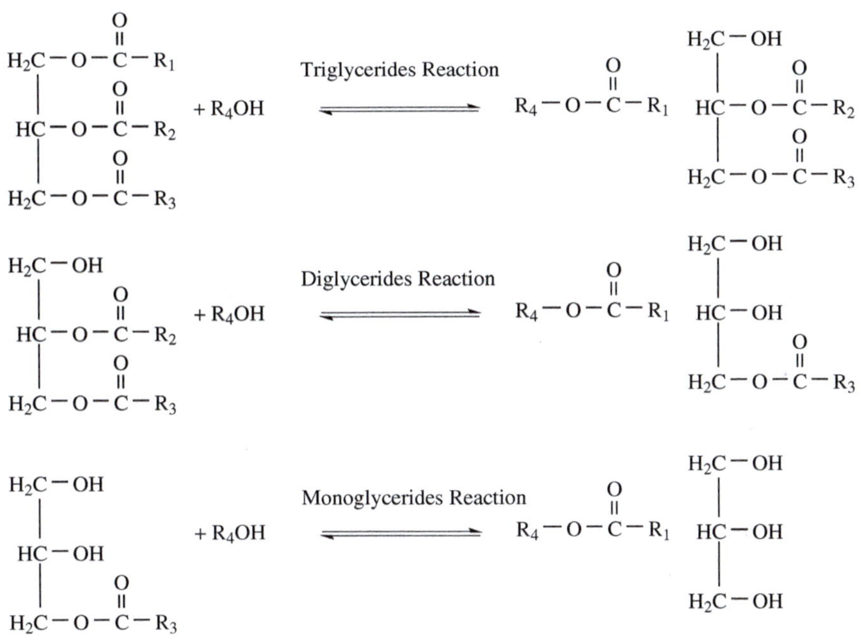

Figure 2.3 Three-step transesterification reactions for biodiesel production. Source: Adapted from (Mansir et al. 2018) with copyright permission from Elsevier.

enzyme-catalyzed (Leung et al. 2010). In acid catalysis, sulfuric acid, hydrochloride acid, or organic sulfonic acid is used. The problems with the acid catalysts are their corrosive properties, increased waste generation, high reaction temperatures, and high costs (Prafulla et al. 2012). Base catalysts such as sodium hydroxide and potassium hydroxide are widely used in base-catalyzed transesterification. Though they are low-cost materials, their high hygroscopic nature reduces the conversion rate, with increased catalyst consumption. Also, saponification rate is higher with their use (Moazeni et al. 2019). In contrast to chemical-based catalysis, the enzyme-catalyzed transesterification process is quite simple. Figures 2.4 and 2.5 show the process design for both chemical-catalyzed and enzyme-catalyzed transesterifications. Figure 2.5 shows a lesser number of process equipments and reduced byproduct generation. Therefore, enzyme catalysts (biocatalysts) are

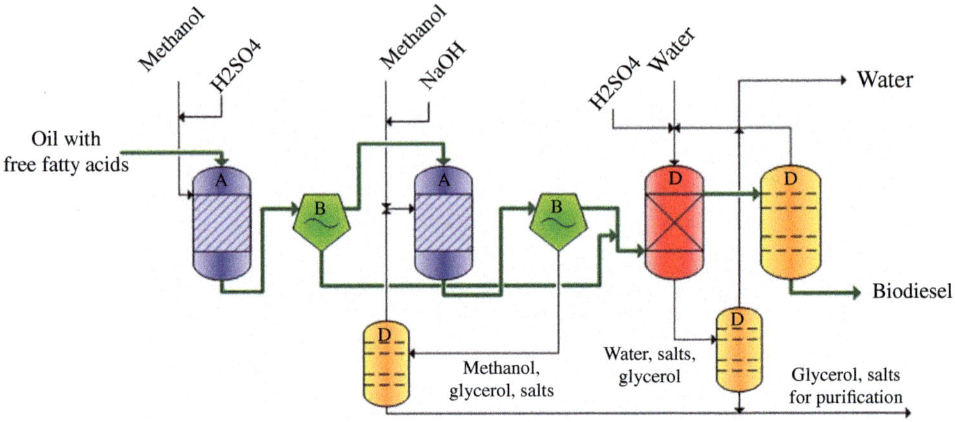

Figure 2.4 Typical process design of the transesterification reaction for biodiesel production through chemical catalysis. A: Reactor, B: Separator, C: Filter, D: Byproduct recovery. Source: Adapted from Fjerbaek et al. (2009) with copyright permission from Wiley and Sons.

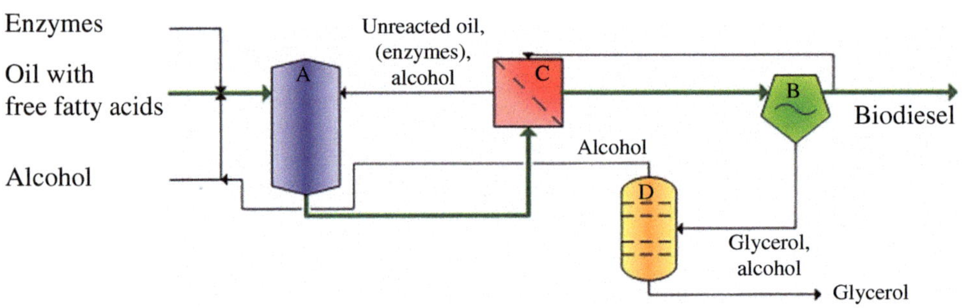

Figure 2.5 Typical process design of the transesterification reaction for biodiesel production through enzymatic catalysis. A: Reactor, B: Separator, C: Filter, D: Byproduct recovery. Source: Adapted from Fjerbaek et al. (2009) with copyright permission from Wiley and Sons.

reportedly proven suitable for the efficient production of biodiesel (Norjannah et al. 2016). Enzyme-catalyzed reactions avoid soap formation and thus offer easy purification and no secondary pollution. The enzymatic biocatalysis for biodiesel production is discussed in details in later sections.

2.4.1 Kinetics of Transesterification

Several kinetic studies of transesterification using lipase enzymes have been reported. The enzyme (lipase)-catalyzed transesterification reaction occurs in two steps. Initially, hydrolysis of ester bond occurs with the release of alcohol moiety. This is followed by esterification using the second substrate. The reaction kinetics of the transesterification is described by the ping-pong bi-bi mechanism (Garcia et al. 1996; Paiva et al. 2000; Pilarek and Szewczyk 2007). In this mechanism, the first substrate triglycerides form a complex with the enzyme to give an intermediate. Then the reaction yields the products of fatty acids and glycerol. In the second step, the generated second substrate (i.e. fatty acids) reacts with the alcohol substrate to give fatty acid esters and the native enzyme is released (Al-Zuhair 2005). The mechanism is given as follows:

$$E + E_{S_s} \rightleftharpoons E \cdot E_{S_s} \rightleftharpoons F \cdot B_p \rightleftharpoons F + B_p \tag{2.1}$$

$$F + A_S \rightleftharpoons F \cdot A_S \rightleftharpoons E \cdot E_{S_p} \rightleftharpoons E + E_{S_p} \tag{2.2}$$

where subscripts "s" and "p" indicate substrate and product, respectively, E is an enzyme, E_S is ester substrate (glycerides from WCOs), F is a fatty acid, B_p is glycerol, A_S is alcohol substrate (methanol or ethanol), and E_{S_p} is the desired biodiesel (fatty acid alkyl esters).

The Michaelis-Menten kinetics was applied to fit experimental data from the transesterification reaction. The initial rate equation from Eqs. (2.1) and (2.2) was given as follows (Dossat et al. 2002; Fjerbaek et al. 2009):

$$v_i = \frac{V_{max}[S][A]}{K_{m_S}[A](1 + [A]/K_{i_A}) + K_{m_A}[S] + [S][A]} \tag{2.3}$$

where v_i is the initial rate, V_{max} is the maximum initial rate, K_{m_S} and K_{m_A} are Michaelis constants for the glyceride substrate and alcohol substrate, K_{i_A} is the inhibitory constant of the alcohol substrate, and $[S]$ and $[A]$ are concentrations of triglycerides and acyl acceptor, respectively.

In practice, other factors that affect the transesterification reaction also need to be considered, including lipase type, quality and constituents of WCOs, and temperature affecting enzyme activity. However, the proceeding kinetics does not give information on intermediate product formation. Also, the enzyme activity is affected by the reaction temperature and hence the equilibrium of the reaction. Therefore, considering the influencing factors, Eq. (2.3) can be made more specific for the type of glyceride by applying the "rake" model. Also, equilibrium limitation can be corrected by adding a reversibility term, as mentioned by Boudart and Djéga-Mariadassou (1984). The effect of temperature can be resolved with the first-order deactivation kinetics of the enzyme. The modified equation is then (Fjerbaek et al. 2009):

$$v_j = \frac{V_{max}(1 - e^{-k_d})[jG][A]}{K_{m_j}[A](1 + [A]/K_{i_A}) + K_{m_A}[jG] + [jG][A]} \times \left(1 - \frac{[B_p][E_{S_p}]/[jG][A]}{K_{eq_j}}\right) \tag{2.4}$$

where j is a specific type of glyceride (tri-, di-, or monoglycerides from WCOs), k_d is the deactivation constant due to temperature effect, and K_{eq_j} is the equilibrium constant for each step.

Transesterification is a complex process and is affected by the individual setup of the experiment. The amount of substrate, reaction conditions, and purity and grade of the enzymes all affect the reaction rate. Also, the type of solvent and immobilization create much difference in terms of mass transfer. Therefore, the kinetics differs from setup to setup, and can be explained by considering a few more parameters added to Eq. (2.4).

2.5 Enzymatic Biocatalysts

As discussed, the chemical catalyst-based reactions require higher reaction temperature (60–250 °C) and therefore have greater energy requirements. Also, post-reaction treatments such as separation of corrosive solvents, glycerol, and catalyst add more cost to the process (Norjannah et al. 2016). In contrast to this, enzymatic catalysis is energy-efficient and generates fewer unwanted products. Therefore, enzymatic catalysts are better alternatives to the chemical-based catalytic systems and offer better selectivity (Christopher et al. 2014). Use of a biocatalyst offers transesterification of a wide range of oil feedstocks with high FFA contents.

This process is environmental friendly, and the biodegradability of the enzymes does not lead to any secondary pollution (Moazeni et al. 2019). Improvement in the stability and the reuse of enzymes is possible through various types of immobilization process (see Section 2.5), which significantly reduces the overall production cost. Immobilization offers easy separation of enzymes from the reaction mixture and helps retain the enzymatic activity for subsequent cycles (Dizge et al. 2009). Lipases are the most widely used enzymes for the transesterification of WCOs in order to produce biodiesel.

2.5.1 Lipases

Lipases (EC 3.1.1.3), also known as triglycerol ester hydrolases, are used for hydrolysis, alcoholysis, esterification, and transesterification processes (Tan et al. 2010). The catalytic activity and stability of the lipases are reported to be excellent in non-aqueous media, and therefore these enzymes are extensively used in organic and ionic liquids (De Los Ríos et al. 2007). The potential of the lipases for biodiesel production has been assessed over the last few decades and they have proven very efficient (Bajaj et al. 2010; Amini et al. 2017). During the transesterification process, lipases convert tri-, di-, and monoglycerides to biodiesel and also transesterify FFA, which is essential from the perspective of yield and production cost (Souza et al. 2016; Vargas et al. 2018). Therefore, lipases can be used in the transesterification process for a wide range of substrate materials.

Lipases are produced by a range of microbes, including bacteria and fungi. Commonly used fungi are *Candida antartica* and *Candida* species 99–125 (Tan et al. 2006; Fan et al. 2012; Narwal and Gupta 2013), while bacteria include *B. subtilis* and *Pseudomonas* species (Norjannah et al. 2016). Depending upon the reaction conditions and the cost of the process, lipases can be applied as biocatalysts for the transesterification process in either extracellular or intracellular form. Their extracellular and intracellular applications are explained in the following sections.

2.5.1.1 Extracellular Lipases

The extracellular lipases are extracted and purified from microbial sources and then applied in the transesterification process. They are initially tested for their optimum performance under suitable reaction conditions before actual production. *Rhizopus oryzae*, *Candida antarctica*, *Mucor miehei*, and *Pseudomonas* sp. are some major producers (Gog et al. 2012). Ali et al. (2017) reported the use of free purified lipase isolated from *P. aeruginosa* for biodiesel production using WCOs in a liquid medium with methanol, with the process optimized using the response surface methodology. The yield was observed as 86% with an optimized oil to methanol ratio of 3.05 : 1 at 44.2 °C for 24 hours (Ali et al. 2017). The extracellular lipases are soluble in the reaction medium and hence their reuse is challenging. Also, the use of free enzyme in the liquid medium hinders the stability of the enzyme, resulting in lower yield. Immobilization of enzyme over some support materials offers better stability and ease of separation from the reaction mixture. Therefore, the process cost and the time required are significantly reduced. The immobilization of enzymes was discussed later in this section. Many researchers have used extracellular lipases extracted from *C. antarctica* for biodiesel production immobilized on magnetic nanoparticles (MNPs) or mesoporous support membranes (Xie and Ma 2010; Chesterfield et al. 2012). In a study by Mehrasbi and colleagues (2017), application of the immobilized lipase showed improved performance and better reusability. The free or extracellular lipase was immobilized on Fe_3O_4 core and silica shell nanoparticles (NPs) functionalized with silanes and retained 97% of the specific activity; silanes are widely used for surface functionalization, forming a silanol bond at the surface and exposing the desired functional groups (Pandey and Pattanayek 2011; Pandey et al. 2013; Hasan and Pandey 2016, 2020). The immobilized lipase was then mixed with WCOs and methanol, with an optimized oil to methanol ratio of 1 : 3. The immobilized lipase gave a 100% yield of biodiesel and reusability for up to six cycles. Different extracellular lipases isolated from various sources have been applied for the transesterification of WCOs, showing good results. Lipases from *Geotrichum* species immobilized by cross-linking in microcrystals have been used for biodiesel production (Yan et al. 2011). Similarly, *Pseudomonas cepacia* lipase has been used for the transesterification of WCOs by immobilization on magnetic NPs (Yu et al. 2013). *Candida* sp. 99–125 isolated lipase was immobilized on the diatom for transesterification of WCO, resulting in a 90% yield (Zhao et al. 2016). Table 2.2 lists different extracellular lipases used in biodiesel production using WCOs.

2.5.1.2 Intracellular Lipases

In this approach, whole-cell microbes producing intracellular lipases are engaged for the transesterification process. This reduces the cost of enzyme extraction and purification and offers easy separation from the reaction mixture (Chen et al. 2017). The biomass-producing lipases are attached on the support and added in the reactor for the transesterification. *Aspergillus* sp. and *Rhizopus* sp. are major microbes producing intracellular lipases used for this process (Pazouki et al. 2010; Balasubramaniam et al. 2012; El-Batal et al. 2016; Touqeer et al. 2020). Chen et al. (2017) have used *Pseudomonas mendocina* cells in the magnetically fluidized bed reactor for the production of biodiesel (Chen et al. 2017). The whole cells-producing lipase was intended to catalyze the transesterification of WCOs in the presence of methanol under optimized reaction conditions. The observed yield of the reactors was 91.8% at 35 °C for 48 hours, and 10 times reusability with the retention of 87.5% of initial yield was found. In a similar study, Balasubramaniam et al. (2012) reported a comparative analysis of biodiesel production by means of purified lipase extracted from *R. oryzae* and the whole cells themselves using WCOs as substrate (Balasubramaniam et al. 2012; Bharathiraja et al. 2015). The whole cell-producing lipase and purified lipase were immobilized in separate calcium alginate beads under pre-optimized reaction conditions of 30 °C and 15 rpm for 24 hours in the presence of a WCO to methanol ratio of 1 : 3. The pure enzyme resulted in higher transesterification (94%) compared to whole cells (84%). However, from an economic perspective, the cost of extraction and purification of lipase from the cells was higher compared to the use of whole cells for biodiesel production. Therefore, the use of intracellular lipase should be preferred in order to minimize production costs

Table 2.2 Different types of bacterial and fungal lipases used for the transesterification of WCOs to produce biodiesel

Microorganism	Type of enzyme production	Alcohol: WCOs	Temp. (°C)	pH	Time (h)	Yield (%)	Reuse cycle	References
Bacteria								
Streptomyces sp. CS273	Extracellular	1 : 1	RT	8.5	48	80	—	Mander et al. (2014)
Streptomyces sp. W007	Extracellular	1 : 3	30		24	95.45	4	Wang et al. (2017)
Thermomyces lanuginosus	Extracellular	6 : 1	45		24	97.8		Sarno et al. (2019)
Thermomyces lanuginosus	Extracellular	6 : 1	65		24	97		Dizge et al. (2009)
Bacillus subtilis	Extracellular	4.5 : 1	25		48	93		Heater et al. (2019)
Bacillus subtilis 1.198	Extracellular	1 : 1	40	6.5	72	90	—	Ying and Chen (2007)
Pseudomonas aeruginosa	Extracellular	3.05 : 1	44.2		24	87		Ali et al. (2017)
Pseudomonas mendocina	Intracellular	3.74 : 1	35		48	91.8	10	Chen et al. (2017)
Pseudomonas sp.	Intracellular	6 : 1	37	7.4	48	70	—	Ali et al. (2011)
Thermomyces lanuginosus	Intracellular	4 : 1	40		48	75	3	Yan et al. (2014b)
Fungi								
Aspergillus niger	Extracellular	4 : 1	45		4	90	5	El-Batal et al. (2016)
Aspergillus terreus AH-F2	Extracellular	6 : 1	37		30	92		Touqeer et al. (2020)
Aspergillus niger	Extracellular	9 : 1	30		10	94	5	Arumugam and Ponnusami (2017)
Candida species 99-125	Extracellular	6 : 1	32.5		60	95.9	—	Mumtaz et al. (2012)
Candida species 99-125	Extracellular	1 : 1	45		100	91.08	—	Chen et al. (2009)
Rhizopus Oryzae	Extracellular	3 : 1	30		24	94		Balasubramaniam et al. (2012)
Candida antarctica	Extracellular	3 : 1	40		24	~90		Liu et al. (2013)
Malassezia globose SMG1-F278N	Extracellular	2 : 1	30		8	98.24	20	Li et al. (2017)

Table 2.2 (Continued)

Microorganism	Type of enzyme production	Alcohol: WCOs	Temp. (°C)	pH	Time (h)	Yield (%)	Reuse cycle	References
Geotrichum sp.	Extracellular	4 : 1	45	4–6	4	85	—-	Yan et al. (2011)
Rhizopus Oryzae	Intracellular	3 : 1	30	5.5	24	83.76		Balasubramaniam et al. (2012)
Rhizopus Oryzae PTCC 5174	Intracellular	3 : 1	35	6.8	72	98.4	—	Pazouki et al. (2011)
Pichia pastoris	Intracellular	4 : 1	40		84	82	3	Yan et al. (2014b)
Pichia pastoris	Intracellular	6 : 1	30		96	87	—	Yan et al. (2014a)
Aspergillus niger KY 401431	Intracellular	3 : 1	30		72	75.5		Amin et al. (2018)
Aspergillus nomius	Intracellular	5 : 1	40		24	95.3	10	Talukder et al. (2013)

and increase reusability. Similarly, Pazouki et al. (2010) reported the use of *R. orazae* as whole-cell catalyst for biodiesel production with a yield of 88% using WCOs. Table 2.2 lists different intracellular lipases used in biodiesel production using WCOs.

2.6 Enzyme Immobilization Techniques

Researchers are looking at various techniques for confining the enzyme within the bulk for easy separation, recovery, and reusability. During enzyme confinement, the base material, termed the "substratum," provides stable support and acts as a carrier of the enzyme for catalysis (Elnashar 2011). However, the size and morphology of the substratum have been found to cause steric hindrance, misconfiguration of enzymes, and diffusion limitations. The efficiency of the immobilized system may be increased by applying engineered substratum surfaces using methods like silanization to tune the surface chemistry. Thus, engineered surfaces can hold the enzymes in a desired orientation in order to achieve maximum activity. Researchers have reported the application of self-assembled monolayers for surface functionalization for different purposes like adsorption of proteins and heavy metals (Jawed et al. 2020; Hasan et al. 2018a). These approaches can be applied for the attachment of proteins like lipases on the nanomaterial surfaces in desired orientations and amounts. Additionally, nanostructured substratum has been reported to minimize steric hindrance by projecting a high surface area and hence improving mass transfer (Ingle et al. 2020b). Ahmed and Sardar (2015) reported improved activity of immobilized enzyme on a nanosubstratum due to Brownian motion, which is negligible in the case of free enzyme. Functionalization of nanomaterials offers anchoring points for the enzyme. Free functional groups interact with the anchors and attach to the surface (Hasan et al. 2018b). The strength of the attachment is proportional to the number and affinity between the interacting groups.

These nanosubstratums are generally composed of different elements, such as iron oxides, carbon, polymers, and composites (Kratošová et al. 2019). The interactions between the substratum and enzyme are classified into physical and chemical types, as shown in Figure 2.6.

Physical interaction employs techniques such as simple adsorption, where the enzyme molecule is immobilized on the surface of the substratum due to ionic or van der Waals forces, or "entrapment," where the enzyme is not strictly restricted to the surface of the substratum, but rather is allowed to move freely in a defined space. On the other hand, chemical interaction deals with the formation of chemical bonds between two entities, making it stronger than physical interaction. Techniques involved in chemical bond formation are classified into two main domains: (i) covalently linked; and (ii) cross-linking. In covalent mode, there is covalent bonding among the enzyme moieties and the substratum, making this the strongest interaction of them all. Finally, apart from the previously mentioned substratum–enzyme-based immobilization techniques, a substratum-free technique has recently been accepted by various scientific communities, where the enzymes are attached to one another using a linker material such as glutaraldehyde (Elnashar 2011).

Table 2.3 lists recent lipase-based immobilization techniques exploited for biodiesel production using nanocatalyst as substratum. The various techniques explored for the immobilization of enzymes and their applications in the production of biodiesel are elaborated on in the next two sections.

2.7 Physical Methods

2.7.1 Adsorption

Adsorption is the most conventional and widely exploited enzyme immobilization technique. It encompasses physical bonding between the substratum and enzyme by van der Waals, hydrogen bonding, hydrophobic interaction, or simple dipole–dipole bonds. During this process, enzymes and nanomaterials are mixed together via simple mixing and separated by centrifugation, rinsed to remove non-adhered enzymes, and further used as a single entity for the catalysis reaction (Sarno and Iuliano 2018). Loss during washing is particularly unavoidable due to the weak force of attachment. Overall, the process is simple, low-cost, and quick, as well as being reversible in nature. This provides the advantages of easy desorption of inactivated enzymes and reuse of the expensive substratum (Nady et al. 2020).

However, this technique often leads to nonspecific binding and overloading of the substratum. In order to overcome this, various surface modification-based attempts have been made in recent decades to improve the bonding between the enzyme and the substratum. In this context, Sarno et al. (2019) immobilized *Thermomyces lanuginosus* lipase over citric acid-modified magnetic Fe_3O_4 for the catalysis of oil to biodiesel. The presence of polar surface groups (i.e. –COOH from citric acid) improved the hydrophilicity of the NPs. Hence, a simple sonication technique was used for the successful mixing of the NPs and enzyme for an initial three minutes, followed by shaking at 200 rpm and 4 °C for an incubation time of three hours. The bionanocatalyst thus prepared showed an excellent biodiesel conversion efficiency of 94% within 24 hours of incubation. In addition, the bionanoassembly exhibited excellent reusability up to 10 cycles (Sarno et al. 2019). Similarly, Xie and Huang (2018)

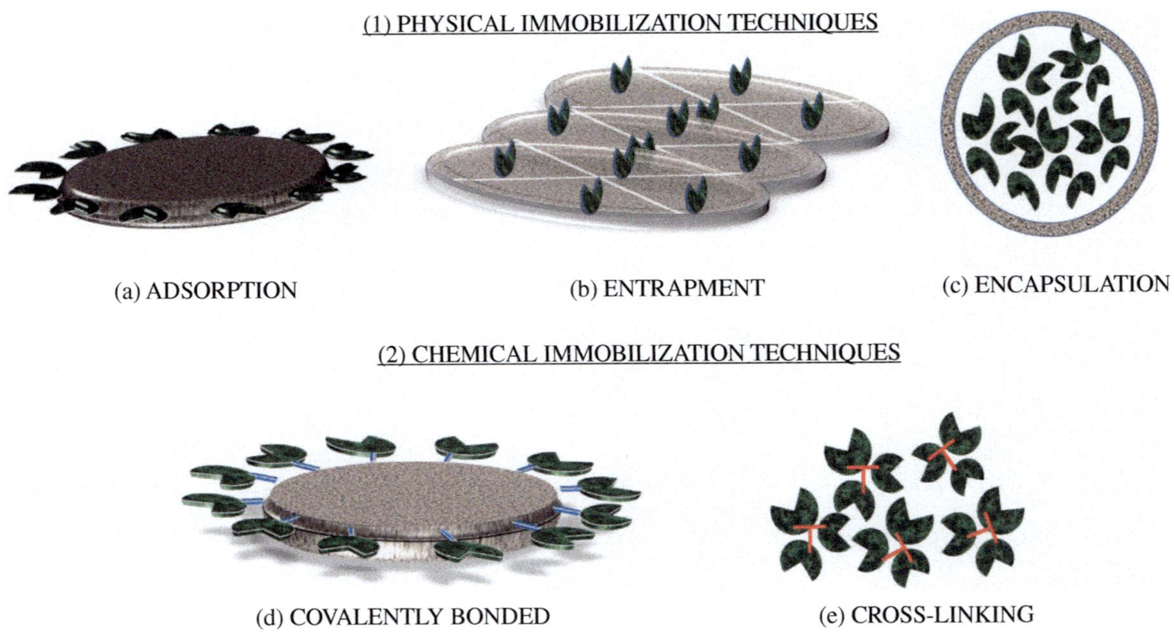

Figure 2.6 Physical and chemical methods for enzyme immobilization.

Table 2.3 Different types of lipase immobilization techniques and their respective yields and reusabilities

Organisms	Oil source	Nanocatalyst	Immobilization technique	Biodiesel yield (%)	Reusability	References
Burkholderia cepacia	Soybean oil	Multi-walled carbon nanotubes modified with polyamidoamine (MMWCNTS-PAMAM)	Covalently bonded	92.8	20	Fan et al. (2017)
Aspergillus oryzae	Soybean oil	Hyperbranched polyglycerol/polyacrylic acid hydrogel	Entrapment	90	5	Ying et al. (2017)
Aspergillus niger	Soybean oil	Sodium titanate nanotubes	Adsorption	89.5	10	Nady et al. (2020)
	Olive oil	Aminopropyl tiethoxysilane (APTES) modified Fe_3O_4 MNPs	Ethyl(dimethylaminopropyl) carbodi-imide – N-hydroxysuccinimide (EDC-NHS) covalent bonding	84.0	5	Jambulingam et al. (2019)
Candida rugosa	Olive oil	$Ca_3(PO_4)_2$ on sulfonated macroporous resins	Encapsulation	89.4	10	Wan et al. (2018)
Candida antarctica	Waste cooking oil	Epoxy-functionalized core-shell MNPs	Covalently bonded	84.0	6	Mehrasbi et al. (2017)
	Soybean oil	MNPs	Cross linking	79.7	5	Modenez et al. (2018)
Candida rugosa	Sunflower oil	Hydrophobic magnetic silica aerogel	Adsorption	72.3	NA	Amirkhani et al. (2019)
Thermomyces lanuginosus	Spent coffee grounds seeds oil	Citric acid (CA) modified Fe_3O_4/Au MNPs	Adsorption	100	5	Sarno and Iuliano (2018)

synthesized graphene oxide-encapsulated Fe_3O_4 NPs, with further surfaces modified with –COOH for the effective affinity of lipase enzyme produced from *Candida rugosa*. The catalyst showed 92.8% biodiesel yield with reusability up to five cycles without loss in enzyme conversion efficacy (Xie and Huang 2018).

Apart from improving the reactivity of the surface using polar functional groups, many researchers have exploited metal chelation techniques for the surface immobilization of proteins. Transition metal ions are known to chelate proteins' –N, –O, and –S functional groups. Such interactions improve the bio-affinity and loading efficiency of the catalyst. In addition, high alkyl groups allow greater enzyme loading capacity. Lipase from *Pseudomonas fluorescens* was studied for biodiesel production using MNPs chemically modified with 3-glycidoxypropyltrimethoxysilane (GOPTS) with 5-aminoisophthalic acid (5-AIPA) functionalization, projecting cobalt ions for improved binding efficiency of lipase enzyme. Thus prepared magnetic nanocatalysts (Co-AGMNPs) were mixed with lipase enzyme in a water bath with constant stirring at 200 rpm. A high conversion efficiency (i.e. 95% of triglycerides of WCOs converted to biodiesel) was achieved in 12 hours of incubation with no significant decrease in activity after 10 repetitive batches (Wang et al. 2019). Hence, stronger bio-affinity on surface modification of substratum led to higher enzyme activity and reusability. Surface chemical modification thus bridges the impaired adsorption capacity of the substratum and hence leads to easy and comparatively stable adsorption of enzymes.

2.7.2 Encapsulation

Encapsulation involves packing enzymes within the grooves of substratum in such a way that nutrients and substrate are easily available to them, and are firmly attached to the substratum to prevent its leakage. This is a type of biomineralization phenomenon in which the enzyme molecule self-assembles on to the substrate in order to overcome instability and conformation-based activity losses (Hou and Ge 2017). It often consists in a semipermeable coating made up of cellulose or nylon or their composites in a size range of 10–100 μm. This acts as a protective layer, allowing the small molecules (substrates and products) to pass through but restricting the washing off of the enzyme and its entities (Elnashar 2011). This configuration holds the enzyme intact with no loss in activity due to an effective mass diffusion phenomenon. However, the membrane often leads to fouling or rupture issues, compromising the enzyme activity in the long run.

Metal organic frameworks (MOFs) have been explored as a selective permeable substratum infrastructure for the immobilization of enzyme. One such microporous MOF is zeolitic imidazolate framework (ZIF-8), which contains coordination bonding between Zn^{2+} ions and 2-methylinidazole. Adnan and colleagues (2018) explored the suitability of this ZIF substratum for the encapsulation of lipase isolated from *Rhizomucor miehei* by forming intermolecular hydrogen and hydrophobic bonds between the imidazole component of the substratum and lipase. An optimized biodiesel conversion efficiency of 95.6% was obtained within eight hours, with almost negligible loss (<15%) in enzyme activity after 10 consecutive cycles (Adnan et al. 2018). Another study revealed the encapsulation of characteristic thermophilic lipase obtained from *Alcaligenes* sp. within zinc acetate and adenine ligands, forming a firm MOF. The coordination bonding between the amine group of lipase and the zinc ions caused the formation of a stable assembly of lipase within the MOF units. This catalytic assembly showed excellent stability at a higher temperature of 90 °C, with 66.3% enzymatic activity achieved where free lipases activity was 11.7%. In addition to excellent stability, the catalyst showed >60% biodiesel conversion with <20% loss in activity over five repetitive cycles (Wrapp et al. 2020). These outcomes facilitate the applicability of this technique in industrial biocatalysis-based biodiesel production applications. Recently, in addition to MOFs, MNPs have been used for the encapsulation of enzyme. These provide the inherent advantage of easy and quick magnetic separation, in contrast to MOFs (Xie and Huang 2018).

2.7.3 Entrapment

Entrapment is another physical enzyme confinement approach, wherein the enzyme molecules are available for 3D movement, reducing the risk of lack of availability of active sites. However, to prevent aggregation within the enzyme molecules, they are confined to the substratum lattice structure, ensuring that there is no leakage from the lattice. Nevertheless, there is no presence of semipermeable membrane, as in the case of encapsulation. Substratum plays a critical role in this technique, as enzyme loading capacity, diffusion of nutrients, and mass transfer depend on the lattice structure. Various polymers with characteristic high tensile strength, nontoxicity, and biodegradability are of prime interest for such immobilization of enzyme. These studies included alginate, sol-gel, agarose, chitosan, and κ-carrageenan (Yagar and Balkan 2017). The entrapment technique has gained enormous attention owing to its rapid, low-cost, simple, and stable characteristics.

Sol-gel-entrapped enzymes have been explored for use in biodiesel production, giving 65–80% yield; however, due to diffusion limitation, researchers have attempted to utilize the porous nature of alginate in the sol-gel preparation (Macario et al. 2009; Zhang et al. 2012). One recent study explored the suitability of biopolymer sodium alginate and polyvinyl alcohol (PVA) for the synthesis of the matrix, in order to entrap lipase for biodiesel production. *P. cepacia* lipase enzyme was immobilized using an interfacial insolubilization process. A homogeneously mixed solution of lipase, PVA, and alginate was prepared and dropwise added to calcium tetraborate (CaB_4O_7) solution. The catalyst exhibited about 75% biodiesel yield within 24 hours of incubation and an overall reusability up to six cycles (Kumar et al. 2020). However, the porosity of sol-gel-synthesized matrix restricts the movement of substrate molecules so that effective reaction cannot take place. The mass diffusion of higher-molecular-weight substrate limits the efficacy of the enzyme, making the entire process lower in biodiesel productivity.

Apart from physical interaction-based immobilization techniques, researchers have also argued for the importance of chemical interaction over existing physical interactions. Cazaban et al. (2018) explored two different modes of immobilization: (i) physical interaction-based *in situ* entrapment; and (ii) chemical covalent immobilization, as shown in Figure 2.7. *T. lanuginosus* lipase was selected for immobilization on biomimetic silica NPs. Improved thermal stability was obtained in both modes, but 88% biodiesel yield was observed in the case of covalent binding, which was 1.6-fold higher than with entrapment (55%). This was due to stronger bonding and greater stability of the enzyme in the case of

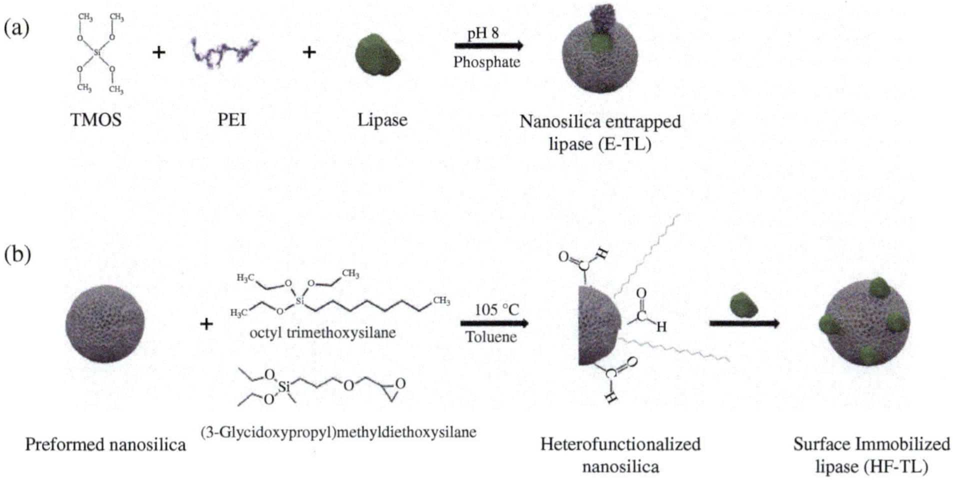

Figure 2.7 Representation of (a) physical entrapment of lipase and (b) chemical covalent bond formation on a biomimetic silica substrate and lipase. Source: Adapted from Cazaban et al. (2018) with copyright permission from Elsevier.

covalent bonding (Cazaban et al. 2018). The advantages of covalent bonding over physical bonding are discussed in detail in the next section.

2.8 Chemical Methods

2.8.1 Covalent Bonding

This technique involves covalent linkage between the enzyme and the substratum, making it the most stable enzyme immobilization technique available. The amino acid groups constituting the enzyme form a covalent bond with the functional groups of the substratum. Such strong attachment prevents the leakage of enzyme even with continuous usage, but it often leads to conformational restriction, decreasing the enzyme activity. Various chemical reactions explored for the covalent attachment of substratum and enzyme are shown in Figure 2.8 (Elnashar 2011).

The major advantage of this technique is the lack of leakage of the enzyme, making it highly efficient and cost-effective. However, the initial cost of manufacture is typically higher due to the chemical reaction required to establish the covalent bond. Nematian et al. (2020a) explored MNPs as a suitable substratum for electrostatic and covalent interactions with the enzyme. The MNPs were chemically modified via functionalization of 3-aminopropyl triethylenesilane (MNP-AP) and its glutaraldehyde derivative (i.e. AP-glutaraldehyde [MNP-AP-GA]) to allow efficient covalent bonding with *Rhizopus oryzae* lipase (ROL) enzyme. It was reported that longer coupling agents enabled better enzyme loading capacity, as MNP-AP-GA exhibited a higher loading capacity of 51.8% compared with bare NP (36.4%). In agreement with this, a higher biodiesel conversion efficiency of ~70% was observed for MNP-AP-GA biocatalyst. Reusability, another important parameter, showed considerable improvement in comparison to other conventional techniques.

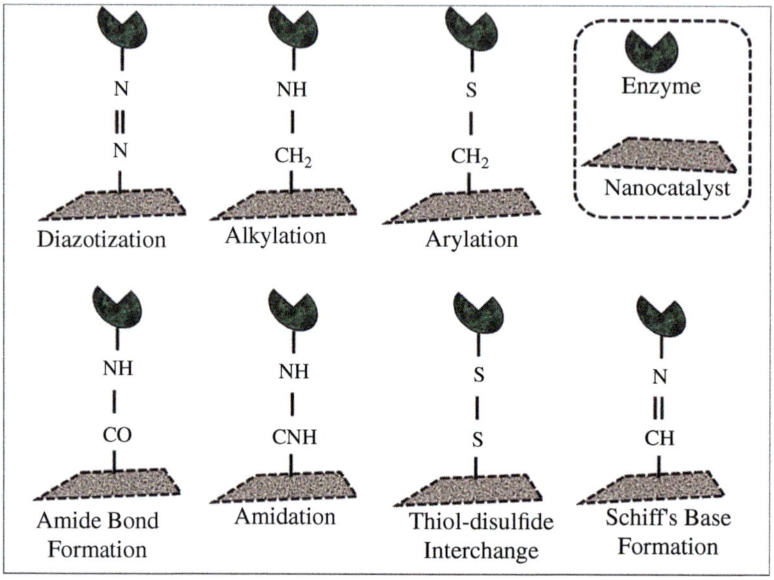

Figure 2.8 Different types of covalent bond formation between substratum and enzyme. Source: Adapted from Elnashar (2011) an open access article.

Covalent bonding has decreased the chances of catalyst wastage due to leakage, with MNP-AP-GA exhibiting two- and threefold improved reusability compared with MNP-AP catalyst and bare MNP, respectively. Hence, a highly functionalized coupling agent along with covalent bonding improves enzyme loading, productivity, and activity retainment even after constant usage (Nematian et al. 2020a). The same group also explored the use of graphene oxide with functionalization of AP and CA groups for covalent and electrostatic bonding with lipase enzyme (MGO-AP-GA). However, the loading capacity was found to be 70.2% with the highest biodiesel conversion of 71.2%. This was ascribed to the lower surface area for the reaction of MNPs as compared with MGO-AP-GA (Nematian et al. 2020b).

Covalent bonding not only showed improvement in conversion efficiency after repetitive reuse but also demonstrated tolerance to higher temperature and pH. Fatima et al. (2020) immobilized lipase enzyme obtained from *Aspergillus terreus* on polydopamine-coated Ceria nanorods (CeO_2@PDA@ *A. terreus* lipase). Schiff base and Michael addition reactions led to bond formation between the $-NH_2$ group from the PDA and the –COOH groups from the lipase, as shown in Figure 2.9. In comparison to free lipase, immobilized lipase revealed tolerance to basic pH and higher temperature (>30 °C), attributed to the strength of covalent bonding. The biodiesel yield was found to be 89.3% and the catalyst was active for five repetitive cycles with no loss in transesterification activity (Fatima et al. 2020). The same researchers explored covalent modification on magnetic Fe_3O_4 NP surfaces for lipase immobilization. A PDA coating on Fe_3O_4 NPs surface was examined for covalent attachment of lipase, as mentioned previously. The maximum biodiesel yield obtained was 92%, owing to the increased surface area for reaction due to smaller NP size (25–30 nm), while reusability was shown for up to five cycles (Touqeer et al. 2020). The

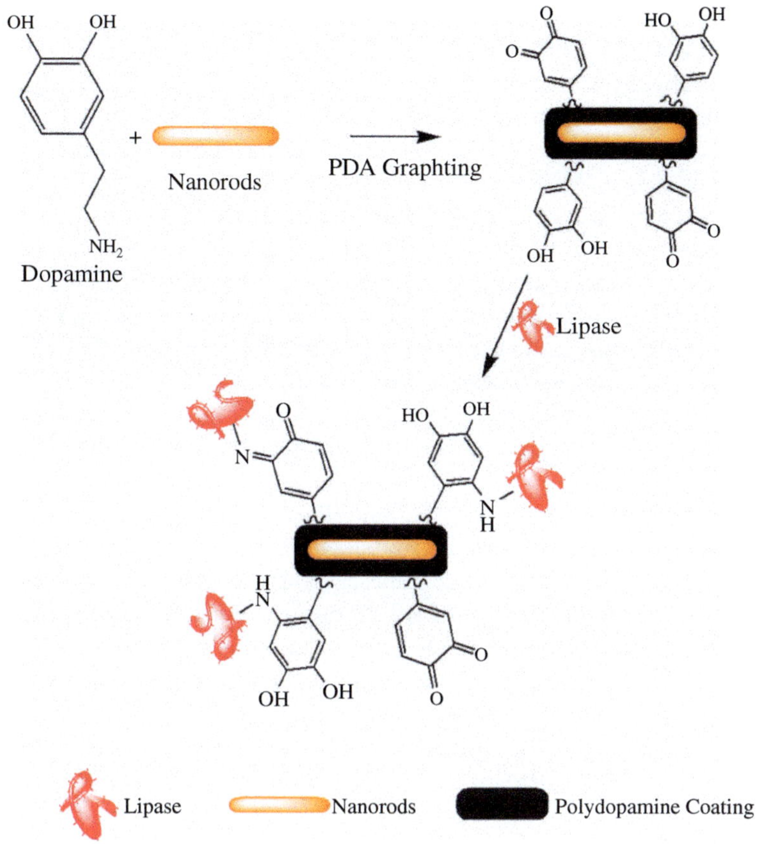

Figure 2.9 Schematic representation of covalent bonding of lipase on polydopamine-coated nanorods. Source: Adapted from Fatima et al. (2020) an open access article.

covalent immobilization technique is thus currently the best method available due to its efficiency, stability, and economic feasibility.

2.8.2 Cross-Linking

This technique involves linking enzymes to one another using specific chemical bonds, without the aid of any substratum. It focuses on restricting enzyme molecules from getting leaked out by bonding them to another molecule of the same enzyme using a spacer or linker support (Elnashar 2011). Such bonding enhances the enzyme's rigidity and conformational stability without compromising its chemical nature (i.e. activity). The support materials are usually derivatives of glutaraldehyde, toluene, or –COOH-terminated compounds. Badoei-dalfard et al. (2019b) investigated the biodiesel production efficiency of Km 12 lipase, a halotolerant enzyme, using the cross-linking technique. Briefly, the lipase enzyme aggregates were cross-linked via glutaraldehyde and then covalently linked to NH_2-functionalized MNPs via amide linkage (mCLEAs-lip). Such covalent linkage improved protein loading on the bionanocatalyst by about 75%. The assembly also imparted overall thermotolerance and conformational and storage stability, attributed to the rigidification of the enzyme due to covalent cross-linking. Though its maximum biodiesel production was found to be slightly decreased (49%), its storage capacity of up to three days and extreme cold tolerance make it suitable for use in colder climates (Badoei-dalfard et al. 2019b). The same group also investigated the suitability of this CLEAs-lip assembly, extracted from *Enterobacter cloacae* MG10, as compared with magnetic NH_2 implanted graphene oxide nanocomposites (maGO-CLEAs-lip). The new assembly exhibited roughly twofold higher biodiesel conversion (78%) within 24 hours, with an enzymatic storage efficiency up to one month (threefold higher than the free lipase assembly). In addition, both the cross-linked catalysts showed reusability up to five cycles. Hence, the effectiveness of this immobilization technique varies with the source enzyme and base material and the nature of the bond formed between them (Badoei-dalfard et al. 2019a).

2.8.3 Summary

Chemical modification techniques are more specific and effective compaed to physical techniques, but their expensive manufacturing and production procedures make them uneconomical for scale-up. In addition, the yield of biodiesel varies not only with the mode of immobilization used, but also with the type of substratum material, oil source, and enzyme source. Further advances are required in the field of designing a suitable catalyst by merging two or more immobilization techniques in order to overcome their individual shortcomings.

2.9 Conclusions

The increasing demand for oil as a fuel and the depletion of petroleum sources call for the use of alternative renewable sources of energy. Biodiesel production using WCOs has become an area of great interest due to its waste-to-wealth strategic approach. Use of WCOs for transesterification prevents pollution caused by their disposal in the environment and provides an alternative source for fuel production. The chemical catalysts used in the transesterification process have various side effects and lead to secondary environmental pollution. Their requirements of low energy and mild reaction conditions, no or reduced waste generation, and the use of cheap biological materials make enzyme catalysts best suited for biodiesel production. Enzyme-catalyzed biodiesel production has shown tremendous success in achieving complete conversion efficiency and improved yields. Lipases are produced intracellularly or extracellularly from different bacterial and fungal species. However, the downstream processing cost is unfeasible due to poor recyclability. Nanomaterials have drawn scrutiny as effective substrata for the immobilization of enzymes due to their improved surface area. Compared with free enzyme, NP-based immobilization offers advantages such as increased loading capacity, conversion efficacy, and reusability. The covalent immobilization technique is the best method due to its efficiency, stability, and economic feasibility. These nanobiocatalysts also allow chemical functionalization, preventing enzyme denaturation and improving overall bioavailability. Hence, they appear promising for the applications in biofuel sectors. However, their applicability varies with the substrate source, enzyme source, and reaction conditions, making the overall process irrational for universal applications. Hence, there is a need to prepare integrated catalysts made up of efficient enzymes and nanosupports and optimize them to work rationally under standard environmental conditions, in order to scale up biodiesel production.

References

Adnan, M., Li, K., Xu, L., and Yan, Y. (2018). X-shaped ZIF-8 for immobilization *Rhizomucor miehei* lipase via encapsulation and its application toward biodiesel production. *Catalysts* 8 (3): 96.

Ahima, R.S. (2020). Global warming threatens human thermoregulation and survival. *Journal of Clinical Investigation* 130 (2): 559–561.

Ahmad, R. and Sardar, M. (2015). Enzyme immobilization: an overview on nanoparticles as immobilization matrix. *Biochemistry and Analytical Biochemistry* 4 (2): 1000178.

Ali, A., Kaur, M., and Mehra, U. (2011). Use of immobilized *Pseudomonas* sp. as whole cell catalyst for the transesterification of used cotton seed oil. *Journal of Oleo Science* 60 (1): 7–10.

Ali, C.H., Qureshi, A.S., Mbadinga, S.M. et al. (2017). Biodiesel production from waste cooking oil using onsite produced purified lipase from *Pseudomonas aeruginosa* FW_SH-1: central composite design approach. *Renewable Energy* 109: 93–100.

Al-Zuhair, S. (2005). Production of biodiesel by lipase-catalyzed transesterification of vegetable oils: a kinetics study. *Biotechnology Progress* 21 (5): 1442–1448.

Amin, H.A., Ahmed, H.M., Mohamed, S.S. et al. (2018). Synthesis of biodiesel from used cooking oil using whole-cell lipase produced by solid state fermentation of *Aspergillus niger* KY401431. *ARPN Journal of Engineering and Applied Sciences* 13: 75–88.

Amini, Z., Ilham, Z., Ong, H.C. et al. (2017). State of the art and prospective of lipase-catalyzed transesterification reaction for biodiesel production. *Energy Conversion and Management* 141: 339–353.

Amirkhani, L., Moghaddas, J., and Jafarizadeh, H. (2019). Optimization of biodiesel production using immobilized *Candida rugosa* lipase on magnetic Fe_3O_4-silica aerogel. *Iranian Journal of Chemistry and Chemical Engineering* 38 (2): 193–201.

Arumugam, A. and Ponnusami, V. (2017). Production of biodiesel by enzymatic transesterification of waste sardine oil and evaluation of its engine performance. *Heliyon* 3 (12): e00486.

Ashraful, A.M., Masjuki, H.H., Kalam, M.A. et al. (2014). Production and comparison of fuel properties, engine performance, and emission characteristics of biodiesel from various non-edible vegetable oils: a review. *Energy Conversion and Management* 80: 202–228.

Asli, H., Ahmadinia, E., Zargar, M., and Karim, M.R. (2012). Investigation on physical properties of waste cooking oil–rejuvenated bitumen binder. *Construction and Building Materials* 37: 398–405.

Badoei-dalfard, A., Karami, Z., and Malekabadi, S. (2019a). Construction of CLEAs-lipase on magnetic graphene oxide nanocomposite: an efficient nanobiocatalyst for biodiesel production. *Bioresource Technology* 278: 473–476.

Badoei-dalfard, A., Malekabadi, S., Karami, Z., and Sargazi, G. (2019b). Magnetic cross-linked enzyme aggregates of Km12 lipase: a stable nanobiocatalyst for biodiesel synthesis from waste cooking oil. *Renewable Energy* 141: 874–882.

Bajaj, A., Lohan, P., Jha, P.N., and Mehrotra, R. (2010). Biodiesel production through lipase catalyzed transesterification: an overview. *Journal of Molecular Catalysis B: Enzymatic* 62 (1): 9–14.

Balasubramaniam, B., Perumal, A.S., Jayaraman, J. et al. (2012). Comparative analysis for the production of fatty acid alkyl esterase using whole cell biocatalyst and purified enzyme from *Rhizopus oryzae* on waste cooking oil (sunflower oil). *Waste Management* 32 (8): 1539–1547.

Bandyopadhyay, K., Chakraborty, C., Chakraborty, S., and Ganguly, S. (2017). Change in physicochemical properties of edible oil during frying: a review. *Pharma Innovation Journal* 6 (10): 196–199.

Bautista, L.F., Vicente, G., Rodriguez, R., and Pacheco, M. (2009). Optimisation of FAME production from waste cooking oil for biodiesel use. *Biomass and Bioenergy* 33 (5): 862–872.

Bharathiraja, B., Jayamuthunagai, J., Praveenkumar, R. et al. (2015). The kinetics of interesterfication on waste cooking oil (sunflower oil) for the production of fatty acid alkyl esters using a whole cell biocatalyst (*Rhizopus oryzae*) and pure lipase enzyme. *International Journal of Green Energy* 12 (10): 1012–1017.

Bouaid, A., Vázquez, R., Martinez, M., and Aracil, J. (2016). Effect of free fatty acids contents on biodiesel quality: pilot plant studies. *Fuel* 174: 54–62.

Boudart, M. and Djéga-Mariadassou, G. (1984). Chapter 3. Kinetics of overall reactions. In: *Kinetics of Heterogeneous Catalytic Reactions* (eds. J.M. Prausnitz and L. Brewer), 77–117. Princeton, NJ: Princeton University Press.

Cazaban, D., Illanes, A., Wilson, L., and Betancor, L. (2018). Bio-inspired silica lipase nanobiocatalysts for the synthesis of fatty acid methyl esters. *Process Biochemistry* 74: 86–93.

Chandel, A.K., Bhatia, L., Garlapati, V.K. et al. (2017). Biofuel policy in Indian perspective: socioeconomic indicators and sustainable rural development. In: *Sustainable Biofuels Development in India* (eds. A.K. Chandel and R.K. Sukumaran), 459–488. Cham: Springer.

Chen, Y., Xiao, B., Chang, J. et al. (2009). Synthesis of biodiesel from waste cooking oil using immobilized lipase in fixed bed reactor. *Energy Conversion and Management* 50 (3): 668–673.

Chen, G., Liu, J., Yao, J. et al. (2017). Biodiesel production from waste cooking oil in a magnetically fluidized bed reactor using whole-cell biocatalysts. *Energy Conversion and Management* 138: 556–564.

Chesterfield, D.M., Rogers, P.L., Al-Zaini, E.O., and Adesina, A.A. (2012). Production of biodiesel via ethanolysis of waste cooking oil using immobilised lipase. *Chemical Engineering Journal* 207: 701–710.

Chincholkar, S., Srivastava, S., Rehman, A. et al. (2005). Biodiesel as an alternative fuel for pollution control in diesel engine. *Asian Journal of Experimental Sciences* 19 (2): 13–22.

Christopher, L.P., Kumar, H., and Zambare, V.P. (2014). Enzymatic biodiesel: challenges and opportunities. *Applied Energy* 119: 497–520.

Datta, P., Tiwari, S., and Pandey, L. (2018). Bioethanol production from waste breads using *Saccharomyces cerevisiae*. In: *Utilization and Management of Bioresources* (ed. S.K. Ghosh), 125–134. Cham: Springer.

de Jong, S., Hoefnagels, R., Wetterlund, E. et al. (2017). Cost optimization of biofuel production–the impact of scale, integration, transport and supply chain configurations. *Applied Energy* 195: 1055–1070.

De Los Ríos, A.P., Hernández-Fernández, F.J., Martínez, F.A. et al. (2007). The effect of ionic liquid media on activity, selectivity and stability of *Candida antarctica* lipase B in transesterification reactions. *Biocatalysis and Biotransformation* 25 (2–4): 151–156.

Demirbas, A. (2008). Biofuels sources, biofuel policy, biofuel economy and global biofuel projections. *Energy Conversion and Management* 49 (8): 2106–2116.

Dizge, N., Aydiner, C., Imer, D.Y. et al. (2009). Biodiesel production from sunflower, soybean, and waste cooking oils by transesterification using lipase immobilized onto a novel microporous polymer. *Bioresource Technology* 100 (6): 1983–1991.

Dossat, V., Combes, D., and Marty, A. (2002). Lipase-catalysed transesterification of high oleic sunflower oil. *Enzyme and Microbial Technology* 30 (1): 90–94.

El-Batal, A.I., Farrag, A.A., Elsayed, M.A., and El-Khawaga, A.M. (2016). Biodiesel production by *Aspergillus niger* lipase immobilized on barium ferrite magnetic nanoparticles. *Bioengineering* 3 (2): 14.

Elnashar, M. (2011). The art of immobilization using biopolymers, biomaterials and nanobiotechnology. In: *Biotechnology of Biopolymers* (ed. M. Elnashar). London: IntechOpen.

Fan, X., Niehus, X., and Sandoval, G. (2012). Lipases as biocatalyst for biodiesel production. In: *Lipases and Phospholipases, Methods in Molecular Biology* (ed. G. Sandoval), 471–483. New York: Humana.

Fan, Y., Su, F., Li, K. et al. (2017). Carbon nanotube filled with magnetic iron oxide and modified with polyamidoamine dendrimers for immobilizing lipase toward application in biodiesel production. *Scientific Reports* 7 (1): 45643.

Fatima, A., Mumtaz, M.W., Mukhtar, H. et al. (2020). Synthesis of lipase-immobilized CeO_2 Nanorods as heterogeneous nano-biocatalyst for optimized biodiesel production from *Eruca sativa* seed oil. *Catalysts* 10 (2): 231.

Felizardo, P., Correia, M.J.N., Raposo, I. et al. (2006). Production of biodiesel from waste frying oils. *Waste Management* 26 (5): 487–494.

Fjerbaek, L., Christensen, K.V., and Norddahl, B. (2009). A review of the current state of biodiesel production using enzymatic transesterification. *Biotechnology and Bioengineering* 102 (5): 1298–1315.

Garcia, T., Coteron, A., Martinez, M., and Aracil, J. (1996). Kinetic modelling of esterification reactions catalysed by immobilized lipases. *Chemical Engineering Science* 51 (11): 2841–2846.

Gharat, N. and Rathod, V.K. (2013). Enzyme catalyzed transesterification of waste cooking oil with dimethyl carbonate. *Journal of Molecular Catalysis B: Enzymatic* 88: 36–40.

Gnanaprakasam, A., Sivakumar, V.M., Surendhar, A. et al. (2013). Recent strategy of biodiesel production from waste cooking oil and process influencing parameters: a review. *Journal of Energy* 2013: 926392.

Gog, A., Roman, M., Toşa, M. et al. (2012). Biodiesel production using enzymatic transesterification – current state and perspectives. *Renewable Energy* 39 (1): 10–16.

Hasan, A. and Pandey, L.M. (2016). Kinetic studies of attachment and re-orientation of octyltriethoxysilane for formation of self-assembled monolayer on a silica substrate. *Materials Science and Engineering: C* 68: 423–429.

Hasan, A. and Pandey, L.M. (2020). Surface modification of Ti6Al4V by forming hybrid self-assembled monolayers and its effect on collagen-I adsorption, osteoblast adhesion and integrin expression. *Applied Surface Science* 505: 144611.

Hasan, A., Saxena, V., and Pandey, L.M. (2018a). Surface functionalization of Ti6Al4V via self-assembled monolayers for improved protein adsorption and fibroblast adhesion. *Langmuir* 34 (11): 3494–3506.

Hasan, A., Pattanayek, S.K., and Pandey, L.M. (2018b). Effect of functional groups of self-assembled monolayers on protein adsorption and initial cell adhesion. *ACS Biomaterials Science & Engineering* 4 (9): 3224–3233.

Heater, B.S., Chan, W.S., Lee, M.M., and Chan, M.K. (2019). Directed evolution of a genetically encoded immobilized lipase for the efficient production of biodiesel from waste cooking oil. *Biotechnology for Biofuels* 12 (1): 165.

Hosseinzadeh-Bandbafha, H., Tabatabaei, M., Aghbashlo, M. et al. (2018). A comprehensive review on the environmental impacts of diesel/biodiesel additives. *Energy Conversion and Management* 174: 579–614.

Hou, M. and Ge, J. (2017). Armoring enzymes by metal–organic frameworks by the coprecipitation method. In: *Methods in Enzymology*, vol. 590 (ed. C.V. Kumar), 59–75. Cambridge, MA: Academic Press.

Ingle, A.P., Chandel, A.K., Pholippini, R. et al. (2020a). Advances in nanocatalysts mediated biodiesel production: a critical appraisal. *Symmetry* 12 (2): 256.

Ingle, A.P., Shende, S., Gupta, I., and Rai, M. (2020b). Recent trends in the development of nano-bioactive compounds and delivery systems. In: *Biotechnological Production of Bioactive Compounds* (eds. M.L. Verma and A.K. Chandel), 409–431. Amsterdam: Elsevier.

Jambulingam, R., Shalma, M., and Shankar, V. (2019). Biodiesel production using lipase immobilised functionalized magnetic nanocatalyst from oleaginous fungal lipid. *Journal of Cleaner Production* 215: 245–258.

Jawed, A., Saxena, V., and Pandey, L.M. (2020). Engineered nanomaterials and their surface functionalization for the removal of heavy metals: a review. *Journal of Water Process Engineering* 33: 101009.

Jegannathan, K.R., Abang, S., Poncelet, D. et al. (2008). Production of biodiesel using immobilized lipase – a critical review. *Critical Reviews in Biotechnology* 28 (4): 253–264.

Knothe, G. and Steidley, K.R. (2009). A comparison of used cooking oils: a very heterogeneous feedstock for biodiesel. *Bioresource Technology* 100 (23): 5796–5801.

Kratošová, G., Holišová, V., Konvičková, Z. et al. (2019). From biotechnology principles to functional and low-cost metallic bionanocatalysts. *Biotechnology Advances* 37 (1): 154–176.

Kumar, D., Das, T., Giri, B.S., and Verma, B. (2020). Preparation and characterization of novel hybrid bio-support material immobilized from *Pseudomonas cepacia* lipase and its application to enhance biodiesel production. *Renewable Energy* 147: 11–24.

Lam, M.K., Lee, K.T., and Mohamed, A.R. (2010). Homogeneous, heterogeneous and enzymatic catalysis for transesterification of high free fatty acid oil (waste cooking oil) to biodiesel: a review. *Biotechnology Advances* 28 (4): 500–518.

Letcher, T.M. (2019). Why do we have global warming? In: *Managing Global Warming* (ed. T.M. Letcher), 3–15. Cambridge, MA: Academic Press.

Leung, D.Y., Wu, X., and Leung, M. (2010). A review on biodiesel production using catalyzed transesterification. *Applied Energy* 87 (4): 1083–1095.

Li, D., Wang, W., Faiza, M. et al. (2017). A novel and highly efficient approach for the production of biodiesel from high-acid content waste cooking oil. *Catalysis Communications* 102: 76–80.

Liu, Y., Wu, H., Yan, Y. et al. (2013). Lipase-catalyzed transesterification for biodiesel production in ionic liquid [Emim] Tfo. *International Journal of Green Energy* 10 (1): 63–71.

Macario, A., Moliner, M., Corma, A., and Giordano, G. (2009). Increasing stability and productivity of lipase enzyme by encapsulation in a porous organic–inorganic system. *Microporous and Mesoporous Materials* 118 (1): 334–340.

Mander, P., Yoo, H.-Y., Kim, S.W. et al. (2014). Transesterification of waste cooking oil by an organic solvent-tolerant alkaline lipase from *Streptomyces* sp. CS273. *Applied Biochemistry and Biotechnology* 172 (3): 1377–1389.

Mansir, N., Teo, S.H., Rashid, U. et al. (2018). Modified waste egg shell derived bifunctional catalyst for biodiesel production from high FFA waste cooking oil. A review. *Renewable and Sustainable Energy Reviews* 82: 3645–3655.

Marchetti, J. and Errazu, A. (2008). Esterification of free fatty acids using sulfuric acid as catalyst in the presence of triglycerides. *Biomass and Bioenergy* 32 (9): 892–895.

Marx, S. (2016). Glycerol-free biodiesel production through transesterification: a review. *Fuel Processing Technology* 151: 139–147.

Mehrasbi, M.R., Mohammadi, J., Peyda, M., and Mohammadi, M. (2017). Covalent immobilization of *Candida antarctica* lipase on core-shell magnetic nanoparticles for production of biodiesel from waste cooking oil. *Renewable Energy* 101: 593–602.

Meunier, S.M., Kariminia, H.-R., and Legge, R.L. (2017). Immobilized enzyme technology for biodiesel production. *Advances in Biofeedstocks and Biofuels* 2: 67–106.

Moazeni, F., Chen, Y.-C., and Zhang, G. (2019). Enzymatic transesterification for biodiesel production from used cooking oil, a review. *Journal of Cleaner Production* 216: 117–128.

Modenez, I.A., Sastre, D.E., Moraes, F.C., and Marques Netto, C.G.C. (2018). Influence of glutaraldehyde cross-linking modes on the recyclability of immobilized lipase B from *Candida antarctica* for transesterification of soy bean oil. *Molecules* 23 (9): 2230.

Mumtaz, M.W., Adnan, A., Mahmood, Z. et al. (2012). Biodiesel from waste cooking oil: optimization of production and monitoring of exhaust emission levels from its combustion in a diesel engine. *International Journal of Green Energy* 9 (7): 685–701.

Nady, D., Zaki, A.H., Raslan, M., and Hozayen, W. (2020). Enhancement of microbial lipase activity via immobilization over sodium titanate nanotubes for fatty acid methyl esters production. *International Journal of Biological Macromolecules* 146: 1169–1179.

Narwal, S.K. and Gupta, R. (2013). Biodiesel production by transesterification using immobilized lipase. *Biotechnology Letters* 35 (4): 479–490.

Nematian, T., Salehi, Z., and Shakeri, A. (2020a). Conversion of bio-oil extracted from *Chlorella vulgaris* micro algae to biodiesel via modified superparamagnetic nano-biocatalyst. *Renewable Energy* 146: 1796–1804.

Nematian, T., Shakeri, A., Salehi, Z., and Saboury, A.A. (2020b). Lipase immobilized on functionalized superparamagnetic few-layer graphene oxide as an efficient nanobiocatalyst for biodiesel production from *Chlorella vulgaris* bio-oil. *Biotechnology for Biofuels* 13 (1): 1–15.

Norjannah, B., Ong, H.C., Masjuki, H. et al. (2016). Enzymatic transesterification for biodiesel production: a comprehensive review. *RSC Advances* 6 (65): 60034–60055.

Otera, J. (1993). Transesterification. *Chemical Reviews* 93 (4): 1449–1470.

Özbay, N., Oktar, N., and Tapan, N.A. (2008). Esterification of free fatty acids in waste cooking oils (WCO): role of ion-exchange resins. *Fuel* 87 (10–11): 1789–1798.

Paiva, A.L., Balcao, V.M., and Malcata, F.X. (2000). Kinetics and mechanisms of reactions catalyzed by immobilized lipases. *Enzyme and Microbial Technology* 27 (3–5): 187–204.

Pandey, L.M. and Pattanayek, S.K. (2011). Hybrid surface from self-assembled layer and its effect on protein adsorption. *Applied Surface Science* 257 (10): 4731–4737.

Pandey, L.M., Pattanayek, S.K., and Delabouglise, D. (2013). Properties of adsorbed bovine serum albumin and fibrinogen on self-assembled monolayers. *Journal of Physical Chemistry C* 117 (12): 6151–6160.

Pazouki, M., Zamani, F., Zamzamian, A.H. et al. (2010). Esterification of free fatty acids by *Rhizopus oryzae* as cell-catalyzed from used cooking oil for biodiesel production. *World Applied Sciences Journal* 8 (6): 719–724.

Pazouki, M., Zamani, F., Zamzamian, S.A.H., and Najafpour, G. (2011). Study on reaction conditions in whole cell biocatalyst methanolysis of pretreated used cooking oil. World Renewable Energy Congress – Sweden, Linköping University Electronic Press, Sweden, pp. 93–100.

Pilarek, M. and Szewczyk, K.W. (2007). Kinetic model of 1,3-specific triacylglycerols alcoholysis catalyzed by lipases. *Journal of Biotechnology* 127 (4): 736–744.

Prafulla, D.P., Veera, G.G., Harvind, K.R. et al. (2012). Biodiesel production from waste cooking oil using sulfuric acid and microwave irradiation processes. *Journal of Environmental Protection* 3: 107–113.

Prasad, S. and Ingle, A.P. (2019). Impacts of sustainable biofuels production from biomass. In: *Sustainable Bioenergy* (eds. M. Rai and A.P. Ingle), 327–346. Amsterdam: Elsevier.

Rodionova, M.V., Poudyal, R.S., Tiwari, I. et al. (2017). Biofuel production: challenges and opportunities. *International Journal of Hydrogen Energy* 42 (12): 8450–8461.

Saravanan, A.P., Mathimani, T., Deviram, G. et al. (2018). Biofuel policy in India: a review of policy barriers in sustainable marketing of biofuel. *Journal of Cleaner Production* 193: 734–747.

Sarno, M. and Iuliano, M. (2018). Active biocatalyst for biodiesel production from spent coffee ground. *Bioresource Technology* 266: 431–438.

Sarno, M., Iuliano, M., and Cirillo, C. (2019). Optimized procedure for the preparation of an enzymatic nanocatalyst to produce a bio-lubricant from waste cooking oil. *Chemical Engineering Journal* 377: 120273.

Saxena, V., Sharma, S., and Pandey, L.M. (2019). Fe(III) doped ZnO nano-assembly as a potential heterogeneous nano-catalyst for the production of biodiesel. *Materials Letters* 237: 232–235.

Shafiee, S. and Topal, E. (2009). When will fossil fuel reserves be diminished? *Energy Policy* 37 (1): 181–189.

Sharma, S. and Pandey, L.M. (2020). Production of biosurfactant by *Bacillus subtilis* RSL-2 isolated from sludge and biosurfactant mediated degradation of oil. *Bioresource Technology* 307: 123261.

Sharma, S., Saxena, V., Baranwal, A. et al. (2018). Engineered nanoporous materials mediated heterogeneous catalysts and their implications in biodiesel production. *Materials Science for Energy Technologies* 1 (1): 11–21.

Sharma, S., Datta, P., Kumar, B. et al. (2019). Production of novel rhamnolipids via biodegradation of waste cooking oil using *Pseudomonas aeruginosa* MTCC7815. *Biodegradation* 30 (4): 301–312.

Singh-Ackbarali, D., Maharaj, R., Mohamed, N., and Ramjattan-Harry, V. (2017). Potential of used frying oil in paving material: solution to environmental pollution problem. *Environmental Science and Pollution Research* 24 (13): 12220–12226.

Souza, L.T., Mendes, A.A., and de Castro, H.F. (2016). Selection of lipases for the synthesis of biodiesel from Jatropha oil and the potential of microwave irradiation to enhance the reaction rate. *BioMed Research International* 2016: 1404567.

Stephens, E., Ross, I.L., Mussgnug, J.H. et al. (2010). Future prospects of microalgal biofuel production systems. *Trends in Plant Science* 15 (10): 554–564.

Surriya, O., Saleem, S.S., Waqar, K. et al. (2015). Bio-fuels: a blessing in disguise. In: *Phytoremediation for Green Energy* (eds. M. Öztürk, M. Ashraf, A. Aksoy and M. Ahmad), 11–54. Dordrecht: Springer.

Talukder, M.M.R., Lee, H.Z.S., Low, R.F. et al. (2013). Potential use of whole cell lipase from a newly isolated *Aspergillus nomius* for methanolysis of palm oil to biodiesel. *Journal of Molecular Catalysis B: Enzymatic* 89: 108–113.

Tan, T., Nie, K., and Wang, F. (2006). Production of biodiesel by immobilized *Candida* sp. lipase at high water content. *Applied Biochemistry and Biotechnology* 128 (2): 109–116.

Tan, T., Lu, J., Nie, K. et al. (2010). Biodiesel production with immobilized lipase: a review. *Biotechnology Advances* 28 (5): 628–634.

Tomei, J. and Helliwell, R. (2016). Food versus fuel? Going beyond biofuels. *Land Use Policy* 56: 320–326.

Touqeer, T., Mumtaz, M.W., Mukhtar, H. et al. (2020). Fe3O4-PDA-lipase as surface functionalized nano biocatalyst for the production of biodiesel using waste cooking oil as feedstock: characterization and process optimization. *Energies* 13 (1): 177.

Uslu, N. and Özcan, M.M. (2018). Determination of the physicochemical changes in the different vegetable oils after fat-product interaction during frying process. *Iranian Journal of Chemistry and Chemical Engineering* 37 (6): 191–199.

Van Gerpen, J. (2005). Biodiesel processing and production. *Fuel Processing Technology* 86 (10): 1097–1107.

Vargas, M., Niehus, X., Casas-Godoy, L., and Sandoval, G. (2018). Lipases as biocatalyst for biodiesel production. In: *Lipases and Phospholipases*, vol. 1835 (ed. G. Sandoval), 377–390. New York: Humana.

Wan, D., Tian, L., Li, X. et al. (2018). A versatile strategy for enzyme immobilization: fabricating lipase/inorganic hybrid nanostructures on macroporous resins with enhanced catalytic properties. *Biochemical Engineering Journal* 139: 101–108.

Wang, Y., Ou, S., Liu, P. et al. (2006). Comparison of two different processes to synthesize biodiesel by waste cooking oil. *Journal of Molecular Catalysis A: Chemical* 252 (1–2): 107–112.

Wang, X., Qin, X., Li, D. et al. (2017). One-step synthesis of high-yield biodiesel from waste cooking oils by a novel and highly methanol-tolerant immobilized lipase. *Bioresource Technology* 235: 18–24.

Wang, J., Li, K., He, Y. et al. (2019). Enhanced performance of lipase immobilized onto Co^{2+}-chelated magnetic nanoparticles and its application in biodiesel production. *Fuel* 255: 115794.

Wrapp, D., Wang, N., Corbett, K.S. et al. (2020). Cryo-EM structure of the 2019-nCoV spike in the prefusion conformation. *Science* 367 (6483): 1260–1263.

Xie, W. and Huang, M. (2018). Immobilization of *Candida rugosa* lipase onto graphene oxide Fe3O4 nanocomposite: characterization and application for biodiesel production. *Energy Conversion and Management* 159: 42–53.

Xie, W. and Ma, N. (2010). Enzymatic transesterification of soybean oil by using immobilized lipase on magnetic nano-particles. *Biomass and Bioenergy* 34 (6): 890–896.

Yagar, H. and Balkan, U. (2017). Entrapment of laurel lipase in chitosan hydrogel beads. *Artificial Cells, Nanomedicine, and Biotechnology* 45 (5): 864–870.

Yahya, S., Razali, F.H., and Harun, F.W. (2019). Physicochemical properties of refined palm cooking oil and used palm cooking oil. *Materials Today: Proceedings* 19: 1166–1172.

Yan, J., Yan, Y., Liu, S. et al. (2011). Preparation of cross-linked lipase-coated micro-crystals for biodiesel production from waste cooking oil. *Bioresource Technology* 102 (7): 4755–4758.

Yan, J., Zheng, X., Du, L., and Li, S. (2014a). Integrated lipase production and in situ biodiesel synthesis in a recombinant *Pichia pastoris* yeast: an efficient dual biocatalytic system composed of cell free enzymes and whole cell catalysts. *Biotechnology for Biofuels* 7 (1): 55.

Yan, J., Zheng, X., and Li, S. (2014b). A novel and robust recombinant *Pichia pastoris* yeast whole cell biocatalyst with intracellular overexpression of a *Thermomyces lanuginosus* lipase: preparation, characterization and application in biodiesel production. *Bioresource Technology* 151: 43–48.

Ying, M. and Chen, G. (2007). Study on the production of biodiesel by magnetic cell biocatalyst based on lipase-producing *Bacillus subtilis*. *Applied Biochemistry and Biotecnology* 137: 793–803.

Ying, H., Zhang, L.-f., Wu, D. et al. (2017). Ionic strength-response hyperbranched polyglycerol/polyacrylic acid hydrogel for the reversible immobilization of enzyme and the synthesis of biodiesel. *Energy Conversion and Management* 144: 303–311.

Yu, C.-Y., Huang, L.-Y., Kuan, I., and Lee, S.-L. (2013). Optimized production of biodiesel from waste cooking oil by lipase immobilized on magnetic nanoparticles. *International Journal of Molecular Sciences* 14 (12): 24074–24086.

Zhang, B., Weng, Y., Xu, H., and Mao, Z. (2012). Enzyme immobilization for biodiesel production. *Applied Microbiology and Biotechnology* 93 (1): 61–70.

Zhao, T., Kim, Y., Kim, Y.S., and Kim, I.-H. (2014). Novel strategy for lipase-catalyzed synthesis of biodiesel using blended alcohol as an acyl acceptor. *Journal of Molecular Catalysis B: Enzymatic* 107: 17–22.

Zhao, K., Di, Q., Cao, X. et al. (2016). Production of biodiesel using immobilized lipase and the characterization of different co-immobilizing agents and immobilization methods. *Sustainability* 8 (9): 764.

3

A Review on the Use of Bio/Nanostructured Heterogeneous Catalysts in Biodiesel Production

Samuel Santos[1], Jaime Puna[1,2], João Gomes[1,2] and Jorge Marchetti[3]

[1] *CERENA – Center for Natural Resources and, Instituto Superior Técnico, Lisbon University, Lisbon, Portugal*
[2] *Chemical Engineering Department, Instituto Superior de Engenharia de Lisboa, Lisbon Polytechnic, Lisbon, Portugal*
[3] *Faculty of Sciences and Technology, Norwegian University of Life Sciences, Ås, Norway*

3.1 Introduction

Research on new renewable energy sources has been growing in recent decades due to the non-renewable nature of fossil fuel resources, the rapid rise in their prices, and an increase in environmental and health concerns (Li et al. 2011b; Safaei-Ghomi et al. 2013; Shi et al. 2017; Ambat et al. 2018). The European Union (EU) aimed to replace 10% of all transportation fuel consumed within its member states with biofuel by the end of 2020 (Schöpe 2008; European Parliament 2009). However, due to concerns over the fuel versus food controversy induced by the use of edible oils (first-generation feedstock), a new directive on the quality of petrol and diesel fuels, and the promotion of energy from renewable sources, the EU has agreed to limit the use of first-generation biofuels (usually based on agricultural crops and edible oils) to a maximum of 7% of total use in order to control the volume of such crops grown solely for energy consumption (European Parliament 2013; Avramovic et al. 2018).

Biodiesel (from the Greek *bio* ("life") + diesel, from Rudolph Diesel) is a preferred replacement for petro-diesel for use in internal combustion engine vehicles, offering benefits such as ready availability, renewability, nontoxicity, reduction of greenhouse gas emissions, improvement of lubrification properties, superior cetane number, and regional development, as well as the social structure benefits it brings to developing countries (Meher et al. 2006; Demirbas 2009; Islam et al. 2014). Compared to the use of fossil fuels, biodiesel combustion results in an over 90% decrease in total unburned hydrocarbons, and a 75–90% decrease in polycyclic aromatic hydrocarbons (Demirbas 2007; Mardhiah et al. 2017). Biodiesel can be mixed with petro-diesel at any weight ratio or percentage, or it can be used without blending with fossil fuels (B100) (Vyas et al. 2010). Its production from biological sources such as vegetable oils, microalgae, animal fats, and waste cooking oils (WCOs) has been thoroughly reviewed (Narasimharao et al. 2007; Zhang et al. 2013; Kouzu et al. 2017; Dai et al. 2019). The most common industrial process used to produce biodiesel from vegetable oils (or fats) involves the use of alkaline homogeneous catalysts. The introduction of a catalyst into a chemical system increases the speed of a reaction in

Nano- and Biocatalysts for Biodiesel Production, First Edition. Edited by Avinash P. Ingle.

one of three ways: by lowering the activation energy of the reaction, acting as a facilitator and bringing the reactive species together more effectively, or creating a higher yield in one of the species when two or more products are formed (Chaturvedi et al. 2012).

Generally, these catalysts are hydroxides such as potassium hydroxide (KOH) and sodium hydroxide (NaOH) or alkoxides of alkali metals such as sodium methoxide (CH_3ONa), potassium methoxide (CH_3OK), sodium propoxide (C_3H_7ONa), and potassium butoxide (C_4H_9OK), to name just a few (Ma et al. 1999; Hu et al. 2019). NaOH and KOH are the most commonly used commercially, due to their ability to perform efficient catalysis on the transesterification reaction under low operation conditions (reaction temperature 60–65 °C, alcohol to oil molar ratio 1 : 1–6 : 1 depending on the alcohol used, catalyst loading 0.5–1.5 wt%, etc.), showing high catalytic activity and hence high yields in the production of methyl esters (around 93–98%) and favorable kinetics, resulting in short reaction times. Another advantage of these catalysts is their low cost and ready market availability (Freedman et al. 1984; Vicente et al. 1998; Ma et al. 1999).

However, the transesterification of triglycerides using alkali homogeneous catalysts depends heavily on the purity of the oil feedstock. Zhang et al. (2003) stated in their work that the alkali process is only effective for a well-refined vegetable oil with less than 0.5 wt% free fatty acid (FFA). Thangaraj et al. (2019) observed that recovery becomes complicated in an alkali-catalyzed transesterification reaction if the oil has more than 1 wt% FFA. Sahoo and Das (2009) and Patel and Sankhavara (2017) demonstrated that homogeneous base catalysts are effective for transesterification of oil having FFA content less than 2 wt%. Leung et al. (2010) established that the maximum amount of FFA acceptable in an alkali-catalyzed system is below 2.5 wt%. Atadashi et al. (2013) claimed that triglyceride feedstocks should not have FFA content exceeding 3 wt%. Regardless of which is correct, an excess of FFA in the oil will promote the formation of soap and water due to the reaction between the FFA and the alkali catalyst. This reaction is called saponification and is undesirable because the soap formed will lower the transesterification yield and hinder the separation of the esters from the glycerol (Vicente et al. 2004; Leung et al. 2010; Fu et al. 2015).

Given the drawbacks of alkaline homogeneous catalysis with regard the purity of the feedstock, it is impossible to use these catalysts in the transesterification of a low-cost oil such as WCO. WCOs are obtained by using and reusing edible vegetable oils such as palm, soybean, sunflower, and corn oil several times in the process of deep-frying food. When compared with unused vegetable oils, the FFA content in WCOs is very different. A chemical pretreatment process can be used to overcome this inconvenience (Wang et al. 2007; Lou et al. 2008; Piker et al. 2016; Ozkan et al. 2019).

Transesterification using acid catalysts is not commonly applied when compared with alkaline-catalyzed transesterification, mainly because it is more time consuming (~4000 times slower) and more energy-demanding due to its need for a higher reaction temperature (usually above 100 °C) (Wang and Yang 2007; Lam et al. 2010; Vyas et al. 2010; Tariq et al. 2012; Boz et al. 2015; Avhad and Marchetti 2016) and it requires a higher alcohol to oil molar ratio. Nonetheless, the ability of acid catalysts to avoid soap formation in the presence of FFAs in the feedstock represents a considerable advantage over homogenous basic alternatives. Acid homogeneous catalysts can be employed in two different ways: via a two-step process comprising chemical treatment of WCO by FFA esterification using acid catalysts followed by transesterification of triglycerides using alkaline catalysts; or via acid-catalyzed transesterification with strong acids. Acid-catalyzed esterification is a chemical pretreatment of acidic feedstocks, which reduce the FFA content via esterification reaction of an alcohol and FFA, producing a fatty-acid methyl ester (FAME) (Boz et al. 2009; Marchetti 2012; Tariq et al. 2012; Mazubert et al. 2013; Avhad and Marchetti 2016; Baskar and Aiswarya 2016). Acid catalysts also have the advantage of being able to catalyze esterification and transesterification simultaneously (López et al. 2005; Lotero et al. 2005; Yan et al. 2009b).

As far as the application of acid catalysts in biodiesel production is concerned, sulfuric acid (H_2SO_4), sulfonic acid (H_2O_3S), ferric sulfate $Fe_2(SO_4)_3$, and hydrochloric acid (HCl) are most preferably utilized (Wang et al. 2007; Zhang et al. 2010; Borges and Díaz 2012; Hu et al. 2012). Among them, H_2SO_4, HCl, and H_2O_3S are most often used in biodiesel production (Guerreiro et al. 2006; Aranda et al. 2008; Guan et al. 2009; Zabeti et al. 2009; Endalew et al. 2011; Ilgen 2011; Lu et al. 2019).

The mechanism of acid-catalyzed transesterification of triglycerides starts with the protonation of the carbonyl group of the ester, leading to carbocation, which following alcohol nucleophilic attack produces a tetrahedral intermediate. This intermediate eliminates glycerol to form a new ester and regenerate the catalyst (Xiao and Gao 2011; Aransiola et al. 2014).

Biocatalysts (i.e. enzymes) have been investigated for decades due to their catalytic power, high degree of specificity, and stereoselectivity (Pollard and Woodley 2007; Carteret et al. 2018; Bilal et al. 2019). These properties helps them to catalyze reactions under mild conditions (low temperature and pressure), making them interesting for a wide variety of industrial applications, including biodiesel production (Garcia-Galan et al. 2011; Bilal et al. 2019). In this regard, enzymatic transesterification using lipases has recently caught the attention of researchers due to their high selectivity toward the desired product.

According to the literature, lipases (triacylglycerol acylhydrolases) found on animals, plants, and microorganisms (Kakugawa et al. 2002; Szczęsna Antczak et al. 2009) that have been used in the production of biodiesel regularly come from microorganisms such as *Aspergillus niger* (Hu et al. 2018), *Burkholderia cepacian* (Shah and Gupta 2008; Liu et al. 2017), *Candida antarctica* (Shimada et al. 2002; Bencze et al. 2016; Wang et al. 2018), *Candida rugose* (Rafiei et al. 2018), *Mucor mihei* (Nelson et al. 1996; Gog et al. 2012), *Pseudomonas cepacian* (Li et al. 2011a), *Pseudomonas fluorescens* (Zhao et al. 2007; Cubides-Roman et al. 2017), *Rhizopus oryzae* (Jun et al. 2012), *Rhizomucor miehei* (Shieh et al. 2003; Adnan et al. 2018), and *Thermomyces lanuginosus* (Du et al. 2005; Xie and Ma 2009). Usually, they catalyze the hydrolysis of triglycerides to glycerol and FFAs at the oil–water interface. Furthermore, lipases can also serve as catalysts for synthetic reactions, such as esterification and transesterification (Aarthy et al. 2014). In comparison to the transesterification of triglycerides using chemical catalysts, biodiesel production by lipase catalysis presents a number of benefits, such as ease of product separation (and consequently, ease of glycerol separation), reduction of the wastewater treatment requirement to a minimum, and a complete lack of side reactions (Tan et al. 2010; Christopher et al. 2014; Zhong et al. 2020).

Another major benefit is related to the possibility of using high-FFA-content oils such as WCO. As conventional homogenous alkaline-catalyzed transesterification cannot achieve high yields of fatty acid alkyl esters (FAAEs) due to the saponification reaction, the use of lipases to transform WCO into biodiesel may be a viable alternative, allowing the conversion of both FFAs and triglycerides into FAAEs, and thus offering a greener route toward biodiesel production under mild reaction conditions (Bajaj et al. 2010; Zhong et al. 2020). However, lipases used in biodiesel production should be nonstereospecific so that all tri-, di-, and monoglycerides and FFAs of different lengths can be converted into FAAEs (Fjerbaek et al. 2009; Tacias-Pascacio et al. 2017; Sánchez-Bayo et al. 2019; Zhong et al. 2020).

3.2 Use of Micro- and Nanostructured Heterogeneous Catalysts in Biodiesel Production

3.2.1 Microstructured Heterogeneous Catalysts

3.2.1.1 Solid Acid Catalysts

Recently, a more "environmentally friendly" approach to biodiesel synthesis has favored the application of solid acid catalysts as replacement for homogeneous acid ones (Helwani et al. 2009). Some peculiar advantages to this class of catalysts include: (i) insensitivity to FFA contents, which ensures the utilization of cheaper and readily available feedstocks without the need for pretreatment; (ii) simultaneous esterification and transesterification reactions (López et al. 2005; Lotero et al. 2005; Yan et al. 2009b; Vyas et al. 2010); (iii) elimination of the biodiesel washing step; (iv) minimized product contamination due to the ease of separation from the reaction medium; (v) tuneability, selectivity, ease of regeneration, and recyclability (Sani et al. 2014); (vi) minimization of corrosion problems even in the presence of acid species (Helwani et al. 2009; Vyas et al. 2010); (vii) applicability in continuous-flow processes (Hillion et al. 2003; Lotero et al. 2005; Chouhan and Sarma 2011); and (viii) the ability to sterically hinder solvation of the catalytic active sites from the action of water, which helps in minimizing catalyst deactivation (Veljković et al. 2006).

Several solid acid catalysts have been studied for use in biodiesel production, including sulfated zirconia, niobium, tantalum oxide and titania (e.g. SO_4^{2-}/ZrO_2, SO_4^{2-}/Ta_2O_5, SO_4^{2-}/Nb_2O_5, and SO_4^{2-}/TiO_2), tungstated zirconia (WO_3/ZrO_2), H-form zeolites, sulfonic ion-exchange resins, sulfonic modified mesostructured silica, sulfonated carbon-based catalysts, heteropoly acids (HPAs), and acidic ionic liquids (ILs) (López et al. 2005, 2008; Kaushik et al. 2008; Xu et al. 2009; Guo et al. 2012; Su and Guo 2014; Yan et al. 2019). Da Conceição et al. (2016) studied the simultaneous esterification and transesterification of macaw palm with a high-FFA-content oil for biodiesel production with ethanol as acylant agent. Sulfated niobium (Nb_2O_5/SO_4) was tested as the heterogeneous catalyst. The biodiesel synthesis was performed at 250 °C with an ethanol to oil molar ratio of 120 : 1, and reaction times were two, three, or four hours. The fatty acid ethyl ester contents for the three reaction times were 91.9%, 98.8%, and 99.2%, respectively. Narkhede et al. (2014) analyzed the effect of using 12-tungstosilic acid anchored to the mesoporous SBA-15 as an acid heterogeneous catalyst to produce biodiesel from WCO. For this synthesis, the different reaction parameters like catalyst loading, methanol to oil molar ratio, reaction temperature, and reaction time were evaluated, as they all influenced the conversion of

WCO into FAME. The maximum conversion achieved for a methanol to oil molar ratio of 8 : 1, a catalyst loading of 0.3 g (6 wt%), a reaction temperature of 65 °C, and a reaction time of eight hours was 86%. Jitputti et al. (2006) studied the transesterification of crude palm kernel oil and crude coconut oil using different solid catalysts. They found that between the several tested acidic and basic solids, only ZnO and SO_4^{2-}/ZrO_2 exhibited the highest activity for the transesterification of both oils. The authors stated that only 1 wt% of sulfated zirconia was needed to catalyze the reaction, resulting in a FAME content higher than 90%. Additionally, they studied the catalyst's recyclability, discovering that it could only be reused after being regenerated and that it could maintain the same catalytic activity for at least four cycles. Boz et al. (2015) produced biodiesel by simultaneous esterification and transesterification of WCO with methanol. Two cationic ion-exchange resins were used: Amberlyst-15 and modified (heat treated at around 493 K) Amberlyst-15. The authors found that the heat treatment decreased the hydrogen ion-exchange capacity of Amberlyst-15 and resulted in the deterioration of the methyl ester yield. Unmodified Amberlyst-15, when applied in a transesterification reaction for nine hours at 338 K under a methanol to oil molar ratio of 12 : 1 and using a catalyst concentration of 3 wt%, could generate a FAME yield of 78%. The team concluded that a high hydrogen ion-exchange capacity was necessary for the catalytic activity of Amberlyst-15 catalysts.

The transesterification mechanism may be explained as follows. First the hydroxyl group of methanol is protonated by Brønsted acid on the catalyst surface, while the protonation of FFAs occurring on the adjacent catalytic site leads to the formation of carbocation. The deprotonation of methanol oxygen produces the nucleophile, which attacks the carbocation to generate a tetrahedral intermediate. The tetrahedral intermediate then eliminates water to form esters (Avhad and Marchetti 2016).

Mass-transport limitation problems, especially for transesterification in heterogeneous acid-catalyzed systems, are one of the major drawbacks of this process. Even though the esterification of FFAs with methanol to FAME is fast on acid catalyst, the transesterification is very slow. Given the high viscosity and poor miscibility of feedstocks with light alcohols, the reaction rate of most of the catalysts is thus hampered. Higher temperature, higher pressure, higher alcohol to oil molar ratio, and longer reaction times are required to obtain considerably high FAME yields. These rigorous conditions are not "friendly" for industrial applications (Lien et al. 2010; Su and Guo 2014).

3.2.1.2 Solid Base Catalysts

A great variety of basic materials have been utilized as heterogeneous catalysts to assist the alcoholysis process, such as: (i) single alkaline earth metal oxides (MgO, BeO, CaO, SrO, BaO, and RaO) and other single metal oxides (SnO_2, ZnO, and CeO_2); (ii) supported alkali metal and alkaline earth metal oxides ($CaZrO_3$, Al_2O_3-SnO, Li/MgO, Al_2O_3/KI, KOH/Al_2O_3, KOH/NaY, and supported alumina/silica supported K_2CO_3); (iii) mixed metal oxides; (iv) hydrotalcites; and (v) anionic ion-exchange resins (Hattori 2001; Kiss et al. 2006; Yan et al. 2009a; Helwani et al. 2009; Shu et al. 2010; Chouhan and Sarma 2011; Refaat 2011; Shuit et al. 2013; Avhad and Marchetti 2016; Kesic et al. 2016; Gardy et al. 2019). Metal oxides are composed of metal cations, which possess Lewis acidity, and oxygen anions, which possess Brønsted basicity. In the methanolysis of oils, these catalysts provide enough adsorptive sites for methanol, in which the O—H bonds readily break into methoxide anions and hydrogen cations. Thus, methoxide anions can react with triglyceride molecules to form corresponding FAMEs (Zabeti et al. 2009).

Among these catalysts, alkaline earth metal oxides are most widely employed as solid base catalysts because of their low cost and high basic strength. Generally, the basic strength of alkaline earth metal oxides is described as follows: MgO < CaO < SrO < BaO (Tanabe and Fukuda 1974). However, BaO is toxic and dissolves in methanol and ethanol, and has an adverse effect on the environment (Endalew et al. 2011; Du et al. 2019). CaO waste materials have potential to be used as solid base catalysts in the transesterification of triglycerides to produce biodiesel because they exhibit high basicity when compared to alkaline homogeneous catalysts, as well as being cheap, noncorrosive, and economically benign, since they come from natural sources such as waste egg and seafood shells (~95% $CaCO_3$) (Yoon et al. 2003; Neves and Mano 2005; Lin et al. 2006; Tsai et al. 2006; Mierzwa-Szymkowiak 2009; Hamester et al. 2012; Safi et al. 2015; Santos et al. 2019). However, the leaching of CaO is a major problem, and the leaching of Ca^{2+} leads to rapid deactivation of the catalyst and formation of soap with FFAs (Hung et al. 2018; Mansir et al. 2018). Meanwhile, CaO can be easily deactivated by poisoning of carbon dioxide and water in the air. In contrast, MgO is not easily leached from the reaction system and maintains considerable activity when reacted with a higher-water-content feedstock oil (Hung et al. 2018).

López et al. (2005) studied the transesterification of triacetin with methanol using several solid acid and base catalysts. Among them was MgO, which performed poorly, achieving only an 18% triacetin conversion after eight hours of reaction at 60 °C. Du et al. (2019) tested a series of carbon-based MgO solid base catalysts with high surface area prepared by sol-gel and calcination. The best results were obtained when urea was added to the MgO precursor and calcined at 800 °C for five hours. This solid showed impressive catalytic performance, and an optimum biodiesel yield of 96.5% was achieved for an ethanol to oil molar ratio of 12 : 1, a catalyst loading of 6 wt%, a reaction temperature of 75 °C, and a reaction time of 60 minutes. Wang et al. (2005a,2005b) tested the transesterification of rapeseed oil at a relatively low temperature of 65 °C, supported by CaO/MgO impregnated on a MgO carrier a $Ca(Ac)_2$ (calcium acetate) solution, followed by its calcination at 700 °C. The catalyst showed higher activity with a glycerol yield of >80% purity (Wang et al. 2005a,2005b). Ilgen and Akin (2009) performed several studies on the effects of methanol/canola oil molar ratio, reaction temperature, KOH loading, and reaction time on the methanolysis of canola oil using MgO-supported KOH catalysts. The best catalytic activity for this reaction was obtained for a 20 wt% KOH loaded on MgO, achieving a FAME yield of 95.05%. Albuquerque et al. (2008) prepared a sample containing 14% of CaO supported on SBA-15, a mesoporous silica-based material, which showed the highest activity among the catalysts tested. The transesterification activity of vegetable oils resulted in a conversion as high as 95% with sunflower oil (after five hours of reaction) and 65% with castor oil (after one hour of reaction). Catarino et al. (2017) produced biodiesel while testing several waste materials rich in Ca as catalysts. Food wastes such as mollusk, shrimp, eggshell, and cuttlebone were used to create heterogeneous catalysts by calcination. All the prepared catalysts were effective in catalyzing the methanolysis of soybean oil. A FAME yield around 96% was obtained after 2.5 hours of reaction at a temperature of 65 °C with a catalyst loading of 5 wt%. WCO was also used as feedstock, and FAME yield decreased to around 65%, with simultaneous production of soap (Catarino et al. 2017). Nakatani et al. (2009) tested the transesterification of soybean oil using CaO from oyster shells as a solid base catalyst. Using a methanol to oil molar ration of 9 : 1, a catalyst loading of 25 wt%, and a rection time of five hours, they achieved a FAME yield of 98.4%. Nair et al. (2012) used CaO from clamshells (*Meretrix meretrix*) for the transesterification of low-FFA WCO. According to their study, for a methanol to oil molar ratio of 6.03 : 1 with a catalyst loading of ~3.0 wt%, a FAME yield of over 97% was possible.

The rate of transesterification by heterogeneous catalysts is generally lower that of homogeneous alkali catalysts due to problems in the diffusion of reactants to the active sites of solid catalysts, which is associated with the three-phase system (oil-methanol-solid catalyst) (Pinto et al. 2005; Islam and Ravindra 2016). On the other hand, solid base catalysts could potentially lead to cheaper production costs by enabling catalyst reuse and creating opportunities to operate in a fixed-bed continuous process (Narasimharao et al. 2007; Kaushik et al. 2008; Boro et al. 2012; Akia et al. 2014).

3.2.2 Nanostructured Heterogeneous Catalysts

Despite the solid phase catalytic systems being intensively studied, their industrial applications are limited. Heterogeneous catalytic methods are usually mass transfer-resistant, time-consuming, and inefficient. Nanocatalysts, on the other hand, have high specific surface areas and high catalytic activities, which can help solve these problems (Chaturvedi et al. 2012). For example, a particle of 30 nm size has 5% of its atoms on its surface, one of 10 nm size has 20% of its atoms on its surface, and one of 3 nm has 50% of its atoms on its surface. Thus, nanoparticles have a much greater surface area per unit of mass when compared with larger particles (Edelstein and Cammarata 1996). According to the International Union of Pure and Applied Chemistry (IUPAC), nanoscale materials, or nanomaterials, are those that have structured components with at least one dimension less than 100 nm (Vert et al. 2012). Recently, there has been a search for new nanomaterials such as nanoclays, nanofibers, carbon nanotubes, nanoporous materials, nanocomposites, and nanoparticles for various applications (Banković–Ilić et al. 2017). Nanocatalysts can be prepared for biodiesel synthesis using a number of methods. The most used are gas condensation, vacuum deposition, chemical deposition, the sol-gel method, impregnation, nanogrinding, and calcination-hydration-dehydration (Rajput 2015; Ambat et al. 2018).

3.2.2.1 Gas Condensation

Nanoparticles can be produced by gas condensation using metals or alloys. The main process involved is controlled evaporation and condensation. In this method, evaporated particles collide with an inert gas molecule. This reduces the overall kinetic energy, leading to the production of nanosized particles in a vacuum. The resulting particles can be collected in a collector attachment to the evaporation device, which is chilled with liquid nitrogen. The main disadvantages of the process are the mismatch between antecedent and source, the difference in the rate of evaporation and choice of temperature, and the long process time (Hahn 1997; Tissue and Yuan 2003; Rajput 2015; Ambat et al. 2018).

3.2.2.2 Vacuum Deposition

Vacuum deposition involves applying a thermal source to evaporate different materials, such as alloys, molecules, and compounds. The substrate molecule is heated in a vacuum at a pressure below 0.1 Pa (Rajput 2015). The technique of chemical vapor deposition (CVD) used for nanocatalyst synthesis involves deposition of the desired materials on to the substrate's surface via chemical reaction. This reaction can be induced in two ways: with high temperatures or plasma. An overhead temperature of 900 °C is required for temperature-induced chemical reaction, whereas in plasma-activated CVD, the temperature is lower, at between 300 and 700 °C. This procedure has the advantage of being cost-effective. Other benefits are the use of proper deposition coatings without any high vacuum and the range of precursors. However, some byproducts of CVD may constitute health risks. CVD is very expensive and is mostly used in electronics industries (Hampden-Smith and Kodas 1995; Frey 2015; Rajput 2015; Ambat et al. 2018).

3.2.2.3 Chemical Deposition

Synthesis of nanocatalyst by chemical precipitation takes place through the reaction of solvent components. In this process, the size can be controlled by the arrested precipitation technique. The dopant is introduced into the main solution prior to precipitation, and the particles formed can be separated using surfactants. Proper quantum arrangement of dried nanomaterial obtained after the separation and cleaning of the precipitant is achieved by polymerization of the nanocluster surface with covering surfactant under ultraviolet light (Wang et al. 2005a,b; Rajput 2015; Ambat et al. 2018).

Feyzi and team (2017) prepared a series of MgO-La_2O_3 nanocatalysts via a co-precipitation method. They premixed aqueous solutions of $MgO(NO_3)_2.6H_2O$ (1 M), and $La(NO_3)_3.6H_2O$ (1 M) with different weight percentages and heated the resulting solution to 343 K. Aqueous N_2CO_3 (1 M) was added dropwise to the nitrate mixture with stirring at 343 K until pH 12.0 was achieved. The resulting precipitate was filtered, washed, dried, and calcinated to give the final catalyst. The results showed that by increasing the weight percent of La, the catalytic activity was increased. Also, the catalyst with Mg/La weight ratio of 60 wt% (La) is an optimal catalyst, and under its optimal conditions (CH_3OH/oil = 18 : 1, 338 K, five hours reaction time, mechanical stirring at 700 rpm), it showed high catalytic activity for biodiesel production, reaching a FAME yield of 97.7% (Feyzi et al. 2017). Wang et al. (2009a) prepared a mesoporous nanosized KF/CaO-MgO catalyst by a co-precipitation method. This catalyst possessed a unique mesoporous structure with pore size ~34 nm and average particle size ~35 nm. The influence of the preparation conditions (e.g. calcination temperature and time, mass ratio of CaO/MgO, and mass ratio KF/CaO-MgO) on the activity of the as-prepared catalyst with regard to biodiesel yield was investigated in detail. With a mass ratio of CaO/MgO of 8 : 2, mass ratio of KF to CaO-MgO of 0.25, calcination temperature of 873 K, and calcination time of three hours, the catalyst performed at its highest activity in the transesterification of rapeseed oil (methanol to oil molar ratio of 12 : 1, catalyst loading of 3% of the total reactants, reaction temperature of 343 K, reaction time of three hours), achieving a 95% FAME yield (Wang et al. 2009a).

3.2.2.4 Sol-Gel Method

The sol-gel method is a liquid phase technique for nanocatalyst production involving hydrolysis reactions, mainly for colloidal particle production. The precursor for synthesizing these colloids consists of ions of metal alkoxides and aloxysilanes. The precipitation of nanoparticles can be achieved with an appropriate quantity of substrate. The reaction requires a low temperature, shaping and inserting can be achieved without much effort, and the process is flexible (Rahman and Padavettan 2012; Rajput 2015; Ambat et al. 2018).

An activated carbon-supported copper oxide (CuO/C) was prepared as follows: copper nanoparticles were synthesized by the sol-gel method and then impregnated into the pore network of the activated carbon support. This solid was calcined at 400 °C for six hours. The obtained (7 wt% CuO/C) catalyst was applied in the esterification of FFA in rubber seed oil, achieving a maximum FFA conversion of ~95% after six hours of reaction at 65 °C with a methanol to oil molar ratio of 10 : 1 and a catalyst loading of 8 wt% (Ong et al. 2014; Avhad and Marchetti 2016). Similarly, nanoparticles of CaO from calcium nitrate and snail shell were successfully synthesized by the sol-gel method and used as active and stable catalysts for biodiesel production. CaO nanocatalyst produced from snail shell showed a stronger basicity, translating to higher catalytic activity. This catalyst achieved a FAME yield of 96% for a transesterification reaction with a reaction temperature of 65 °C, methanol to oil molar ratio of 12 : 1, catalyst loading of 8 wt%, and reaction time of six hours (Safaei-Ghomi et al. 2013; Gupta and Agarwal 2016).

3.2.2.5 Impregnation

In the impregnation method of nanocatalyst synthesis, the desired component-containing liquid encounters a solid, and the appropriate component from the liquid is deposited on the solid's surface. During this process, adsorption of diverse species occurs at different rates. The materials are selectively adsorbed to the surface by means of various forces, including hydrogen bonds, van der Waals forces, Coulomb forces, and ion exchange. Impregnation is a simple, well-studied, and low-cost method, but it is difficult to control the size of the particles. The properties of the nanocatalysts can be modified by adjusting synthesis parameters such as concentration, reducing agent, and calcination temperature (Jiang 2006; Rajput 2015; Ambat et al. 2018).

Bet-Moushoul et al. (2016) synthesized five different types of CaO-based catalyst-supported gold nanoparticles (AuNPs), including commercial CaO, eggshell, mussel shell, calcite, and dolomite, using the impregnation method. CaO was added to AuNP solution and stirred at room temperature for one hour. All solutions lost their color by deposition of AuNPs on CaO. This method allowed for an average amount of AuNPs to be deposited on the catalyst (0.012 ± 0.18 wt%). The catalysts obtained were then used in the transesterification of triglycerides. The influence of the reaction parameters on the catalytic activity was also assessed, and optimal conditions were found to be a methanol to oil molar ratio of 9 : 1, reaction temperature of 65 °C, catalyst loading of 3 wt%, and reaction time of three hours. Under these conditions, the best FAME yield reached for all catalysts ranged between 94% and 97%, with calcite/AuNPs the most active. Reusability tests were also performed, and after the 10th run there was no significant loss of activity for any catalysts (Bet-Moushoul et al. 2016).

Kaur and Ali (2011) prepared lithium-impregnated CaO in nanoparticle form using a wet impregnation method. X-ray diffraction (XRD) and transmission electron microscopy (TEM) were used to confirm that the solid was in fact in nanoparticle form. Basic strength was measured by Hammett indicators. 1.75 wt% Li-CaO nanocatalyst was used as a solid catalyst in the transesterification of Karanja and jatropha oil, containing 3.4 and 8.3 wt% of FFA, respectively. The reactions were performed at 65 °C using a methanol to oil molar ratio of 12 : 1 and a catalyst loading of 5 wt% (oil), and achieved conversions for Karanja and jatropha oils of >99% FFA in one and two hours, respectively (Kaur and Ali 2011). In another study, two types of efficient and novel sulfonated ZrO_2-TiO_2 nanorods solid acid catalysts, designed as ZrO_2-TiO_2-SO_3H nanorods and ZrO_2-TiO_2@SO_4^{2-} nanorods, were synthesized to produce biodiesel by esterification and transesterification of palmitic acid. The ZrO_2-TiO_2 nanorods were prepared by a wet-chemical deposition precipitation method. $ZrOCl_2.8H_2O$ was dissolved in deionized water and mixed with TiO_2 nanorods. NH_4OH was added to the solution dropwise until pH 11.0 was reached. The mixture was ultrasonicated to form a homogeneous mix, which afterward was dried and calcinated. ZrO_2-TiO_2@SO_4^{2-} nanorods were prepared through the impregnation method by immersing the obtained ZrO_2-TiO_2 nanorods in a sulfuric acid solution. The mixture was separated by centrifugation followed by calcination. ZrO_2-TiO_2-SO_3H nanorods were prepared via a post-grafting method through a substitution reaction between the hydroxyl-groups on the surface of the ZrO_2-TiO_2 nanorods and KH-580. Biodiesel synthesis conditions were a methanol to oil molar ratio of 20 : 1, reaction temperature of 100 °C, catalyst loading of 1.5 wt%, and reaction time of five hours. Although the activity of ZrO_2-TiO_2@SO_4^{2-}nanorods was better (biodiesel yield of 95.3%) than that of ZrO_2-TiO_2-SO_3H nanorods (biodiesel yield of 93.1%), the stability of the latter was superior, maintaining high catalytic activity after five successive uses (biodiesel yield 85.1%). The authors concluded that ZrO_2-TiO_2-SO_3H nanorods were more competent solid acid catalysts, with high catalytic activity and excellent reusability (Fan et al. 2019).

3.2.2.6 Nanogrinding

Nanogrinding is a simple comminution process capable of producing ultra-fine particles (<10 μm) with high specific surface areas. It can alter the size, structure, composition, and morphological characteristics of materials by carefully selecting appropriate parameters. Panjaitan et al. (2017) prepared CaO with specific surface areas ranging from 4.5 to 62.5 m^2g^{-1} by calcinating scallop shells and then applying methanol-assisted nanogrinding. They found that three distinct phases were formed on the surfaces of these catalysts: calcium methoxide, calcium hydroxide, and calcium oxide. The methanol-assisted nanogrinding process was carried out using a planetary ball mill, which was equipped with two zirconia pots of 80 cm^3 volume, each containing 100 g yttria-stabilized zirconia milling beads (3 mm diameter). The transesterification of soybean oil was carried out at 65 °C with a methanol to oil molar ratio of 9 : 1 and a catalyst concentration of 4 wt% (oil). The highest catalytic activity in a heterogeneous methanolysis reaction was observed for CaO-ss (scallop shell) samples that had undergone partial surface conversions to $Ca(OH)_2$.

3.2.2.7 Calcination-Hydration-Dehydration

This method is normally used for waste materials rich in calcium (i.e. eggshells, seafood shells, and bones). The first step consists in calcining the catalytic precursors in order to obtain the CaO from calcium carbonate ($CaCO_3$). Afterward, the obtained oxides are refluxed in water for several hours to allow hydration and consequent formation of calcium hydroxides ($Ca(OH)_2$), giving an aqueous slurry. These slurries are then filtrated and dried overnight in an oven at 105 °C. Finally, the solids are ground into a fine powder and once again calcined to allow dehydration to take place, transforming the hydroxides in oxides.

Pandit and Fulekar (2017) utilized waste eggshells for the synthesis of highly active heterogeneous calcium oxide (CaO) nanocatalyst in order to transesterify dry biomass (microalgae) into methyl esters (biodiesel). The CaO nanocatalyst was prepared using the calcination-hydration-dehydration technique. This involved washing the eggshells with warm tap water to remove unwanted materials on the surface and then rinsing and drying them. The eggshells were then crushed using a blender and sieved, and the resultant powder was calcinated. The CaO obtained was refluxed in water at 60 °C for six hours and dried in a hot-air oven overnight. The solid dried particles were further ground by a blender and dehydrated by calcination to 870 °C for three hours to convert the calcium hydroxide into CaO and carbon dioxide. The obtained solid showed nanoparticles of 75 nm size and 16.4 m^2g^{-1} surface area. The optimal reaction conditions were obtained using a response surface methodology study: a methanol to oil ratio of 10 : 1 wt/vol, a reaction temperature of 70 °C, a catalyst loading of 1.7 wt%, and a reaction time of 3.6 hours. Under these conditions, the catalyst was able to achieve 86.41% of microalgae biodiesel yield.

Reddy et al. (2016) synthesized a highly active CaO nanocatalyst from *Polymedosa erosa* seashells with a diameter of 66 ± 3 nm and a surface area of 90.61 m^2/g, also using a calcination-hydration-dehydration technique. The transesterification of crude jatropha oil was preceded by an acid pretreatment in the form of esterification with H_2SO_4 in order to minimize the FFA content to a value of 1%. The biodiesel production took place using the CaO nanocatalyst, with a methanol to oil molar ratio ranging from 4 : 1 to 6 : 1, a catalyst loading ranging from 1 to 3 wt%, a reaction temperature of 60 °C, and a reaction time of 60–180 minutes. The authors stated that under the optimal conditions and using the synthesized CaO nanocatalyst they were able to reach a FAME yield of 98.54%.

3.3 Enzymatic Catalysis

3.3.1 Heterogeneous Biocatalysts (Lipases) and Their Immobilization

Biocatalysts are living (biological) systems, or their parts, that are able to catalyze chemical reactions and processes. However, they have some disadvantages. The use of free enzymes as catalysts is not economically viable because they are not always as stable as needed under operational conditions and they are restricted to a one-time usage. Moreover, enzyme separation from the reaction mixture is very complicated, making catalyst recycling very much impossible (Garcia-Galan et al. 2011; Tran and Balkus 2011; Rodrigues et al. 2013). This limits the large-scale application of enzymatic methods (Tan et al. 2010).

One way to overcome these drawbacks involves immobilizing the enzyme on to supports. Enzyme immobilization on solid carriers can increase both thermal and chemical stability and protects the enzyme molecules from denaturation (Sánchez-Bayo et al. 2019). Also, immobilized enzymes can be described as being physically enclosed or located in a certain defined space while retaining their catalytic activities, with the advantage they may be used repeatedly and continuously (Jegannathan et al. 2008; Zhang et al. 2015). To date, a series of immobilization techniques have been developed, among which the most used are physical adsorption, entrapment, covalent bonding, and cross-linking (Dalla-Vecchia et al. 2004; Ricardi et al. 2018; Zdarta et al. 2018; Bilal et al. 2019).

3.3.1.1 Physical Adsorption

The process of physically adsorbing a protein on to a supporting material is regarded as very simple and is one of the most used methods. In adsorption, the enzyme is immobilized through low-energy bonds, such as van der Waals or hydrophobic interactions or hydrogen or ionic bonds with the exterior surface of the supporting material (Villeneuve et al. 2000; Brady and Jordaan 2009). De Paola et al. (2009) performed a factor analysis for the enzymatic transesterification of waste oil for biodiesel production. The chosen biocatalyst was a lipase extracted from *Rhizomocur miehei* that was immobilized on an ion-exchange macroporous hydrophilic resin with irregular shape, with dimensions ranging between 0.02 and 0.06 cm, bulk density of 0.4 g/cm^3, and porosity of 0.53 ± 0.02. The developed model predictions were in good agreement with experimental data, with yields of 90% after eight hours of reaction (De Paola et al. 2009).

Yagiz and coworkers (2007) used hydrotalcite and four other types of zeolites (13-X, 5A, FM-8, and AW-300) as immobilization materials for lipases using physical adsorption. The amount of lipase adsorbed on hydrotalcite was highest at pH 8.5 and 4 °C. The immobilized lipase was able to retain, after the seventh reuse cycle, 36% of its initial activity at 45 °C and 14% at 55 °C for the transesterification of WCO. Comparing the obtained results, the free lipase achieved a yield of 95% and the immobilized lipase in hydrotalcite a yield of 92.8%. However, considering the amount of lipase in the reaction, the immobilized lipase showed higher activity. The main reason for this may be the fact that hydrotalcite itself is also an active site for transesterification , being used as a solid base heterogeneous catalyst in transesterification reactions of edible oils. For the other types of support, the transesterification reaction showed no conversion of triglycerides to FAAEs.

In another report, biodiesel was produced by alcoholysis of *Jatropha curcas* oil, a non-edible oil, using biocatalysts (lipase). The best yield achieved was 98% (w/w) for *Pseudomonas cepacia* immobilized by adsorption on celite at 50 °C in the presence of 4–5% (w/w) water following eight hours of reaction. Further, it was found that this biocatalyst could be recycled four times without any loss of activity (Shah and Gupta 2007).

Katiyar and Ali (2012) studied the preparation of MCM-41 as support for the immobilization of *Candida rugosa* lipase by the physical adsorption technique. They observed maximum lipase immobilization (250 mg/g) on the support at pH 6, and used the resulting immobilized lipase as a biocatalyst for the transesterification of the cotton seed oil with methanol. After optimization of all reaction conditions (pH, temperature, and methanol to oil molar ratio), a maximum yield of FAME (98.3%) was achieved.

In physical adsorption, however, enzyme leaching is a real problem, limiting the use of the immobilized enzyme under different reaction conditions (Hartmann and Jung 2010; Liu et al. 2020). One solution might be to employ entrapment to restrict the enzyme inside the support framework (Safaryan et al. 2017; Souza et al. 2017).

3.3.1.2 Entrapment

Entrapment involves the capture of a lipase typically within a polymeric network, such as an organic polymer or microcapsules of polymers, that allows the substrate and products to pass through but retains the lipase. Usually, lipase entrapment is performed *in situ*; that is, the lipase is placed into the monomeric phase solution, which, as polymerization develops, leads to its entrapment (Villeneuve et al. 2000; Brady and Jordaan 2009; Zhao et al.

2015). Compared to physical adsorption, lipase immobilization by entrapment is much more stable, protecting the enzyme from the surrounding environment (hydrophobic solvents, mechanical sheer, and gas bubbles) (Lalonde and Margolin 2002)

Moreno-Pirajàn and Giraldo (2011) studied the enzymatic transesterification of palm oil with methanol and ethanol. The lipase obtained from *Candida rugosa* was investigated in immobilized form, entrapped within activated carbon as a support. This lipase resulted in the highest yield of FAAEs among the four tested. Under optimal reaction conditions, FAME and FAEE formations of 70 and 85 mol% were achieved after one hour of reaction (Moreno-Pirajàn and Giraldo 2011). Noureddini et al. (2005) analyzed the enzymatic transesterification of soybean oil with methanol and ethanol using nine different lipases. A lipase from *P. cepacian* gave the highest yield of FAAEs. This lipase was further investigated in immobilized form within a chemically inert and hydrophobic sol-gel support. Under optimal reaction conditions, FAME and FAAE yields were 67% and 65% molar after one hour of reaction (Noureddini et al. 2005).

Kuan et al. (2013) entrapped lipase from *P. cepacian* in a polyallylamine-mediated biometric silica. This biocatalyst was applied to the production of biodiesel with soybean oil and WCO as feedstock. For WCO transesterification, the optimal conditions were a reaction temperature of 43.3 °C, substrate molar ratio of 5 : 1, and 38% (w/w) *n*-hexane. The predicted and experimental values of conversion were 68% and 67%, respectively (Kuan et al. 2013). Moreover, another study investigated the lipase-catalyzed synthesis of alkyl esterified from tallow and grease using *P. cepacian* lipase immobilized within a phyllosilicate sol-gel matrix. The matrix-immobilized lipase effectively converted grease and tallow into ethyl esters at >95% when using ethanol. The matrix-immobilized lipase was also easily recovered and could be reused at least five times without losing any catalytic activity (Hsu et al. 2001). Unlike covalent bonding, entrapment uses a relatively simple procedure while maintaining the immobilized lipase's catalytic activity and stability (Kennedy et al. 1990).

3.3.1.3 Covalent Bonding

Lipase immobilization by covalent bonding to a solid support depends on an irreversible bonding of the lipase to the support matrix (Tan et al. 2010). This bond consists in a chemical reaction between the active amino acid residue in the exterior of the active catalytic site of the lipase (usually thiol and amine groups) and the active groups of the support matrix (Brena and Batista-Viera 2006; Stoycheva et al. 2011). This results in strong interactions between the lipase and the support, making enzyme leaching to the reaction moiety during the catalytic process very rare, since the activity and stability of the immobilized enzyme are directly related to the properties of the support, including pore size, chemical stability, and binding affinity to the enzyme (Mateo et al. 2006; Tan et al. 2010; Ricardi et al. 2018). Multipoint covalent bonding is the most efficient immobilization technique available in terms of thermal and operational lipase stability (Mendes et al. 2011).

Tang et al. (2019) studied the immobilization of lipase from *Candida cylindracea* via covalent bonding with alkaline Ca-bentonite modified with glutamic acid. The immobilized lipase obtained demonstrated a significantly improved catalytic activity. At 50 °C, free lipase retained only 6% of its initial activity after six hours of reaction. However, immobilized lipase (i.e. the Glu-Ca-Bent-lipase) managed to retain 50% of its activity after eight hours. The biocatalyst prepared for biodiesel production presented an improvement, achieving a transesterification reaction conversion of 99%, where free lipase achieved only 52.8%. Reusability-wise, said biocatalyst kept the reaction conversion at 56.2% after being reused for five times, while free lipase was inactive upon two uses (Tang et al. 2019).

Rodrigues et al. (2016) evaluated the feasibility of biodiesel production by transesterification of Jatropha oil with methanol, catalyzed by noncommercial *sn*-1,3-regioslective lipases. Heterologous *Rhizopus oryzae* lipase (rROL) was immobilized by covalent bonding on different supports and tested. Transesterification reaction was performed at 30 °C, with seven stepwise methanol additions. For all the tested biocatalysts, an average FAME conversion of 51–65% was obtained after four hours of reaction (Rodrigues et al. 2016). Yücel (2011) immobilized microbial lipase from *T. lanuginosus* on to olive pomace via covalent bonding. The immobilized support was used to produce biodiesel from pomace oil with methanol. Under the optimized conditions, the maximum biodiesel yield was 93% at 25 °C after 24 hours of reaction. The immobilized enzyme retained its activity during 10 repeated batch reactions (Yücel 2011).

Another report demonstrated the immobilization of microbial lipases from *T. lanuginosus* and *P. fluorescens* by multipoint covalent bonding on Toyopearl AF-amino-650M resin. The active and thermostable biocatalyst thus prepared was used to catalyze the transesterification of babasu and palm oils with ethanol. Compared with free lipase, *T. lanuginosus*-immobilized biocatalyst presented higher hydrolytic activity and thermal stability (27–31 times more stable). The highest conversion into biodiesel was found in the transesterification of palm oil catalyzed by both *T. lanuginosus* and *P. fluorescens* biocatalysts, with transesterification yield ranging from 93.5% to 100% (Mendes et al. 2011).

3.3.1.4 Cross-Linking

The cross-linking technique is based on the immobilization of lipase via the formation of intermolecular cross-linkages with the aid of a cross-linker. Usually, cross-linkers are bi- or multifunctional reagents such as glutaraldehyde or bisdiazobenzidine. Generally, this technique is support-free and involves the aggregation to form a 3D structure. The first step is the precipitation of the enzyme aggregate via the use of acetone. The aggregates are then cross-linked using a cross-linker to form a more robust structure (Murty et al. 2002; Jegannathan et al. 2008; Zhao et al. 2015). Cross-linking can represent an improvement on covalent bonding, because the enzyme can be cross-linked to the support through the cross-linker. This procedure usually includes a preliminary step involving the immobilization on an ion-exchange resin, followed by treatment with a buffered solution of cross-linker to form a Schiff base (Villeneuve et al. 2000; Liu et al. 2020).

Abdulla and Ravindra (2013) studied the immobilization of lipase from *B. cepacian*. The lipase was first cross-linked with glutaraldehyde and then entrapped into a hybrid matrix of equal proportions of alginate and κ-carrageenan. Afterward, this biocatalyst was applied in the transesterification of *J. curcas* L. oil. Under optimal conditions, 100% fatty acid ethyl esters could be achieved. The immobilized lipase was stable and capable of retaining 73% relative transesterification activity after six cycles of reuse (Abdulla and Ravindra 2013).

Similarly, a novel immobilization method was developed for lipase from *T. lanuginosus* within hydrophilic polyurethane foams, using polygluturaldehyde as a cross-linking agent. The immobilized lipase was used as a biocatalyst to produce biodiesel with canola oil and methanol, and the effects of enzyme loading, alcohol to oil molar ratio, water concentration, and temperature on the transesterification reaction were investigated. The maximum FAME yield achieved was 90%, with enzymatic activity maintained after 10 batches. The immobilized lipase proved to be stable and lost little activity when subjected to repeated uses (Dizge and Keskinler 2008).

Kumari et al. (2006) investigated the use of immobilized lipase from *P. cepacian* as catalyst in the transesterification reaction of *Madhuca indica* oil. One of the best results was obtained using cross-linked enzyme aggregates (CLEAs) of the studied lipase. While free lipase powder could achieve a conversion yield of 98% in six hours of reaction, after process optimization, CLEAs of *P. cepacian* gave a conversion of 92% in 2.5 hours. Both methods used around the same amount of lipase (Kumari et al. 2006).

Han and Kim (2011) explored the methanol-tolerant lipase M37 from *Photobacterium lipolyticum* and its immobilization by the CLEA method. The cross-linked lipase could produce biodiesel from olive oil and alcohols such as methanol and ethanol. Regarding its physical stability and reusability, the cross-linked lipase could potentially be used as a catalyst in organic synthesis, including biodiesel production (Han and Kim 2011).

3.3.2 Nano(Bio)Catalysts: Immobilization of Enzymes on Nanosupports

Conventional immobilized enzyme materials are typically micron sized (Ansari and Husain 2012). Recently, many researchers have been investigating the use of nanomaterials as support matrixes for enzymes, due to their intrinsic large surface-to-area-to-volume ratio, allowing for high enzyme loading and enhanced mass transfer (Hwang and Gu 2013). Immobilization of lipase on functionalized nanomaterials such as nanoparticles, nanotubes, nanofibers, and nanocomposites has very promising applications, particularly that of nano-immobilized lipase in packed-bed reactors, resulting in multiple reuse and effective protection from enzyme denaturation in biodiesel production (Verma et al. 2016; Kim et al. 2018).

3.3.2.1 Nanoparticles

Nanoparticles of both inorganic and organic nature have been extensively studied recently as possible supports for enzyme immobilization (Jia et al. 2003; Kim et al. 2006). Nanoparticles can be divided into two types: magnetic and nonmagnetic.

3.3.2.1.1 Nonmagnetic Nanoparticles

Silica is one of the most used inorganic support materials for enzyme immobilization. Its characteristics, including high thermal and chemical resistance, good mechanical properties, low cost, nontoxicity, and good compatibility, make it suitable for several applications. Silica supports show good adsorption properties due to their high surface area and porosity, providing effective enzyme immobilization and minimal diffusion limitations (Jesionowski 2001; Jesionowski and Krysztafkiewicz 2002; Adlercreutz 2013; Liese and Hilterhaus 2013). Macario et al. (2013) demonstrated the immobilization of lipase from *Rhizomucor miehei* in porous inorganic silica and its use as biocatalyst in the transesterification of triolein with methanol for biodiesel production. This achieved a substrate conversion of 98%, with a biodiesel yield of 89% after three hours of reaction at 37 °C with a methanol to oil molar ration of 6 : 1.

Babaki et al. (2016) studied the enzymatic production of biodiesel by methanolysis of canola oil. Mesoporous SBA-15 nanoparticles were prepared, characterized, and functionalized by 3-glycidiloxypropyl trimethysilane. Lipases from *C. antarctica*, *T. lanuginosus*, and *R. miehei* were covalently immobilized on to SBA-epoxy. Thermal stability and the influence of methanol concentration on the catalytic activity were also evaluated. The immobilized lipases showed rather good operational stability, being able to retain most of their activity after several cycles of reaction (Babaki et al. 2016).

Nanometals represent another type of inorganic nanoparticle. They usually include metal oxides such as zirconia, titania, and alumina, and other metals like gold and silver. These supports are known for their high stability, mechanical resistance, and good adsorption capacity. They are inert under various reaction conditions, facilitating their application as supports for the immobilization of various classes of enzymes (Cipolatti et al. 2014; Zdarta et al. 2018).

Dumri and Hung Anh (2014) evaluated the immobilization of a novel lipase (EC3.1.1.3) on to polydopamine-functionalized silver nanoparticles. Biodiesel production yield was assessed through gas chromatography, showing 95% yield for the immobilized lipase after six hours of reaction at 40 °C. In terms of reusability, the immobilized lipase was still active after seven batches, with subsequent conversion rate of soybean oil decreased by only 27% (Dumri and Hung Anh 2014). Similarly, zirconia nanoparticles were modified with carboxylic acid with a long alkyl chain and shown to significantly enhance the activity of immobilized lipases. Comparing catalytic activity results for free *P. cepacian* lipase and the immobilized lipase on unmodified zirconia, the initial activity for the immobilized lipase was almost double that of the free one. If the best modified zirconia is considered, then the initial activity becomes over 16 times higher than for the free lipase (Chen et al. 2009).

Organic nanoparticles, such as chitosan, can be divided into two classes: synthetic materials (mainly polymers) and renewable materials obtained from natural sources (biopolymers) (Zhi et al. 2006; Zdarta et al. 2018). Chitosan is a biopolymer with vast potential as an enzyme immobilization support (Krajewska 2004). Its nanoparticles possess bioactivity that is harmless for humans and presents outstanding physical–chemical, antimicrobial, and biological properties, which makes them viable for a number of applications, ranging from tissue engineering, pharmaceutical, and food packaging to biosensing, wastewater treatment, and enzyme immobilization (Verma et al. 2020). Chitosan is also commercially interesting due to its low cost, being obtained from the shells of shellfish (mainly crab, shrimp, lobster, and krill) and waste from the seafood processing industry (Krajewska 2004; Cipolatti et al. 2014).

Ghadi et al. (2015) synthetized chitosan magnetic core-shell nanoparticles in order to use them as supports for *B. cepacian* lipase immobilization on the transesterification of soybean oil. The catalytic activity of the synthetized biocatalyst was compared to that of free lipase. Results showed better biological activity while the enzyme was immobilized on the support, due to the polymeric and hydrophilic properties of chitosan (Ghadi et al. 2015).

Xie and Wang (2012) immobilized a lipase on to magnetic chitosan nanoparticles in order to prepare a biocatalyst for the transesterification of soybean oil. This immobilized lipase reached a FAME conversion yield of 87% under the optimized conditions of methanol to oil molar ratio of 4 : 1 with a three-step addition of methanol at 35 °C for 30 hours. Furthermore, it could be reused up to four times without significant loss of catalytic activity. Among nanoparticles, the nonmagnetic supports for enzyme immobilization are well dispersed in the reaction solution. This creates a problem for recovery of the biocatalysts to be reused as they are difficult to separate from the reaction mixture, requiring high-speed centrifugation (Zhao Chen et al. 2008).

3.3.2.1.2 Magnetic Nanoparticles

One possible solution to overcome the problem of nanoparticle recovery is attachment of the enzyme molecules to magnetic iron oxide nanoparticles (magnetite, Fe_3O_4) and simple separation of the biocatalyst with the use of an external magnetic field (Cao et al. 2012). In lipase immobilization, magnetite is of great interest due to its low toxicity, biocompatibility, and large surface area. The abundance of hydroxyl groups on its surface allows effortless functionalization and strong covalent bonding with the enzyme (Li et al. 2013; Vaghari et al. 2016). However, bare magnetite always tends to aggregate due to magnetic dipole–dipole attractions (Xie and Zang 2016). According to the literature, functionalization of the magnetic nanoparticles of Fe_3O_4 can lead to an improvement in their dispersity and chemical stability (Rosenholm et al. 2011).

López et al. (2014) tested magnetic nanoparticles for immobilization in two robust types of nanobiocatalysts: lipases covalently bonded on to the surfaces of magnetic nanoparticles functionalized with amino groups, and magnetic CLEAs. For this purpose, lipase B of *C. antarctica* was used. The catalytic activity of the nanobiocatalysts was evaluated in the transesterification reaction of olive oil at 40 °C with a 2-propanol to oil molar ratio of 6 : 1. In terms of catalytic activity, both nanobiocatalysts showed similar results for the transesterification reaction, with the magnetic CLEAs having slightly higher numbers. Biodiesel conversion-wise, the magnetic CLEAs presented higher yields (30.3%) when compared with immobilized lipase (20.5%). In terms of stability, after 10 cycles of reuse, both nanobiocatalysts had their biodiesel conversion reduced to above 50% (López et al. 2014).

Jambulingam et al. (2019) produced biodiesel using a lipase immobilized on to functionalized magnetic nanoparticles. In this study, APTES ((3-aminopropyl)triethoxysilane)-functionalized magnetic Fe_3O_4 nanoparticles were tested in the transesterification of *Aspergillus niger* fungal lipid. The catalytic activity of the nanobiocatalyst allowed production of biodiesel with a yield of 85.3% (w/w) under optimal reaction conditions. The nanobiocatalyst showed catalytic activity after five cycles without any loss of performance (Jambulingam et al. 2019).

3.3.2.2 Carbon Nanotubes

Due to their formidable mechanical properties, large surface area, small size, unique morphology, and chemical stability, carbon nanotubes, whether single-walled carbon nanotubes (SWCNTs) or multiwalled carbon nanotubes (MWCNTs), are very promising and effective supports for the immobilization of enzymes (Shim et al. 2002; Yang et al. 2015; Kim et al. 2018; Szelwicka et al. 2019). However, their small size, one of their best properties, is also one of their main disadvantages, because it makes recovery very difficult after completion of a reaction. One solution is to combine carbon nanotubes with magnetic nanoparticles (Tan et al. 2012). Further, carbon nanotubes have few active sites, which can only be introduced by functionalizing them with dendrimer molecules (Zhao et al. 2011; Zhong et al. 2020). Bencze et al. (2016) used SWCNTs as supports for the covalent immobilization of *C. antarctica* lipase B. The obtained nanobiocatalysts with low diffusional limitation were tested in batches in the ethanolysis of sunflower oil. The most efficient and stable among them achieved an conversion yield of 83.4% after four hours of reaction at 35 °C, retaining over 90% of its original activity after 10 cycles (Bencze et al. 2016).

Fan et al. (2016) grafted magnetic multiwalled carbon nanotubes (mMWCNTs) on to a polyamidoamine (PAMAM) dendrimer, in order to combine magnetic properties with a large surface functionalized with amino groups. The obtained support was used in the immobilization of *Rhizomucor miehei* lipase. The authors found recovery to be quite high, and the corresponding esterification activity of the immobilized enzyme to be 27-fold higher than that of the free lipase. Said immobilized lipase was employed as a biocatalyst for biodiesel production from WCO in a *tert*-butanol solvent system. Biodiesel conversion reached 94% under optimal conditions, and the immobilized lipase could be easily recovered without showing any decrease in conversion rates after 10 cycles of reuse. Given these results, the immobilized lipase is a catalyst with high stability and exceptional reusability for biodiesel production (Fan et al. 2016).

3.3.2.3 Nanofibers

Some problems that certain nanoparticles and carbon nanotubes can suffer from are mass-transfer limitations and difficulties in recycling due to good dispersion (Wang et al. 2009b). A promising alternative support for enzyme immobilization capable of overcoming these problems is nanofibers (Kim et al. 2018). Nanofibers are extensively utilized as supports for lipase immobilization due to their essential functional groups and uniform diameter, their very high surface area-to-volume ratios, and their ease of separation from the reaction mixture (Li et al. 2007; Wang et al. 2009b). They are commonly prepared by self-assembly, electrospinning, template synthesis, and phase separation. Among such preparation methods, electrospinning is the most universal and highly effective. Electrospun nanofibers present preferred dispersion and functional groups for lipase immobilization and almost infinite length (Ansari and Husain 2012; Ahmed et al. 2015).

Li et al. (2011a) fabricated a polyacrylonitrile nanofibrous membrane by electrospinning and activated it by amidination to immobilize *P. cepacian* lipase. The lipase was then employed as a biocatalyst for the transesterification of soybean oil to biodiesel by adding methanol as solvent. Under optimal reaction conditions, the biodiesel conversion of soybean oil was 90% after 24 hours. In terms of reusability, the immobilized lipase still retained 91% of its initial catalytic activity after 10 cycles (Li et al. 2011a). Another study demonstrated the production of butyl-biodiesel using polyacrylonitrile nanofibers obtained by electrospinning with *P. cepacian* immobilized through physical adsorption. About 80% conversion to butyl-biodiesel was achieved after 24 hours by suspending the biocatalyst at 2.4 mg/ml with a *n*-butanol to rapeseed oil molar ratio of 3 : 1 at 40 °C. A further 24 hours of reaction resulted in 94% conversion. The immobilized lipase continued to be catalytically active for 27 days, with a 15% reduction in conversion yield (Sakai et al. 2010).

3.3.2.4 Nanocomposites

In light of the unique characteristics and properties of inorganic and organic nanoparticles, researchers have been investigating the possibility of maximizing the benefits by combining the two. The resulting biocatalyst may exhibit properties not observed for either of the individual components (Yang et al. 2012). An example of a precursor for a nanocomposite is the magnetic nanoparticle. Such nanomaterials are nonporous and can cause damage to native nanoparticles through erosion of reacting agents (Lu et al. 2007). To avoid this environmental reaction, the encapsulation or coating of nanoparticles with the desired organic/inorganic molecules presents an enhancement to their function as nanosupports, thus creating nanocomposites or hybrid materials (Long et al. 2004; Sen et al. 2006, 2010). This adds to nanoparticles' high versatility for surface modifications, making them an ideal nanosupport for lipase immobilization (Chaubey et al. 2009).

Tran et al. (2012) immobilized a lipase from *Burkholderia* sp. on magnetic nanoparticles in order to catalyze biodiesel synthesis. Core-shell nanoparticles were produced by coating Fe_3O_4 core with silica shell. They were then treated with dimethyl octadecyl [3-(trimethoxysilyl)propyl] ammonium chloride. The immobilized lipase was applied as biocatalyst in the transesterification of olive oil with methanol to produce FAMEs, attaining a FAME conversion yield of over 90% within 30 hours of reaction, with 11 wt% of catalyst loading. The immobilized lipase could be used for 10 cycles without significant loss in its catalytic activity (Tran et al. 2012).

Mehrasbi et al. (2017) covalently immobilized a lipase from *C. antarctica* on functionalized magnetic nanoparticles to catalyze biodiesel synthesis. Core-shell nanoparticles were synthesized by coating Fe_3O_4 core with silica shell ($Fe_3O_4@SiO_2$). The nanoparticles were functionalized with (3-glycidoxypropyl)trimethoxysilane (GPTMS) and utilized as immobilization support. When compared with the free lipase, the immobilized lipase presented higher thermal stability and methanol tolerance. It was then employed in the transesterification of WCO with methanol. Under optimum conditions, the biodiesel conversion yield was almost 100%, with the immobilized lipase showing good reusability, retaining 100% of its initial catalytic activity after six cycles of reaction (Mehrasbi et al. 2017). All these studies clearly indicate that various nanobiocatalysts can be effectively used in the production of biodiesel.

3.4 Conclusions

This chapter presents a review of several heterogeneous catalysts for biodiesel production, along with their methods of synthesis. The application of these catalysts could enhance the production process. Microstructured heterogeneous catalysts offer solutions to some of the concerns present in the conventional process with homogenous catalysts; specifically, concerns about the washing/purification of the final products of transesterification and reusability. However, they have some drawbacks of their own, such as being mass transfer-resistant, time-consuming, and inefficient. With their higher surface areas and enhanced catalytic activity, nanostructured catalysts exhibit the same advantages as microstructured one, with analogous performance to homogeneous catalysts. Moreover, it has been confirmed that various nanostructures can be effectively used as solid supports for the immobilization of enzymes, in order to develop most efficient biocatalysts for biodiesel production. The development of magnetic-supported biocatalysts is of great importance because of their nontoxic nature and reusability over multiple cycles of transesterification, making the process eco-friendly and economically viable.

References

Aarthy, M., Saravanan, P., Gowthaman, M.K. et al. (2014). Enzymatic transesterification for production of biodiesel using yeast lipases: an overview. *Chemical Engineering Research and Design* 92: 1591–1601.

Abdulla, R. and Ravindra, P. (2013). Immobilized *Burkholderia cepacia* lipase for biodiesel production from crude *Jatropha curcas* L. oil. *Biomass and Bioenergy* 56: 8–13.

Adlercreutz, P. (2013). Immobilisation and application of lipases in organic media. *Chemical Society Reviews* 42: 6406–6436.

Adnan, M., Li, K., Xu, L., and Yan, Y. (2018). X-shaped Zif-8 for immobilization *Rhizomucor miehei* lipase via encapsulation and its application toward biodiesel production. *Catalysts* 8: 96.

Ahmed, F.E., Lalia, B.S., and Hashaikeh, R. (2015). A review on electrospinning for membrane fabrication: challenges and applications. *Desalination* 356: 15–30.

Akia, M., Yazdani, F., Motaee, E. et al. (2014). A review on conversion of biomass to biofuel by nanocatalysts. *Biofuel Research Journal* 1: 16–25.

Albuquerque, M.C.G., Jiménez-Urbistondo, I., Santamaría-González, J. et al. (2008). CaO supported on mesoporous silicas as basic catalysts for transesterification reactions. *Applied Catalysis A: General* 334: 35–43.

Ambat, I., Srivastava, V., and Sillanpää, M. (2018). Recent advancement in biodiesel production methodologies using various feedstock: a review. *Renewable and Sustainable Energy Reviews* 90: 356–369.

Ansari, S.A. and Husain, Q. (2012). Potential applications of enzymes immobilized on/in nano materials: a review. *Biotechnology Advances* 30: 512–523.

Aranda, D.A.G., Santos, R.T.P., Tapanes, N.C.O. et al. (2008). Acid-catalyzed homogeneous esterification reaction for biodiesel production from palm fatty acids. *Catalysis Letters* 122: 20–25.

Aransiola, E.F., Ojumu, T.V., Oyekola, O.O. et al. (2014). A review of current technology for biodiesel production: state of the art. *Biomass and Bioenergy* 61: 276–297.

Atadashi, I.M., Aroua, M.K., Abdul Aziz, A.R., and Sulaiman, N.M.N. (2013). The effects of catalysts in biodiesel production: a review. *Journal of Industrial and Engineering Chemistry* 19: 14–26.

Avhad, M.R. and Marchetti, J.M. (2016). Innovation in solid heterogeneous catalysis for the generation of economically viable and ecofriendly biodiesel: a review. *Catalysis Reviews: Science and Engineering* 58: 157–208.

Avramovic, J., Velikovic, A., and Veljkovic, V. (2018). Challenges in biodiesel industry: socio-economic, occupational health, and policy issues. *Safety Engineering* 8: 79–84.

Babaki, M., Yousefi, M., Habibi, Z. et al. (2016). Enzymatic production of biodiesel using lipases immobilized on silica nanoparticles as highly reusable biocatalysts: effect of water, t-butanol and blue silica gel contents. *Renewable Energy* 91: 196–206.

Bajaj, A., Lohan, P., Jha, P.N., and Mehrotra, R. (2010). Biodiesel production through lipase catalyzed transesterification: an overview. *Journal of Molecular Catalysis B: Enzymatic* 62: 9–14.

Banković–Ilić, I.B., Miladinović, M.R., Stamenković, O.S., and Veljković, V.B. (2017). Application of nano CaO-based catalysts in biodiesel synthesis. *Renewable and Sustainable Energy Reviews* 72: 746–760.

Baskar, G. and Aiswarya, R. (2016). Trends in catalytic production of biodiesel from various feedstocks. *Renewable and Sustainable Energy Reviews* 57: 496–504.

Bencze, L.C., Bartha-Vári, J.H., Katona, G. et al. (2016). Nanobioconjugates of *Candida antarctica* lipase B and single-walled carbon nanotubes in biodiesel production. *Bioresource Technology* 200: 853–860.

Bet-Moushoul, E., Farhadi, K., Mansourpanah, Y. et al. (2016). Application of CaO-based/Au nanoparticles as heterogeneous nanocatalysts in biodiesel production. *Fuel* 164: 119–127.

Bilal, M., Zhao, Y., Noreen, S. et al. (2019). Modifying bio-catalytic properties of enzymes for efficient biocatalysis: a review from immobilization strategies viewpoint. *Biocatalysis and Biotransformation* 37: 159–182.

Borges, M.E. and Díaz, L. (2012). Recent developments on heterogeneous catalysts for biodiesel production by oil esterification and transesterification reactions: a review. *Renewable and Sustainable Energy Reviews* 16: 2839–2849.

Boro, J., Deka, D., and Thakur, A.J. (2012). A review on solid oxide derived from waste shells as catalyst for biodiesel production. *Renewable and Sustainable Energy Reviews* 16: 904–910.

Boz, N., Degirmenbasi, N., and Kalyon, D.M. (2009). Conversion of biomass to fuel: transesterification of vegetable oil to biodiesel using KF loaded nano-γ-Al_2O_3 as catalyst. *Applied Catalysis B: Environmental* 89: 590–596.

Boz, N., Degirmenbasi, N., and Kalyon, D.M. (2015). Esterification and transesterification of waste cooking oil over Amberlyst 15 and modified Amberlyst 15 catalysts. *Applied Catalysis B: Environmental* 165: 723–730.

Brady, D. and Jordaan, J. (2009). Advances in enzyme immobilisation. *Biotechnology Letters* 31: 1639–1650.

Brena, B.M. and Batista-Viera, F. (2006). Immobilization of enzymes. In: *Immobilization of Enzymes and Cells* (ed. J.M. Guisan), 15–30. Totowa, NJ: Humana.

Cao, M., Li, Z., Wang, J. et al. (2012). Food related applications of magnetic iron oxide nanoparticles: enzyme immobilization, protein purification, and food analysis. *Trends in Food Science & Technology* 27: 47–56.

Carteret, C., Jacoby, J., and Blin, J.L. (2018). Using factorial experimental design to optimize biocatalytic biodiesel production from mucor miehei lipase immobilized onto ordered mesoporous materials. *Microporous and Mesoporous Materials* 268: 39–45.

Catarino, M., Ramos, M., Dias, A.P.S. et al. (2017). Calcium rich food wastes based catalysts for biodiesel production. *Waste and Biomass Valorization* 8: 1699–1707.

Chaturvedi, S., Dave, P.N., and Shah, N.K. (2012). Applications of nano-catalyst in new era. *Journal of Saudi Chemical Society* 16: 307–325.

Chaubey, A., Parshad, R., Taneja, S.C., and Qazi, G.N. (2009). *Arthrobacter* sp. lipase immobilization on magnetic sol-gel composite supports for enantioselectivity improvement. *Process Biochemistry* 44: 154–160.

Chen, Y.Z., Ching, C.B., and Xu, R. (2009). Lipase immobilization on modified zirconia nanoparticles: studies on the effects of modifiers. *Process Biochemistry* 44: 1245–1251.

Chouhan, A.P.P.S.S. and Sarma, A.K.K. (2011). Modern heterogeneous catalysts for biodiesel production: a comprehensive review. *Renewable and Sustainable Energy Reviews* 15: 4378–4399.

Christopher, L.P., Kumar, H., and Zambare, V.P. (2014). Enzymatic biodiesel: challenges and opportunities. *Applied Energy* 119: 497–520.

Cipolatti, E.P., Silva, M.J.A., Klein, M. et al. (2014). Current status and trends in enzymatic nanoimmobilization. *Journal of Molecular Catalysis B: Enzymatic* 99: 56–67.

Cubides-Roman, D.C., Pérez, V.H., de Castro, H.F. et al. (2017). Ethyl esters (biodiesel) production by *Pseudomonas fluorescens* lipase immobilized on chitosan with magnetic properties in a bioreactor assisted by electromagnetic field. *Fuel* 196: 481–487.

da Conceição, L.R.V., Carneiro, L.M., Rivaldi, J.D., and de Castro, H.F. (2016). Solid acid as catalyst for biodiesel production via simultaneous esterification and transesterification of macaw palm oil. *Industrial Crops and Products* 89: 416–424.

Dai, Y.M., Hsieh, C.H., Lin, J.H. et al. (2019). Biodiesel production using bauxite in low-cost solid base catalyst precursors. *Catalysts* 9: 1–9.

Dalla-Vecchia, R., da Graça Nascimento, M., and Soldi, V. (2004). Aplicações sintéticas de lipases imobilizadas em polímeros. *Química Nova* 27: 623–630.

De Paola, M.G., Ricca, E., Calabrò, V. et al. (2009). Factor analysis of transesterification reaction of waste oil for biodiesel production. *Bioresource Technology* 100: 5126–5131.

Demirbas, A. (2007). Importance of biodiesel as transportation fuel. *Energy Policy* 35: 4661–4670.

Demirbas, A. (2009). Progress and recent trends in biodiesel fuels. *Energy Conversion and Management* 50: 14–34.

Dizge, N. and Keskinler, B. (2008). Enzymatic production of biodiesel from Canola oil using immobilized lipase. *Biomass and Bioenergy* 32: 1274–1278.

Du, W., Xu, Y.Y., Liu, D.H., and Li, Z.B. (2005). Study on acyl migration in immobilized lipozyme TL-catalyzed transesterification of soybean oil for biodiesel production. *Journal of Molecular Catalysis B: Enzymatic* 37: 68–71.

Du, L., Li, Z., Ding, S. et al. (2019). Synthesis and characterization of carbon-based MgO catalysts for biodiesel production from Castor oil. *Fuel* 258: 116122.

Dumri, K. and Hung Anh, D. (2014). Immobilization of lipase on silver nanoparticles via adhesive polydopamine for biodiesel production. *Enzyme Research* 2014: 389739.

Edelstein, A.S. and Cammarata, R.C. (1996). *Nanomaterials: Synthesis, Properties and Applications*. New York: Taylor & Francis.

Endalew, A.K., Kiros, Y., and Zanzi, R. (2011). Inorganic heterogeneous catalysts for biodiesel production from vegetable oils. *Biomass and Bioenergy* 35: 3787–3809.

European Parliament. (2009). European Parliament Directive 2009/30/EC. *Official Journal of the European Union*, L140/88–L140/113.

European Parliament. (2013). Fuel Quality Directive and Renewable Energy Dir II.

Fan, Y., Wu, G., Su, F. et al. (2016). Lipase oriented-immobilized on dendrimer-coated magnetic multi-walled carbon nanotubes toward catalyzing biodiesel production from waste vegetable oil. *Fuel* 178: 172–178.

Fan, M., Si, Z., Sun, W., and Zhang, P. (2019). Sulfonated ZrO_2-TiO_2 nanorods as efficient solid acid catalysts for heterogeneous esterification of palmitic acid. *Fuel* 252: 254–261.

Feyzi, M., Hosseini, N., Yaghobi, N., and Ezzati, R. (2017). Preparation, characterization, kinetic and thermodynamic studies of MgO-La_2O_3 nanocatalysts for biodiesel production from sunflower oil. *Chemical Physics Letters* 677: 19–29.

Fjerbaek, L., Christensen, K.V., and Norddahl, B. (2009). A review of the current state of biodiesel production using enzymatic transesterification. *Biotechnology and Bioengineering* 102: 1298–1315.

Freedman, B., Pryde, E.H., and Mounts, T.L. (1984). Variables affecting the yields of fatty esters from transesterified vegetable oils. *Journal of the American Oil Chemists' Society* 61: 1638–1643.

Frey, H. (2015). Chemical vapor deposition (CVD). In: *Handbook of Thin-Film Technology* (eds. H. Frey and H.R. Khan), 225–252. Cham: Springer.

Fu, J., Chen, L., Lv, P. et al. (2015). Free fatty acids esterification for biodiesel production using self-synthesized macroporous cation exchange resin as solid acid catalyst. *Fuel* 154: 1–8.

Garcia-Galan, C., Berenguer-Murcia, Á., Fernandez-Lafuente, R., and Rodrigues, R.C. (2011). Potential of different enzyme immobilization strategies to improve enzyme performance. *Advanced Synthesis and Catalysis* 353: 2885–2904.

Gardy, J., Rehan, M., Hassanpour, A. et al. (2019). Advances in nano-catalysts based biodiesel production from non-food feedstocks. *Journal of Environmental Management* 249: 109316.

Ghadi, A., Tabandeh, F., Mahjoub, S. et al. (2015). Fabrication and characterization of core-shell magnetic chitosan nanoparticles as a novel carrier for immobilization of *Burkholderia cepacia* lipase. *Journal of Oleo Science* 64: 423–430.

Gog, A., Roman, M., Toşa, M. et al. (2012). Biodiesel production using enzymatic transesterification- current state and perspectives. *Renewable Energy* 39: 10–16.

Guan, G., Kusakabe, K., Sakurai, N., and Moriyama, K. (2009). Transesterification of vegetable oil to biodiesel fuel using acid catalysts in the presence of dimethyl ether. *Fuel* 88: 81–86.

Guerreiro, L., Castanheiro, J.E., Fonseca, I.M. et al. (2006). Transesterification of Soybean oil over sulfonic acid functionalised polymeric membranes. *Catalysis Today* 118: 166–171.

Guo, F., Fang, Z., Xu, C.C., and Smith, R.L. (2012). Solid acid mediated hydrolysis of biomass for producing biofuels. *Progress in Energy and Combustion Science* 38: 672–690.

Gupta, J. and Agarwal, M. (2016). Preparation and characterizaton of CaO nanoparticle for biodiesel production. *AIP Conference Proceedings* 1724: 020066.

Hahn, H. (1997). Gas phase synthesis of nanocrystalline materials. *Nanostructured Materials* 9: 3–12.

Hamester, M.R.R., Balzer, P.S., and Becker, D. (2012). Characterization of calcium carbonate obtained from oyster and mussel shells and incorporation in polypropylene. *Materials Research* 15: 204–208.

Hampden-Smith, M.J. and Kodas, T.T. (1995). Chemical vapor deposition of metals: part 1. An overview of CVD processes. *Chemical Vapor Deposition* 1: 8–23.

Han, J.Y. and Kim, H.K. (2011). Transesterification using the cross-linked enzyme aggregate of *Photobacterium lipolyticum* lipase M37. *Journal of Microbiology and Biotechnology* 21: 1159–1165.

Hartmann, M. and Jung, D. (2010). Biocatalysis with enzymes immobilized on mesoporous hosts: the status quo and future trends. *Journal of Materials Chemistry* 20: 844–857.

Hattori, H. (2001). Solid base catalysts: generation of basic sites and application to organic synthesis. *Applied Catalysis A: General* 222: 247–259.

Helwani, Z., Othman, M.R., Aziz, N. et al. (2009). Solid heterogeneous catalysts for transesterification of triglycerides with methanol: a review. *Applied Catalysis A: General* 363: 1–10.

Hillion, G., Delfort, B., Le Pennec, D. et al. (2003). Biodiesel production by a continuous process using a heterogeneous catalyst. *ACS Division of Fuel Chemistry, Preprints* 48: 636–638.

Hsu, A.F., Jones, K., Marmer, W.N., and Foglia, T.A. (2001). Production of alkyl esters from tallow and grease using lipase immobilized in a phyllosilicate sol-gel. *Journal of the American Oil Chemists' Society* 78: 585–588.

Hu, S., Wen, L., Wang, Y. et al. (2012). Gas-liquid countercurrent integration process for continuous biodiesel production using a microporous solid base KF/CaO as catalyst. *Bioresource Technology* 123: 413–418.

Hu, Y., Dai, L., Liu, D., and Du, W. (2018). Rationally designing hydrophobic UiO-66 support for the enhanced enzymatic performance of immobilized lipase. *Green Chemistry* 20: 4500–4506.

Hu, Y., Gong, M., Feng, S. et al. (2019). A review of recent developments of pre-treatment technologies and hydrothermal liquefaction of microalgae for bio-crude oil production. *Renewable and Sustainable Energy Reviews* 101: 476–492.

Hung, C.-H., Chen, C.-S., Sheu, H.-S., and Chang, J.-R. (2018). Deactivation and rejuvenation of pellet MgO/SiO_2 catalysts for transesterification of soybean oil with methanol to biodiesel: roles of MgO morphology change in catalysis. *Industrial & Engineering Chemistry Research* 57: 456–469.

Hwang, E.T. and Gu, M.B. (2013). Enzyme stabilization by nano/microsized hybrid materials. *Engineering in Life Sciences* 13: 49–61.

Ilgen, O. (2011). Dolomite as a heterogeneous catalyst for transesterification of Canola oil. *Fuel Processing Technology* 92: 452–455.

Ilgen, O. and Akin, A.N. (2009). Transesterification of Canola oil to biodiesel using MgO loaded with KOH as a heterogeneous catalyst. *Energy and Fuels* 23: 1786–1789.

Islam, A. and Ravindra, P. (2016). *Biodiesel Production with Green Technologies*, 1–133. Berlin: Springer.

Islam, A., Taufiq-Yap, Y.H., Chan, E.-S.S. et al. (2014). Advances in solid-catalytic and non-catalytic technologies for biodiesel production. *Energy Conversion and Management* 88: 1200–1218.

Jambulingam, R., Shalma, M., and Shankar, V. (2019). Biodiesel production using lipase immobilised functionalized magnetic nanocatalyst from oleaginous fungal lipid. *Journal of Cleaner Production* 215: 245–258.

Jegannathan, K.R., Abang, S., Poncelet, D. et al. (2008). Production of biodiesel using immobilized lipase – a critical review. *Critical Reviews in Biotechnology* 28: 253–264.

Jesionowski, T. (2001). Preparation of colloidal silica from sodium metasilicate solution and sulphuric acid in emulsion medium. *Colloids and Surfaces A: Physicochemical and Engineering Aspects* 190: 153–165.

Jesionowski, T. and Krysztafkiewicz, A. (2002). Preparation of the hydrophilic/hydrophobic silica particles. *Colloids and Surfaces A: Physicochemical and Engineering Aspects* 207: 49–58.

Jia, H., Zhu, G., and Wang, P. (2003). Catalytic behaviors of enzymes attached to nanoparticles: the effect of particle mobility. *Biotechnology and Bioengineering* 84: 406–414.

Jiang, S.P. (2006). A review of wet impregnation – an alternative method for the fabrication of high performance and nano-structured electrodes of solid oxide fuel cells. *Materials Science and Engineering A* 418: 199–210.

Jitputti, J., Kitiyanan, B., Rangsunvigit, P. et al. (2006). Transesterification of crude palm kernel oil and crude coconut oil by different solid catalysts. *Chemical Engineering Journal* 116: 61–66.

Jun, S.-H., Lee, J., Chan Kim, B. et al. (2012). Highly efficient enzyme immobilization and stabilization within meso-structured onion-like silica for biodiesel production. *Chemistry of Materials* 24: 924–929.

Kakugawa, K., Shobayashi, M., Suzuki, O., and Miyakawa, T. (2002). Purification and characterization of a lipase from the glycolipid-producing yeast Kurtzmanomyces Sp. I-11. *Bioscience Biotechnology and Biochemistry* 66: 978–985.

Katiyar, M. and Ali, A. (2012). Immobilization of Candida rugosa lipase on MCM-41 for the transesterification of cotton seed oil. *Journal of Oleo Science* 61: 469–475.

Kaur, M. and Ali, A. (2011). Lithium ion impregnated calcium oxide as nano catalyst for the biodiesel production from Karanja and Jatropha oils. *Renewable Energy* 36: 2866–2871.

Kaushik, N., Kumar, K., and Kumar, S. (2008). Potential of *Jatropha curcas* for biofuels. *Journal of Biobased Materials and Bioenergy* 1: 301–314.

Kennedy, J.F., Melo, E.H.M., and Jumel, K. (1990). Immobilized enzyme and cells. *Chemical Engineering Progress* 86: 81–89.

Kesic, Z., Lukic, I., Zdujic, M. et al. (2016). Calcium oxide based catalysts for biodiesel production: a review. *Chemical Industry and Chemical Engineering Quarterly* 22: 391–408.

Kim, J., Grate, J.W., and Wang, P. (2006). Nanostructures for enzyme stabilization. *Chemical Engineering Science* 61: 1017–1026.

Kim, K.H., Lee, O.K., and Lee, E.Y. (2018). Nano-immobilized biocatalysts for biodiesel production from renewable and sustainable resources. *Catalysts* 8: 68.

Kiss, A.A., Dimian, A.C., and Rothenberg, G. (2006). Solid acid catalysts for biodiesel production – towards sustainable energy. *Advanced Synthesis and Catalysis* 348: 75–81.

Kouzu, M., Fujimori, A., Suzuki, T. et al. (2017). Industrial feasibility of powdery CaO catalyst for production of biodiesel. *Fuel Processing Technology* 165: 94–101.

Krajewska, B. (2004). Application of chitin- and chitosan-based materials for enzyme immobilizations: a review. *Enzyme and Microbial Technology* 35: 126–139.

Kuan, I.C., Lee, C.C., Tsai, B.H. et al. (2013). Optimizing the production of biodiesel using lipase entrapped in biomimetic silica. *Energies* 6: 2052–2064.

Kumari, V., Shah, S., and Gupta, M.N. (2006). Preparation of biodiesel by lipase-catalyzed transesterification of high free fatty acid containing oil from *Madhuca indica*. *Energy & Fuels* 21: 368–372.

Lalonde, J. and Margolin, A. (2002). Immobilization of enzymes. In: *Enzyme Catalysis in Organic Synthesis* (eds. K. Drauz and H. Waldmann), 163–184. Weinheim: Wiley-VCH Verlag GmbH.

Lam, M.K., Lee, K.T., and Mohamed, A.R. (2010). Homogeneous, heterogeneous and enzymatic catalysis for transesterification of high free fatty acid oil (waste cooking oil) to biodiesel: a review. *Biotechnology Advances* 28: 500–518.

Leung, D.Y.C., Wu, X., and Leung, M.K.H. (2010). A review on biodiesel production using catalyzed transesterification. *Applied Energy* 87: 1083–1095.

Li, S.-F., Chen, J.-P., and Wu, W.-T. (2007). Electrospun polyacrylonitrile nanofibrous membranes for lipase immobilization. *Journal of Molecular Catalysis B: Enzymatic* 47: 117–124.

Li, S.-F., Fan, Y.-H., Hu, R.-F., and Wu, W.-T. (2011a). *Pseudomonas cepacia* lipase immobilized onto the electrospun PAN nanofibrous membranes for biodiesel production from soybean oil. *Journal of Molecular Catalysis B: Enzymatic* 72: 40–45.

Li, Y., Qiu, F., Yang, D. et al. (2011b). Preparation, characterization and application of heterogeneous solid base catalyst for biodiesel production from soybean oil. *Biomass and Bioenergy* 35: 2787–2795.

Li, X.-S., Zhu, G.-T., Luo, Y.-B. et al. (2013). Synthesis and applications of functionalized magnetic materials in sample preparation. *TrAC Trends in Analytical Chemistry* 45: 233–247.

Lien, Y.-S., Hsieh, L.-S., and Wu, C.S.J. (2010). Biodiesel synthesis by simultaneous esterification and transesterification using oleophilic acid catalyst. *Industrial & Engineering Chemistry Research* 49: 2118–2121.

Liese, A. and Hilterhaus, L. (2013). Evaluation of immobilized enzymes for industrial applications. *Chemical Society Reviews* 42: 6236–6249.

Lin, A.Y.M., Meyers, M.A., and Vecchio, K.S. (2006). Mechanical properties and structure of *Strombus gigas*, *Tridacna gigas*, and *Haliotis rufescens* sea shells: a comparative study. *Materials Science and Engineering C* 26: 1380–1389.

Liu, L.H., Shih, Y.H., Liu, W.L. et al. (2017). Enzyme immobilized on nanoporous carbon derived from metal–organic framework: a new support for biodiesel synthesis. *ChemSusChem* 10: 1364–1369.

Liu, J., Ma, R.-T., and Shi, Y.-P. (2020). Recent advances on support materials for lipase immobilization and applicability as biocatalysts in inhibitors screening methods – a review. *Analytica Chimica Acta* 1101: 9–22.

Long, J.W., Logan, M.S., Rhodes, C.P. et al. (2004). Nanocrystalline iron oxide aerogels as mesoporous magnetic architectures. *Journal of the American Chemical Society* 126: 16879–16889.

López, D.E., Goodwin, J.G., Bruce, D.A., and Lotero, E. (2005). Transesterification of triacetin with methanol on solid acid and base catalysts. *Applied Catalysis A: General* 295: 97–105.

López, D.E., Goodwin, J.G., Bruce, D.A., and Furuta, S. (2008). Esterification and transesterification using modified-zirconia catalysts. *Applied Catalysis A: General* 339: 76–83.

López, C., Cruz-Izquierdo, Á., Picó, E.A. et al. (2014). Magnetic biocatalysts and their uses to obtain biodiesel and biosurfactants. *Frontiers in Chemistry* 2: 1–11.

Lotero, E., Liu, Y., Lopez, D.E. et al. (2005). Synthesis of biodiesel via acid catalysis. *Industrial and Engineering Chemistry Research* 44: 5353–5363.

Lou, W.-Y., Zong, M.-H., and Duan, Z.-Q. (2008). Efficient production of biodiesel from high free fatty acid-containing waste oils using various carbohydrate-derived solid acid catalysts. *Bioresource Technology* 99: 8752–8758.

Lu, A.H., Salabas, E.L., and Schüth, F. (2007). Magnetic nanoparticles: synthesis, protection, functionalization, and application. *Angewandte Chemie International Edition* 46: 1222–1244.

Lu, P., Wang, K., and Gong, J. (2019). Optimization of salix carbonation solid acid catalysts for one-step synthesis by response surface method. *Applied Sciences (Switzerland)* 9: 1518.

Ma, F., Hanna, M.A., and Fangrui, M. (1999). Biodiesel production: a review 1. *Bioresource Technology* 70: 1–15.

Macario, A., Verri, F., Diaz, U. et al. (2013). Pure silica nanoparticles for liposome/lipase system encapsulation: application in biodiesel production. *Catalysis Today* 204: 148–155.

Mansir, N., Teo, S.H., Rashid, U., and Taufiq-Yap, Y.H. (2018). Efficient waste gallus domesticus shell derived calcium-based catalyst for biodiesel production. *Fuel* 211: 67–75.

Marchetti, J.M. (2012). A summary of the available technologies for biodiesel production based on a comparison of different feedstock's properties. *Process Safety and Environmental Protection* 90: 157–163.

Mardhiah, H.H., Ong, H.C., Masjuki, H.H.H. et al. (2017). A review on latest developments and future prospects of heterogeneous catalyst in biodiesel production from non-edible oils. *Renewable and Sustainable Energy Reviews* 67: 1225–1236.

Mateo, C., Fernandes, B., van Rantwijk, F. et al. (2006). Stabilisation of oxygen-labile nitrilases via co-aggregation with poly(ethyleneimine). *Journal of Molecular Catalysis B: Enzymatic* 38: 154–157.

Mazubert, A., Poux, M., and Aubin, J. (2013). Intensified processes for FAME production from waste cooking oil: a technological review. *Chemical Engineering Journal* 233: 201–223.

Meher, L.C., Vidya Sagar, D., and Naik, S.N. (2006). Technical aspects of biodiesel production by transesterification – a review. *Renewable and Sustainable Energy Reviews* 10: 248–268.

Mehrasbi, M.R., Mohammadi, J., Peyda, M., and Mohammadi, M. (2017). Covalent immobilization of *Candida antarctica* lipase on core-shell magnetic nanoparticles for production of biodiesel from waste cooking oil. *Renewable Energy* 101: 593–602.

Mendes, A.A., Giordano, R.C., de LC Giordano, R., and de Castro, H.F. (2011). Immobilization and stabilization of microbial lipases by multipoint covalent attachment on aldehyde-resin affinity: application of the biocatalysts in biodiesel synthesis. *Journal of Molecular Catalysis B: Enzymatic* 68: 109–115.

Mierzwa-Szymkowiak, D. (2009). Chemical composition and structure of the shell of Cepaea Vindobonensis (Férussac, 1821) (Gastropoda: Pulmonata: Helicidae). *Folia Malacologica* 16: 17–20.

Moreno-Pirajàn, J.C. and Giraldo, L. (2011). Study of immobilized Candida Rugosa lipase for biodiesel fuel production from palm oil by flow microcalorimetry. *Arabian Journal of Chemistry* 4: 55–62.

Murty, V.R., Bhat, J., and Muniswaran, P.K.A. (2002). Hydrolysis of oils by using immobilized lipase enzyme: a review. *Biotechnology and Bioprocess Engineering* 7: 57–66.

Nair, P., Singh, B., Upadhyay, S.N., and Sharma, Y.C. (2012). Synthesis of biodiesel from low FFA waste frying oil using calcium oxide derived from *Mereterix mereterix* as a heterogeneous catalyst. *Journal of Cleaner Production* 29–30: 82–90.

Nakatani, N., Takamori, H., Takeda, K., and Sakugawa, H. (2009). Transesterification of soybean oil using combusted oyster shell waste as a catalyst. *Bioresource Technology* 100: 1510–1513.

Narasimharao, K., Lee, A., and Wilson, K. (2007). Catalysts in production of biodiesel: a review. *Journal of Biobased Materials and Bioenergy* 1: 1–12.

Narkhede, N., Brahmkhatri, V., and Patel, A. (2014). Efficient synthesis of biodiesel from waste cooking oil using solid acid catalyst comprising 12-tungstosilicic acid and SBA-15. *Fuel* 135: 253–261.

Nelson, L.A., Foglia, T.A., and Marmer, W.N. (1996). Lipase-catalyzed production of biodiesel. *JAOCS, Journal of the American Oil Chemists' Society* 73: 1191–1195.

Neves, N.M. and Mano, J.F. (2005). Structure/mechanical behavior relationships in crossed-lamellar sea shells. *Materials Science and Engineering C* 25: 113–118.

Noureddini, H., Gao, X., and Philkana, R.S. (2005). Immobilized *Pseudomonas cepacia* lipase for biodiesel fuel production from soybean oil. *Bioresource Technology* 96: 769–777.

Ong, H.R., Khan, M.R., Chowdhury, M.N.K. et al. (2014). Synthesis and characterization of CuO/C catalyst for the esterification of free fatty acid in rubber seed oil. *Fuel* 120: 195–201.

Ozkan, S., Puna, J.F., Gomes, J.F. et al. (2019). Preliminary study on the use of biodiesel obtained from waste vegetable oils for blending with hydrotreated kerosene fossil fuel using calcium oxide (CaO) from natural waste materials as heterogeneous catalyst. *Energies* 12: 4306.

Pandit, P.R. and Fulekar, M.H. (2017). Egg shell waste as heterogeneous nanocatalyst for biodiesel production: optimized by response surface methodology. *Journal of Environmental Management* 198: 319–329.

Panjaitan, F.R., Yamanaka, S., and Kuga, Y. (2017). Soybean oil methanolysis over scallop shell-derived CaO prepared via methanol-assisted dry nano-grinding. *Advanced Powder Technology* 28: 1627–1635.

Patel, R.L. and Sankhavara, C.D. (2017). Biodiesel production from Karanja oil and its use in diesel engine: a review. *Renewable and Sustainable Energy Reviews* 71: 464–474.

Piker, A., Tabah, B., Perkas, N., and Gedanken, A. (2016). A green and low-cost room temperature biodiesel production method from waste oil using egg shells as catalyst. *Fuel* 182: 34–41.

Pinto, A.C., Guarieiro, L.L.N., Rezende, M.J.C. et al. (2005). Biodiesel: An Overview. *Journal of the Brazilian Chemical Society* 16: 1313–1330.

Pollard, D.J. and Woodley, J.M. (2007). Biocatalysis for pharmaceutical intermediates: the future is now. *Trends in Biotechnology* 25: 66–73.

Rafiei, S., Tangestaninejad, S., Horcajada, P. et al. (2018). Efficient biodiesel production using a lipase@ZIF-67 nanobioreactor. *Chemical Engineering Journal* 334: 1233–1241.

Rahman, I.A. and Padavettan, V. (2012). Synthesis of silica nanoparticles by sol-gel: size-dependent properties, surface modification, and applications in silica-polymer nanocomposites – a review. *Journal of Nanomaterials* 2012: 132424.

Rajput, N. (2015). Methods of preparation of nanoparticles – a review. *International Journal of Advances in Engineering & Technology* 7: 1806–1811.

Reddy, A.N.R., Saleh, A.A., Saiful Islam, M. et al. (2016). Biodiesel production from crude Jatropha oil using a highly active heterogeneous nanocatalyst by optimizing transesterification reaction parameters. *Energy and Fuels* 30: 334–343.

Refaat, A.A. (2011). Biodiesel production using solid metal oxide catalysts. *International Journal of Environmental Science and Technology* 8: 203–221.

Ricardi, N.C., de Menezes, E.W., Valmir Benvenutti, E. et al. (2018). Highly stable novel silica/chitosan support for β-galactosidase immobilization for application in dairy technology. *Food Chemistry* 246: 343–350.

Rodrigues, R.C., Ortiz, C., Berenguer-Murcia, Á. et al. (2013). Modifying enzyme activity and selectivity by immobilization. *Chemical Society Reviews* 42: 6290–6307.

Rodrigues, J., Canet, A., Rivera, I. et al. (2016). Biodiesel production from crude Jatropha oil catalyzed by non-commercial immobilized heterologous *Rhizopus oryzae* and *Carica papaya* lipases. *Bioresource Technology* 213: 88–95.

Rosenholm, J.M., Zhang, J., Sun, W., and Gu, H. (2011). Large-pore mesoporous silica-coated magnetite core-shell nanocomposites and their relevance for biomedical applications. *Microporous and Mesoporous Materials* 145: 14–20.

Safaei-Ghomi, J., Ghasemzadeh, M.A., and Mehrabi, M. (2013). Calcium oxide nanoparticles catalyzed one-step multicomponent synthesis of highly substituted pyridines in aqueous ethanol media. *Scientia Iranica* 20: 549–554.

Safaryan, S.M., Yakovlev, A.V., Pidko, E.A. et al. (2017). Reversible sol–gel–sol medium for enzymatic optical biosensors. *Journal of Materials Chemistry B* 5: 85–91.

Safi, B., Saidi, M., Daoui, A. et al. (2015). The use of seashells as a fine aggregate (by sand substitution) in self-compacting mortar (SCM). *Construction and Building Materials* 78: 430–438.

Sahoo, P.K. and Das, L.M. (2009). Process optimization for biodiesel production from Jatropha, Karanja and Polanga oils. *Fuel* 88: 1588–1594.

Sakai, S., Liu, Y., Yamaguchi, T. et al. (2010). Production of butyl-biodiesel using lipase physically-adsorbed onto electrospun polyacrylonitrile fibers. *Bioresource Technology* 101: 7344–7349.

Sánchez-Bayo, A., Morales, V., Rodríguez, R. et al. (2019). Biodiesel production (FAEEs) by heterogeneous combi-lipase biocatalysts using wet extracted lipids from microalgae. *Catalysts* 9: 1–15.

Sani, Y.M., Daud, W.M.A.W., and Abdul Aziz, A.R.R. (2014). Activity of solid acid catalysts for biodiesel production: a critical review. *Applied Catalysis A: General* 470: 140–161.

Santos, S., Nobre, L., Gomes, J. et al. (2019). Soybean oil transesterification for biodiesel production with micro-structured calcium oxide (CaO) from natural waste materials as a heterogeneous catalyst. *Energies* 12: 1–9.

Schöpe, M. (2008). Renewable energy directive. *European Wind Energy Conference and Exhibition 2008* 1: 32–38.

Sen, T., Sebastianelli, A., and James Bruce, I. (2006). Mesoporous silica–magnetite nanocomposite: fabrication and applications in magnetic bioseparations. *Journal of the American Chemical Society* 128: 7130–7131.

Sen, T., Bruce, I.J., and Mercer, T. (2010). Fabrication of novel hierarchically ordered porous magnetic nanocomposites for bio-catalysis. *Chemical Communications* 46: 6807–6809.

Shah, S. and Gupta, M.N. (2007). Lipase catalyzed preparation of biodiesel from Jatropha oil in a solvent free system. *Process Biochemistry* 42: 409–414.

Shah, S. and Gupta, M.N. (2008). The effect of ultrasonic pre-treatment on the catalytic activity of lipases in aqueous and non-aqueous media. *Chemistry Central Journal* 2: 1–9.

Shi, M., Zhang, P., Fan, M. et al. (2017). Influence of crystal of Fe_2O_3in magnetism and activity of nanoparticle CaO@$Fe_2O_3$for biodiesel production. *Fuel* 197: 343–347.

Shieh, C.-J., Liao, H.-F., and Lee, C.-C. (2003). Optimization of lipase-catalyzed biodiesel by response surface methodology. *Bioresource Technology* 88: 103–106.

Shim, M., Wong Shi Kam, N.J., Chen, R. et al. (2002). Functionalization of carbon nanotubes for biocompatibility and biomolecular recognition. *Nano Letters* 2: 285–288.

Shimada, Y., Watanabe, Y., Sugihara, A., and Tominaga, Y. (2002). Enzymatic alcoholysis for biodiesel fuel production and application of the reaction to oil processing. *Journal of Molecular Catalysis B: Enzymatic* 17: 133–142.

Shu, Q., Gao, J., Nawaz, Z. et al. (2010). Synthesis of biodiesel from waste vegetable oil with large amounts of free fatty acids using a carbon-based solid acid catalyst. *Applied Energy* 87: 2589–2596.

Shuit, S.H., Yee, K.F., Lee, K.T. et al. (2013). Evolution towards the utilisation of functionalised carbon nanotubes as a new generation catalyst support in biodiesel production: an overview. *RSC Advances* 3: 9070.

Souza, R.L., Faria, E.L.P., Figueiredo, R.T. et al. (2017). Protic ionic liquid applied to enhance the immobilization of lipase in sol–gel matrices. *Journal of Thermal Analysis and Calorimetry* 128: 833–840.

Stoycheva, M., Montero, G., Toscano, L., and Valdez, B. (2011). The immobilized lipases in biodiesel production. In: *Biodiesel - Feedstocks and Processing Technologies* (eds. M. Stoycheva and G. Montero), 397–410. Horwich: InTech.

Su, F. and Guo, Y. (2014). Advancements in solid acid catalysts for biodiesel production. *Green Chemistry* 16: 2934–2957.

Szczęsna Antczak, M., Kubiak, A., Antczak, T., and Bielecki, S. (2009). Enzymatic biodiesel synthesis–key factors affecting efficiency of the process. *Renewable Energy* 34: 1185–1194.

Szelwicka, A., Boncel, S., Jurczyk, S., and Chrobok, A. (2019). Exceptionally active and reusable nanobiocatalyst comprising lipase non-covalently immobilized on multi-wall carbon nanotubes for the synthesis of diester plasticizers. *Applied Catalysis A: General* 574: 41–47.

Tacias-Pascacio, V.G., Virgen-Ortíz, J.J., Jiménez-Pérez, M. et al. (2017). Evaluation of different lipase biocatalysts in the production of biodiesel from used cooking oil: critical role of the immobilization support. *Fuel* 200: 1–10.

Tan, T., Lu, J., Nie, K. et al. (2010). Biodiesel production with immobilized lipase: a review. *Biotechnology Advances* 28: 628–634.

Tan, H., Feng, W., and Ji, P. (2012). Lipase immobilized on magnetic multi-walled carbon nanotubes. *Bioresource Technology* 115: 172–176.

Tanabe, K. and Fukuda, Y. (1974). Basic properties of alkaline earth metal oxides and their catalytic activity in the decomposition of diacetone alcohol. *Reaction Kinetics and Catalysis Letters* 1: 21–24.

Tang, A., Zhang, Y., Wei, T. et al. (2019). Immobilization of *Candida cylindracea* lipase by covalent attachment on glu-modified bentonite. *Applied Biochemistry and Biotechnology* 187: 870–883.

Tariq, M., Ali, S., and Khalid, N. (2012). Activity of homogeneous and heterogeneous catalysts, spectroscopic and chromatographic characterization of biodiesel: a review. *Renewable and Sustainable Energy Reviews* 16: 6303–6316.

Thangaraj, B., Solomon, P.R., Muniyandi, B. et al. (2019). Catalysis in biodiesel production – a review. *Clean Energy*.

Tissue, B. and Yuan, H. (2003). Structure, particle size, and annealing of gas phase-condensed Eu^{3+}:Y_2O_3 nanophosphors. *Journal of Solid State Chemistry* 171: 12–18.

Tran, D.N. and Balkus, K.J. (2011). Perspective of recent progress in immobilization of enzymes. *ACS Catalysis* 1: 956–968.

Tran, D.-T., Chen, C.-L., and Chang, J.-S. (2012). Immobilization of Burkholderia sp. lipase on a ferric silica nanocomposite for biodiesel production. *Journal of Biotechnology* 158: 112–119.

Tsai, W.T., Yang, J.M., Lai, C.W. et al. (2006). Characterization and adsorption properties of eggshells and eggshell membrane. *Bioresource Technology* 97: 488–493.

Vaghari, H., Jafarizadeh-Malmiri, H., Mohammadlou, M. et al. (2016). Application of magnetic nanoparticles in smart enzyme immobilization. *Biotechnology Letters* 38: 223–233.

Veljković, V.B., Lakićević, S.H., Stamenković, O.S. et al. (2006). Biodiesel production from tobacco (*Nicotiana tabacum* L.) seed oil with a high content of free fatty acids. *Fuel* 85: 2671–2675.

Verma, M.L., Puri, M., and Barrow, C.J. (2016). Recent trends in nanomaterials immobilised enzymes for biofuel production. *Critical Reviews in Biotechnology* 36: 108–119.

Verma, M.L., Kumar, S., Das, A. et al. (2020). Chitin and chitosan-based support materials for enzyme immobilization and biotechnological applications. *Environmental Chemistry Letters* 18: 315–323.

Vert, M., Doi, Y., Hellwich, K.H. et al. (2012). Terminology for biorelated polymers and applications (IUPAC recommendations 2012). *Pure and Applied Chemistry* 84: 377–410.

Vicente, G., Coteron, A., Martinez, M., and Aracil, J. (1998). Application of the factorial design of experiments and response surface methodology to optimize biodiesel production. *Industrial Crops and Products* 8: 29–35.

Vicente, G., Martı′ nez, M., and Aracil, J. (2004). Integrated biodiesel production: a comparison of different homogeneous catalysts systems. *Bioresource Technology* 92: 297–305.

Villeneuve, P., Muderhwa, J.M., Graille, J., and Haas, M.J. (2000). Customizing lipases for biocatalysis: a survey of chemical, physical and molecular biological approaches. *Journal of Molecular Catalysis B: Enzymatic* 9: 113–148.

Vyas, A.P., Verma, J.L., and Subrahmanyam, N. (2010). A review on FAME production processes. *Fuel* 89: 1–9.

Wang, L. and Yang, J. (2007). Transesterification of soybean oil with nano-MgO or not in supercritical and subcritical methanol. *Fuel* 86: 328–333.

Wang, G., Yan, S., Zhou, C., and Liang, B. (2005a). Preparation and activity measurement of CaO/MgO catalyst for biodiesel fuel production. *Zhongguo Youzhi* 30 (10): 66–69.

Wang, L.K., Vaccari, D.A., and Shammas NK, L.Y. (2005b). Chemical precipitation. In: *Physicochemical Treatment Processes* (eds. L.K. Wang, Y.-T. Hung and N.K. Shammas), 141–197. Totowa, NJ: Humana.

Wang, Y., Pengzhan Liu, S.O., and Zhang, Z. (2007). Preparation of biodiesel from waste cooking oil via two-step catalyzed process. *Energy Conversion and Management* 48: 184–188.

Wang, Y., Hu, S.Y., Guan, Y.P. et al. (2009a). Preparation of mesoporous nanosized KF/CaO-MgO catalyst and its application for biodiesel production by transesterification. *Catalysis Letters* 131: 574–578.

Wang, Z.-G., Wan, L.-S., Liu, Z.-M. et al. (2009b). Enzyme immobilization on electrospun polymer nanofibers: an overview. *Journal of Molecular Catalysis B: Enzymatic* 56: 189–195.

Wang, L., Liu, X., Jiang, Y. et al. (2018). Biocatalytic pickering emulsions stabilized by lipase-immobilized carbon nanotubes for biodiesel production. *Catalysts* 8: 587.

Xiao, G. and Gao, L. (2011). First generation biodiesel. In: *Biofuel Production – Recent Developments and Prospects* (ed. M.A., S. dos Bernardes), 45–64. Horwich: InTech.

Xie, W. and Ma, N. (2009). Immobilized lipase on Fe_3O_4 nanoparticles as biocatalyst for biodiesel production. *Energy & Fuels* 23: 1347–1353.

Xie, W. and Wang, J. (2012). Immobilized lipase on magnetic chitosan microspheres for transesterification of soybean oil. *Biomass and Bioenergy* 36: 373–380.

Xie, W. and Zang, X. (2016). Immobilized lipase on core–shell structured Fe_3O_4–MCM-41 nanocomposites as a magnetically recyclable biocatalyst for interesterification of soybean oil and lard. *Food Chemistry* 194: 1283–1292.

Xu, L., Wang, Y., Yang, X. et al. (2009). Simultaneous esterification and transesterification of soybean oil with methanol catalyzed by mesoporous Ta_2O_5/SiO_2-$[H_3PW_{12}O_{40}/R]$ (R = Me or Ph) hybrid catalysts. *Green Chemistry* 11: 314–317.

Yagiz, F., Kazan, D., and Akin, A.N. (2007). Biodiesel production from waste oils by using lipase immobilized on hydrotalcite and zeolites. *Chemical Engineering Journal* 134: 262–267.

Yan, S., Kim, M., Salley, S.O., and Ng, K.Y.S. (2009a). Oil transesterification over calcium oxides modified with lanthanum. *Applied Catalysis A: General* 360: 163–170.

Yan, S., Salley, S.O., and Simon Ng, K.Y. (2009b). Simultaneous transesterification and esterification of unrefined or waste oils over ZnO-La_2O_3 catalysts. *Applied Catalysis A: General* 353: 203–212.

Yan, G.X., Wang, A., Wachs, I.E., and Baltrusaitis, J. (2019). Critical review on the active site structure of sulfated zirconia catalysts and prospects in fuel production. *Applied Catalysis A: General* 572: 210–225.

Yang, X.Y., Tian, G., Jiang, N., and Su, B.L. (2012). Immobilization technology: a sustainable solution for biofuel cell design. *Energy and Environmental Science* 5: 5540–5563.

Yang, N., Chen, X., Ren, T. et al. (2015). Carbon nanotube based biosensors. *Sensors and Actuators B: Chemical* 207: 690–715.

Yoon, G.L., Kim, B.T., Kim, B.O., and Han, S.H. (2003). Chemical-mechanical characteristics of crushed oyster-shell. *Waste Management* 23: 825–834.

Yücel, Y. (2011). Biodiesel production from pomace oil by using lipase immobilized onto olive pomace. *Bioresource Technology* 102: 3977–3980.

Zabeti, M., Wan Daud, W.M.A., and Aroua, M.K. (2009). Activity of solid catalysts for biodiesel production: a review. *Fuel Processing Technology* 90: 770–777.

Zdarta, J., Meyer, A.S., Jesionowski, T., and Pinelo, M. (2018). A general overview of support materials for enzyme immobilization: characteristics, properties, practical utility. *Catalysts* 8: 92.

Zhang, Y., Dubé, M.A., McLean, D.D., and Kates, M. (2003). Biodiesel production from waste cooking oil: 2. Economic assessment and sensitivity analysis. *Bioresource Technology* 90: 229–240.

Zhang, J., Chen, S., Yang, R., and Yan, Y. (2010). Biodiesel production from vegetable oil using heterogenous acid and alkali catalyst. *Fuel* 89: 2939–2944.

Zhang, X.L., Yan, S., Tyagi, R.D., and Surampalli, R.Y. (2013). Biodiesel production from heterotrophic microalgae through transesterification and nanotechnology application in the production. *Renewable and Sustainable Energy Reviews* 26: 216–223.

Zhang, Y., Ge, J., and Liu, Z. (2015). Enhanced activity of immobilized or chemically modified enzymes. *ACS Catalysis* 5: 4503–4513.

Zhao Chen, Y., Ting Yang, C., Bun Ching, C., and Xu, R. (2008). Immobilization of lipases on hydrophobilized zirconia nanoparticles: highly enantioselective and reusable biocatalysts. *Langmuir* 24: 8877–8884.

Zhao, X., El-Zahab, B., Brosnahan, R. et al. (2007). An organic soluble lipase for water-free synthesis of biodiesel. *Applied Biochemistry and Biotechnology* 143: 236–243.

Zhao, G., Li, Y., Wang, J., and Zhu, H. (2011). Reversible immobilization of glucoamylase onto magnetic carbon nanotubes functionalized with dendrimer. *Applied Microbiology and Biotechnology* 91: 591–601.

Zhao, X., Qi, F., Yuan, C. et al. (2015). Lipase-catalyzed process for biodiesel production: enzyme immobilization, process simulation and optimization. *Renewable and Sustainable Energy Reviews* 44: 182–197.

Zhi, J., Wang, Y., Lu, Y. et al. (2006). In situ preparation of magnetic chitosan/Fe_3O_4 composite nanoparticles in tiny pools of water-in-oil microemulsion. *Reactive and Functional Polymers* 66: 1552–1558.

Zhong, L., Feng, Y., Wang, G. et al. (2020). Production and use of immobilized lipases in/on nanomaterials: a review from the waste to biodiesel production. *International Journal of Biological Macromolecules* 152: 207–222.

4

Calcium-Based Nanocatalysts in Biodiesel Production

Priti R. Pandit and Archit Mohapatra

Gujarat Biotechnology Research Centre, Gandhinagar, Gujarat, India

4.1 Introduction

Global energy demand is increasing with increasing industrialization and population growth, leading to greater dependence on fossil fuels. According to the World Energy Forum, worldwide reserves such as coal, natural gas and fossil-based oil will be exhausted in less than 10 decades (Lee et al. 2011). Moreover, the combustion of fossil fuel has raised various environmental issues, including greenhouse gas (GHG) emissions, which significantly contribute to climate change, global warming, and many incurable diseases (Lam and Lee 2012). It has become progressively more important to develop abatement techniques and adopt policies to reduce the impact of GHGs.

Therefore, it is necessary to develop new technologies in order to discover alternative sources of fuel that can mitigate environmental problems and act as renewable sources of energy. In this context, biofuel can be an important substitute for fossil fuel. Biofuels may be solid, liquid, or gaseous in form and are predominantly derived from the biological materials. Solid biofuel is mainly produced from lignified cellulose (wood). Liquid and gaseous biofuels include bioethanol, biodiesel, and methane from anaerobic digestion (Yu et al. 2012). Currently, biodiesel is receiving worldwide attention as a renewable, nontoxic, and biodegradable fuel source. Furthermore, it has a high flash point, good lubricant properties, and low gas emissions, with significantly lower sulfur content when compared to ultra-sulfur diesel (ULSD) (Yuan et al. 2020). It can also be blended with commercially available diesel to improve its fuel properties (Van Gerpen 2005).

Biodiesel can be produced through various techniques, such as thermal cracking, micro-emulsification, ultrasonic reaction, microwaves techniques, and enzyme-based methods (Tan et al. 2015). Commercially, the transesterification method is used for the production of biodiesel. This involves the reaction of oils such as animal fats, vegetable oils, used cooking oils, and algae oil with alcohols: mainly methanol and ethanol (Yuan et al. 2020). It is catalyzed by different type of catalysts, including acid and base homogeneous catalysts, acid and base heterogeneous catalysts, and enzymatic catalysts. Currently, homogeneous base catalysis is the most common option commercially for the production

Nano- and Biocatalysts for Biodiesel Production, First Edition. Edited by Avinash P. Ingle.

of biodiesel. Initially, hydroxides and alkoxides such as NaOH, KOH, $NaOCH_3$, $KOCH_3$, $Ca(OH)_2$, LiOH, and $NaOCH_2CH_3$ were the main transesterification catalysts (Galadima and Muraza 2014).

However, this existing technology for biodiesel production on an industrial scale is facing many challenges, including difficulty over separation of the product the inability to recover and reuse the catalyst at the end of the process. The catalysts used are commercially expensive and lead to the production of large quantities of waste. To overcome these challenges, carbon-derived heterogeneous catalysts can be used as promising alternatives to replace the homogeneous ones. The heterogeneous catalysts are noncorrosive, environmentally friendly, easily separated from the product, and easy to reuse, requiring no cleansing with water (Marinkovi et al. 2016). Among them, calcium oxide (CaO) has been most widely used, due to its high activity, low cost, ease of availability, eco-friendliness, low solubility in organic solvents, and noncorrosive nature. CaO can be regenerated and reused due to its high basicity and is widely used as a catalyst for the transesterification reaction for different feedstocks (Bankovi et al. 2017).

Kaur and Ali (2014a,b) used waste cottonseed oil as a feedstock for transesterification. In the literature, other similar studies have proved CaO is a highly promising catalyst for biodiesel preparation (Tang et al. 2008; Tan et al. 2017). Despite its many advantages, however, the application of CaO is restricted by the requirement for a long reaction time and a high molar ratio of methanol to oil (Li et al. 2016). Therefore, in order to improve the catalytic activity and increase the reaction rate of existing CaO catalysts, they have recently been prepared in their nano form and functionalized with metal oxides.

A nanocatalyst is a nanomaterial with an average particle size in the range of 1–100 nm. Nanocatalysts play an important role in biodiesel production because of their highly specific surface area, high catalytic efficiency, resistance to saponification, and good rigidity. Many researchers have proved the potentiality of CaO nanocatalysts in biodiesel production (Manikandan and Rajasekaran 2013).

This chapter provides an overview of the CaO nanocatalysts, including recent advances in their synthesis and characterization, the possibility of utilizing waste material for their preparation, their applications in the conversion of different oil sources for biodiesel production, and the factors affecting production such as temperature, stirring rate, alcohol to oil ratio, reaction time, and catalyst loading. Finally, the assessment of their stability and reusability is also described.

4.2 Nanocatalysts

Over the last few decades, nanotechnology has emerged as a most promising technology with various applications in numerous fields. "Nanotechnology" is a broad term typically used to describe materials and phenomena at nanometer scale (i.e. on the scale of 1 billionth to several tens of billionths of a meter). However, it specifically implies not only the miniaturization but also the precise manipulation of atoms and molecules to design and control the properties of a nanomaterial. Materials that are nanoscale in two dimensions include nanotubes and nanowires, while those with three dimensions include particles such as colloids and quantum dots (Chaturvedi et al. 2012). One of the most important applications of nanotechnology is in catalysis. Nanocatalysts combine the characteristics of both homogenous and heterogeneous catalysts, while reducing the disadvantages of both. In a homogenous catalyst, the staring material and catalytic substance exist in the same phase, which increases its selectivity and activity. However, the application of homogenous catalyst is limited due to the difficulty of separating it from the end product (Hamilton et al. 2014). In heterogeneous catalysts, on the other hand, the staring materials and the catalytic substance are in different phases, improving separation (Olveira et al. 2014) but reducing surface area and catalytic activity, thus necessitating a high dosage of catalyst for any given reaction. To overcome this latter problem, the catalyst surface to volume ratio should be increased by decreasing the size of the catalytically active material. This can be achieved by engineering the catalyst on the nanoscale (Hussain et al. 2011).

Two approaches are usually used in nanocatalysis: the bottom-up strategy used for synthesis of well-controlled (in terms of shape and size) nanoparticles and the top-down strategy used to obtain more selective nanocatalysts. The important features to pursue are 100% selective reactions, extremely high activity, low energy consumption, high yield, minimization of waste, nontoxicity, long lifetime of the catalyst system, and enhanced possibilities for isolating and reusing the nanocatalyst (Venkat-Reddy et al. 2006; Wang et al. 2009). These can be achieved by controlling the catalyst's size, shape, surface composition, and electronic structure, as well as thermal and chemical nanocomponent stability. Nanocatalysts with different shapes (e.g. cube, triangle, platelet, rod) have been shown to have different activities and selectivities, as well as different stabilities in catalytic processes (Philippot and Serp 2013). Heterogeneous nanocatalysts, with advantages such as high pore size, large surface area, eco-friendly nature, and high catalytic efficiency, represent attractive potential catalysts for biodiesel production.

In the literature, various nanocatalysts have been reported for biodiesel production, including mixed oxide KF-ZnO_2 (Xie and Huang 2006), sulfated Fe_2O_3–SiO_2 (Anuradha et al. 2014), WO_3/SnO_2 (Xie and Wang 2013), TiO_2-ZnO (Madhuvilakku and Piraman 2013), and CaO/Fe_3O_4@SiO_2 (Feyzi and Norouzi 2016). Among them, CaO-based nanocatalysts are the most attractive for use in transesterification reactions, having high basicity, low solubility in methanol, the ability to be synthesized from low-cost material such as limestone or calcium hydroxide, high specific surface area, large porosity, and degenerative activity (Tahvildari et al. 2015).

4.3 CaO-Based Nanocatalysts for Biodiesel Production

CaO-based catalyst has been widely used as a solid base nanocatalyst due to its low cost and high catalytic activity. The calcium-based nanocatalyst is prepared in two steps: first, by using different methods such as the sol-gel method, co-precipitation method, thermal decomposition method, and impregnation; second, by calcinating the prepared nanocatalyst at high temperature in order to achieve high catalytic activity. In this way, the specific surface area and active site of the CaO nanocatalyst increase, so that they can be effectively employed in the transesterification reaction for biodiesel production. An overview of the methods of synthesis and properties of different CaO-based nanocatalysts is given in Table 4.1.

Table 4.1 Methods of synthesis and properties of different CaO-based nanocatalysts.

Catalyst	Synthesis method	Diameter (nm)	Surface area (m^2/g)	Reusability	References
KF/CaO	KF-impregnated calcium oxide by impregnation method followed by calcination at 600 °C for three hours	30–100	109	16 times, after which yield obtained is 91%	Wen et al. (2010)
Al_2Co_3-CaO	CaO-doped Al_2O_3 using aluminum isopropoxide and calcium nitrate as a precursor, stirring (two hours), drying for 18 hours at 120 °C, and calcination for six hours at 500 °C	20–30 (XRD)	70–82.74	3 times, checked for 25% catalyst	Sivaprakash et al. (2020)
$MgFe_2O_4$-CaO	Precipitation of $MgSO_4$ and $FeCl_3.6H_2O$ (1 : 2 ratio), washing and drying, calcination for three hours at 700 °C	49.76 (XRD)	3.72	4 times, loss in activity after every run	Liu et al. (2016)
Li-CaO	Lithium-impregnated calcium oxide by impregnation method	50–70 (TEM)	1.7	—	Kaur and Ali (2011)
CaO/Fe_3O_4	Magnetic solid base catalysts	49	—	—	Liu et al. (2010)
$KF/CaO-Fe_3O_4$	Fe_3O_4 prepared by Co-precipitation, followed by mixing of Fe_3O_4 and CaO then impregnation of KF, calcination for three hours at 600 °C	50	20.08	14 times, then loss in activity	Hu et al. (2011a)
CaO-K_2CO_3	CaO nanocatalyst impregnated with K_2CO_3 followed by drying for 16 hours at 120 °C and calcination for three hours at 500 °C	50–100	12.85	5 times, then activity decreases	Degirmenbasi et al. (2014)
$CuFe_2O_4$-CaO	Eggshells washed, dried (four hours at 800 °C), and calcinated, then chemically precipitated with a solution of $CuFe_2O_4$	50	—	—	Seffati et al. (2019)
CaO	Sol–gel method	3.3	67.78	—	Bharti et al. (2019)

Nano-crystalline CaO	Sonication-assisted reverse emulsion-based, followed by calcination for two hours at 1000 °C	18.81%	97.63	3 times, with loss in the activity after every run	Badnore et al. (2018)
$Ca/Fe_3O_4@SiO_2$	Combination of sol-gel and wetness impregnation	51	189.2	—	Feyzi et al. (2013)
CaO	*Polymesoda erosa* shells hydrated then dried and calcinated	—	—		Anr et al. (2016)
				9 times	
CaO-/CaN, CaO/	Sol-gel method	—	—	5 times	Gupta and Agarwal, (2016)
CaO-MgO	Sol-gel method for CaO and sol-gel self-combustion for MgO	—	—	6 times	Tahvildari et al. (2015)
Calcium aluminate on to Fe_3O_4	$Ca(OH)_2$ prepared by calcination and then hydrated, Na-meta-aluminate solution prepared, nano-Fe_3O_4, NaOH, and $Ca(OH)_2$ mixed, stirred (five hours at 80 °C), washed, dehydrated, and dried (12 hours), followed by calcination for six hours at 600 °C	31.51 (XRD)	25.89	5 times	Tang et al. (2012)
Nano CaO	Nano-CaO catalyst derived from eggshell	—	—	—	Siva and Marimuthu (2015)
Cao	Commercially available	37.3	22.25	10 time	Zhao et al. (2013)
Cao	Fresh calcite ($CaCO_3$) thermally decomposed then hydrated for six hours at 60 °C and filtrated, before being heated at 120 °C overnight and calcinated for three hours at 600 °C	42.2	25	5 times	Yoosuk et al. (2010)

(Continued)

Table 4.1 (Continued)

Catalyst	Synthesis method	Diameter (nm)	Surface area (m^2/g)	Reusability	References
LiC/CaO	Li_2CO_3 impregnated in aqueous solution, stirred (2 hours), dried, and heated (24 hours at 120 °C)	70 (XRD)	—	—	Kumar and Ali (2010)
K/CaO	KOH impregnated in aqueous solution, stirred (3 h), and dried (24 hours at 180 °C)	50 (TEM)	5.84	3 times, with partial loss of activity	Kumar and Ali (2012)
Zn/CaO	$Zn(CH_3COO)_2$ impregnated from aqueous solution then stirred (three hours), air dried, and calcinated (12 hours at 550 °C)	13 (TEM)	10.95	5 times, can be recovered	Kumar and Ali (2013)
Zr/CaO	$ZrOCl_2 \cdot 8H_2O$ wet impregnated from aqueous solution, then stirred (four hours), dried, heated (24 hours at 120 °C), and calcinated (700 °C)	30 (TEM)	1.86	4 times, loss in activity after every subsequent run	Kaur and Ali (2014a,2014b)
CaO/graphite oxide	Aqueous solution of $Ca(NO)_3$ and graphite oxide prepared, stirred (two hours), dried (80 °C, atmospheric pressure), and calcinated for one hour at 800 °C	27 (XRD)	437	4 times, then activity lost to 18.6%	Zu et al. (2011)
CaO	*Musa balbisiana* colla underground stem ash dried, burned to ashes in thin pieces, and dried for two hours at 550 °C	<100 (XRD)	38.71	2 times, with 20% decreases from next run	Sarma et al. (2014)
CaO-Au	CaO prepared from wastes then nanogold-CaO prepared using the impregnation method	50 (TEM)	—	10 times, then activity reduces	Bet-moushoul et al. (2016)
$CaO/Fe_3O_4@SiO_2$	Sol-method and wet implementation method	48–52 (SEM)	—	Several runs	Feyzi and Norouzi (2016)

4.3.1 Synthesis and Characterization of CaO-Based Nanocatalysts Using Waste Material

Recently, many researchers have turned their attention toward development of a CaO-based nanocatalyst as a valuable catalytically active material for biodiesel production using various raw waste materials such as shells (chicken egg, mud crab, mollusk, mussel, oyster, and clam) (Viriya-empikul et al. 2010; Niju et al. 2014), biomass fly ashes, bones, and lime (Kouzu et al. 2008). The utilization of wastes as a raw material for the synthesis of nanocatalyst will help resolve the problem of waste management and produce an economically viable catalyst with high efficiency. Generally, CaO nanocatalysts are prepared by the calcination-hydration-dehydration method.

Sarma et al. (2014) investigated the preparation, structure, composition, and use of ash obtained from the underground stem of *Musa balbisiana* Colla (banana tree) as a nanocatalyst for the transesterification of jatropha oil with a high percentage of free fatty acids (FFAs). They reported a very effective transesterification conversion of 98% of fatty acid methyl esters (FAMEs) under high temperature (275 °C) and high pressure (4.2 MPa) with a 10% CaO nanocatalyst load. Anr et al. (2016) successfully synthesized a highly active heterogeneous CaO nanocatalyst from *Polymesoda erosa* seashells through a calcination-hydration-dehydration method. Further, the nanocatalyst thus synthesized was evaluated for its efficacy in the two-step transesterification of jatropha oil for biodiesel production. It was reported that the nanocatalyst possessed high surface area (90.61 m^2/g) and low particle sizes (66 nm), which contributed to enhance the transesterification reaction rate. A 98.54% biodiesel yield was obtained under optimal conditions of 0.02 : 1 (w/w) catalyst ratio, 133.1 minutes reaction time, and 5.15 : 1 mol/mol ratio of methanol to pretreated oil. Moreover, it was also observed that after a sixth cycle of reuse, an average of 95.8% biodiesel yield was obtained. Table 4.2 summarizes different waste materials used for the synthesis of CaO nanocatalyst and their application in biodiesel production using various feeds.

Similarly, Sirisomboonchai et al. (2015) utilized waste scallop shell as a catalyst for transesterification of waste cooking oil (WCO). It was observed that CaO obtained from scallop shell showed higher catalytic activity as compared with commercial CaO. An 86% biodiesel yield was obtained using 5 wt% catalysts for a reaction time of two hours and a temperature of 65 °C. Viriya-empikul et al. (2010) synthesized CaO catalyst using waste shells of egg, golden apple snail, and *Meretrix venus* for the transesterification of palm oil and obtained >90% biodiesel yield. Khemthong et al. (2012) prepared waste eggshell CaO heterogeneous catalyst for biodiesel production using microwaves. Compared to conventional heating, it was reported that a high yield of FAME could be achieved (96.7%) through microwave-assisted heating by employing 15% catalyst and a reaction time of four minutes at 900 W microwave power. However, Suryaputra et al. (2013) synthesized waste capiz shells for the transesterification of palm oil for biodiesel production.

Recently, Foroutan et al. (2020) investigated the application of waste edible oil (WEO) for biodiesel production using CaO/MgO nanocatalyst derived from waste chicken eggshells. The highest biodiesel yield of 98.37% was achieved with a reaction time of 7.08 hours, reaction temperature of 69.37 °C, methanol to oil molar ratio of 16.7 : 1, and catalyst concentration of 4.57 wt%. Meanwhile, Chung et al. (2019) demonstrated the comparative efficacy of

Table 4.2 Synthesis of CaO nanocatalyst from different waste materials and their application in biodiesel production using various feedstocks.

Catalyst used for transesterification	Source of nanocatalyst synthesis	Feedstock	Temperature	Alcohol/oil ratio	Catalyst Concentration (wt%)	Reaction time (h)	Biodiesel yield %	References
CaO	Lime stone	Soybean oil	64.7	12 : 1	—	1	93	Kouzu et al. (2008)
CaO	Red mud	Soybean oil	65	24 : 1	4	3	94	Liu et al. (2013)
CaO	Oyster shell	Soybean oil	65	6 : 1	25	5	98.4	Nakatani et al. (2009)
CaO	Mussel shell	Soybean oil	60	24 : 1	12	8	94.1	Rezaei et al. (2013)
Porous CaO	—	Soybean oil	65	9 : 1	3	3	98.72	Bai et al. (2009)
CaO	Lime mud	Refine peanut oil	64	15 : 1	6	3	94.4	Li et al. (2014)
Calcined $Ca(NO_3)_2$	Dolomite rock	Palm kernel oil	60	15 : 1	10	3	99.9	Ngamcharussrivichai et al. (2010)
Calcined Dolomite rock	Dolomite rock	Palm kernel oil	60	30 : 1	6	3	98	Ngamcharussrivichai et al. (2010)
Calcined chicken eggshell	Chicken eggshell	Palmolein oil	60	12 : 1	10	2	94.1	Viriya-empikul et al. (2012)
Calcined golden apple snail shell	Golden apple snail shell	Palmolein oil	60	15 : 1	10	2	93.2	Viriya-empikul et al. (2012)
Calcined *Meretrix venus* shell	*Meretrix venus* shell	Palmolein oil	60	15 : 1	10	2	92.3	Viriya-empikul et al. (2012
Calcined freshwater mussel shell	Freshwater mussel shell	Chinese tallow oil	70	12 : 1	5	1.5	90	Hu et al. (2011b)

CaO	Quail eggshell	Palm oil	65	12 : 1	1	2	98	Cho and Seo (2010)
CaO	Capiz shell	Palm oil	60	8 : 1	3	6	92.83	Suryaputra et al. (2013)
CaO		Canola oil	64.85	9 : 1	—	8	97.67 ± 1.7	Degirmenbasi et al. (2014)
CaO	Snail shell	Canola oil	65	12 : 1	8	6	96	Gupta and Agarwal (2016)
CaO	Eggshell	Sunflower oil	60	9 : 1	3	3	97.75 ± 0.02	Correia et al. (2014)
CaO-based catalyst-supported gold nanoparticles	Waste mussel and eggshells	Sunflower oil	65	9 : 1	3	3	90–97	Bet-moushoul et al. (2016)
CaO	Quail eggshell	Sunflower oil	60	10.5 : 1	2	2	99.00 ± 0.02	Correia et al. (2017)
CaO	—	Fried oil	65	6 : 1	5	3	92	Viola et al. (2012)
Crushed and powdered shell	Eggshell	Algae oil	55	9 : 1	2	—	96.3	Siva and Marimuthu (2015)
CaO	Eggshell	Algae biomass	75	12 : 1	1.6	3.5	86.41	Pandit and Fulekar (2017)

both CaO and KOH catalyst for transesterification, observing that the utilization of waste chicken eggshell-derived CaO had lesser ecotoxicity compared to KOH because the calcination process was not complex and had less of an impact on climate change. Therefore, utilization of waste as a nanocatalyst is not only economically viable but also eco-friendly.

4.3.2 CaO Nanocatalysts Supported with Metal Oxides for Biodiesel Production

Currently, use of various alkali and alkaline-based metal oxides as catalysts is under investigation for biodiesel production (Avhad and Marchetti 2015). The alkaline earth metals and hydroxides have been arranged according to increasing order of basic strength: Mg < Ca < Sr < Ba. Among them, CaO is most extensively studied as a heterogeneous catalyst due to its various beneficial properties, as already discussed. However, naked CaO nanocatalysts have some limitations, such as formation of hydrogen bonds with methanol, which complicates downstream processing and increases production costs. In order to overcome these challenges, CaO nanocatalysts are usually doped or immobilized with certain materials like oxides of alkali and alkaline earth metals. In many reports, it has been demonstrated that functionalization of CaO catalysts with metal oxides can improve catalytic activity and increase reaction rate. However, these reported CaO-based catalysts are unstable and suffer from leaching of active sites with methanol due to their partial solubility in it (Kesic et al. 2016).

Li et al. (2016) successfully functionalized CaO catalyst with strontium oxide (SrO) due to its high basicity and insolubility in the methanol and FAMEs. This nanocatalyst was prepared using four different methods: wet impregnation, co-precipitation, solid mixing, and improved co-precipitation. The maximum conversion of FAMEs (98.31%) from palm oil was obtained under optimum conditions of 5 wt. % catalyst and 9 M ratio of methanol to oil at 65 °C with a reaction time of 30 minutes. It was also reported that the improved co-precipitation method was ideal for the strongest interaction between Sr and Ca. Lin et al. (2012) successfully synthesized a magnetic CaO/alpha Fe-fiber nanocatalyst using an organic gel thermal decomposition approach normally used for the transesterification of rapeseed oil. They recorded an 85% biodiesel yield after 20 cycles of transesterification. Due to strong magnetic behavior, the catalyst was easily separated from the reaction mixture with the application of a magnetic field and reused. Moreover, many research groups have reported the immobilization of CaO with different carriers in order to produce a highly active catalyst for transesterification reactions (Table 4.3). The different carriers, such as MgO, Al_2O_3, and zeolite, were immobilized on CaO and used for biodiesel production (Kesic et al. 2016). Al_2O_3 loaded with CaO and MgO catalyst was prepared and successfully applied to the transesterification of lipid of *Nannochloropsis oculata* microalgae. It was observed that the naked CaO and MgO nanocatalyst showsed less activity than CaO/Al_2O_3 nanocatalyst. The higher activity of CaO-loaded Al_2O_3 was due to higher basic site concentration and basic strength (Umdu et al. 2009).

Zu et al. (2011) synthesized a graphite oxide-supported CaO nanocatalyst and successfully applied it to biodiesel production using soybean oil as a promising feedstock. The nanocatalyst thus synthesized showed high catalytic activity and easily regenerated via a simple heat-treatment process. It was concluded that the oxygen-containing groups (e.g. hydroxyl,

Table 4.3 Synthesis of CaO nanocatalyst supported by different alkaline metals and metal oxide.

Catalyst	Feedstock	Catalyst concentration (% wt to oil)	Alcohol: oil ratio	Temperature (°C)	Reaction time (h)	Biodiesel yield (%)	References
CaO/Au	Sunflower oil	3	9 : 1	65	3	90–97	Bet-moushoul et al. (2016)
CaO-ZnO	Sunflower oil	3	12 : 1	60	1.5	>90	
CaO-ZnO	Sunflower oil	2	10 : 1	60	4	97.5	Kesic et al. (2016)
$CaO/Fe_3O_4@SiO_2$	Sunflower oil	6	15 : 1	65	5	97	Feyzi and Norouzi (2016)
CaO (16 wt%)-ZnO	Sunflower oil	1.3	12 : 1	60	2	>90	
CaO–NiO	Waste cottonseed oil	5	15 : 1	65	4	≥99	Kaur and Ali (2014a)
K/CaO	Waste cottonseed oil	7.5	12 : 1	65	1.25	≥99	Kumar and Ali (2012)
$CaO–CeO_2$	Refined palm oil	5	20 : 1	85	3	95	Thitsartarn and Kawi (2011)
$CaO-ZrO_2$	Waste cooking oil	10	30 : 1	65	2	92.1	Dehkordi and Ghasemi (2012)
$CaO-Al_2O_3$	Palm oil	6	12 : 1	65	5	98.64	
$CaO-La_2O_3$	*Jatropha curcas* oil	4	24 : 1	65	6	86.51	Lee et al. (2015)
CaO/Fe_3O_4	Jatropha curcas oil	2	15 : 1	70	4	99	Liu et al. (2010)
CaO-MgO	Jatropha curcas oil	3	25 : 1	120	3	90	Taufiq-yap et al. (2011)
$KF-CaO-Fe_3O_4$	Stillingia oil	4	12 : 1	65	3	>95	Hu et al. (2011a)
Al_2O_3-CaO	*P. murex* plant tissue	25	14 : 1	80	6	60	Sivaprakash et al. (2020)
Li-Cao	Karanja oil and jatropha oil	5	12 : 1	65	1	99	Kaur and Ali (2011)
K/CaO	Karanja oil	7.5	12 : 1	65	2	≥99	Kumar and Ali (2012)
CaO/Fe_3O_4	Jatropha oil	2	15 : 1	70	1.33	95	
K/CaO	Jatropha oil	7.5	12 : 1	65	2.5	≥99	Kumar and Ali (2012)
CaO/MgO	Jatropha oil	2	18 : 1	64.5	3.5	92	Tahvildari et al. (2015)
K_2CO_3/CaO	Canola oil	3	9 : 1	65	8	97.67	Degirmenbasi et al. (2014)
KF/(CaO/MgO)	Rapeseed oil	3	12 : 1	70	3	95	Kawashima et al. (2008)

epoxide groups, and carboxyl groups) present on the surface of graphite oxide act as effective anchoring centers for the active phase of CaO, resulting in a stable and active catalyst. Moreover, Kumar and Ali (2010) investigated the influence of different alkali elements such as Li, Na, and K on the catalytic performance of CaO nanocatalysts and found that Li-doped CaO nanocatalyst had comparatively higher efficiency. However, it was proposed that the highest activity of Li/CaO might be due to the formation of strong basic sites with CaO. The authors further concluded that nanocatalyst has a potential for converting the cheap feedstock into biodiesel.

There are some reports suggesting that CaO can be mixed with magnetic material to form nano-magnetic solid base catalysts. Hu et al. (2011a) evaluated the catalytic efficacies of KF/CaO-Fe_3O_4, KF/MgO-Fe_3O_4, and KF/SrO-Fe_3O_4 nanocatalysts for the transesterification of Stillingia oil. They also investigated the influence of KF loading and calcination temperature in achieving maximum catalytic activity. The results obtained for KF/CaO-Fe_3O_4 nanocatalyst showed that the biodiesel yield could be improved from 75% to 95% when the KF loading was increased from 15 to 25 wt%. The increased catalytic activity was due to the increase in the active sites on the catalyst surface with increasing KF loading. The authors further revealed that the calcination temperature played an important role in increasing the catalytic activity of the catalyst: at higher calcination temperatures, new catalytic active sites were generated due to the interaction between the support and the active ingredient. Similar trends were also observed with both KF/MgO-Fe_3O_4 and KF/SrO-Fe_3O_4 nanocatalysts.

Zhang et al. (2014) synthesized the magnetic solid catalyst CaO/$CoFe_2O_4$ using a hydrothermal method and employed it in the production of biodiesel. Their results showed that CaO/$CoFe_2O_4$ catalyst had better catalytic activity, weaker hydroscopicity, and stronger wettability than CaO/$ZnFe_2O_4$ and CaO/$MnFe_2O_4$, and produced 87.4% biodiesel yield. Moreover, CaO/$CoFe_2O_4$ had stronger magnetic strength, and could easily be separated by applying a magnetic field after transesterification and reused for subsequent cycles of transesterification in biodiesel production. In another similar study, Tang et al. (2012) evaluated the potential of Ca/Al Fe_2O_4 magnetic nanocatalyst in biodiesel production. They compared the two catalysts with and without magnetic nanoparticles. The magnetic composite catalyst (i.e. Ca/Al Fe_2O_4) resulted in a higher product yield of 98.71% in the transesterification of rapeseed oil due to its small size, higher specific surface area, and higher catalytic activity. Furthermore, the Ca/Al Fe_2O_4 catalyst was easily separated from the reaction mixture by application of an external magnetic field like a solid magnet. It was observed that the recovery rate of this catalyst was higher than that of a calcium acuminate catalyst. The introduction of microwave and ultrasound irradiation in the transesterification process allowed the reaction to be completed in a shorter time.

Mootabadi et al. (2010) investigated the ultrasonic-assisted transesterification of palm oil in the presence of CaO, SrO, and BaO under different operating conditions with 20 kHz ultrasonic cavitation. The biodiesel yield was increased from 5.5% to 77.3%, 48.2% to 95.2%, and 67.3% to 95.2% for CaO, SrO, and BaO catalyst, respectively, after 60 minutes of reaction time. The ultrasonic waves generated intense microturbulence (oscillation of liquid) in the medium and led to more efficient mixing of the biomass and solvent (without induction of shear stress), while shockwaves caused rupture of the cell wall. This enhances transport process rates and eliminates mass transfer resistance in heterogeneous systems

(Naveena et al. 2015). Nayebzadeh et al. (2018) proposed a simple and cost-effective method for the biodiesel production under microwave irradiation. They reported the successful synthesis of carbonated alumina doped by CaO using a microwave combustion method. The results showed that 98.8% of canola oil was converted to biodiesel under optimum conditions (270 W microwave power, 15 : 1 M ratio of methanol to oil, 4 wt% of catalyst, and 30 minutes' reaction time). Zhang et al. (2010) concluded that absorption of microwave irradiation by carbon groups enhanced the reaction rate and consequently decreased the reaction time.

4.4 Effects of Different Parameters on Biodiesel Production

As far as production of biodiesel is concerned, it is well known that different reaction parameters such as reaction time, temperature, methanol to oil molar ratio, and catalyst load play a pivotal role in the success of the transesterification reaction (Seffati et al. 2019). In order to obtain maximum yield, it is very important to optimize these parameters by changing only one at a time. Different reports are available investigating the production of biodiesel by changing parameters, as already mentioned (Naureen et al. 2014; Esmaeili and Foroutan 2018), and the following equation has been derived:

$$\textit{Biodiesel conversion}\ (\%) = \textit{Weight of biodiesel produced} \times 100 \div \textit{Weight of oil}$$

4.4.1 Reaction Time

With evolution, the reaction time is known to vary depending upon the other parameters required for the conversion of feedstock into biodiesel. In the case of heterogeneous multiple systems, specific surface area and mass transfer are directly affected by the reaction time. With increasing reaction time, productivity increases until an equilibrium is reached, before decreasing again due to the reversible nature of the reaction (Rashtizadeh et al. 2014). Buasri et al. (2013) suggested that when the optimum reaction time is provided, the yield obtained is maximal, whereas when insufficient time (less time for the interaction) or too much time (overproduction of product) is provided, the yield obtained is comparatively less.

4.4.2 Temperature

The performance of the catalyst depends upon the temperature at which the system performs the conversion of biodiesel. It has been found that with increasing temperature, production increases; however, at too high a temperature, methanol, which is widely used in such reactions, becomes evaporated (~70 °C). Evaporation of methanol forms bubbles, which may inhibit the process and reduce the contact time between the methanol and the oil, resulting in a decrease in the biodiesel yield (Zhang et al. 2010; Qiu et al. 2011; Suryaputra et al. 2013). Further, at high temperatures, the viscosity of the oil and diffusion resistance both decrease, causing an increase in the intrinsic rate constant (Badnore et al. 2018). Many researchers have shown that proper yields can be obtained at a lower temperature (i.e. 35 °C) applied for a longer period of time.

4.4.3 Methanol to Oil Molar Ratio

The transesterification reaction mainly depends upon the ratio of methanol to oil (Kim et al. 2004). Different feedstocks or oils are used in the production of biodiesel, along with either methanol or ethanol. It has been observed that the stoichiometric ratio should be less than the molar ratio, as this moves the rate of reaction toward completion of product and helps in the regeneration of the catalyst active site by washing (Feyzi and Norouzi 2016). A higher methanol to oil ratio will hinder the recovery of product, as it may emulsify after washing (Demirbas 2009; Sharma and Singh 2009).

4.4.4 Catalyst Load

Different catalysts with different weight percentages are used for conversion through transesterification. The main function of the catalyst is to provide a surface area on which the reaction can take place, which increases with increasing catalyst concentration. Production remains constant and the catalyst does not further contribute to the reaction as it is limited by the availability of reactant, and increasing reactant increases the viscosity of the blend, decreasing the contact between the catalyst active site and the reactant (Bet-moushoul et al. 2016). Buasri et al. (2013) suggested that when optimized amounts of catalyst are loaded, maximum yield is obtained, while when higher or lower amounts are loaded, production is less than optimal due to unfavorable mass transfer.

4.5 Reusability and Leaching of Nanocatalysts

CaO nanocatalyst has been widely utilized in the transesterification reaction for biodiesel production. Besides having promising catalytic activity and selectivity for biodiesel production, the heterogeneous catalyst can be recovered and reused multiple times for such reactions. Following completion of the reaction, the catalyst is recovered and regenerated by filtration, washing with methanol or n-hexane, and then drying or calcination (Bankovi et al. 2017). The recovered catalyst can be reused for transesterification under the same experimental conditions until partial or complete loss of its activity.

Catalytic activity can be reduced by prolonged reuse of the catalyst due to leaching: loss of its active sites. Venkat-Reddy et al. (2006) investigated the reusability of CaO nanocatalyst and reported its deactivation after eight cycles with soybean oil and three with poultry fat. The authors concluded that the deactivation was due to the presence of organic impurities or adventitious moisture and enolate formation by the deprotonation of the carbon alpha to the carboxy group in the FAMEs or triglyceride. Nayebzadeh et al. (2018) reported a reduction in activity of 9.2%, 16.6%, 15.6%, and 27.8% after three cycles for carbonated alumina doped with CaO nanocatalyst prepared by sorbitol, triethylene glycol, starch, and citric acid, respectively, as fuel. Thangaraj et al. (2019) explained in a review paper that while CaO showed a higher catalytic activity during transesterification, after a few cycles of reuse the catalyst leached out the calcium ion, affecting the reaction product. The leaching of calcium ion was caused by the presence of moisture, which hydrolyzed the calcium diglyceride. Moreover, during the transesterification reaction, the formation of soluble calcium ions directly dissociated with CaO. Leached calcium ions reacted with the oil, leading to saponification and deactivating the effective function of the catalyst.

Liu et al. (2013) studied the reusability and stability of CaO/KF nanocatalyst and reported a significant decrease in biodiesel yield from 99.9% to 96.6% from the first to the fourth cycle. The authors explained that the leakage of K and Ca with water and menthol, as well as the occupation of active sites by oleic acid, was the main reason for the decrease in biodiesel yield. Zhao et al. (2013) observed a significant loss of activity after the 15th cycle of reuse and concluded that the loss was due to a change in physical properties and the formation of $Ca(OCH_3)_2$, $CaCO_3$, and a calcium glycerol complex.

$CaO/CoFe_2O_4$ nanocatalyst used for the transesterification of soybean oil maintained its activity over five cycles with 80% biodiesel yield (Zhang et al. 2014). A nano $KF/CaO\text{-}Fe_3O_4$ magnetic nanocatalyst with porous structure could be reused up to 14 times without loss of catalytic activity (Hu et al. 2011a). CaO/graphite nanocatalyst, after regeneration, lost its activity significantly after four consecutive reaction cycles (Zu et al. 2011). The magnetic nanocomposite $Ca/Al/Fe_3O_4$ catalyst maintained its overall catalytic activity for five successive cycles of transesterification after regenerations by calcination (Tang et al. 2012).

4.6 Conclusions

Numerous reports have proved that CaO base nanocatalyst is a potential heterogeneous catalyst that can be used to enhance the rate of transesterification for biodiesel production. CaO catalyst shows high specific surface area and high basicity and can be obtained from cheap natural or waste sources. Proper preparation, activation, and regeneration should be applied in obtaining it in sufficiently active and stable form. However, the reuse and recoverability of CaO catalyst in successive cycles faces challenges due to a reduction in catalytic sites caused by deposition of organic matter and leaching of CaO in the reaction medium. Hence, more research work is needed to discover new cost-effective precursors and carriers for the synthesis of nano-CaO catalyst, to reduce the leaching of active sites, and to improve reusability.

Moreover, modification with an appropriate metal oxide on the surface of CaO nanocatalyst may contribute significantly to the enhancement of catalytic activity. However, CaO nanocatalyst transesterification systems require more improvement to the existing technology and the development of new techniques for biodiesel production. Furthermore, new reactor types such as microwave, ultrasound irradiance, and supercritical fluid extraction may help improve the biodiesel yield. Optimization of reaction processes by application of statistical analysis and investigation of the kinetics of transesterification reactions through development of a kinetic model will further enhance the downstream processing of biodiesel production.

References

Anr, R., Saleh, A.A., Islam, M.S. et al. (2016). Biodiesel production from crude jatropha oil using a highly active heterogeneous nanocatalyst by optimizing transesterification reaction parameters. *Energy and Fuels* 30: 334–343.

Anuradha, S., Raj, K., Vijayaraghavan, V.R., and Viswanathan, B. (2014). Sulphated Fe2O3-TiO2 catalysed transesterification of soybean oil to biodiesel. *Indian Journal of Chemistry – Section A Inorganic, Physical, Theoretical and Analytical Chemistry* 53: 1493–1499.

Avhad, M.R. and Marchetti, J.M. (2015). A review on recent advancement in catalytic materials for biodiesel production. *Renewable and Sustainable Energy Reviews* 50: 696–718.

Badnore, A.U., Jadhav, N.L., Pinjari, D.V., and Pandit, A.B. (2018). Efficacy of newly developed nano-crystalline calcium oxide catalyst for biodiesel production. *Chemical Engineering and Processing Process Intensification* 133: 312–319.

Bai, H.X., Shen, X.Z., Liu, H.X., and Liu, Y.S. (2009). Synthesis of porous CaO microsphere and its application in catalyzing transesterification reaction for biodiesel. *Transactions of the Nonferrous Metals Society of China* 19 (Suppl. 3): s674–s677.

Bankovi, I.B., Miladinovi, M.R., Stamenkovi, O.S., and Veljkovi, V.B. (2017). Application of nano CaO-based catalysts in biodiesel synthesis. *Renewable and Sustainable Energy Reviews* 72: 746–760.

Bet-moushoul, E., Farhadi, K., Mansourpanah, Y. et al. (2016). Application of CaO-based/Au nanoparticles as heterogeneous nanocatalysts in biodiesel production. *Fuel* 164: 119–127.

Bharti, P., Singh, B., and Dey, R.K. (2019). Process optimization of biodiesel production catalyzed by CaO nanocatalyst using response surface methodology. *Journal of Nanostructure in Chemistry* 9: 269–280.

Buasri, A., Chaiyut, N., Loryuenyong, V. et al. (2013). Calcium oxide derived from waste shells of mussel, cockle, and scallop as the heterogeneous catalyst for biodiesel production. *The Scientific World Journal* 2013: 460923.

Chaturvedi, S., Dave, P.N., and Shah, N.K. (2012). Applications of nano-catalyst in new era. *Journal of Saudi Chemical Society* 16: 307–325.

Cho, Y.B. and Seo, G. (2010). High activity of acid-treated quail eggshell catalysts in the transesterification of palm oil with methanol. *Bioresource Technology* 101: 8515–8519.

Chung, Z.L., Tan, Y.H., San Chan, Y. et al. (2019). Life cycle assessment of waste cooking oil for biodiesel production using waste chicken eggshell derived CaO as catalyst via transesterification. *Biocatalysis and Agricultural Biotechnology* 21: 101317.

Correia, L.M., Saboya, R.M., de Sousa Campelo, N. et al. (2014). Characterization of calcium oxide catalysts from natural sources and their application in the transesterification of sunflower oil. *Bioresource Technology* 151: 207–213.

Correia, L.M., Cecilia, J.A., Rodriguez-Castellon, E. et al. (2017). Relevance of the physicochemical properties of Calcined quail eggshell (CaO) as a catalyst for biodiesel production. *Journal of Chemistry* 2017: 5679512.

Degirmenbasi, N., Boz, N., and Kalyon, D.M. (2014). Biofuel production via transesterification using sepiolite-supported alkaline catalysts. *Applied Catalysis B: Environmental* 150: 147–156.

Dehkordi, A.M. and Ghasemi, M. (2012). Transesterification of waste cooking oil to biodiesel using Ca and Zr mixed oxides as heterogeneous base catalysts. *Fuel Processing Technology* 97: 45–51.

Demirbas, A. (2009). Production of biodiesel fuels from linseed oil using methanol and ethanol in non-catalytic SCF conditions. *Biomass and Bioenergy* 33: 113–118.

Esmaeili, H. and Foroutan, R. (2018). Optimization od biodiesel production from goat tallow using alkaline catalysts and combinig them with diesel. *Chemistry & Chemical Technology* 12: 120–126.

Feyzi, M. and Norouzi, L. (2016). Preparation and kinetic study of magnetic Ca/Fe3O4 @ SiO2 nanocatalysts for biodiesel production. *Renewable Energy* 94: 579–586.

Feyzi, M., Hassankhani, A., and Rafiee, H.R. (2013). Preparation and characterization of Cs/Al/Fe3O4 nanocatalysts for biodiesel production. *Energy Conversion and Management* 71: 62–68.

Foroutan, R., Mohammadi, R., Esmaeili, H. et al. (2020). Transesterification of waste edible oils to biodiesel using calcium oxide@magnesium oxide nanocatalyst. *Waste Management* 105: 373–383.

Galadima, A. and Muraza, O. (2014). Biodiesel production from algae by using heterogeneous catalysts: a critical review. *Energy* 78: 72–83.

Gupta, J. and Agarwal, M. (2016). Preparation and characterizaton of CaO nanoparticle for biodiesel production. *AIP Conference Proceedings* 1724: 020066.

Hamilton, M.L., Haslam, R.P., Napier, J.A., and Sayanova, O. (2014). Metabolic engineering of *Phaeodactylum tricornutum* for the enhanced accumulation of omega-3 long chain polyunsaturated fatty acids. *Metabolic Engineering* 22: 3–9.

Hu, S., Guan, Y., Wang, Y., and Han, H. (2011a). Nano-magnetic catalyst KF/CaO–Fe3O4 for biodiesel production. *Applied Energy* 88: 2685–2690.

Hu, S., Wang, Y., and Han, H. (2011b). Utilization of waste freshwater mussel shell as an economic catalyst for biodiesel production. *Biomass and Bioenergy* 35: 3627–3635.

Hussain, S.T., Ali, S.A., Bano, A., and Mahmood, T. (2011). Use of nanotechnology for the production of biofuels from butchery waste. *International Journal of Physical Science* 6: 7271–7279.

Kaur, M. and Ali, A. (2011). Lithium ion impregnated calcium oxide as nano catalyst for the biodiesel production from karanja and jatropha oils. *Renewable Energy* 36: 2866–2871.

Kaur, M. and Ali, A. (2014a). Potassium fluoride impregnated CaO/NiO: an efficient heterogeneous catalyst for transesteri fi cation of waste cottonseed oil. *European Journal of Lipid Science and Technology* 116: 80–88.

Kaur, N. and Ali, A. (2014b). Kinetics and reusability of Zr/CaO as heterogeneous catalyst for the ethanolysis and methanolysis of *Jatropha crucas* oil. *Fuel Processing Technology* 119: 173–184.

Kawashima, A., Matsubara, K., and Honda, K. (2008). Development of heterogeneous base catalysts for biodiesel production. *Bioresource Technology* 99: 3439–3443.

Kesic, Z., Lukic, I., Zdujic, M. et al. (2016). Calcium oxide based catalysts for biodiesel production: a review. *Chemical Industry and Chemical Engineering Quarterly* 22: 391–408.

Khemthong, P., Luadthong, C., Nualpaeng, W. et al. (2012). Industrial eggshell wastes as the heterogeneous catalysts for microwave-assisted biodiesel production. *Catalysis Today* 190: 112–116.

Kim, J.H., Kang, S.B., Kim, J.M. et al. (2004). Transesterification of vegetable oil to biodiesel using heterogeneous bas catalyst. *Catalysis Today* 93–95: 315–320.

Kouzu, M., Kasuno, T., Tajika, M., and Sugimoto, Y. (2008). Calcium oxide as a solid base catalyst for transesterification of soybean oil and its application to biodiesel production. *Fuel* 87: 2798–2806.

Kumar, D. and Ali, A. (2010). Nanocrystalline lithium ion impregnated calcium oxide as heterogeneous catalyst for transesterification of high moisture containing cotton seed oil. *Energy and Fuels* 24: 2091–2097.

Kumar, D. and Ali, A. (2012). Nanocrystalline K–CaO for the transesterification of a variety of feedstocks: structure, kinetics and catalytic properties. *Biomass and Bioenergy* 46: 459–468.

Kumar, D. and Ali, A. (2013). Transesterification of low-quality triglycerides over a Zn/CaO heterogeneous catalyst: kinetics and reusability studies. *Energy & Fuels* 27: 3758–3768.

Lam, M.K. and Lee, K.T. (2012). Microalgae biofuels: a critical review of issues, problems and the way forward. *Biotechnology Advances* 30: 673–690.

Lee, H.V., Yunus, R., Juan, J.C., and Tau, Y.H. (2011). Process optimization design for jatropha-based biodiesel production using response surface methodology. *Fuel Processing Technology* 92: 2420–2428.

Lee, H.V., Juan, J.C., and Tau, Y.H. (2015). Preparation and application of binary acid-base CaO-La2O3 catalyst for biodiesel production. *Renewable Energy* 74: 124–132.

Li, H., Niu, S., Lu, C. et al. (2014). Use of lime mud from paper mill as a heterogeneous catalyst for transesterification. *Science China Technological Sciences* 57: 438–444.

Li, H., Niu, S., Lu, C., and Li, J. (2016). Calcium oxide functionalized with strontium as heterogeneous transesterification catalyst for biodiesel production. *Fuel* 176: 63–71.

Lin, L., Li, X., Cui, F., and Zhou, H. (2012). Transesterification of rapeseed oil to biodiesel on CaO/α-Fe hollow fiber catalyst: optimization by response surface methodology. *Bioenergy Research* 5: 949–957.

Liu, C., Lv, P., Yuan, Z. et al. (2010). The nanometer magnetic solid base catalyst for production of biodiesel. *Renewable Energy* 35: 1531–1536.

Liu, Q., Xin, R., Li, C. et al. (2013). Application of red mud as a basic catalyst for biodiesel production. *Journal of Environmental Sciences* 25: 823–829.

Liu, Y., Zhang, P., Fan, M., and Jiang, P. (2016). Biodiesel production from soybean oil catalyzed by magnetic nanoparticle MgFe2O4@CaO. *Fuel* 164: 314–321.

Madhuvilakku, R. and Piraman, S. (2013). Biodiesel synthesis by TiO2–ZnO mixed oxide nanocatalyst catalyzed palm oil transesterification process. *Bioresource Technology* 150: 55–59.

Manikandan, G. and Rajasekaran, R. (2013). Transesterification of algal oil using nano CaO catalyst. *International Journal of Chemical Sciences* 11: 591–597.

Marinkovi, D.M., Stankovi, M.V., Veli, A.V. et al. (2016). Calcium oxide as a promising heterogeneous catalyst for biodiesel production: current state and perspectives. *Renewable and Sustanaible Energy Reviews* 56: 1387–1408.

Mootabadi, H., Salamatinia, B., Bhatia, S., and Abdullah, A.Z. (2010). Ultrasonic-assisted biodiesel production process from palm oil using alkaline earth metal oxides as the heterogeneous catalysts. *Fuel* 89: 1818–1825.

Nakatani, N., Takamori, H., Takeda, K., and Sakugawa, H. (2009). Transesterification of soybean oil using combusted oyster shell waste as a catalyst. *Bioresource Technology* 100: 1510–1513.

Naureen, R., Tariq, M., Yusoff, I. et al. (2014). Synthesis, spectroscopic and chromatographic studies of sunflower oil biodiesel using optimized base catalyzed methanolysis. *Saudi Journal of Biological Sciences* 22: 332–339.

Naveena, B., Armshaw, P., and Pembroke, T.J. (2015). Ultrasonic intensification as a tool for enhanced microbial biofuel yields. *Biotechnology for Biofuels* 8: 140.

Nayebzadeh, H., Haghighi, M., and Saghatoleslami, N. (2018). Fabrication of carbonated alumina doped by calcium oxide via microwave combustion method used as nanocatalyst in biodiesel production: influence of carbon source type. *Energy Conversion and Management* 171: 566–575.

Ngamcharussrivichai, C., Nunthasanti, P., Tanachai, S., and Bunyakiat, K. (2010). Biodiesel production through transesterification over natural calciums. *Fuel Processing Technology* 91: 1409–1415.

Niju, S., Begum, K.M.M.S., and Anantharaman, N. (2014). Modification of egg shell and its application in biodiesel production. *Journal of Saudi Chemical Society* 18: 702–706.

Olveira, S., Forster, S.P., and Seeger, S. (2014). Nanocatalysis: academic discipline and industrial realities. *Journal of Nanotechnology* 2014: 324089.

Pandit, P.R. and Fulekar, M.H. (2017). Egg shell waste as heterogeneous nanocatalyst for biodiesel production: optimized by response surface methodology. *Journal of Environmental Managment* 198: 319–329.

Philippot, K. and Serp, P. (2013). Concepts in nanocatalysis. *Nanomaterials in Catalysis* 1: 1–54.

Qiu, F., Li, Y., Yang, D. et al. (2011). Heterogeneous solid base nanocatalyst: preparation, characterization and application in biodiesel production. *Bioresource Technology* 102: 4150–4156.

Rashtizadeh, E., Farzaneh, F., and Talebpour, Z. (2014). Synthesis and characterization of Sr3Alo6 nanocomposite as catalyst for biodiesel production. *Bioresource Technology* 154: 32–37.

Rezaei, R., Mohadesi, M., and Moradi, G.R. (2013). Optimization of biodiesel production using waste mussel shell catalyst. *Fuel* 109: 534–541.

Sarma, A.K., Kumar, P., Aslam, M., and Chouhan, A.P. (2014). Preparation and characterization of *Musa balbisiana* Colla underground stem nano-material for biodiesel production under elevated conditions. *Catalysis Letters* 144: 1344–1353.

Seffati, K., Honarvar, B., Esmaeili, H., and Esfandiari, N. (2019). Enhanced biodiesel production from chicken fat using CaO/CuFe2 nanocatalyst and its combination with diesel to improve fuel properties. *Fuel* 235: 1238–1244.

Sharma, Y.C. and Singh, B. (2009). Development of biodiesel: current scenario. *Renewable and Sustainable Energy Reviews* 13: 1646–1651.

Sirisomboonchai, S., Abuduwayiti, M., Guan, G., and Samart, C. (2015). Biodiesel production from waste cooking oil using calcined scallop shell as catalyst. *Energy Conversion and Management* 95: 242–247.

Siva, S. and Marimuthu, C. (2015). Production of biodiesel by transesterification of algae oil with an assistance of Nano-CaO catalyst derived from egg shell. *International Journal of ChemTech Research* 7: 2112–2116.

Sivaprakash, G., Mohanrasu, K., Ravindran, B. et al. (2020). Integrated approach: Al2O3-CaO nanocatalytic biodiesel production and antibacterial potential silver nanoparticle synthesis from *Pedalium murex* extract. *Journal of King Saud University-Science* 32: 1503–1509.

Suryaputra, W., Winata, I., Indraswati, N., and Ismadji, S. (2013). Waste capiz (*Amusium cristatum*) shell as a new heterogeneous catalyst for biodiesel production. *Renewable Energy* 50: 795–799.

Tahvildari, K., Anaraki, Y.N., Fazaeli, R. et al. (2015). The study of CaO and MgO heterogenic nano-catalyst coupling on transesterification reaction efficacy in the production of biodiesel from recycled cooking oil. *Journal of Environmental Health Science and Engineering* 13: 73.

Tan, Y.H., Abdullah, M.O., and Nolasco-Hipolito, C. (2015). The potential of wsate cooking oil-based biodiesel using heterogenous catalyst derived from various calcined eggshell coupled with an reduction and engine performance. *Renewable and Sustainable Energy Reviews* 47: 589–603.

Tan, Y.H., Abdullah, M.O., Nolasco-Hipolito, C., and Ahmad Zauzi, N.S. (2017). Application of RSM and Taguchi methods for optimizing the transesterification of waste cooking oil catalyzed by solid ostrich and chicken-eggshell derived CaO. *Renewable Energy* 114: 437–447.

Tang, Z.X., Claveau, D., Corcuff, R. et al. (2008). Preparation of nano-CaO using thermal-decomposition method. *Materials Letters* 62: 2096–2098.

Tang, S., Wang, L., Zhang, Y. et al. (2012). Study on preparation of Ca/Al/Fe3O4 magnetic composite solid catalyst and its application in biodiesel transesterification. *Fuel Processing Technology* 95: 84–89.

Taufiq-yap, Y.H., Lee, H.V., Yunus, R., and Juan, J.C. (2011). Transesterification of non-edible jatropha curcas oil to biodiesel using binary Ca–Mg mixed oxide catalyst: effect of stoichiometric composition. *Chemical Engineering Journal* 178: 342–347.

Thangaraj, B., Solomon, P.R., Muniyandi, B. et al. (2019). Catalysis in biodiesel production – a review. *Clean Energy* 3: 2–23.

Thitsartarn, W. and Kawi, S. (2011). An active and stable CaO–CeO2 catalyst for transesterification of oil to biodiesel. *Green Chemistry* 13: 3423–3430.

Umdu, E.S., Tuncer, M., and Seker, E. (2009). Transesterification of *Nannochloropsis oculata* microalga's lipid to biodiesel on Al2O3 supported CaO and MgO catalysts. *Bioresource Technology* 100: 2828–2831.

Van Gerpen, J. (2005). Biodiesel processing and production. *Fuel Processing Technology* 86: 1097–1107.

Venkat-Reddy, C.R., Oshel, R., and Verkade, J.G. (2006). Room-temperature conversion of soybean oil and poultry fat to biodiesel catalyzed by nanocrystalline calcium oxides. *Energy and Fuels* 20: 1310–1314.

Viola, E., Blasi, A., Valerio, V. et al. (2012). Biodiesel from fried vegetable oils via transesterification by heterogeneous catalysis. *Catalysis Today* 179: 185–190.

Viriya-empikul, N., Krasae, P., Puttasawat, B. et al. (2010). Waste shells of mollusk and egg as biodiesel production catalysts. *Bioresource Technology* 101: 3765–3767.

Viriya-empikul, N., Krasae, P., Nualpaeng, W. et al. (2012). Biodiesel production over Ca-based solid catalysts derived from industrial wastes. *Fuel* 92: 239–244.

Wang, Y., Hu, S.Y., Guan, Y.P. et al. (2009). Preparation of mesoporous nanosized KF/CaO–MgO catalyst and its application for biodiesel production by transesterification. *Catalysis Letters* 131: 574–578.

Wen, L., Wang, Y., Lu, D. et al. (2010). Preparation of KF/CaO nanocatalyst and its application in biodiesel production from Chinese tallow seed oil. *Fuel* 89: 2267–2271.

Xie, W. and Huang, X. (2006). Synthesis of biodiesel from soybean oil using heterogeneous KF/ZnO catalyst. *Catalysis Letters* 107: 53–59.

Xie, W. and Wang, T. (2013). Biodiesel production from soybean oil transesterification using tin oxide-supported WO3 catalysts. *Fuel Processing Technology* 109: 150–155.

Yoosuk, B., Udomsap, P., Puttasawat, B., and Krasae, P. (2010). Improving transesterification acitvity of CaO with hydration technique. *Bioresource Technology* 101: 3784–3786.

Yu, X., Zhao, P., He, C. et al. (2012). Isolation of a novel strain of *Monoraphidium* sp. and characterization of its potential application as biodiesel feedstock. *Bioresource Technology* 121: 256–262.

Yuan, S., Periasamy, L.A.P., Ming, C. et al. (2020). Biodiesel synthesis using natural solid catalyst derived from biomass waste – a review. *Journal of Industrial and Engineering Chemistry* 81: 41–60.

Zhang, S., Zu, Y., Fu, Y. et al. (2010). Rapid microwave-assisted transesterification of yellow horn oil to biodiesel using a heteropolyacid solid catalyst. *Bioresource Technology* 101: 931–936.

Zhang, P., Han, Q., Fan, M., and Jiang, P. (2014). Magnetic solid base catalyst CaO/CoFe2O4 for biodiesel production: influence of basicity and wettability of the catalyst in catalytic performance. *Applied Surface Science* 317: 1125–1130.

Zhao, L., Qiu, Z., and Stagg-williams, S.M. (2013). Transesterification of canola oil catalyzed by nanopowder calcium oxide. *Fuel Process Technolology* 114: 154–162.

Zu, Y., Tang, J., Zhu, W. et al. (2011). Graphite oxide-supported CaO catalysts for transesterification of soybean oil with methanol. *Bioresource Technology* 102: 8939–8944.

5

Titanium Dioxide-Based Nanocatalysts in Biodiesel Production

Elijah Olawale Ajala, Mary Adejoke Ajala and Harvis Bamidele Saka

Department of Chemical Engineering, University of Ilorin, Ilorin, Kwara State, Nigeria

5.1 Introduction

Titanium dioxide bulk particles (TiO_2 BPs) are white solid inorganic materials that have found application in a variety of industries, including plastics, cosmetics, automobile and building coatings, textiles, pharmaceuticals and food (Akinnawo 2019). They are widely used and well researched due to their appealing characteristics, including stability, biocompatibility, non-toxicity, electrical and optical properties (Nyamukamba et al. 2016). The high physical and chemical stability as well as the high refractive index of TiO_2 make it widely sought after for in-depth research. Hence, it has been reported for use in several fields, including sensors, solar cells, air cleaning, photocatalysts, wastewater treatment and self-cleaning applications. It had been modified with other relevant materials and has played significant roles in a number of processes due to its high conductivity and stability in alkaline and acidic media (Bagheri et al. 2014).

The TiO_2 BPs exist in the crystalline forms of brookite, rutile, and anatase (Dubey et al. 2019). In mineral ores, they additionally exist as ilmenite, leucoxene, perovskite and sphene. The Ilmenite, leucoxene and rutile have particular economic importance (Sarapää et al. 2015). Among the three crystalline forms, anatase and rutile phases of TiO_2 are most in demand for industrial applications (Dubey et al. 2019). To obtain rutile, both anatase and brookite can be modified through thermal treatment at temperatures between 450 and 1200°C. The rutile form is the most stable of all the phases at all temperatures and pressures below 60 kbar. It is worthy of note that the transformation temperature to rutile is dependent on the preparation conditions, nature and structure of precursor (Kumar et al. 1993; Nyamukamba et al. 2016). These crystalline phase transformations of TiO_2 BPs are considered of utmost importance due to their capability to modify the properties and efficiency of the catalyst (Oi et al. 2016).

Due to these characteristics, TiO_2 is a qualified substance for the development of photocatalysts, catalyst supports and heterogeneous catalysts. Other suitable characteristics for these purposes include its long-term photostability, high effectiveness in mineralizing toxic non-biodegradable environmental contaminants and stability in acidic and oxidative environments (Bagheri et al. 2014). The TiO_2 BPs can be developed into various TiO_2-based

Nano- and Biocatalysts for Biodiesel Production, First Edition. Edited by Avinash P. Ingle.

catalysts for wide application. These catalysts are divided into organic, inorganic, mixed and complex types. The organic and inorganic compounds of TiO_2-based catalysts are components of the complex catalysts used for esterification and transesterification reactions (Carlucci et al. 2019). This is due to the ability of TiO_2 BPs to act as alternative materials in the synthesis of solid-base and solid-acid (heterogeneous) catalysts, given their high surface area, which stabilizes catalysts in its mesoporous structure (Carlucci et al. 2019). The TiO_2 BPs used for the synthesis of heterogeneous catalysts have been proven to improve catalyst performance and modulate catalyst activities for many reactions, including dehydrogenation, hydrodesulfurisation, water gas shift and thermal catalytic decomposition. However, there are some challenges associated with the use of TiO_2 BPs as heterogeneous catalysts, including their small specific surface area, low quantum efficiency and low adsorption ability. Other drawbacks are the cost of synthesis, challenges in the separation of catalyst from the reaction medium and inadequacy for continuous processing in large-scale production (Bagheri et al. 2014). The necessary improvements in the synthesis of TiO_2-based catalysts are offered by the use of nanotechnology.

The nanotechnology of different materials, including metallic materials, has garnered significant attention in all aspects of science and engineering. Most of the time, metallic materials are synthesized into nanoparticles by physical and chemical methods. However, these methods have some drawbacks, including toxicity, high cost and energy demands. These have motivated researchers to consider the novel approach to the development of nanostructured materials which is green synthesis (Nadeem et al. 2018). A nanostructure is beneficial to the catalyst development and the technology improves activity, lifetime and resistance to poisoning and offers other novel abilities. Such improvements and novelties cannot always be attained when catalysts are synthesized by other methods (Hussain et al. 2011). This development has created great interest among scientific researchers, as nanotechnology is fascinating and unique, offering characteristics such as a high surface area to volume ratio (Chenari et al. 2016). It has assisted in the modification of nanostructures of metallic compounds such as TiO_2 BPs into spheroidal nanocrystallites, nanoparticles, nanorods, nanowires, nanotubes, nanosheets and nanofibers. These various shapes and structures have found numerous applications, especially in photocatalysis and photovoltaics (Zhang et al. 2018).

Madhuvilakku and Piraman (2013) highlight the benefits of nanocatalysts in the production of biodiesel. They combine the benefits of the two basic classes of catalyst, homogenous and heterogeneous, by eliminating their respective challenges. They can easily be separated from product mixtures and reused, under simple operational procedures; can lead to less pollution. They have emerged with improved characteristics suitable for water purification, fuel cell development, energy storage, medicine and biodiesel production. However, factors such as catalyst size, surface composition, catalyst geometry, aggregation attitude, surface functionality and the physical and chemical environment affect catalytic performance (Akubude et al. 2019). These challenges can be overcome through the use of appropriate materials such as TiO_2 BPs in nanocatalyst synthesis.

Titanium dioxide nanoparticles (TiO_2 NPs) specifically are of research interest as they are among the most widely used nanocatalysts in photocatalysis, heterogeneous catalysis and catalyst support. This is due to their desirable properties such as acidity, low toxicity, high surface area, low cost, high availability, high reactivity, good redox selectivity, chemical stability and high reusability (Gardy et al. 2017). The TiO_2 NPs can combine with catalyst supports to enhance catalytic activity and optimize selectivity in the desired applications (Oi et al. 2016). Modification of TiO_2 NPs by doping with metals and non-metals leads to novel properties as the dopants improve the physical and optoelectronic properties of the NPs, such as in carbon-modified TiO_2 NPs and scandium-modified TiO_2 powders (Chenari et al. 2016). The TiO_2 NPs-modified catalysts have high activity for various reduction and oxidation reactions at low pressures and temperatures and can be suitable for biodiesel production. Hence, nanosynthesis is a novel technology that can be employed to synthesize TiO_2-based heterogeneous catalysts for use in biodiesel production. Nanotechnology and nanoscience improve these catalysts by controlling the pore structure and modifying the support (Bagheri et al. 2014).

Therefore, the modified TiO_2 NPs thus have excellent properties and easy recoverability in the production of biodiesel. However, a wide range of synthetic routes need to be followed in developing surface-modified NPs, in order to improve their surface acidity/basicity for biodiesel production. These catalysts would be suitable for the production of biodiesel using cheap feedstock with high free fatty acids (FFAs) via simultaneous esterification and transesterification without soap formation (Gardy et al. 2019).

The main aim of this chapter is to discuss the emerging role of TiO_2 NPs in biodiesel production. Various other important aspects, such as the natural occurrence of titanium, precursors for synthesis of the TiO_2 NPs and methods of synthesis are also described.

5.2 Natural Occurrences of Titania

5.2.1 Rutile

Rutile is the second most abundant mineral of titania, containing about 98% TiO_2 in the ore. It exists as a tetrahedral crystal structure, usually in twins or rosettes crystals. The structure is made up of parallel chains of octahedrons, which are composed of a titanium ion surrounded by six oxygen atoms, as shown in Figure 5.1a. Finely powdered rutile is a shining

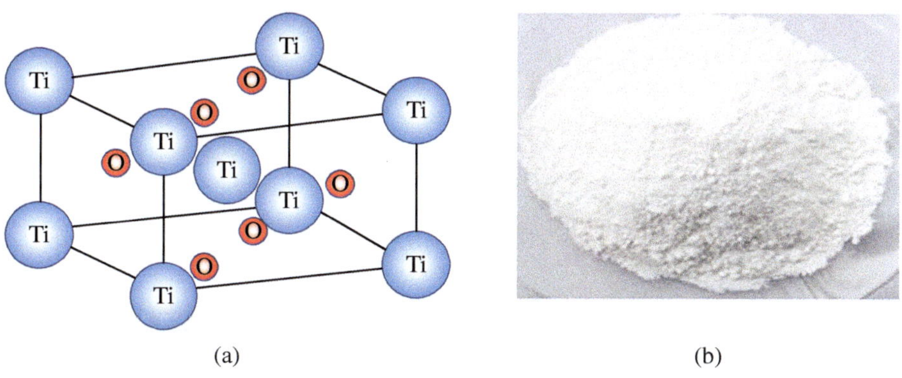

Figure 5.1 (a) Crystalline structure of rutile. (b) Fine powder of rutile.

white color, as shown in Figure 5.1b. Rutile naturally occurs in igneous and metamorphic rocks, usually in small quantities alongside other impurities such as iron oxide, vanadium oxide, and chromium oxide.

5.2.2 Anatase

Anatase is made up of edge- and corner-sharing octahedra shapes which outline a 3D structure as shown in Figure 5.2a (Scheinost 2004). It is a meta-stable mineral form of TiO_2, which in its natural forms exists mostly as a black solid as shown in Figure 5.2b. Sometimes, anatase appears white or colorless, especially in its pure form (Bordbar et al. 2017). These crystals are found in soils as a product of titanite ($CaTiSiO_5$) and Ti-bearing silicates through the process of weathering (Scheinost 2004).

5.2.3 Rhombic Brookite

Brookite is the ortho-rhombic form of TiO_2 that occurs in many natural polymorphic forms (minerals with the same composition but different structure), as shown in Figure 5.3a. Brookite is usually brown, sometimes yellowish or reddish-brown, as shown in Figure 5.3b. It is the most difficult titania phase to prepare in thin-film form. Brookite-containing films have been reported for vapor-deposited films (Moret et al. 2000).

5.2.4 Kaolin Clays

Kaolin clays are generally known as aluminum silicate ($(OH)_8Si_4Al_4O_{10}$) and are composed of 46.54% SiO_2, 39.5% Al_2O_3 and 13.96% H_2O (Yang and Hwang 2013). Kaolinites formed by hydrothermal weathering of feldspar and the muscovite process in granitic and pegmatitic rocks have been observed to have a low titanium composition, while kaolin

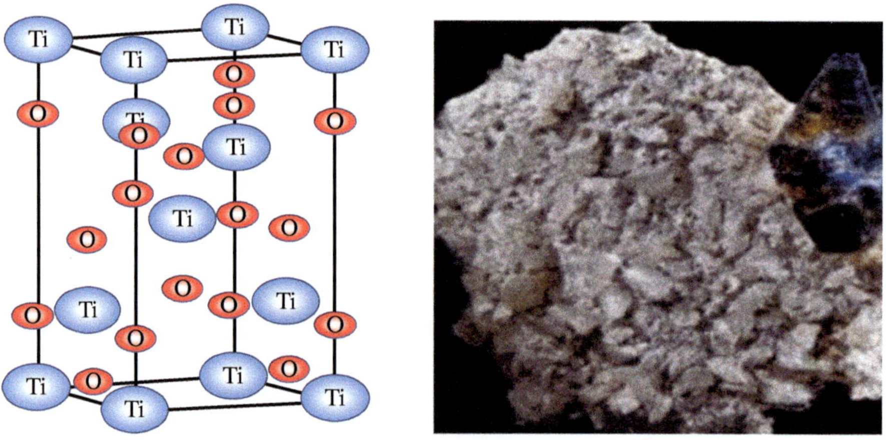

(a) (b)

Figure 5.2 (a) Crystalline structure anatase. (b) Image of anatase.

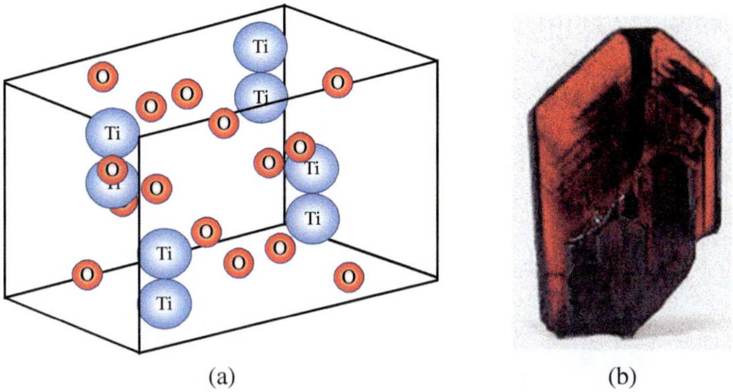

Figure 5.3 (a) Crystalline structure of rhombic brookite. (b) Image of rhombic brookite.

deposits derived from biotite-rich granites, granite gneiss and schists have a relatively high one (Weaver 1976). Sedimentary kaolins contain 1.3–3.5% of TiO_2, unlike other types of kaolin which have just 0.5%. The TiO_2 in sedimentary kaolin has been identified as the anatase phase, while other minerals such as leucoxene and brookite are detected in small quantities (Maynard et al. 1969). Most of the titanium in kaolinites is present in the anatase phase with little amounts of rutile phase. The rutile-phased titanium is mostly inherited, while the anatase phase is formed authigenically through the chemical weathering of the titanium-containing minerals. Titanium aggregates that are not attracted by magnets have been reported to have relatively high kaolinite contents (Weaver 1976). The structure of kaolinite is shown in Figure 5.4a, while its image is depicted in Figure 5.4b.

To remove titanium mineral from kaolin clays, excessive amounts of dispersing agents are usually required in order to treat the clay and produce maximum deflocculation of the particles. In the second stage of the process, reflocculation takes pace, which makes the kaoline particle attain a high degree of suspension and great stability. The titanium mineral particles, mostly in anatase phase, are thus liberated and can be separated by sedimentation (Maynard et al. 1969).

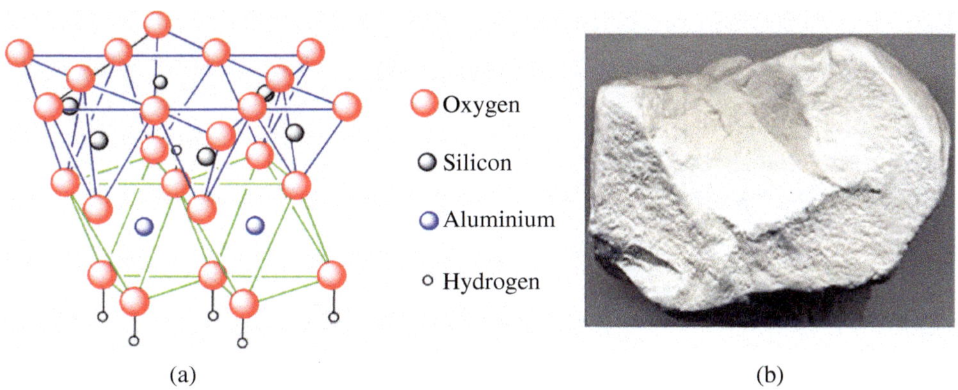

Figure 5.4 (a) Structure of kaolinite clay. (b) Image of clay.

Figure 5.5 (a) Image of ilmenites. (b) Structure of ilmenites.

5.2.5 Ilmenites or Manaccanite

Ilmenite ($FeTiO_3$) has been used as a source of about 92% of all titanium minerals consumption globally (Woodruff et al. 1802). This is because titanium is considered the most abundant among its dark minerals (Weaver 1976). Therefore, ilmenite can serve as a precursor to the development of TiO_2 as it is mostly found in large quantities, deposited as sedimentary heavy mineral concentrates in coastal regions or on ocean beaches. In smaller quantities, it is located in alluvial environments such as riverbeds and deltas (Sarapää et al. 2015). The ilmenite forms as a major mineral in igneous rocks arranged and concentrated into layers by a route called magmatic segregation. It is a metallic to submetallic mineral that is majorly iron black, as shown in Figure 5.5a. It has a similar isostructure to hematite, with one-half of the octahedral center occupied by Fe2þ and the other half by Ti4þ, as shown in Figure 5.5b. In soils, it is relatively unstable due to easy weathering, producing pseudorutile and mixtures of rutile, anatase and Fe oxides (Scheinost 2004).

The production of TiO_2 from ilmenite ore involves mixing with sulfuric acid to eliminate the iron oxide group. The iron sulfate formed as a byproduct is then crystallized and removed by filtration to produce the titanium salt in the digestion solution. This then undergoes some other processes to yield TiO_2 (Speight 2017). However, the process is challenging as the recovery of TiO_2 from an ilmenite mixture of magnetite grains is not economically viable.

5.3 Precursors Used for the Synthesis of TiO_2 NPs

Both artificial and natural sources serve as precursors for the production of TiO_2 in the environment. Artificial sources include titanium tetrachloride, titanium isopropoxide and titanium butoxide, among many others. Natural sources are found mostly in rutile (TiO_2), river water, kaolin clays and ilmenite ($FeTiO_3$), as described in the previous section. Other natural sources are geikielite ($MgTiO_2$), pyrophanite ($MnTiO_2$), anatase, brookite and akaogiite. These compounds are mostly found in metamorphic, platonic rocks and beach sands (Bordbar et al. 2017).

5.3.1 Titanium Tetrachloride

Titanium tetrachloride (tetrachlorotitanium or titanium chloride) is an inorganic compound with the chemical formula $TiCl_4$ and molecular formula 189.679 g/mol. It exhibits isotopes that are abundant in nature as shown in Figure 5.6. It is an imperative precursor in the production of TiO_2 NPs. It tends to show stunning opaque clouds of TiO_2 and hydrated hydrogen chloride when brought in contact with humid air. Its boiling point, density and melting point are 136.4°C, 1.73 g/cm^3 and −24°C respectively. It is a volatile liquid with a strong odor. Eqs. (5.1) and (5.2) illustrate the synthesis of TiO_2 through hydrolysis and oxidation of titanium tetrachloride (Kang and Okabe 2014).

$$TiCl_4 + 2H_2O \quad TiO_2 + 4HCl \text{ (Hydrolysis)} \tag{5.1}$$

$$TiCl_4 + O_2 \quad TiO_2 + 2Cl_2 \text{ (Oxidation)} \tag{5.2}$$

5.3.2 Titanium Tetraisopropoxide

Titanium tetraisopropoxide (TTIP) is a chemical compound with the chemical formula $Ti\{OCH(CH_3)_2\}_4$ and a molecular formula of 284.219 g/mol. Its appearance ranges from colorless to yellowish. Its boiling point, density and melting point are 232°C, 0.96 g/cm^3 and 17°C respectively. Eq. (5.3) illustrates the synthesis of TiO_2 through hydrolysis of titanium isopropoxide. This reaction is usually employed in the sol-gel synthesis technique to produce TiO_2-based materials in the form of thin films or powders (Ritala and Leskela 1993).

$$Ti\{OCH(CH_3)_2\}_4 + 2H_2O \quad TiO_2 + 4(CH_3)_2CHOH \text{ (Hydrolysis)} \tag{5.3}$$

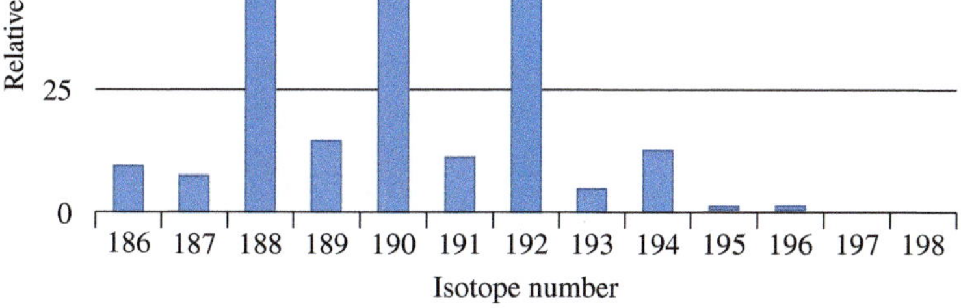

Figure 5.6 Schematic presentation of different isotopes of $TiCl_4$ and their relative abundance.

5.3.3 Titanium Butoxide

Titanium butoxide is a chemical compound with the chemical formula $Ti(OCH_2CH_2CH_2CH_3)_4$ and a molecular formula of 340.32164 g/mol. It possesses a weak alcohol-like odor and its boiling point, density and melting point are 312°C, 0.998 g/cm^3 and −55°C respectively. It is readily hydrolyzed to form TiO_2 or pyrolyzed to form TiO_2 as shown in Eqs. (2.4) and (2.5) (Dittert et al. 2019).

$$Ti(OCH_2CH_2CH_2CH_3)_4 + 2H_2O \ TiO_2 + 4HOCH_2CH_2CH_2CH_3 \ \text{(Hydrolysis)} \quad (5.4)$$

$$Ti(OCH_2CH_2CH_2CH_3)_4 + TiO_2 + (CH_2CH_2CH_2CH_3)_2O \ \text{(Pyrolysis)} \quad (5.5)$$

5.4 Methods for the Synthesis of TiO_2 NPs

Generally, NPs can be synthesized either by a top-down or a bottom-up approach. The synthesis of TiO_2 NPs can further be classified as physical, chemical and biological methods (Akinnawo 2019). These methods usually involve a complicated process, particular equipment and alkoxide precursors. The cost of raw materials and complex preparation procedure have constrained the use of TiO_2 NPs in large-scale industries (Zhang et al. 2018).

The low temperature of process plays a significant role in the synthesis of crystalline TiO_2 NPs, allowing a higher degree of control over the developed polymorph. It also influences the intra- or interparticle porosity. It is made possible through a solution-based approach in the absence of thermal treatment, with special attention paid to the polymorphs developed, crystal morphology, surface area and particle dimensions (Bagheri et al. 2014). The procedures for synthesis of TiO_2 NPs are further divided as:

(1) **Physical methods:** Including methods such as ball milling, laser ablation/photablation and sputtering.
(2) **Chemical methods:** Including microemulsion, precipitation, sol-gel, hydrothermal, solvothermal, electrochemical/deposition, sonochemical and direct oxidation methods.
(3) **Biological methods:** Green synthesis methods involving the use of microbes and enzymes.

5.4.1 Physical Methods

5.4.1.1 Ball Milling

Ball milling involves grinding nanotubes into extremely fine powders. The collision between the tiny rigid balls in a concealed container generates localized areas of high pressure (Jalalah et al. 2013). Usually, ceramic, flint and stainless steel are used. This method can be used alone or in combination with other methods. Farbod and Khademalrasool (2011) documented the synthesis of TiO_2 NPs via the combination of sol-gel and ball milling techniques. The TiO_2 was synthesized from tetra n-butyl titanate by the sol-gel method. After synthesis, a yellowish precipitate was obtained, which was dried and divided into two parts. The first part was ball milled with a lab-scale planetary mill for 30 minutes using a ball to powder mass ratio of 10 : 1, while the second part was not milled. It was shown that milling of dry gel is necessary to prepare nanosized particles (Jalalah et al. 2013).

5.4.1.2 Laser Ablation/Photoablation

Laser ablation or photoablation is a technique involving the removal of materials from the surface of a solid by irradiating it with a laser beam (Moret et al. 2000). The material is heated by the absorbed laser energy and sublimates at low laser flux. Singh et al. (2009) reported the synthesis of colloidal TiO_2 NPs by pulsed laser ablation (PLA) in water in the absence of surfactants. A slice of highly purified titanium rod was settled on the bottom of a glass vessel containing 30 ml double-distilled water and allowed to irradiate with a focused output of 1064 nm of pulsed Nd:YAG laser (Spectra-Physics, USA) operating at 35 mJ/pulse energy, 10 ns pulse width and 10 Hz frequency for 60 minutes. The vessel was manually translated in the X–Y plane to achieve homogeneous ablation of the target surface. Lambda 35 and LS 55 Perkin Elmer spectrophotometers were employed for the recording of UV-visible absorption and photoluminescence spectra respectively, of a colloidal solution of TiO_2 NPs. A similar case was also reported by Asadnajafi et al. (2019).

In another study, TiO_2 NPs were manufactured by PLA of a high purity titanium plate in de-ionized water (Singh et al. 2009). A titanium plate was placed at the bottom of an open glass cylindrical vessel filled with 80 ml of de-ionized water. Before the experiment, the titanium plate and containers were thoroughly cleaned in acetone, alcohol and de-ionized water. The NPs were manufactured using the fundamental beam of an Nd:YAG pulsed laser operating at a wavelength of 1064 nm. The laser pulse width was 7 ns, with 10 Hz repetition rate at 1, 2, and 3 J/cm^2 for samples a, b, and c respectively, in order to produce NPs of different sizes.

5.4.1.3 Sputtering

Sputtering is usually understood as sputter deposition. It is a physical means of synthesizing TiO_2 NPs in which atoms are transformed from the solid to the gas phase by bombardment with energetic ions from inert gas. Dreesen et al. (2009) documented the deposition of TiO_2 NPs on carbon-coated copper transmission electron microscopy grids by reactive DC magnetron sputtering with the substrate at ground potential. A 15 cm distance between the target and substrate was set and a temperature around 400°C was maintained throughout the deposition period. A base pressure of 10^{-4} N/m^2 was set in the plasma chamber using a 300 l/min turbopump (vessel size: 0.05 m^3). The titanium target was covered by an oxide layer as the oxygen flow was higher than a critical value above which there would be deposition of TiO_2 (Navío et al. 1998).

5.4.2 Chemical Methods

5.4.2.1 Microemulsion

A microemulsion is a thermodynamically stable system consisting of oil, water, amphiphilic surfactant and co-surfactants. These surfactants are used to stabilize small droplets of liquids or two immiscible liquids consisting of microdomains (Muhammad et al. 2019).

This scenario provides a micro-heterogeneous medium for the creation of NPs. The configuration of particles in such systems is restricted by the behavior of interdroplet exchange of reactants and by the reactant distribution in the droplets. The surfactant-stabilized microcavities provide a restrictive effect that helps immobilize the particles and also limit their nucleation, growth and agglomeration. Microemulsions of NPs occur in two forms: W/O (water in oil) and O/W (oil in water) which are similar to aqueous and nonaqueous mediums respectively. One of the advantages of this method is that the size of the particles can be varied based on the ratio of surfactant to water. However, there have been only a few reports of controlled TiO_2 NP synthesis by microemulsions for biodiesel production.

5.4.2.2 Precipitation

Physical transformation due to change of temperature, pH, solvent evaporation, reactant concentration or chemical process as a result of the addition of bases or acids or the use of complex-forming agents allows the formation of precipitate from a homogeneous liquid phase. Generally, the formation of a new solid phase in a liquid medium occurs by two elemental processes that happen either sequentially or simultaneously (Muhammad et al. 2019). These can include nucleation and growth or agglomeration of the particles. Nucleation tends to occur when the smallest elementary particles of the new phase are formed and stabilized under the precipitation conditions. The kinetics of nucleation and particle growth inside homogeneous solutions can be adjusted by the controlled release of the anions and cations. Careful control of the kinetics of the precipitation can result in monodisperse NPs (Bagheri et al. 2014). By controlling the factors that determine precipitation processes such as the pH and concentration of reactants/ions, nanocatalysts with narrow size distributions can be produced. For the development of TiO_2, co-precipitation involves precipitation of hydroxides by the addition of a solution like NaOH, NH_4OH or urea to a raw material. This is followed by thermal treatment such as calcination to crystallize TiO_2 nanocatalyst. Lower processing temperatures result in better metal dispersion and very minimal particle size (Muhammad et al. 2019).

5.4.2.3 Sol-Gel

Dubey et al. (2019) reported the use of the sol-gel method for the synthesis of TiO_2 NPs using titanium isopropoxide (TTIP, $Ti[OCH(CH_3)_2]_4$) as a precursor. First, 150 ml ethanol was added to 10 ml of de-ionized water in a beaker and mixed under constant agitation. Then, 9 ml of TTIP was added to the mixture and combined at the maintained agitation speed and a constant temperature of 85°C (via paraffin oil bath) for 4 hours. The gel formed was oven-dried at 60°C and calcined at 400°C for 3 hours. This method has become widely used for the synthesis of TiO_2 NPs due to its ease of operation, scalability of production and moderate reaction conditions.

Sol-gel is based on inorganic polymerization reactions. In the sol-gel process, inorganic metal salts or metal-organic compounds (monomers) such as metal alkoxides are usually transformed into a colloidal solution or sol, that acts as the precursor for an integrated network. Generally, this process passes through four major steps: hydrolysis, polycondensation, drying and thermal decomposition (Muhammad et al. 2019). Hydrolysis of the precursors of the metal or non-metal alkoxides takes place with water or alcohols. It has also been reported that acid or base helps in the hydrolysis of the precursor. After condensation of the solution to a gel, the solvent is removed. Both drying and thermal decomposition at higher temperature are needed to decompose the organic precursor. After these steps, the materials begins to solidify. The size of the solid particles can be adjusted or tuned by varying the solution composition, pH and temperature during the reaction. The sol-gel method has exceptional potential through glycosylated precursor to make TiO_2 a prime candidate for heterogeneous catalyst support (Bagheri et al. 2014).

5.4.2.4 Hydrothermal

Hydrothermal synthesis involves water, both as a catalyst and occasionally as a component of solid phases at an elevated temperature greater than 100°C. It is an attractive technique for producing monodispersed and highly homogeneous titanium nanocatalysts (Muhammad et al. 2019). Detailed research has been conducted on the mild hydrothermal synthesis of TiO_2 particles and the influence of various independent variables such as experimental duration, temperature, solvent type, pressure (percentage fill), pH and the starting charge on the resultant product. A small autoclave of the Morey type provided with Teflon liners is usually used in the synthesis of TiO_2 NPs. The conditions selected are pressure <100 bars and temperature ≤200°C. Employing Teflon liners helps in achieving pure and homogeneous TiO_2 NPs. Solutions of HNO_3, KOH, NaOH, HCl, H_2SO_4 and HCOOH were treated so that they would mineralize and it was found that HNO_3 was better for obtaining monodispersed TiO_2 NPs with homogeneous composition under the previously stated experimental conditions. This might not hold if the experimental conditions were adjusted.

5.4.2.5 Solvothermal

This technique is similar to the hydrothermal method except that organic solvents are employed in place of water. It explains how under supercritical conditions, organic solvents such as methanol, 1,4 butanol and toluene are transformed into a chemical reaction. In most cases, successive thermal treatment is essential to crystallizing the final material. The crystalline phase, surface chemistry and particle morphology of the solvothermal-derived TiO_2 can be easily tuned by controlling the reaction temperature, pressure, composition of precursor, solvent property and aging time. The solvothermal process allows the product to be free from alien anions because the organic solution, having a low relative permittivity, is free from ionic species. However, this is not the case in the hydrothermal technique (Muhammad et al. 2019). Also, solvothermal processing offers a superb chemical homogeneity and the likelihood of deriving distinctive stable structures at low reaction temperatures. The TiO_2 nanocatalyst prepared via this technique may exhibit a better photocatalytic activity than those of commercial TiO_2 products.

5.4.2.6 Electrochemical/Deposition

Electrochemical synthesis offers an all-round, low-temperature approach to the fabrication of advanced thin films such as epitaxial, superlattice, quantum dot and nonporous films. Varying the electrolysis parameters, such as potential, current density, temperature and pH, can easily control the attributes and qualities of the films produced. Both anodic and cathodic approaches may be utilized in depositing the TiO_2 nanocatalyst. However, only the anodic electro-synthesis variant has emerged for application purposes (Gupta and Tripathi 2012).

5.4.2.7 Sonochemical

The sonochemical approach has been utilized in the synthesis of various nanostructured catalytic materials, including colloids, oxides, alloys, carbides and high-surface-area transition metals. In the sonochemical process, ultrasonic irradiations ensure the principle of acoustic cavitation, that is, rapid formation, growth and disintegration of unstable bubbles in liquids. These conditions elevate the local pressure and temperature up to 20 MPa and 5000 K, with a cooling rate of 1010 Ks^{-1} (Suslick 1986).

5.4.2.8 Direct Oxidation

Direct oxidation of titanium foils by hydrogen peroxide (H_2O_2) can easily produce TiO_2 NPs with a rod-like array at a low temperature of 80°C. More importantly, the preparation process is simple to manage as reported by Song et al. (2014). The titanium foil is first degreased by sonication through the utilization of sound energy to agitate particles in acetone, 2-propanol and ethanol. It is subsequently rinsed with ultrapure water and dried in a nitrogen stream. Song et al. (2014) immersed their foil in a 30 ml H_2O_2 (30 wt.%) solution and kept the mixture at 80°C for 72 hours. The sample was then mixed and rinsed with plenty of ultrapure water before undergoing thermal treatment in a muffle furnace at a temperature of between 300 and 500°C. Oxygen-containing organic molecules, such as ethanol and acetaldehyde can also be used as sources of oxygen to oxidize titanium foils at high temperatures in order to generate TiO_2 NPs.

5.4.3 Biological Methods

The many adverse effects associated with the chemical and physical methods of NP synthesis heralded the advent of the biological method, including (i) handling of hazardous chemicals, (ii) low conversion/yield of material, (iii) high cost and energy consumption and (iv) difficulty and wastefulness of purification. The biological approach involves the use of plant/leaf extracts, microorganisms and enzymes to synthesize nanoparticles as shown in Figure 5.7. These agents of biosynthesis have been adjudged a simple and viable alternative to the use of chemical and physical procedures (Abu-dalo et al. 2019). Thus, green chemistry of NP synthesis mitigates or eliminates the use and generation of hazardous substances in the design, manufacture and application of chemical products (Hussain et al. 2016). This method is increasingly attractive as it establishes that many organisms can manufacture inorganic materials either intra- or extracellularly.

5.4.3.1 Green Synthesis Using Plant Extracts

The green synthesis of NPs using plant extracts has attracted the attention of researchers due to its simplicity and viability. Compared with chemical and physical methods, green synthesis is an environmentally benign alternative (Huang et al. 2007). Even among the biological methods, utilization of plant extracts for synthesis appears the best means of developing NPs (Abu-dalo et al. 2019). Plant extracts contain many biomolecules such as proteins, polysaccharides, amino acids, organic acids, vitamins and phytochemicals (Hossain et al. 2019). The phytochemicals, which are novel secondary metabolites are polyphenols, flavonoids, terpenoids, alkaloids, tannins and alcoholic compounds. These act as reducing and stabilizing agents in the green synthesis of nanoparticles (Madkour 2017).

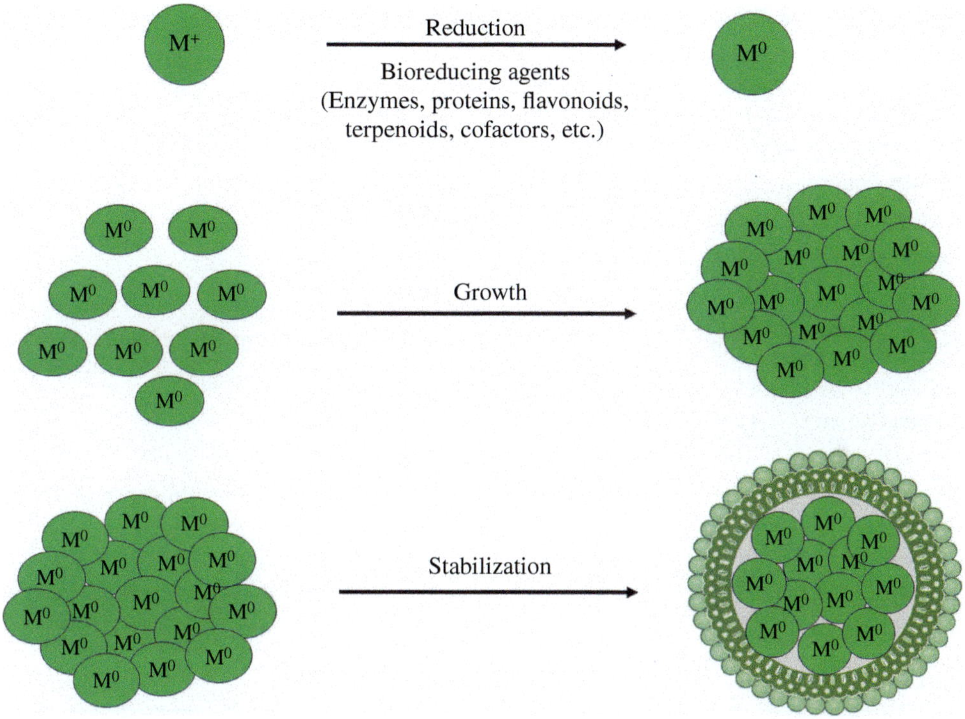

Figure 5.7 Schematic illustration of biological synthesis of NPs. (Source: Adapted and modified from Madkour 2017; an open access article).

The green synthesis of NPs has been reported to be easy, reliable, safe and scalable to industrial applications (Nadeem et al. 2018). It avoids or reduces overdependency on toxic chemicals, simple and cost-effective. It further connects nanotechnology with plants (Abu-dalo et al. 2019).

Chatterjee et al. (2016) synthesized TiO_2 NPs using a green synthesis approach involving green gram seed extract. The 20 ml extract was mixed with 80 ml of 1 mM TiO_2 solution and stirred for 24 hours at room temperature. The resulting mixture was filtered, the residue was dried and calcined to obtain TiO_2 NPs. Rajakumar et al. (2012) reported the use of *Eclipta prostrata* leaf extract to prepare TiO_2 NPs with a particle size in the range of 36–38 nm. The TiO_2 NPs produced were spherical and easily synthesized at room temperature using titanium hydroxide $TiO(OH)_2$ solution as a precursor. Sankar et al. (2014) also synthesized TiO_2 NPs of 124 nm particle size using *Azadirachta indica* leaf extract. *A. indica* leaf extract was used to develop TiO_2 NPs by adding 10 ml of it to 90 ml of 5 mM aqueous TiO_2. The TiO_2 NPs formed were purified by repeated centrifugation at 7826 × g for 20 minutes using double-distilled water. The pellet obtained was dispersed in de-ionized water and heated at 60 °C for 1 hour to obtain TiO_2 NPs (Thakur et al. 2019). *Echinacea purpurea* Herba has also been reported for the synthesis of TiO_2 NPs (Dobrucka 2017). Patidar and Jain (2017) identified a list of plant extracts used for the synthesis of TiO_2 NPs, which included *A. indica, Camellia sinensis, Nyctanthes arbor-tristis, Coriandrum, Nelumbo nucifera* and *Ocimum sanctum*. These extracts are compatible with green

Table 5.1 Different plants, their parts and precursors for the synthesis of TiO_2 NPs.

Plants	Part	TiO_2 precursor	Size (nm)	Sources
Calotropis gigantea	Flower	$TiO(OH)_2$	10.52	Marimuthu et al. (2013)
Psidium guajava	Leaf	$TiO(OH)_2$	32.58	Santhoshkumar et al. (2014)
Azadirachta indica	Leaf	TiO_2 BPs	25–87	Thakur et al. (2019)
Solanum xanthocarpum berry	Fruit	$Ti(OCH(CH_3)_2)_4$	50–100	Muthukumar et al. (2017)
Citrus limon	Fruit	TiO_2BPs	20–200	Hossain et al. (2019)
Syzygium cumini	Leaf	TTIP	22	Sethy et al. (2020)
Jatropha curcas	Leaf	$TiCl_4$	10–20	Goutam et al. (2018)
Moringa oleifera	Leaf	TTIP	12.22	Patidar and Jain (2017)
Glycosmis cochinchinensis	Leaf	$TiO(OH)_2$	40	Rosi and Kalyanasundaram (2018)
Curcuma longa	Plant	TiO_2 BPs	76.36	Abdul Jalill et al. (2016)
Trigonella foenum graecum	Leaf	$TiOSO_4$	20–90	Subhapriya and Gomathipriya (2018)
Hibiscus rosa sinensis	Flower	TTIP	43.3	Rao et al. (2014)

chemistry principles, which give them their phytochemical properties. These and many other plant extracts have been reported in the literature in relation to the development of TiO_2 NPs. The NPs were found to be efficient and effective in the synthesis of TiO_2-based nanocatalysts for various applications. Other plants utilized for the synthesis of TiO_2 NPs are listed in Table 5.1, along with the TiO_2 precursors used and the performance of each with respect to the nanosize achieved.

5.4.3.2 Microbial Synthesis

Biological systems such as prokaryotic and eukaryotic microorganisms have been exploited for the synthesis of different metal NPs, particularly transition metals such as Au, Ag, Cd, Pt, Zn and Fe, under operating conditions of mild temperature and pressure (Palomo 2019). Microorganisms have the ability to produce NPs from a precursor both intracellularly and extracellularly. The cell-free extract of microbes obtained after incubation for one to two days in a rotary shaker under optimal operating conditions of pH, temperature and medium composition is usually used for extracellular synthesis. After proper shaking of the culture medium, biomass is removed by centrifugation and the cell-free filtrate (supernatant) is recovered to synthesize the NPs. Intracellular synthesis of NPs, on the other hand, is carried out after culturing the microorganism for a certain optimum growth period and then centrifuging it to obtain biomass. The biomass is then thoroughly washed using sterile water and dissolved in sterile water that contains a filter-sterilized solution of metal salt. At the end of the incubation period, the biomass is removed by repeated cycles of ultrasonication, washing and centrifugation. It can then be used to disintegrate the cell wall and release any NPs produced (Singh et al. 2016). This method, therefore, employs microorganisms to metabolize a precursor of a metallic salt such as titanium chloride to TiO_2 NPs.

Different microorganisms have been exploited for the synthesis of TiO_2 NPs. Those synthesized by microbes have been found to have various shapes and sizes, which enhance their functionality as good catalysts. A major challenge with the use of microorganisms in the synthesis of NPs concerns the deployment of pathogenic strains. These strains are disease-causing microorganism that can infect their handler. The application of non-pathogenic strains of microorganisms would eliminate the risk in this process, making it suitable for commercial purposes (Nadeem et al. 2018). Significant polydispersity is usually observed in microorganism-derived NPs, since the size of the mineral particles is dependent on cell growth, the composition of the incubation solution and the growth conditions of the minerals (Salman et al. 2014). However, mono-dispersed products can be produced by proper regulation of synthetic parameters such as concentration, pH and temperature. Furthermore, microorganisms can accumulate and detoxify heavy metals via various reductase enzymes. These can reduce metal salts to metal NPs with a narrow size distribution.

Bacterial metabolites have also been employed to develop TiO_2 NPs by bio-reduction and stabilization. *Bacillus mycoides* was used for the synthesis of spherical TiO_2 NPs in a size range of 40–60 nm. *Aeromonas hydrophila* was reported to produce TiO_2 NPs of 28–54 nm size (Nadeem et al. 2018). Feyel et al. (2006) documented the use of *Lactobacillus* sp. as a green and cost-effective method for the biosynthesis of TiO_2 NPs, which had an average size of around 30 nm. *Staphylococcus aureus* was reported to have the ability to synthesize TiO_2 NPs. Moreover, it was also reported that bacterium *Lactobacillus* can bio-synthesize TiO_2 NPs through the combined action of oxido-reductases and glucose at mild pH. It is worthy of note that the TiO_2 NPs synthesized from bacterial system have excellent poly-dispersity and stability compared to NPs produced from plant extracts and fungal systems. The bacterial system also has the advantages of easy handling and hereditary manipulation (Malarkodi et al. 2013). However, pathogenicity and high energy consumption may hamper its commercialization (Nadeem et al. 2018).

The utilization of fungi in the synthesis of metallic NPs has gained wide acceptance due to its numerous superior advantages over bacterial systems. Fungi have been adjudged to be applicable at commercial scale and to be economically viable. The TiO_2 NPs developed by fungi were found to have various shapes and sizes, due to the intrinsic ability of the fungi to reduce bulk salt to an elemental or ionic state via either enzymes or metabolites. To date, several fungi have been reported to synthesize a variety of nanoparticles, including TiO_2 NPs (e.g. *Aspergillus flavus*). In addition, extracts of yeast (e.g. *Saccharomyces cerevisiae*) were responsible for the synthesis of TiO_2 NPs of 12.6 nm size (Nadeem et al. 2018). Table 5.2 shows various microbial systems and precursors used for the synthesis of TiO_2 NPs.

5.4.3.3 Enzyme-Mediated Synthesis

Enzymes have been successfully applied in the synthesis of different types of NPs including TiO_2 NPs (Palomo 2019). The utilization of enzymes for the preparation of NPs is considered a novel approach with growing interest due to its simple and eco-friendly nature. However, selection of suitable enzymes with the right size and shape is extremely important (Palomo and Filice 2016). Enzymes obtained from different sources such as bacteria, yeast and fungi, have been successfully used in the preparation of NPs

Table 5.2 Types of microbes and TiO_2 precursors used to achieve the nanosize of TiO_2 NPs.

Microorganism	TiO_2 precursor	Nanosize (nm)	Sources
Staphylococcus aureus	TTIP	20	Landage et al. (2020)
Halomonas elongate	$TiO(OH)_2$	104	Taran et al. (2018)
Lactobacillus sp.	$TiO(OH)_2$	10–70	Jha and Prasad (2010)
Aeromonas hydrophila	$TiO(OH)_2$	40.50	Jayaseelan et al. (2013)
Bacillus mycoides	TTIP	40–60	Órdenes-Aenishanslins et al. (2014)
Lactobacillus johnsonii	TiO_2BPs	4–9	Al-Zahrani et al. (2018)
Planomicrobium sp.	TiO_2BPs	8.89	Malarkodi et al. (2013)
Aspergillus flavus	TiO_2BPs	62–74	Rajakumar et al. (2012)

(Palomo 2019). In the enzymatic synthesis of TiO_2 NPs, the hydrogen ion concentration present in the reaction medium plays a significant role in selecting the size of the resulting NPs. Therefore, the pH and hydrogen ion concentration of the medium used for TiO_2 NP synthesis decide the catalytic efficiency of the enzyme (Raliya and Tarafdar 2014). Enzymatic synthesis does not required any additional external reducing agent, because the enzyme has the reducing ability to actively develop NPs that remain catalytically active at the end of the synthesis (Palomo and Filice 2016). Moreover, some reports propose the synthesis of TiO_2 NPs using different enzymes (Sumerel et al. 2003; Palomo 2019).

5.5 Methods for the Synthesis of TiO_2-Based Nanocatalysts

Wet and dry impregnation are commonly used to develop TiO_2-based nanocatalysts. This is because the chemical compound to be used is usually functionalized with TiO_2 NPs to achieve a synergetic effect on the chemical and physical properties of the nanocatalyst. However, these methods may be employed along with techniques such as sol-gel and precipitation. The major difference between the wet and the dry impregnation methods is in the volume of solution used to contact a solid phase or support (Deraz 2018).

5.5.1 Wet Impregnation

Wet impregnation requires an excess volume of the solution during operation. Afterward, the solid catalyst is carefully separated from the solution and dried until it reaches a constant weight. Factors such as temperature and concentration of precursor usually affect the efficiency and quality of the TiO_2 nanocatalyst developed (Deraz 2018). Alsharifi et al. (2017) employed wet impregnation to develop a TiO_2-Li nanocatalyst for biodiesel production. A known quantity of lithium nitrate was suspended in de-ionized water using an ultrasonic bath. The TiO_2 was slowly added into the lithium nitrate solution to form a slurry, which was agitated for 4 hours on a magnetic stirrer at room temperature. The mixture was then allowed to evaporate to dryness using a magnetic stirrer hot plate at 120°C for 12 hours. The resulting sample was calcined (450, 600, or 750°C) in a muffle furnace under airflow. The process was used to achieve Li-impregnated TiO_2 nanocatalyst for use in transesterification reactions to produce biodiesel.

5.5.2 Dry Impregnation

During dry impregnation, the volume of the solution used is usually somewhat less than or equal to the pore volume of the active solid phase. After the catalyst has been impregnated into the support or an added active solid phase, it is dried and calcinated before use (Deraz 2018).

5.6 TiO_2-Based Nanocatalysts for Biodiesel Production

The TiO_2-based nanocatalysts have been broadly utilized for the production of biodiesel using various grades of vegetable oil and animal fat as shown in Table 5.3. The table demonstrates the suitability of various catalysts for biodiesel production, showing their excellent percentage yields. This is due to their excellent thermal and chemical stability in acidic and oxidative environments (Muhammad et al. 2019). The TiO_2 NPs have great properties, including non-toxicity, high effectiveness and photocatalytic stability, which make them suitable for the synthesis of nanocatalysts for various applications. Recently, researchers have demonstrated that the TiO_2 NPs improves the performance of catalysts by allowing the modulation of catalytic activities. This makes it a primary candidate for heterogeneous catalyst support (Muhammad et al. 2019). Furthermore, the TiO_2 NPs has been found to crystallize into either anatase or rutile, enabling great control over its activities as a support (Tang et al. 2016). A few substances that can be supported by the TiO_2 NPs include metals, bimetals and metal oxides which have been developed into various catalysts for biodiesel production using vegetable oils/animal fats. These showed great performance when used with any triglyceride acid (TGA) source. Table 5.3 shows the performance yields and operating conditions for various TiO_2-based nanocatalysts, homogeneous and heterogeneous catalysts for the production of biodiesel using different types of TGAs. The yields reported for the TiO_2-based nanocatalysts compared well with those produced by other catalysts. Compounds/elements that can be combined with the TiO_2 NPs to develop TiO_2-based nanocatalysts have been extensively identified and discussed as followed.

5.6.1 Sulfated TiO_2 Nanocatalysts

Sulfonation of metal oxides generally shows great performance due to the addition of sulfate ions to the metal oxides, which enhances their acidic properties and lowers their deactivation process. Nanocatalysts developed via sulfonation of TiO_2 NPs have been reported to perform better due to enhancement of their acid strength by the loading of SO_4^{2-} groups on the surface of the TiO_2 (Carlucci et al. 2019). Carlucci et al. (2019) confirmed that sulfated TiO_2 nanocatalyst showed greater performance compared to other sulfated metal oxides, as the TiO_2 NPs had better acid strength. Examples include TiO_2-SO_4^{2-}, SO_4^{2-}/TiO_2-ZrO_2, SO_4^{2-}/TiO_2-SiO_2, $SO_4^{2-}/TiO_2/La^{3+}$ and TiO_2-HNO_3. Zhou

Table 5.3 Comparative appraisal of the performance of TiO_2-based nanocatalysts, homogeneous and heterogeneous catalysts for biodiesel production.

Catalyst		TGA source	Process conditions			Performance	
Type	**Name**	**Vegetable/animal fat**	**Temp. (°C)**	**Reaction time (h)**	**Catalyst loading (wt%)**	**Biodiesel yield (%)**	**References**
TiO_2-based	$TiO_2/PrSO_3H$	Waste cooking oil	60	9	4.5	98.3	Gardy et al. (2017)
Heterogeneous	Fe_2O_3-MnO-SO_4^{2}/ZrO_2		180	**4**	3	96.5	Alhassan et al. (2015)
Homogeneous	KOH		45	2	0.55	93.6	Kolhe et al. (2017)
TiO_2-based	TiO_2–ZnO	Palm oil	60	5	9.2	92.2	Madhuvilakku and Piraman (2013)
Heterogeneous	Montmorillonite KSF		190	3	3	79.6	Kansedo et al. (2009)
Homogeneous	H_2SO_4/NaOH		64.5/56	2.5/2	2.5/0.65	90	Anguebes-Franseschi et al. (2016)
TiO_2-based	TiO_2–MgO	Waste cooking oil	160	10	6	92.3	Wen et al. (2010)
Heterogeneous	CaO		50	1.5	1	96	Degfie et al. (2019)
Homogeneous	KOH		62.4	0.03	1.16	98.26	Mohadesi et al. (2019)
TiO_2-based	Ba_2TiO_4	Microalgae	65	3	3.5	98.41	Singh et al. (2019)
Heterogeneous	K-pumice		60	2	10	77	Cercado et al. (2017)
Homogeneous	H_2SO_4/chloroform		90	0.7	0.6/4 (ml/wt)	95.38	Boopathy et al. (2020)
TiO_2-based	Li/TiO_2	Canola oil	55	3	5	98	Alsharifi et al. (2017)
Heterogeneous	Na/BaO		50	4	2	97.5	D'Cruz et al. (2007)
TiO_2-based	SO_4^{-2}/TiO_2	Jatropha oil	140	24	4	85.3	Chen et al. (2018)
Heterogeneous	Transition metal glycerolates		150	6	6	99.7	Lau et al. (2016)
Homogeneous	KOH		60	1	1	90	Folaranmi (2013)
TiO_2-based	CaO-TiO_2	Palm oil	65	1	0.5	76.67	Mohamad et al. (2018)
Heterogeneous	KF/ZnO		65	9.72	5.52	89.23	Hameed et al. (2009)
Homogeneous	KOH		60	0.5	1.5	61.2%	Aworanti et al. (2019)

et al. (2013) demonstrated the application of SO_4^{2-}/TiO_2-ZrO_2 in biodiesel production using high-FFA waste cooking oil (WCO). The optimum yield of 96.68% biodiesel was obtained at a catalyst loading of 5 wt% oil, reaction temperature of 120°C and reaction time of 8 hours. The catalyst showed evidence of great reusability over three cycles at 80% yield of biodiesel. The SO_4^{2-}/ZrO_2-TiO_2/La^{3+} heterogeneous catalyst was also reported for biodiesel production by Li et al. (2010), obtaining high catalytic activity with an efficient percentage yield of >90%. The catalyst was reused for five cycles and maintained high conversion efficiency. The sulfated TiO_2 NPs is a good catalyst with great performance and reusability for biodiesel production from high-FFA vegetable oil/animal fat.

5.6.2 Alkaline TiO_2 Nanocatalysts

The TiO_2 NPs can also be modified with alkaline earth metals to provide a solid catalyst for biodiesel production. Wen et al. (2010) reported the use of TiO_2-MgO (alkaline earth metal) nanocatalyst that was synthesized by sol-gel method, in this regard. The addition of titanium resulted in the substitution of titanium ions for Mg ions in the magnesia lattice, improving stability and maintaining catalytic activity. Reusability of the catalyst was an added advantage, as its regeneration improved specific surface area and average pore diameter compared to fresh/unused TiO_2-MgO NPs. The TiO_2-MgO NPs were tested for biodiesel production using WCO, giving a yield of 92.3% with a reaction time of 6 hours, a reaction temperature of 160°C, and a catalyst loading of 10% (w/w oil). Regeneration of TiO_2-MgO NPs after the fourth cycle gave an improved yield of 93.8%. These findings justify that the TiO_2-MgO NPs could be used as catalysts for industrial-scale production of biodiesel.

The potential of TiO_2 impregnated on calcium oxide (CaO) to produce CaO-TiO_2 nanocatalyst for biodiesel production was investigated by Mohamad et al. (2018). They found out that the impregnation of TiO_2 NPs improved the surface area and strength of basic sites, increasing the catalytic activity of the TiO_2-CaO nanocatalyst. The functionality of the TiO_2-CaO nanocatalyst in biodiesel production using palm oil yielded 93.33% biodiesel. This demonstrates the significant importance of TiO_2 NPs to improving catalyst activity.

5.6.3 Co-Transition TiO_2 Nanocatalysts

The modification of TiO_2NPs with different co-transition elements and their use as solid catalysts for biodiesel production are reported in the literature. Madhuvilakku and Piraman (2013) synthesized TiO_2-ZnO NPs by substituting titanium ions on the zinc lattice to creates defects, which were responsible for the stability of the catalytic activity. The TiO_2-ZnO NPs achieved a 92% yield of biodiesel with a reaction time of 5 hours, a reaction temperature of 60°C, and a catalyst quantity of 200 mg (Table 5.3). The nanocatalyst shows great potential in large-scale production of biodiesel as it significantly improves yield compared to other ZnO nanocatalysts.

TiO_2-Au nanocatalyst is another co-transition TiO_2 catalyst found to be suitable as a heterogeneous catalyst for biodiesel production. The particle size and properties of TiO_2 NPs usually enhance the catalytic activities of Au-based heterogeneous catalysts. The TiO_2 NPs were reported to be a good support for the Au heterogeneous catalyst due to their strong interaction with metal supports, thermal and chemical stability as well as acid-base properties. To develop effective Au-supported TiO_2 NPs, various operating parameters must be properly controlled which includes, (i) incorporation of impurities; (ii) pH value; (iii) drying pretreatment; (iv) precipitation; (v) Au-cluster size/morphology; (vi) surface area of catalyst conditions; (vii) nature of support; (viii) acid-base treatment of support; (ix) loading of TiO_2 NPs; (x) CO adsorption; (xi) catalytic reaction conditions; and (xii) chemical/electronic state in catalyst (Tang et al. 2016).

TiO_2-Co is another co-transition TiO_2 nanocatalyst that has been found to be applicable in biodiesel production. It is worthy of note that the rutile phase of titanium is the most suitable to dope with cobalt in order to form nanocatalyst, which is not the case for other metals (Larsson et al. 1996). It should be noted that cobalt supported on TiO_2 NPs is significantly connected to the TiO_2 crystal phase and the loading of Co^{3+} ions on the catalyst support (Yu et al. 2013). A Co/TiO_2 heterogeneous catalyst synthesized with >15% rutile phase was found to be four times more efficient than anatase phase (Takanabe et al. 2005).

TiO_2-CuO is also an example of co-transition TiO_2 nanocatalyst which has been found to be effective for biodiesel production. The interaction between the Cu and the support materials has been reported to have a very significant effect, especially with TiO_2 supports. The CuO heterogeneous catalyst is highly active in CO oxidation and is cost-effective compared to other noble metal heterogeneous catalysts (Hao et al. 2006). The CuO also possesses high chemical and thermal stability.

Ruthenium oxide-supported TiO_2 (TiO_2-RuO_2) heterogeneous nanocatalyst has also become attractive for use in the oxidation process (Blondeel et al. 1983). Generally, RuO_2/TiO_2 heterogeneous nanocatalyst is prepared by the impregnation of Ru salt followed by thermal treatment, with limited control over the properties of RuO_2 species formation. A green preparation of calibrated RuO_2 NPs has recently been developed and analyzed for the synthesis of RuO_2/TiO_2 heterogeneous catalyst and is expected to exhibits excellent oxidation activity (Swider et al. 1997).

Palladium-supported TiO_2 (TiO_2-Pd) is another example of a co-transition TiO_2 nanocatalyst. Palladium is mostly supported with anatase phase of TiO_2, which possesses excellent catalytic activity vis-á-vis methanol electro-oxidation. Reports have also claimed that the Pd-supported TiO_2 anatase heterogeneous catalyst exhibits greater activity than TiO_2 rutile in selective hydrogenation reactions, due to the superior metal supporting behavior of TiO_2 anatase. Furthermore, the treatment of Pd/TiO_2 anatase solid catalyst by an H_2 reduction at 200°C produces better effects than that of Pd/TiO_2 in rutile phase (Shahreen et al. 2013).

5.6.4 Alkali TiO_2 Nanocatalysts

Alkali metal can be used to modify TiO_2 NPs in order to develop effective nanocatalysts for biodiesel production. Sodium titanate was doped with potassium, an alkali metal, to synthesize a NaK-TiO_2 nanocatalyst for biodiesel production (Martínez-Klimova et al. 2015). Alsharifi et al. (2017) developed a TiO_2-based catalyst by doping with an alkali metal, lithium, to form TiO_2-Li nanocatalyst. The insertion of lithium improved the reactivity of the TiO_2 NPs by enhancing their surface properties. This increased the efficiency of the TiO_2-Li nanocatalyst without altering the TiO_2 NP structure. The performance efficiency of the TiO_2-Li nanocatalyst shows a maximum 98% yield of biodiesel. However, its reusability decreases after repeated runs due to filling of pores by organic materials and leaching of Li^+.

5.6.5 Bimetallic TiO_2 Nanocatalysts

5.6.5.1 TiO_2-Pd-Ni

Spherical TiO_2 NPs were used to amalgamate the Pd-Ni-supported TiO_2 electro-catalyst for methanol oxidation (Ju et al. 2012). The methanol oxidation mechanism of the Pd-Ni/TiO_2 heterogeneous catalyst mainly results from the high catalytic activity of the hybrid system without UV light illumination. Therefore, Pd-Ni/TiO_2 catalyst may become a promising material for a direct-methanol fuel cell.

5.6.5.2 TiO_2-Au-Cu

A TiO_2-Au-Cu heterogeneous nanocatalyst was studied in reactions like total oxidation of methane, water, gas shift and propane as well as epoxidation of propane, and found to be highly effective. The Au and Cu complement each other in terms of surface stability, electronic properties and oxygen mobility (Bracey et al. 2009). The interaction between the two in the Au-Cu/TiO_2 bimetallic heterogeneous catalyst was analyzed for methanol oxidation, and it was found that catalytic activity and selectivity were greater than with the monolithic catalyst (Bracey et al. 2009).

5.7 Other TiO_2 Nanocomposite Catalysts

Abu-dalo et al. (2019) described another approach to the development of TiO_2 nanocomposites using plant material, combining pristine pomegranate peel (PPP) with TiO_2 NPs. First, the peels were separated and shredded into small pieces. Then, they were thoroughly washed with tapwater and rinsed using de-ionized water. The prepared sample was submerged in hot de-ionized water, boiled for 30 minutes, and poured into clean glassware. It was then oven-dried at a temperature of 102°C for 6 hours and pulverized into a fine powder using an electric blender. The powder was re-dried in an oven at 102°C for 30 minutes and stored to prepare a solution of PPP extract. The TiO_2 NPs were then prepared using titanium isopropoxide (IV) ($Ti(OCH(CH_3)_2)_4$), 97% as a starting solution diluted in isopropyl alcohol of 99.5%. The mixture was vigorously mixed for 10 hours at room temperature to precipitate TiO_2 NPs by adding alkaline distilled water of pH 8. The formed precipitate was washed with distilled water and centrifuged. The residue was calcined at a temperature range of 200–1100°C. Ten grams of the PPP extract was weighed into an Erlenmeyer flask and 100 ml of distilled water was added to form a solution. The solution was agitated continuously at 1500 HZ and a temperature of 102°C for 15 minutes. Then 0.1 g TiO_2 was weighed into a separate Erlenmeyer flask and 100 ml of distilled water was added. The mixture was heated at 102°C and stirred for 15 minutes. The TiO_2 and PPP extract solutions were mixed and reheated at 102°C under constant agitation for 15 minutes. The final mixture was sonicated for 15 minutes at 50°C and oven-dried in a glass petri dish at a temperature of 102°C. The nanocomposite was pulverized into a fine powder by an electric blender and oven-dried at 102°C for 30 minutes. The sample was then dried again using nitrogen gas for 15 minutes in order to remove any moisture from the powder. This process gave a TiO_2-PPP nanocatalyst using plant peel extract.

5.8 Conclusion

Ilmenite is the richest natural source of TiO_2 and is suitable for the synthesis of TiO_2 NPs. Among the various methods of synthesis proposed, the biological methods are preferred due to their easy handling, non-toxicity and eco-friendly nature. Synthesis of NPs using plant extracts is referred to as green synthesis which has been found to be effective and suitable for the development of the TiO_2 NPs. The TiO_2 NPs can be combined with a good number of other elements/materials to form nanocatalysts, improving their functionality in biodiesel production. Wet and dry impregnation methods have been used for this purpose. Performance evaluation of TiO_2-based nanocatalysts in biodiesel production reveals that they are reliable and reusable. They are also economically viable and can easily be scaled up for large-scale production.

References

Abdul Jalill, R.D., Nuaman, R.S., and Abd, A.N. (2016). Biological synthesis of titanium dioxide nanoparticles by *Curcuma longa* plant extract and study its biological properties. *World Scientific News* 49 (2): 204–222.

Abu-dalo, M., Jaradat, A., Albiss, B.A., and Al-rawashdeh, N.A.F. (2019). Green synthesis of TiO_2 NPs/*Pristine pomegranate* peel extract nanocomposite and its antimicrobial activity for water disinfection. *Journal of Environmental Chemical Engineering* 7 (5): 103370.

Akinnawo, S. (2019). Synthesis, modification, applications and challenges of titanium dioxide nanoparticles. *Research Journal of Nanoscience and Engineering* 3 (4): 10–22.

Akubude, V.C., Nwaigwe, K.N., and Dintwa, E. (2019). Production of biodiesel from microalgae via nanocatalyzed transesterification process: a review. *Materials Science for Energy Technologies* 2 (2): 216–225.

Alhassan, F.H., Rashid, U., and Taufiq-yap, Y.H. (2015). Synthesis of waste cooking oil-based biodiesel via effectual recyclable bi-functional $Fe_2O_3.MnO.SO_4^{2-}/ZrO_2$ nanoparticle solid catalyst. *Fuel* 142: 38–45.

Alsharifi, M., Znad, H., Hena, S., and Ang, M. (2017). Biodiesel production from canola oil using novel Li/TiO_2 as a heterogeneous catalyst prepared via impregnation method. *Renewable Energy* 114: 1077–1089.

Al-Zahrani, H., El-Waseif, A., and El-Ghwas, D. (2018). Biosynthesis and evaluation of TiO_2 and ZnO nanoparticles from in vitro stimulation of *Lactobacillus johnsonii*. *Journal of Innovations in Pharmaceutical and Biological Sciences* 5 (1): 16–20.

Anguebes-Franseschi, F., Córdova-Quiroz, A., Cerón-Bretón, J. et al. (2016). Optimization of biodiesel production from African crude palm oil (*Elaeis guineensis*) with high concentration of free fatty acids by a two-step transesterification process. *Open Journal of Ecology* 6 (1): 13–21.

Asadnajafi, S., Shahidi, S., and Dorranian, D. (2019). *In situ* synthesis and exhaustion of nano TiO2 on fabric samples using laser ablation method. *Journal of The Textile Institute* 111 (1): 122–128.

Aworanti, O.A., Ajani, A.O., Agarry, S.E. et al. (2019). Effect of process variables on the transesterification process of palm oil sludge to biodiesel. *Biotechnology Journal International* 23 (2): 1–14.

Bagheri, S., Muhd Julkapli, N., and Bee Abd Hamid, S. (2014). Titanium dioxide as a catalyst support in heterogeneous catalysis. *Scientific World Journal* 2014: 727496.

Blondeel, G., Harriman, A., Porter, G. et al. (1983). Design, preparation, and characterization of RuO_2/TiO_2 colloidal catalytic surfaces active in photo-oxidation of water. *Journal of Physical Chemistry* 87 (14): 2629–2636.

Boopathy, A.B., Jayakumar, T., Chinnasamy, S. et al. (2020). Biomass and lipid production potential of an Indian marine algal isolate *Tetraselmis striata* BBRR1. *Energies* 13 (341): 1–12.

Bordbar, H., Yousefi, A.A., and Abedini, H. (2017). Production of titanium tetrachloride ($TiCl_4$) from titanium ores: a review. *Polyolefins Journal* 4 (2): 149–173.

Bracey, C.L., Ellis, P.R., and Hutchings, G.J. (2009). Application of copper–gold alloys in catalysis: current status and future perspectives. *Chemical Society Reviews* 38 (8): 2231–2243.

Carlucci, C., Degennaro, L., and Luisi, R. (2019). Titanium dioxide as a catalyst in biodiesel production. *Catalysts* 9 (75): 1–25.

Cercado, A.P.I., Ballesteros, F.C., and Capareda, S.C. (2017). Biodiesel from microalgae using synthesized novel alkali catalyst. *Applied Ecology and Environmental Research* 15 (4): 1859–1870.

Chatterjee, A., Nishanthini, D., Sandhiya, N., and Abraham, J. (2016). Biosynthesis of titanium dioxide nanoparticles using *Vigna radiata*. *Asian Journal of Pharmaceutical and Clinical Research* 9 (4): 85–88.

Chen, C., Cai, L., Shangguan, X. et al. (2018). Heterogeneous and efficient transesterification of *Jatropha curcas* L. seed oil to produce biodiesel catalysed by nano-sized SO_4^{2-}/TiO_2. *Royal Society Open Science* 5: 181331.

Chenari, H.M., Seibel, C., Hauschild, D. et al. (2016). Titanium dioxide nanoparticles: synthesis, X-ray line analysis and chemical composition study. *Materials Research* 19 (6): 1319–1323.

D'Cruz, A., Kulkarni, M.G., Meher, L.C., and Dalai, A.K. (2007). Synthesis of biodiesel from canola oil using heterogeneous base catalyst. *Journal of the American Oil Chemists' Society* 84: 937–943.

Degfie, T.A., Mamo, T.T., and Mekonnen, Y.S. (2019). Optimized biodiesel production from waste cooking oil (WCO) using calcium oxide (CaO) nano-catalyst. *Scientific Reports* 9 (1): 1–8.

Deraz, N.M. (2018). The comparative jurisprudence of catalysts preparation methods: I. precipitation and impregnation methods. *Journal of Industrial and Environmental Chemistry* 2 (1): 19–21.

Dittert, B., Wiessner, M., Angerer, P. et al. (2019). Tailored Al_2O_3-Al_2TiO_5-TiO_2 composite ceramics from different titanium precursors. *Archives of Metallurgy and Materials* 64 (4): 1277–1286.

Dobrucka, R. (2017). Synthesis of titanium dioxide nanoparticles using *Echinacea purpurea* herba. *Iranian Journal of Pharmaceutical Research* 16 (2): 756–762.

Dreesen, L., Colomer, J.F., Limage, H. et al. (2009). Synthesis of titanium dioxide nanoparticles by reactive DC magnetron sputtering. *Thin Solid Films* 518: 112–115.

Dubey, R.S., Venkata, K., and Singh, S. (2019). Experimental studies of TiO_2 nanoparticles synthesized by sol-gel and solvothermal routes for DSSCs application. *Results in Physics* 14: 102390.

Farbod, M. and Khademalrasool, M. (2011). Synthesis of TiO_2 nanoparticles by a combined sol-gel ball milling method and investigation of nanoparticle size effect on their photocatalytic activities. *Powder Technology* 214: 344–348.

Feyel, S., Schroder, D., and Schwarz, H. (2006). Gas-phase oxidation of isomeric butenes and small alkanes by vanadium-oxide and hydroxide cluster cations. *Journal of Physical Chemistry A* 110 (8): 2647–2654.

Folaranmi, J. (2013). Production of biodiesel (B100) from Jatropha oil using sodium hydroxide as catalyst. *Journal of Petroleum Engineering* 2013: 1–6.

Gardy, J., Hassanpour, A., Lai, X. et al. (2017). Biodiesel production from used cooking oil using a novel surface functionalised TiO_2 nano-catalyst. *Applied Catalysis B: Environmental* 207: 297–310.

Gardy, J., Rehan, M., Hassanpour, A. et al. (2019). Advances in nano-catalysts based biodiesel production from non-food feedstocks. *Journal of Environmental Management* 249: 109316.

Goutam, S.P., Saxena, G., Singh, V. et al. (2018). Green synthesis of TiO_2 nanoparticles using leaf extract of *Jatropha curcas L.* for photocatalytic degradation of tannery wastewater. *Chemical Engineering Journal* 336: 386–396.

Gupta, S.M. and Tripathi, M. (2012). A review on the synthesis of TiO_2 nanoparticles by solution route. *Central European Journal of Chemistry* 10 (2): 279–294.

Hameed, B.H., Lai, L.F., and Chin, L.H. (2009). Production of biodiesel from palm oil (*Elaeis guineensis*) using heterogeneous catalyst: an optimized process. *Fuel Processing Technology* 90 (4): 606–610.

Hao, X.Y., Zhang, Y.Z., Wang, J.W. et al. (2006). A novel approach to prepare MCM-41 supported CuO catalyst with high metal loading and dispersion. *Microporous and Mesoporous Materials* 88 (1–3): 38–47.

Hossain, A., Abdallah, Y., Ali, M.A. et al. (2019). Lemon-fruit-based green synthesis of zinc oxide nanoparticles and titanium dioxide nanoparticles against soft rot bacterial pathogen *Dickeya dadantii*. *Biomolecules* 9 (12): 863.

Huang, J., Li, Q., Sun, D. et al. (2007). Biosynthesis of silver and gold nanoparticles by novel sundried *Cinnamomum camphora* leaf. *Nanotechnology* 18: 105104.

Hussain, S.T., Ali, S.A., Bano, A., and Mahmood, T. (2011). Use of nanotechnology for the production of biofuels from butchery waste. *International Journal of the Physical Sciences* 6 (31): 7271–7279.

Hussain, I., Singh, N.B., Singh, A. et al. (2016). Green synthesis of nanoparticles and its potential application. *Biotechnology Letters* 38: 545–560.

Jalalah, M., Faisal, M., Bouzid, H. et al. (2013). Dielectric and photocatalytic properties of sulfur-doped TiO_2 nanoparticles prepared by ball milling. *Materials Research Bulletin* 48: 3351–3356.

Jayaseelan, C., Rahuman, A.A., Roopan, S.M. et al. (2013). Biological approach to synthesize TiO_2 nanoparticles using *Aeromonas hydrophila* and its antibacterial activity. *Spectrochimica Acta – Part A: Molecular and Biomolecular Spectroscopy* 107: 82–89.

Jha, A.K. and Prasad, K. (2010). Biosynthesis of metal and oxide nanoparticles using *Lactobacilli* from yoghurt and probiotic spore tablets. *Biotechnology Journal* 5 (3): 285–291.

Ju, J., Shi, Y., and Wu, D. (2012). TiO_2 nanotube supported Pd-Ni catalyst for methanol electro-oxidation. *Powder Technology* 230: 252–256.

Kang, J. and Okabe, T.H. (2014). Production of titanium dioxide directly from titanium ore through selective chlorination using titanium tetrachloride. *Mining and Materials Processing Institute of Japan* 55 (3): 591–598.

Kansedo, J., Lee, K.T., and Bhatia, S. (2009). Biodiesel production from palm oil via heterogeneous transesterification. *Biomass and Bioenergy* 33 (2): 271–276.

Kolhe, N.S., Gupta, A.R., and Rathod, V.K. (2017). Production and purification of biodiesel produced from used frying oil using hydrodynamic cavitation. *Resource-Efficient Technologies* 3 (2): 198–203.

Kumar, K.N.P., Keizer, K., Burggraaf, A.J. et al. (1993). Synthesis and textural properties of unsupported and supported rutile (TiO_2) membranes. *Journal of Materials Chemistry* 3 (9): 923–929.

Landage, K.S., Arbade, G.K., Khanna, P., and Bhongale, C.J. (2020). Biological approach to synthesize tio_2 nanoparticles using *Staphylococcus aureus* for antibacterial and anti-biofilm applications. *Journal of Microbiology & Experimentation* 8 (1): 36–43.

Larsson, P.O., Andersson, A., Wallenberg, L.R., and Svensson, B. (1996). Combustion of CO and toluene; characterisation of copper oxide supported on titania and activity comparisons with supported cobalt, iron, and manganese oxide. *Journal of Catalysis* 163 (2): 279–293.

Lau, P., Kwong, T., and Yung, K. (2016). Effective heterogeneous transition metal glycerolates catalysts for one-step biodiesel production from low grade non-refined jatropha oil and crude aqueous bioethanol. *Nature Publishing Group* March: 1–10.

Li, Y., Zhang, X.D., Sun, L. et al. (2010). Fatty acid methyl ester synthesis catalyzed by solid superacid catalyst $SO_4^{2-}/ZrO^{2-}/TiO_2/La^{3+}$. *Applied Energy* 87 (1): 156–159.

Madhuvilakku, R. and Piraman, S. (2013). Biodiesel synthesis by TiO_2–ZnO mixed oxide nanocatalyst catalyzed palm oil transesterification process. *Bioresource Technology* 150: 55–59.

Madkour, H.L. (2017). Ecofriendly green biosynthesized of metallic nanoparticles: bio-reduction mechanism, characterization and pharmaceutical applications in biotechnology industry. *Global Drugs and Therapeutics* 3 (1): 1–11.

Malarkodi, C., Chitra, K., Rajeshkumar, S. et al. (2013). Novel eco-friendly synthesis of titanium oxide nanoparticles by using *Planomicrobium sp.* and its antimicrobial evaluation. *Der Pharmacia Sinica* 4 (3): 59–66.

Marimuthu, S., Rahuman, A.A., Jayaseelan, C. et al. (2013). Acaricidal activity of synthesized titanium dioxide nanoparticles using *Calotropis gigantea* against *Rhipicephalus microplus* and *Haemaphysalis bispinosa*. *Asian Pacific Journal of Tropical Medicine* 6 (9): 682–688.

Martínez-Klimova, E., Hernández-Hipólito, P., and Klimova, T.E. (2015). Biodiesel production with nanotubular sodium titanate doped with potassium as a catalyst. *Materials Research Society* 1 (6): 415–420.

Maynard, R.N., Millman, N., and Iannicelli, H. (1969). A method for removing titanium dioxide impurities from kaolin. *Clays and Clay Minerals* 17: 59–62.

Mohadesi, M., Aghel, B., Maleki, M., and Ansari, A. (2019). Production of biodiesel from waste cooking oil using a homogeneous catalyst: study of semi-industrial pilot of microreactor. *Renewable Energy* 136: 677–682.

Mohamad, M., Ngadi, N., Wong, S. et al. (2018). Synthesis and characterization of CaO-TiO_2 for transesterification of vegetable palm oil. *International Journal of Engineering, Transactions B: Applications* 31 (8): 1326–1333.

Moret, M.P., Zallen, R., Vijay, D.P., and Desu, S.B. (2000). Brookite-rich titania films made by pulsed laser deposition. *Thin Solid Film* 366: 8–10.

Muhammad, T.N., Muhammad, A.A., and Azam, A. (2019). Synthesis and applications of nano-TiO_2: a review. *Environmental Science and Pollution Research* 26 (4): 3262–3291.

Muthukumar, S., Kalyanasundharam, S., and Rosy, K.S.P.J. (2017). Biosynthesis, characterization of tio_2 nanoparticles by using *Solanum xanthocarpum* berry extract and their biomedical and photocatalytic activity. *International Journal for Research in Applied Science & Engineering Technology (IJRASET)* 5 (XII): 738–746.

Nadeem, M., Tungmunnithum, D., Hano, C. et al. (2018). The current trends in the green syntheses of titanium oxide nanoparticles and their applications. *Green Chemistry Letters and Reviews* 11 (4): 492–502.

Navío, J.A., Colón, G., Trillas, M. et al. (1998). Heterogeneous photocatalytic reactions of nitrite oxidation and Cr(VI) reduction on iron-doped titania prepared by the wet impregnation method. *Applied Catalysis B: Environmental* 16 (2): 187–196.

Nyamukamba, P., Okoh, O., Tichagwa, L., and Greyling, C. (2016). Preparation of titanium dioxide nanoparticles immobilized on polyacrylonitrile nanofibres for the photodegradation of methyl orange. *International Journal of Photoenergy* 2016: 1–9.

Oi, L.E., Choo, M.Y., Lee, H.V. et al. (2016). Recent advances of titanium dioxide (TiO_2) for green organic synthesis. *RSC Advances* 6 (110): 108741–108754.

Órdenes-Aenishanslins, N.A., Saona, L.A., Durán-Toro, V.M. et al. (2014). Use of titanium dioxide nanoparticles biosynthesized by *bacillus mycoides* in quantum dot sensitized solar cells. *Microbial Cell Factories* 13 (1): 1–10.

Palomo, J.M. (2019). Nanobiohybrids: a new concept for metal nanoparticles synthesis. *Chemical Communications* 55: 9583–9589.

Palomo, J.M. and Filice, M. (2016). Biosynthesis of metal nanoparticles: novel efficient heterogeneous nanocatalysts. *Nanomaterials* 6 (84): 1–16.

Patidar, V. and Jain, P. (2017). Green synthesis of TiO_2 nanoparticle using *Moringa oleifera* leaf extract. *International Research Journal of Engineering and Technology (IRJET)* 4 (3): 470–473.

Rajakumar, G., Rahuman, A.A., Roopan, S.M. et al. (2012). Fungus-mediated biosynthesis and characterization of TiO_2 nanoparticles and their activity against pathogenic bacteria. *Spectrochimica Acta Part A: Molecular and Biomolecular Spectroscopy* 91: 23–29.

Raliya, R. and Tarafdar, J.C. (2014). Biosynthesis and characterization of zinc, magnesium and titanium nanoparticles: an eco-friendly approach. *International Nano Letters* 4 (93): 1–10.

Rao, K.G., Rao, K.V., Rajendar, V. et al. (2014). Green synthesis of TiO_2 nanoparticles using hibiscus flower extract. *International Conference on Emerging Technologies in Mechanical Sciences* 2014: 79–82.

Ritala, M. and Leskela, M. (1993). Titanium isopropoxide as a precursor in atomic layer epitaxy of titanium dioxide thin films. *Chemistry Materials* 5 (8): 1174–1181.

Rosi, H. and Kalyanasundaram, S. (2018). Synthesis, characterization, structural and optical properties of titanium-dioxide nanoparticles using *Glycosmis cochinchinensis* leaf extract and its photocatalytic evaluation and antimicrobial properties. *World News of Natural Sciences* 17: 1–15.

Salman, J.A.S., Ibrahem, K.H., and Ali, F.A. (2014). Effect of culture media on biosynthesis of titanium dioxide nanoparticles using *Lactobacillus crispatus*. *International Journal of Advanced Research* 2 (5): 1014–1021.

Sankar, R., Rizwana, K., Shivashangari, S.K., and Ravikumar, V. (2014). Ultra-rapid photocatalytic activity Azadirachtaindica engineered colloidal titanium dioxide nanoparticles. *Applied Nanoscience* 5 (6): 731–736.

Santhoshkumar, T., Rahuman, A.A., Jayaseelan, C. et al. (2014). Green synthesis of titanium dioxide nanoparticles using *Psidium guajava* extract and its antibacterial and antioxidant properties. *Asian Pacific Journal of Tropical Medicine* 7 (12): 968–976.

Sarapää, O., Kärkkäinen, N., Ahtola, T., and Al-Ani, T. (2015). High-tech metals in Finland. In: *Mineral Deposits of Finland* (eds. W. Maier, R. Lahtinen and H. O'Brien). Amsterdam: Elsevier.

Scheinost, A.C. (2004). Metal oxides. In: *Encyclopedia of Soils in the Environment*, vol. 4 (ed. D. Hillel). Amsterdam: Elsevier.

Sethy, N.K., Arif, Z., Mishra, P.K., and Kumar, P. (2020). Green synthesis of TiO_2 nanoparticles from *Syzygium cumini* extract for photo-catalytic removal of lead (Pb) in explosive industrial wastewater. *Green Processing and Synthesis* 9 (1): 171–181.

Shahreen, I., Chase, G.G., Turinske, A.J. et al. (2013). NO decomposition by CO over Pd catalyst supported on TiO_2 nanofibers. *Chemical Engineering Journal* 13 (1): 340–349.

Singh, S.C., Swarnkar, R.K., and Gopal, R. (2009). Synthesis of titanium dioxide nanomaterial by pulsed laser ablation in water. *Journal of Nanoscience and Nanotechnology* 9: 5367–5371.

Singh, P., Kim, Y.J., Zhang, D., and Yang, D.C. (2016). Biological synthesis of nanoparticles from plants and microorganisms. *Trends in Biotechnology* 34 (7): 588–599.

Singh, R., Kumar, A., and Sharma, Y.C. (2019). Biodiesel synthesis from microalgae (*Anabaena* PCC 7120) by using barium titanium oxide (Ba_2TiO_4) solid base catalyst. *Bioresource Technology* 287: 121357.

Song, H., Dai, M., Wan, X. et al. (2014). Synthesis of a Ni2P catalyst supported on anatase-TiO2 whiskers with high hydrodesulfurization activity, based on triphenylphosphine. *Catalysis Communications* 43: 151–154.

Speight, J.G. (2017). Industrial inorganic chemistry. In: *Environmental Inorganic Chemistry for Engineers* (ed. D. Speight). Oxford: Butterworth-Heinemann.

Subhapriya, S. and Gomathipriya, P. (2018). Green synthesis of titanium dioxide (TiO_2) nanoparticles by *Trigonella foenum-graecum* extract and its antimicrobial properties. *Microbial Pathogenesis* 116: 215–220.

Sumerel, J.L., Yang, W., Kisailus, D. et al. (2003). Biocatalytically templated synthesis of titanium dioxide. *Chemistry of Materials* 15: 4804–4809.

Suslick, K.S. (1986). The site of sonochemical reaction. *IEEE Transactions on Ultrasonics, Ferroelectrics, and Frequency Control* 33 (2): 143–147.

Swider, K.E., Merzbacher, C.I., Hagans, P.L., and Rolison, D.R. (1997). Synthesis of ruthenium dioxide-titanium dioxide aerogels: redistribution of electrical properties on the nanoscale. *Chemistry of Materials* 9 (5): 1248–1255.

Takanabe, K., Nagaoka, K., Nariai, K., and Aika, K.I. (2005). Titania supported cobalt and nickel bimetallic catalysts for carbon dioxide reforming of methane. *Journal of Catalysis* 232 (2): 268–275.

Tang, H., Liu, F., Wei, J. et al. (2016). Ultrastable hydroxyapatite/titanium-dioxide-supported gold nanocatalyst with strong metal–support interaction for carbon monoxide oxidation. *Angewandte Chemie – International Edition* 55 (36): 10606–10611.

Taran, M., Rad, M., and Alavi, M. (2018). Biosynthesis of TiO_2 and ZnO nanoparticles by *Halomonas elongata* IBRC-M 10214 in different conditions of medium. *BioImpacts: BI* 8 (2): 81–89.

Thakur, B.K., Kumar, A., and Kumar, D. (2019). Green synthesis of titanium dioxide nanoparticles using *Azadirachta indica* leaf extract and evaluation of their antibacterial activity. *South African Journal of Botany* 124: 223–227.

Weaver, C.E. (1976). The nature of TiO_2 in kaolinite. *Clays and Clay Minerals* 24: 215–218.

Wen, Z., Yu, X., Tu, S.T. et al. (2010). Biodiesel production from waste cooking oil catalyzed by tio_2-mgo mixed oxides. *Bioresource Technology* 101 (24): 9570–9576.

Woodruff, L.G., Bedinger, G.M., and Piatak, N.M. (1802). Critical mineral resources of the United States – economic and environmental geology and prospects for future supply. In: *Titanium* (eds. K.J. Schulz, J.J.H. DeYoung, R.R. Seal and D.C. Bradley). Reston, VA: U.S. Geological Survey.

Yang, W. and Hwang, W. (2013). A study on kaolin and titanium dioxide affecting physical properties of electrocoating. *Corrosion Science and Technology* 12 (5): 203–208.

Yu, S., Ma, Y., Zhi, Y. et al. (2013). Synthesis of cobalt-based catalyst supported on TiO_2 nanotubes and its Fischer–Tropsch reaction. *Integrated Ferroelectrics* 147 (1): 59–66.

Zhang, X., Wang, L., Hu, W. et al. (2018). The synthesis of titanium dioxide nanoparticles from titanium slag and its use for low temperature SCR catalyst. *IOP Conference Series: Earth and Environmental Science* 208: 1–4.

Zhou, H., Cao, Y., and Li, J. (2013). Research of preparation of SO_4^{2-}/TiO_2-ZrO_2 and its application on synthesis of biodiesel from waste cooking oil. *Applied Mechanics and Materials* 316–317: 906–910.

6

Zinc-Based Nanocatalysts in Biodiesel Production

Avinash P. Ingle

Biotechnology Centre, Department of Agricultural Botany, Dr. Panjabrao Deshmukh Krishi Vidyapeeth, Akola, Maharashtra, India

6.1 Introduction

Nowadays, rapid and continuous depletion in fossil fuel resources and environmental concerns caused by their burning (e.g. emission of greenhouse gases, climate change, and global warming) are global problems (Ingle et al. 2020a,b). These challenges create a pressing need for the search of novel, renewable, eco-friendly, and environmentally sustainable alternative fuels. In this context, production of clean-energy alternatives such as biofuels has attracted considerable attention from the global scientific community (Ingle et al. 2020b). According to available reports, ~19% growth is expected in the renewable energy sector by 2023, along with a rise in biofuel production of 15% (Bhagat et al. 2020).

Biofuels are renewable energy sources usually produced using contemporary processes from biomass, organic matter, or waste. Currently, they are the largest source of renewable energy used in various sectors, including the transportation sector, where they are blended with existing petroleum fuels such as gasoline (petrol) and diesel. Among the biofuels, ethanol is produced on the largest scale (~66% of all biofuels produced), followed by biodiesel and hydrotreated vegetable oil (Mohammed et al. 2018). Infrastructure and technologies are less developed for these latter fuels than for bioethanol. Thus, more studies are required so that effective technologies can be developed for the industrial production of biodiesel. Biodiesel is a mono alkyl ester of long-chain fatty acids conventionally produced through processes like transesterification and esterification of a variety of feedstocks (biomass, vegetable oils, microbial oils, animal fats, waste oils, etc.) using different kinds of catalysts (Sandoval et al. 2017).

The catalysts usually used for biodiesel production are categorized as heterogeneous and homogeneous acid or base catalysts. The large-scale production of biodiesel through transesterification mainly involves the use of homogeneous base catalysts like sodium hydroxide and potassium hydroxide (Cong et al. 2020); sodium hydroxide is more often used as it is comparatively cheaper (Gupta et al. 2020). In addition to homogeneous catalysts, heterogeneous ones – particularly base catalysts like alkaline earth metal oxides – have been found to have promising applications in biodiesel production, with a report claiming a 98% yield

Nano- and Biocatalysts for Biodiesel Production, First Edition. Edited by Avinash P. Ingle.

of fatty acid methyl ester (FAME) (biodiesel) with their use (Colombo et al. 2017). Another category is the biocatalysts (i.e. enzymes), many of which have been shown to have high potential in the transesterification of low-quality and cheap feedstock (Hama and Kondo 2013). In addition to their ability to catalyze feedstock with a high free fatty acid (FFA) content, they may be recovered from reaction mixtures and reused over multiple cycles when immobilized on any suitable solid support. Moreover, they require only a low reaction temperature – another most important advantage (Hamd 2016). However, all homogeneous, heterogeneous, and biocatalysts have certain disadvantages. Homogeneous catalysts are cheap but require high reaction temperatures, and the application of acid catalysts has a corrosive effect on reactors. Heterogeneous catalysts are comparatively better but still suffer from the same issues. The most common drawbacks associated with the use of biocatalysts are low biodiesel yield, high input costs due to the price of the enzymes, and the possibility of inactivation by alcohols like methanol and glycerol, especially when the catalyst is recycled over several uses (Gog et al. 2012; Cerveró et al. 2014).

In this context, and given the catalytic applications of nanomaterials, nanotechnology is now one of the most explored areas in contemporary research (Grunes et al. 2003). The unique properties possessed by nano-based catalysts (e.g. high surface area due to their nanoscale size) have attracted a great deal of attention within the scientific community. To date, numerous nano-based catalysts have been developed and used for the production of biodiesel from different feedstocks (Rai et al. 2019; Ingle et al. 2020c). Some of the most important nanocatalysts and nanobiocatalysts have been already covered in this book. In the present chapter, special emphasis is given to the role of zinc oxide-based nanocatalysts in the production of biodiesel.

6.2 Feedstocks Used for Biodiesel Production

The total input cost in the production of biodiesel mainly includes the costs for raw materials, equipment, labor, and other utilities, with the raw materials (or feedstock) making up the bulk of this (>80%). Therefore, selection of appropriate and cheap raw materials will help in reducing the total input cost without negatively affecting the quality of biodiesel produced. Considering these aspects, a number of feedstocks have been successfully exploited for industrial biodiesel production. Some are briefly discussed in this section.

6.2.1 Vegetable Oils

Both edible and non-edible oils produced from different vegetables or oil seeds have been preferentially used in the production of biodiesel. However, extensive use of edible oils for biodiesel production may result in a shortage of oils for consumption and lead to a hike in oil prices. Therefore, use of non-edible oils is recommended (Gui et al. 2008). The most commonly used edible oils includes rapeseed oil, sesame oil, peanut oil, rice oil, sunflower oil, coconut oil, soybean oil, corn oil, and hazelnut oil (Alptekin and Canakci 2008; Sanli and Canakci 2008; Karmakar et al. 2010; Visser et al. 2011). Non-edible oils include palm oil jatropha oil, castor oil, *Moringa oleifera* seed oil, neem oil, jojoba oil, cottonseed oil, and sea mango oil (Pramanik 2003; Gui et al. 2008; Nabi et al. 2009; Ogbunugafor et al. 2011; Visser et al. 2011; Salamatinia et al. 2013). Table 6.1 shows a list of important edible and non-edible oils used for biodiesel production.

Table 6.1 Important edible and non-edible oils used for biodiesel production.

Group	Oil source
Major oils	Coconut (copra), corn (maize), cottonseed, canola (a variety of rapeseed), olive, peanut (groundnut), safflower, sesame, soybean, and sunflower
Nut oils	Almond, cashew, hazelnut, macadamia, pecan, pistachio, and walnut
Other edible oils	Amaranth, apricot, argan, artichoke, avocado, babassu, bay laurel, beech nut, ben, Borneo tallow nut, carob pod (algaroba), cohune, coriander seed, false flax, grape seed, hemp, kapok seed, lallemantia, lemon seed, macauba fruit (*Acrocomia sclerocarpa*), meadowfoam seed, mustard, okra seed (hibiscus seed), perilla seed, pequi, (*Caryocar brasiliensis* seed), pine nut, poppy seed, prune kernel, quinoa, ramtil (*Guizotia abyssinica* seed or Niger pea), rice bran, tallow, tea (camellia), thistle (*Silybum marianum* seed), and wheat germ
Non-edible oils	Babassu tree, copaiba, honge, jatropha or ratanjyote, jojoba, karanja or honge, mahua, milk bush, nagchampa, neem, petroleum nut, rubber seed tree, silk cotton tree, and tall
Other oils	Castor, radish, and tung

Source: Adapted from Sani et al. 2013; an open access article.

6.2.2 Microbial Oils

Algal species have high oil contents and therefore represent promising feedstocks for biodiesel production. They are non-edible and therefore have no conflict with food consumption. Moreover, it is possible to grow algae with very low input cost, making them ideal feedstocks as far as biodiesel production is concerned (Chen et al. 2018). The algal species most commonly used are *Chlorella protothecoides* (Chen et al. 2012), *Chlorella emersonii*, *Chlorella salina*, and *Chlorella vulgaris* (Talebi et al. 2013), and *Nannochloropsis oculate*, *Phaeodactylum tricornutum*, and *Scenedesmus dimorphus* (Islam et al. 2013).

6.2.3 Animal Fats

Animal fats are a rich source of oils that can acts as potential feedstocks for biodiesel production. The most studied include chicken fat (Seffati et al. 2019), goat fat (Chakraborty and Sahu 2014), duck fat (Liu and Wang 2013), lamb, cow, and pork fats (Banković–Ilić et al. 2014), and mutton fat (Mutreja et al. 2011). Among these, goat and mutton fat are most preferentially used in biodiesel production.

6.2.4 Waste Oils

As discussed, the majority of the cost involved in biodiesel production is associated with the feedstock. In this context, utilization of waste edible oils (WEOs) is of great interest. WEOs have the benefit of being environmentally sustainable. It has been estimated that more than 15 million tons of WEOs are generated annually around the globe. Utilization of such oils would help manage their disposal and make the process more economically viable (Kulkarni and Dalai 2006).

6.2.5 Biomass

The biomass obtained from various sources, such as agriculture and forestry, can be used for biodiesel production. Several reports claim a potential use for lignocellulosic biomass in the effective production of biodiesel. Among the most important biomasses used to date are corn stover (Kim et al. 2015) and corncob (Li et al. 2015).

6.3 Conventional Methods of Biodiesel Production

A number of different technologies have been proposed for the production of biodiesel. Chief among them are pyrolysis and transesterification (Sani et al. 2013).

6.3.1 Pyrolysis

Pyrolysis, also referred to as catalytic cracking, mainly involves heating organic matter in the absence of air to produce a gas, a liquid, and a solid (Pramanik 2003). Heating, alone or in combination with catalyst, is usually used to break vegetable oils or animal fats into smaller constituents, such as olefins and paraffins, which show similar properties to petro-diesel. Hence, these products are referred to as diesel-like fuels.

6.3.2 Transesterification

As already mentioned, transesterification is the most preferentially used method for the commercial production of biodiesel. It is usually performed by heating an oil to a specific temperature in the presence of alcohol and a catalyst, thereby restructuring its chemical makeup and thus reducing its high viscosity (Sani et al. 2013). The transesterification of triglyceride molecules occurs through successive reactions. Initially, FFA is neutralized by the triglyceride from the alcohol, resulting in the formation of 1 mol of glycerol and 3 mol of alkyl esters (for each mole of triglyceride converted) at the completion of the net reaction. The reaction mixture is then separated into three layers: biodiesel on top, a soapy substance in the middle, and glycerol at the bottom (Fukuda et al. 2001). The yield of biodiesel depends upon many factors, one of the chief ones being the choice of catalyst. A number of catalyst-based methods of transesterification are briefly discussed in this section.

6.3.2.1 Homogeneous Acid and Base (Alkali)-Catalyzed Transesterification

To date, a number of homogeneous acid catalysts (H_2SO_4, HCl, H_3PO_4, etc.) have been successfully exploited in the transesterification of a variety of oils, but this method is not routinely used because of its poor performance, high cost, and environmental impacts (due to its corrosive nature) (Al-widyan and Al-shyoukh 2000; Yong et al. 2006). Further, homogeneous alkali- or base-catalyzed reactions are about 4000 times faster than homogeneous acid-catalyzed ones. Base catalysts such as sodium and potassium hydroxide are commonly used due to their inexpensive nature and high catalytic performance (Vicente et al. 2004). However, the homogeneous catalyzed transesterification approaches have

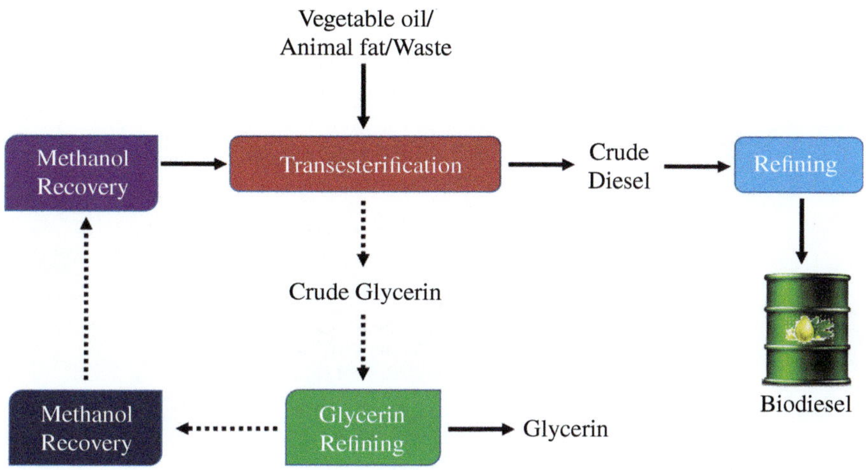

Figure 6.1 Schematic illustration of the general procedure of biodiesel production via transesterification.

certain disadvantages, such as high sensitivity toward FFAs, problems of saponification, tedious and expensive separation of product, wastewater generation, and high energy consumption.

6.3.2.2 Heterogeneous Acid and Base (Alkali)-Catalyzed Transesterification

Heterogeneous catalyzed transesterification has a number of advantages over its homogeneous equivalent. Heterogeneous acid catalysts (i.e. solid acid catalysts) have the ability to catalyze both esterification and transesterification reactions simultaneously, without any requirement for pretreatment of the feedstock used. Therefore, this approach is considered cost-effective (Lotero et al. 2005). Heterogeneous base or alkali catalysts (i.e. solid basic catalysts) are also very effective in biodiesel production and are economically viable due to their low consumption and ability to be reused and regenerated (Sani et al. 2013). These catalysts possess a few disadvantages, however, such as mass transfer (diffusion), loss of catalytic activity in the presence of water, and expensive downstem processing. Figure 6.1 shows a schematic illustration of the general procedure of biodiesel production through transesterification.

6.3.2.3 Enzymatic Transesterification

The concept of enzymatic transesterification has been proposed to overcome the problems associated with homogeneous and heterogeneous catalyzed transesterification (Raman et al. 2008). Applications of biocatalysts (i.e. enzymes) in free and immobilized form enhance the rate of reaction. However, immobilization on any solid support increases the stability of the enzyme and the quality of the product, in addition to providing easy recovery and reuse (Nielsen et al. 2008). A very high cost and the chances of enzyme inactivation and product contamination are the major issues associated with enzymatic transesterification. In this context, nanotechnology-based catalysts would be more effective. The ideal process of enzymatic biodiesel production is illustrated in Figure 6.2.

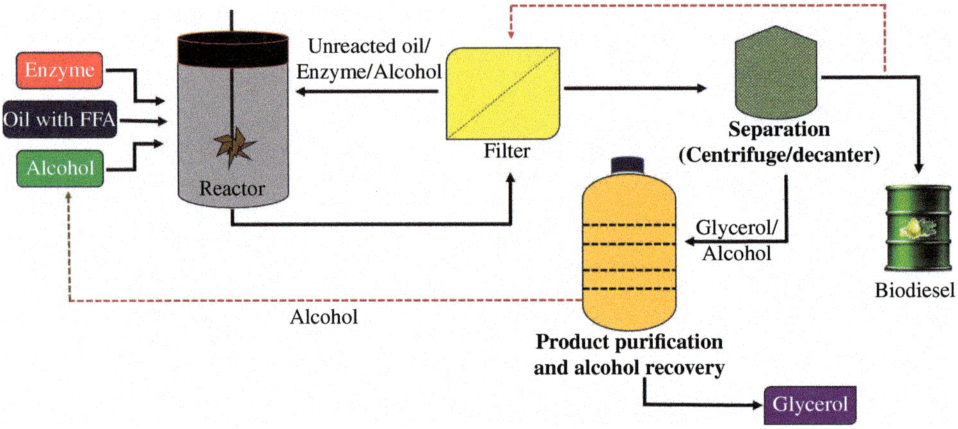

Figure 6.2 Design for the ideal process of enzymatic biodiesel production.

6.4 Nanotechnology in Biodiesel Production

Nanotechnology has great potential to solve the challenges associated with conventional technologies used in biorefinery, particularly in biodiesel production. As already mentioned, biodiesel is a renewable fuel that can be produced from different sources, such as vegetable oils, microbial oils, animal fats, and waste oils, through transesterification using numerous catalysts (Ingle et al. 2020c). Nanotechnology usually deals with the self-assembly of atoms, molecules, or molecular groups into nanostructures or nanocomposites with the aim of creating materials having a size in the range of 1–100 nm and with novel properties and functions (Kumar et al. 2019).

Nanomaterials are considered the building blocks of nanotechnology, and these functional materials are reported to have a wide range of applications in biodiesel industries. Among them, nanoparticles are most commonly used as nanocatalysts due to their outstanding properties resulting from their nano-size. Application of nanocatalysts in biodiesel production has several advantages over that of conventional catalysts, which mainly include significant catalytic performance, good selectivity, high reactivity, high yield, reduction of the reaction time, and recovery of reactants following esterification and transesterification processes (Ingle et al. 2020c). To date, a variety of nanocatalysts (metal-oxide based nanocatalysts, magnetic nanocatalysts, nanohydrotalcites, nanozeolites, etc.) have been developed and used in biodiesel production. Among them, the metal oxide-based nanocatalysts are considered the most promising as far as biodiesel conversion from different feedstocks is concerned.

6.5 Zinc-Based Nanocatalysts in Biodiesel Production

Zinc is mostly used in the form of zinc oxide (ZnO), which is reported to have highly efficient catalytic activity in various processes, including esterification and transesterification. A number of reports are available demonstrating the efficacy of ZnO as a promising catalyst for biodiesel production. In recent years, much effort has been made to enhance its catalytic ability, and it has been observed that when the same ZnO catalyst is used in its nanoform it shows comparatively higher catalytic activity. Therefore, scientists across the world are currently utilizing ZnO nanocatalysts for improved production of biodiesel using a variety of feedstocks.

In this context, Madhuvilakku and Piraman (2013) proposed the synthesis of stable and active heterogeneous ZnO and mixed metal oxide TiO_2-ZnO nanocatalysts for the development of an easy transesterification process and enhanced biodiesel production. Thus developed catalysts were used in the transesterification of palm oil, where it was found that the TiO_2-ZnO nanocatalysts exhibited comparatively higher catalytic activity than the ZnO. The authors showed that the catalytic performance of these Zn-based nanocatalysts mainly depended on factors such as concentration of catalyst used, methanol to oil molar ratio, reaction temperature, and reaction time. A maximum 98% conversion of palm oil was achieved at the optimum reaction parameters with 200 mg of catalyst loading. The authors concluded that, between the two nanocatalysts, TiO_2-ZnO was the better potential candidate for large-scale biodiesel production from palm oil at ambient reaction conditions, due to its better catalytic efficacy.

Gurunathan and Ravi (2015a,2015b) developed Cu-doped ZnO nanocatalysts with a size of about 80 nm and studied their catalytic efficacy in the transesterification of waste cooking oil (WCO) and neem oil, with promising results. In the case of WCO, the maximum yield (97.71% w/w) of biodiesel was obtained under optimum reactions conditions of 8 : 1 methanol to oil molar ratio, 12% (w/w) nanocatalyst concentration, 55 °C reaction temperature, and 50 minutes reaction time, whereas a 73.95% yield was recorded for the transesterification of neem oil at 10 : 1 methanol to oil molar ratio, 10% (w/w) nanocatalyst concentration, 55 °C reaction temperature, and one hour reaction time. Being heterogeneous, the catalysts in both cases were recovered and reused for six successive cycles of transesterification. Moreover, the results indicated that the final yield of biodiesel depended on a number of factors, chief among them are the type and concentration of nanocatalyst and the type of feedstock used. Hence, different yields of biodiesel were recorded for two different feedstocks using the same Cu-doped ZnO nanocatalyst.

In another similar study, production of biodiesel using crystalline copper (Cu)-doped ZnO nanocatalyst was demonstrated using WCO as feedstock. A promising transesterification rate was recorded, which confirmed the significant catalytic efficacy of such nanocatalysts. A maximum yield (98%) of biodiesel was achieved with reactions conditions of 5 : 1 methanol to oil molar ratio, 4% (w/w) nanocatalyst loading, 60 °C reaction temperature, and 40 minutes reaction time (Sandhya et al. 2015).

Baskar and Soumiya (2016) proposed a co-precipitation method for the synthesis of ferromagnetic (iron (II)-doped) ZnO nanocatalysts and evaluated their catalytic efficacy in the esterification of castor oil. From the findings obtained, the authors claimed a maximum 91% yield of biodiesel up to the fourth cycle of reaction. However, after the fourth cycle, a reduction in catalytic efficacy was recorded, which resulted in a decline in biodiesel yield (87%). From this study, it can be concluded that easy recovery and reuse of such Zn-based heterogeneous nanocatalysts for multiple cycles of esterification would help reduce input costs and make biodiesel production economically viable. Another report from the same group on the use of the same nanocatalysts for the production of biodiesel via esterification

Table 6.2 Different Zn-based nanocatalysts used in the production of biodiesel from different feedstocks.

Type of catalysts	Feedstock	Catalyst loading (w%)	Reaction time (h)	Reaction temperature (°C)	Methanol to oil molar ratio	Yield/ conversion (%)	References
Cu-doped ZnO	Waste cooking oil	4	40 min	60	5 : 1	98	Sandhya et al. (2015)
	Waste cooking oil	12	50	55	8 : 1	97.71	Gurunathan and Ravi (2015a)
	Neem oil	10	1	55	10 : 1	73.95	Gurunathan and Ravi (2015b)
Ferromagnetic (iron (II)-doped) ZnO	Castor oil	14	55 min	50	12 : 1	91	Baskar and Soumiya (2016)
Iron (II)-doped ZnO	Pongamia oil	12	55 min	50	10 : 1	93	Baskar et al. (2016)
Heteropoly acid-coated ZnO	*Madhuca indica* oil	—	5	55 ± 5	—	95	Thangaraj and Piraman (2016)
Mn-doped ZnO	Mahua oil	8	50 min	50	7 : 1	97	Baskar et al. (2017a)
Ni-doped ZnO	Castor oil	11	1	55	8 : 1	95.20	Baskar et al. (2017b)
Ag-ZnO nanoparticles	Fromsimarouba oil					84.5	Nagaraju et al. (2017)
Co-doped ZnO	*Mesua ferrea* oil	2.5	3	60	9 : 1	98.03	Borah et al. (2019)

of Pongamia oil demonstrated multiple recovery and reuse of ZnO-based nanocatalysts. A maximum yield of 93% was recorded with a stirring rate of 250 rpm and a calcination temperature of 700 °C (Baskar et al. 2016).

Similarly, in two different studies, synthesis of manganese (Mn)- and nickel (Ni)-doped ZnO nanoparticles was reported via co-precipitation. These nanocatalysts were subsequently used in the production of biodiesel from Mahua and castor oils (Baskar et al. 2017a,b). The observations recorded from these comparative production studies revealed that Mn-doped ZnO nanocatalyst showed higher catalytic efficacy, as evinced by a higher biodiesel yield of 97% under optimum reaction conditions of 7 : 1 methanol to oil molar ratio, 8% catalyst loading, 50 °C reaction temperature, and 50 minutes reaction time. Ni-doped ZnO nanocatalyst showed a 95.20% biodiesel yield with an 8 : 1 methanol to oil molar ratio, 11% catalyst loading, 55 °C reaction temperature, and one hour reaction time.

Varghese et al. (2018) demonstrated a comparative study on ultrasonic-assisted catalytic transesterification using a ZnO nanocatalyst and the conventional stirring method with WCO as feedstock. In addition, they investigated the effects of different reaction parameters (e.g. catalyst loading amount, oil to methanol ratio, reaction time and temperature) in order to optimize the transesterification conditions. The results clearly suggested that ultrasonic-assisted catalytic transesterification using ZnO nanocatalyst was better than the conventional approach, as evidenced from a 96% FAME conversion yield. Moreover, the density and viscosity (which are considered the most important properties of biodiesel) of the thus produced FAME compared favorably with those of the standard biodiesel. The data recorded clearly indicate that WCO can be effectively used as a potential feedstock for the production of biodiesel suitable for use in diesel engines.

A recent study by Borah et al. (2019) demonstrated the synthesis of cobalt (Co)-doped ZnO nanocatalyst and evaluated its catalytic potential in the production of biodiesel from non-edible oil of *Mesua ferrea*. The outcomes confirmed significant catalytic activity, with a maximum biodiesel conversion of 98.03% under optimum reaction conditions of 9 : 1 methanol to oil molar ratio, 2.5 wt% catalyst loading, 60 °C reaction temperature, and three hour reaction time.

All these studies clearly suggest that Zn-based nanocatalysts can be potentially be used for the enhanced production of biodiesel from various sources. Table 6.2 shows the different ZnO-based nanocatalysts used in the transesterification of a variety of feedstocks for biodiesel production.

6.6 Conclusions

The use of fossil fuels in transportation and other sectors has numerous hazardous effects for the health of environment and human beings. In these circumstances, the use of biofuels as a renewable and sustainable energy source is the only alternative. Although conventional technologies have been developed and routinely used for the industrial production of biofuels, including biodiesel, they have several limitations, such as high input costs, energy-intensiveness, and large time requirements. The search for newer technologies to overcome these challenges is thus of utmost importance. In this context, nanotechnology-based solutions (i.e. the use of nanomaterials in the form of nanocatalysts) offer an economic and environment friendly alternative. Zn-based nanocatalysts have tremendous potential to revolutionize the current status quo in the biodiesel industries, making processes more simple, effective, and productive. The author strongly believes that further extensive studies are required to provide a clear understanding of their negative effects, if any.

References

Alptekin, E. and Canakci, M. (2008). Determination of the density and the viscosities of biodiesel–diesel fuel blends. *Renewable Energy* 33 (12): 2623–2630.

Al-widyan, M.I. and Al-shyoukh, A.O. (2000). Experimental evaluation of the transesterification of waste palm oil into biodiesel. *Bioresource Technology* 85: 253–256.

Banković-Ilić, I.B., Stojković, I.J., Stamenković, O.S. et al. (2014). Waste animal fats as feedstocks for biodiesel production. *Renewable and Sustainable Energy Reviews* 32: 238–254.

Baskar, G. and Soumiya, S. (2016). Production of biodiesel from castor oil using iron (II) doped zinc oxide nanocatalyst. *Renewable Energy* 98: 101–107.

Baskar, G., Soumiya, S., and Aiswarya, R. (2016). Biodiesel production from Pongamia oil using magnetic composite of zinc oxide nanocatalyst. *International Journal of Modern Science and Technology* 1: 129–137.

Baskar, G., Gurugulladevi, A., Nishanthini, T. et al. (2017a). Optimization and kinetics of biodiesel production from Mahua oil using manganese doped zinc oxide nanocatalyst. *Renewable Energy* 103: 641–646.

Baskar, G., Aberna, I., and Aiswarya, R. (2017b). Biodiesel production using castor oil using heterogeneous Ni doped ZnO nanocatalyst. *Bioresource Technology* 250: 793–798.

Bhagat, R., Panakkal, H., Gupta, I., and Ingle, A.P. (2020). Recent advances in the production of biodiesel using lignocellulosic biomass. In: *Lignocellulosic Biorefining Technologies* (eds. A.P. Ingle, A.K. Chandel and S.S. Da Silva), 69–86. Oxford: Wiley-Blackwell.

Borah, M.J., Devi, A., Borah, R., and Deka, D. (2019). Synthesis and application of co doped ZnO as heterogeneous nanocatalyst for biodiesel production from non-edible oil. *Renewable Energy* 133: 512–519.

Cerveró, J.M., Álvarez, J.R., and Luque, S. (2014). Novozym 435-catalyzed synthesis of fatty acid ethyl esters from soybean oil for biodiesel production. *Biomass and Bioenergy* 61: 131–137.

Chakraborty, R. and Sahu, H. (2014). Intensification of biodiesel production from waste coal tallow using infrared radiation: process evaluation through response surface methodology and artificial neural network. *Applied Energy* 114: 827–836.

Chen, Y.H., Huang, B.Y., Chiang, T.H., and Tang, T.C. (2012). Fuel properties of microalgae (*Chlorella protothecoides*) oil biodiesel and its blends with petroleum diesel. *Fuel* 94: 270–273.

Chen, J., Li, J., Dong, W. et al. (2018). The potential of microalgae in biodiesel production. *Renewable and Sustainable Energy Reviews* 90: 336–346.

Colombo, K., Ender, L., and Barros, A.A.C. (2017). The study of biodiesel production using CaO as a heterogeneous catalytic reaction. *Egyptian Journal of Petroleum* 26: 341–349.

Cong, W.J., Wang, Y.T., Li, H. et al. (2020). Direct production of biodiesel from waste oils with a strong solid base from alkalized industrial clay ash. *Applied Energy* 264: 114735.

Fukuda, H., Kondo, A., and Noda, H. (2001). Biodiesel fuel production by transesterification of oils. *Journal of Bioscience and Bioengineering* 92: 405–416.

Gog, A., Roman, M., Toşa, M. et al. (2012). Biodiesel production using enzymatic transesterification: current state and perspectives. *Renewable Energy* 39: 10–16.

Grunes, J., Somorjai, G.A., Sciences, M. et al. (2003). Catalysis and nanoscience. *Chemical Communications* 8: 2257–2260.

Gui, M.M., Lee, K.T., and Bhatia, S. (2008). Feasibility of edible oil vs. non-edible oil vs. waste edible oil as biodiesel feedstock. *Energy* 33 (11): 1646–1653.

Gupta, J., Agarwal, M., and Dalai, A.K. (2020). An overview on the recent advancements of sustainable heterogeneous catalysts and prominent continuous reactor for biodiesel production. *Journal of Industrial and Engineering Chemistry* 85: 58–77.

Gurunathan, B. and Ravi, A. (2015a). Biodiesel production from waste cooking oil using copper doped zinc oxide nanocomposite as heterogeneous catalyst. *Bioresource Technology* 188: 124–127.

Gurunathan, B. and Ravi, A. (2015b). Process optimization and kinetics of biodiesel production from neem oil using copper doped zinc oxide heterogeneous nanocatalyst. *Bioresource Technology* 190: 424–428.

Hama, S. and Kondo, A. (2013). Enzymatic biodiesel production: an overview of potential feedstocks and process development. *Bioresource Technology* 135: 386–395.

Hamd, H.T. (2016). Kinetic processes simulation for production of the biodiesel using enzyme as catalyst. *Journal of Natural Science Research* 6 (3): 17–22.

Ingle, A.P., Philippini, R.R., de Souza Melo, Y.C., and da Silva, S.S. (2020a). Acid-functionalized magnetic nanocatalysts mediated pretreatment of sugarcane straw: an eco-friendly and cost-effective approach. *Cellulose* 27: 7067–7078.

Ingle, A.P., Philippini, R.R., and da Silva, S.S. (2020b). Pretreatment of sugarcane bagasse using two different acid-functionalized magnetic nanoparticles: a novel approach for high sugar recovery. *Renewable Energy* 150: 957–964.

Ingle, A.P., Chandel, A.K., Philippini, R. et al. (2020c). Advances in nanocatalysts mediated biodiesel production: a critical appraisal. *Symmetry* 12: 256.

Islam, M., Magnusson, M., Brown, R. et al. (2013). Microalgal species selection for biodiesel production based on fuel properties derived from fatty acid profiles. *Energies* 6 (3): 5676–5702.

Karmakar, A., Karamkar, S., and Mukherjee, S. (2010). Properties of various plants and animals feedstocks for biodiesel production. *Bioresource Technology* 101 (19): 7201–7210.

Kim, I., Seo, Y.H., Kim, G.Y., and Han, J.I. (2015). Co-production of bioethanol and biodiesel from corn Stover pretreated with nitric acid. *Fuel* 143: 285–289.

Kulkarni, M.G. and Dalai, A.K. (2006). Waste cooking oil – an economical source for biodiesel production: a review. *Industrial and Engineering Chemistry* 45 (9): 2901–2913.

Kumar, L.R., Ram, S.K., and Tyagi, R.D. (2019). Application of nanotechnology in biodiesel production. In: *Biodiesel Production: Technologies, Challenges, and Future Prospects* (eds. R.D. Tyagi, R.Y. Surampalli, T.C. Zhang, et al.), 397–419. Reston, VA: American Society of Civil Engineers.

Li, W., Li, Q., Zheng, L. et al. (2015). Potential biodiesel and biogas production from corncob by anaerobic fermentation and black soldier fly. *Bioresource Technology* 194: 276–282.

Liu, K. and Wang, R. (2013). Biodiesel production by transesterification of duck oil with methanol in the presence of alkali catalyst. *Petroleum and Coal* 55 (1): 68–72.

Lotero, E., Liu, Y., Lopez, D.E. et al. (2005). Synthesis of biodiesel via acid catalysis. *Industrial & Engineering Chemistry Research* 44 (14): 5353–5363.

Madhuvilakku, R. and Piraman, S. (2013). Biodiesel synthesis by TiO_2-ZnO mixed oxide nanocatalyst catalyzed palm oil transesterification process. *Bioresource Technology* 150: 55–59.

Mohammed, N.I., Kabbashi, N.A., Alade, A.O., and Sulaiman, S. (2018). Advancement in the utilization of biomass-derived heterogeneous catalysts in biodiesel production. *Green and Sustainable Chemistry* 8: 74–91.

Mutreja, V., Singh, S., and Ali, A. (2011). Biodiesel from mutton fat using KOH impregnated MgO as heterogeneous catalysts. *Renewable Energy* 36 (8): 2253–2258.

Nabi, M.N., Rahman, M.M., and Akhter, M.S. (2009). Biodiesel from cotton seed oil and its effect on engine performance and exhaust emissions. *Applied Thermal Engineering* 29 (11–12): 2265–2270.

Nagaraju, G., Udayabhanu, Shivaraj et al. (2017). Electrochemical heavy metal detection, photocatalytic, photoluminescence, biodiesel production and antibacterial activities of Ag-ZnO nanomaterial. *Materials Research Bulletin* 94: 54–63.

Nielsen, P.M., Brask, J., and Fjerbaek, L. (2008). Enzymatic biodiesel production: technical and economical considerations. *European Journal of Lipid Science and Technology* 110: 692–700.

Ogbunugafor, H.A., Eneh, F.U., and Ozumba, A.N. (2011). Physico-chemical and antioxidant properties of *Moringa oleifera* seed oil. *Pakistan Journal of Nutrition* 10 (5): 409–414.

Pramanik, K. (2003). Properties and use of jatropha curcas oil and diesel fuel blends in compression ignition engine. *Renewable Energy* 28 (2): 239–248.

Rai, M., Ingle, A.P., Pandit, R. et al. (2019). Emerging role of nanobiocatalysts in hydrolysis of lignocellulosic biomass leading to sustainable bioethanol production. *Catalysis Reviews* 61 (1): 1–26.

Raman, J.K., Sariah, A., Denis, P. et al. (2008). Production of biodiesel using immobilized lipase- a critical review. *Critical Reviews in Biotechnology* 28: 253–264.

Salamatinia, B., Hashemizadeh, I., and Ahmad-Zuhairi, A. (2013). Alkaline earth metal oxide catalysts for biodiesel production from palm oil elucidation of process behaviors and modeling using response surface methodology. *Iranian Journal of Chemistry and Chemical Engineering* 32 (1): 113–126.

Sandhya, R., Velavan, R., and Ravichandran, J. (2015). Transesterification of waste cooking oil catalysed by crystalline copper doped zinc oxide nanocatalyst. *Journal of Advances in Chemistry* 12: 5798–5808.

Sandoval, G., Casas-Godoy, L., Bonet-Ragel, K. et al. (2017). Enzyme-catalyzed production of biodiesel as alternative to chemical-catalyzed processes: advantages and constraints. *Current Biochemical Engineering* 4: 109–141.

Sani, Y.M., Daud, W.M.A.W., and Abdul-Aziz, A.R. (2013). Biodiesel feedstock and production technologies: successes, challenges and prospects. In: *Biodiesel: Feedstocks, Production and Applications* (ed. Z. Fang). London: IntechOpen.

Sanli, H. and Canakci, M. (2008). Effects of different alcohol and catalyst usage on biodiesel production from different vegetable oils. *Energy & Fuels* 22 (4): 2713–2719.

Seffati, K., Honarvar, B., Esmaeili, H., and Esfandiari, N. (2019). Enhanced biodiesel production from chicken fat using CaO/$CuFe_2O_4$ nanocatalyst and its combination with diesel to improve fuel properties. *Fuel* 235: 1238–1244.

Talebi, A.F., Mohtashami, S.K., Tabatabaei, M. et al. (2013). Fatty acids profiling: a selective criterion for screening microalgae strains for biodiesel production. *Algal Research* 2 (3): 258–267.

Thangaraj, B. and Piraman, S. (2016). Heteropoly acid coated ZnO nanocatalyst for *Madhuca indica* biodiesel synthesis. *Biofuels* 7: 13–20.

Varghese, R., Henry, J.P., and Irudayaraj, J. (2018). Ultrasonication-assisted transesterification for biodiesel production by using heterogeneous ZnO nanocatalyst. *Environmental Progress & Sustainable Energy* 37: 1176–1182.

Vicente, G., Martinez, M., and Aracil, J. (2004). Integrated biodiesel production: a comparison of different homogeneous catalysts systems. *Bioresource Technology* 92: 297–305.

Visser, E.M., Oliveira Filho, D., Martins, M.A., and Steward Brian, L. (2011). Bioethanol production potential from Brazilian biodiesel co-products. *Biomass and Bioenergy* 35 (1): 489–494.

Yong, W., Shiyi, O., Pengzhan, L. et al. (2006). Comparison of two different processes to synthesize biodiesel by waste cooking oil. *Journal of Molecular Catalysis A: Chemical* 252: 107–112.

7

Carbon-Based Nanocatalysts in Biodiesel Production

Rahul Bhagat[1], Harris Panakkal[1], Indarchand Gupta[1] and Avinash P. Ingle[2]

[1] *Department of Biotechnology, Government Institute of Science, Aurangabad, Maharashtra, India*
[2] *Biotechnology Centre, Department of Agricultural Botany, Dr. Panjabrao Deshmukh Krishi Vidyapeeth, Akola, Maharashtra, India*

7.1 Introduction

With rapid expansion in the global human population and continuous development in technology, energy demand is increasing. This demand is met mainly by fossil fuels. The International Energy Agency (IEA) has predicted that the peak in world oil production will be between the years 1996 and 2035 (Akia et al. 2014). In the transportation sector, diesel is the fuel of choice. Exhaust emissions during transportation gravely affect human health and play a key role in degrading air quality (Mofijur et al. 2016). They also contribute to serious problems like global warming. Moreover, fossil fuel resources are very limited and will become exhausted in the near future. Hence, there is a constant search for a better, renewable alternative. Many renewable sources of energy have been tested, including solar energy, hydropower, wind energy, ocean thermal energy, geothermal energy, and tidal energy. However, the use of biofuel is now gaining wide attention, mainly due to demand from the automobile sector.

Biofuel is defined as either a biodiesel or a renewable diesel. In both its pure form and blended with petrodiesel, it is becoming accepted as a good alternative to traditional diesel. It produces lower carbon dioxide, hydrocarbon, and particulate matter emissions after combustion (Hosseinzadeh-Bandba et al. 2018), and possesses the additional advantages of renewability, ready availability, portability, high efficiency, and safe handling (Banković-Ilić et al. 2014). The use of biodiesel thus represents a greener alternative to traditional diesel. Its production cost is also lower than that of diesel, giving it a high potential to become widely distributed and commercialized. In 2016, the global biodiesel supply was 30.8 billion liters, and it was predicted to soon increase to 41.4 billion (Muniyappan et al. 2019). Some countries, such as Brazil, the United States, and member-states of the European Union, have agreed on targets of the use of 20%, 25%, and 10% biodiesel blends, respectively, by the year 2020. Moreover, the US Navy plans to convert approximately 50% of its entire energy resources to alternative fuels, predominantly biodiesel (Fazal et al. 2019).

Nano- and Biocatalysts for Biodiesel Production, First Edition. Edited by Avinash P. Ingle.

Renewable feedstocks are used in the production of biodiesel, with yields depending on the feedstock composition. Many studies have shown that biomass can be used to produce both liquid and gaseous fuel. It is a general observation that biodiesel production from biomass involves many types of catalytic process, making the selection of catalyst likewise very important. Biodiesel production is generally performed through transesterification of triglycerides from biomass or other feedstocks with methanol in the presence of a good catalyst (Vahid et al. 2017). Large numbers of catalysts have been identified that can be used in effective biodiesel production. However, there is a major drawback associated with most of them: they show difficulties in their separation at the end of the reaction, meaning they are not reusable. This adds significantly to production costs, affecting the cost-efficiency of the biodiesel product and thus making it less acceptable to the end consumer and to society at large. The scientific community across the globe is therefore looking for reusable alternative catalytic sources. Nanotechnology offers a potential option in the form of nanocatalysts.

Today, much research is focused on identifying novel nanocatalysts and their potential for use in biofuel production. They can be very useful for improving the quality of the product and attaining optimum operating conditions. Their high surface area and highly catalytic surfaces make them favorable for contact with the substrate, which prominently increases the efficiency of the transesterification reaction (Borah et al. 2019). This chapter focuses specifically on the use of carbon-based nanocatalysts.

7.2 Feedstocks Used for Biodiesel Production

The source oil has a huge effect on biodiesel quality. Any change in the environment or the physiology of the plant will affect the quality of the oil. Before selecting a biodiesel feedstock, it is essential to know its price, availability, oil content, and oil production per unit area. The choice of good-quality feedstock will help in producing high-quality biodiesel with an acceptable cost of production. According to directive 2018/2001 of the European Union, acceptable feedstocks are categorized into two lists: List A, which mainly includes vegetable oils and algae, and List B, which includes animal fats and waste cooking oils (WCOs) (European Union 2018; Godri Pollitt et al. 2019).

7.2.1 Vegetable Oils

Vegetable oils are classified into edible and non-edible oils. The biodiesel obtained from edible oils is typically termed first-generation biodiesel. It constitutes >90% of various biodiesel feedstocks, while plant seed oils are major contributors to various vegetable oils. To date, more than 350 crop species have been recognized as potential feedstocks for biodiesel production (Muniyappan et al. 2019), including palm oil, rapeseed oil, sunflower oil, soybean oil, and coconut oil (Borah et al. 2018). It is reported that, compared to other plants, the palm seed contains the highest percentage of oil (Fazal et al. 2010). On the other hand, soybean can be produced with no or nearly no nitrogen, making it a cost-effective feedstock, as there is no need to use nitrogen fertilizer (Pimentel and Patzek 2005). Sunflower is the fifth largest oilseed crop cultivated across the globe. Compared to soybean, it provides a higher yield of oil per unit area. It also requires less fertilizer and water. However, due to its high iodine number and low oxidation stability, sunflower oil is not suitable for use as a fuel (Li and Khanal 2016). Coconut oil is another source used in biodiesel production. It largely consists of saturated fatty acids and has a high biodiesel yielding capacity. It has been observed that biodiesel derived from coconut can augment the mileage of automobile by 1–2 km and decrease emission levels by 60% (Li and Khanal 2016).

The high price of edible vegetable oils and their use in human food restrict their use for biodiesel production. Hence, non-edible oils are the preferred feedstock choice (Borah et al. 2018). Second-generation biodiesels are produced from non-edible oils including jatropha, lineseed oil, neem oil, and karanja oil (Fazal et al. 2019). Jatropha oil has the highest acid value among them, which has the potential to cause corrosion of the fuel system material. Jatropha seeds contains approximately 20–60% oil, consisting mostly of unsaturated fatty acids. Karanja is cultivated in the United States, China, and countries of South East Asia. It yields approximately 25–40% oil. However, linseed usually contains a high amount of unsaturated fatty acids and less saturated ones (Singh et al. 2020). Neem is generally found in South Asian countries. After 15 years, neem seed can produce approximately 20–30% oil, consisting mainly of unsaturated fatty acids (Singh et al. 2020). Non-edible oil plants can grow in arid to semi-arid regions, which are favorable for the cultivation of food plants. They are not currently grown in large areas. Publicity for their use in biodiesels could thus help stimulate local economies (Singh et al. 2020).

7.2.2 Algae

Micro- and macroalgae are potential sources of biodiesel synthesis. Microalgae are considered to be the feedstocks for third-generation biodiesel. They have gained wide attention in the past two decades as they can generate a huge amount of lipids that are suitable for biodiesel (Chandra et al. 2020). Of the two types, microalgae are preferred due to their high yielding capacity, making them the more economically viable source of biodiesel. They can be grown in both open and closed pond systems. In an open pond system, water temperature, environmental conditions, and lighting are not controlled. It is beneficial in terms of being less expensive and providing greater production capacity. A closed pond system, on the other hand, involves growing algae in a greenhouse. It requires a smaller area and offers controlled conditions for cultivation, keeping the algae protected from both living and non-living contamination. Like non-edible plants, algae can be grown in both arid and semi-arid conditions, and their per unit area yield is supposed to be higher than that of tropical oil plants (Singh et al. 2020).

More than 25 000 species of microalgae have been identified, of which only 15 are used for biodiesel production. The vast majority have not been tested in this regard. Microalgae mainly consist of diatoms, blue-green algae, green algae, and golden brown algae. They are photosynthetic organisms with a high oil content (up to 80%) in dry biomass and a doubling time of 24 hours, and are easy to grow (Borugadda and Goud 2012). *Chlorella vulgaris* is one of the preferred choices for biodiesel production. It is rich in lipid content and easy to cultivate (Nematian et al. 2020). Other popular algae include *Chlorella protothecoides*, *Botrycoccus* sp., *Phaeodactylum tricronutum*, *Chlorococcum infusarium*, and *Nannochloropsis oceanica* (Nematian et al. 2020). Many can be grown in saline water, even in polluted water bodies and oceans, making them beneficial to cultivate under any conditions (Brennan and Owende 2010).

7.2.3 Animal Fats

Animal processing facilities are the main source of animal fat. Fat largely contains triglycerides, proteins, water, and several minerals. It is collected as waste from the fishery and meat industries, the agri-food industry, slaughterhouses, and dairies, and also from restaurant grease trap waste. It is generally obtained from fish, cattle, chicken, and pigs. The fatty acid wastes produced therein are categorized as yellow and brown grease. However, animal fat usually contains yellow grease, due to the advanced state of oxidation (Abishek et al. 2014; Ndiaye et al. 2020).

Animal fat is an important feedstock for biodiesel as it is cheap. The biodiesel produced from animal fat containing saturated fatty acids has certain benefits, such as short delay in ignition, stability under oxidation, and good calorific value. Due to frequent scandals related to animal illness, animal fats are not preferred for use as foods. Hence, they can be used for biodiesel production. For example, beef tallow is a key feedstock for biodiesel production. Yellow grease and lard, the main byproducts of the leather and meat industries, are also used for this purpose (Jayathilakan et al. 2011). However, there is a major limitation related to their irregular supply, hence animal fat is not primarily used for biodiesel production. Moreover, the high percentage of saturated fatty acids makes them unsuitable for use under low temperature condition, and they can't be used at all in cold countries (Adewale et al. 2015). The high impurity and acidity of fatty effluents from the food industry further make them difficult to convert into biodiesel (Ndiaye et al. 2020).

7.2.4 Waste Cooking Oils

The cost of biodiesel is majorly dependent on the cost of feedstock. Compared to animal fats and vegetable oils, WCOs collected from household waste represent a less expensive substitute. Their use in biodiesel production can help reduce the overall production cost of biodiesel. In addition, waste oils can be collected from the food industry, non-food industries, and restaurants. Reports suggest that WCOs from soybean, palm, coconut, rapeseed, and other edible sources can be promisingly used for biodiesel production. As per Canakci (2007), the annual collection of waste fat from fast-food establishments and restaurants in the United States is approximately 2.5 billion pounds. Furthermore, the Kyoto has made use of WCOs from various sources to generate biodiesel for its buses. However, there is a problem associated with the use of waste oil for biodiesel: it contains large quantities of free fatty acids (FFAs). It thus cannot be converted to biodiesel in the presence of alkaline catalysts, as it forms their salt (i.e. soap), preventing the separation of the biodiesel from glycerine. Furthermore, these oils require extra processing in order to produce clean biodiesel (Knothe and Steidley 2009).

7.3 Conventional Heterogeneous Catalysts

As mentioned earlier, biodiesel is usually obtained via transesterification of vegetable oils or animal fats. To accelerate the reaction, distinct catalysts are used that can be categorized into homogeneous and heterogeneous types. Between them, heterogeneous catalysts have emerged as the more desirable, overcoming many of the disadvantages associated with the use of the homogeneous type. They are capable of producing high yields of biodiesel, they are noncorrosive, they generate less impurities in the final product (leading to a reduction in the purification cost), they can be easily recovered and reused (leading to industrial advantage), and as they involve less water they are environment friendly, leading them to be counted among the new green catalysts being developed to reduce reaction time and cost compared with existing catalysts.

Many heterogeneous catalysts, both acid and base, have been studied by many researchers in the context of the transesterification reaction. In comparison to base catalysts, acid catalysts are of greater interest, since they are least sensitive to the FFA and moisture content of feedstock and are able to carry out esterification and transesterification simultaneously (Barot et al. 2017). Some of the solid acid catalysts studied for transesterification are sulfonic acid functionalized mesoporous SBA-15, KIT-6, Cs-exchanged heteropolyacids, tungstate zirconia, zirconium phosphate, Nafion/Silica composite (SAC-13), and sulfated zirconia (Barot et al. 2017). It has been found that sulfated zirconia displays significant activity toward the transesterification of model triglyceride triacetin under mild reaction conditions, with a maximum 96% yield of monoalkyl ester.

However, it has also been reported that acid catalysts usually entail a high cost for their preparation, require more severe reaction conditions, may undergo deactivation, have low activity, and require high reaction temperature and longer reaction time as compared to base heterogeneous catalysts (Babak et al. 2013; de Lima et al. 2016). Some of the basic heterogeneous catalysts capable of transesterification studied to date include hydrotalcites (HTs, hydroxycarbonates of magnesium and aluminum), alkaline earth metal oxides such as calcium and magnesium oxides, and various types of support materials such as ion-exchange resins, mixed oxides, and zeolites.

Hydrotalcites are minerals belonging to the family of anionic clays, containing anionic species in the interlamellar space. These materials have the following general formula: $[M^{II}{}_{1-x}\, M^{III}\, x\,(OH)_2]^{x+}\, A_{x/n}{}^{n-}\,.\, nH_2O$ (M^{II} = Mg, Fe, Co, or Zn; M^{III} = metallic trivalent cations; A^{n-} = anions; $x = M_{II}/(M_{II} + M_{III})$, ranges from 0.1 to 0.5) (de Lima et al. 2016). Hydrotalcite catalysts were synthesized by the co-precipitation method with a Mg : Al atomic ratio of 2 : 1 and calcination at 507 °C; when used for the transesterification of soybean oil at 60 °C, 5 wt% catalyst loading, and 12 : 1 methanol to oil molar ratio over six hours, they gave a yield of 92.5% methyl esters (de Lima et al. 2016).

Mesoporous silica nanoparticles have recently garnered substantial attention as a means of supporting hydroxides and oxides, especially in catalysis. SBA-15 and MCM-41, well known as hexagonally ordered mesoporous materials, are considered exceptional supports for heterogeneous catalysts owing to their large pore size, high surface area, and thermal stability. These materials are composed of siloxane groups in the bulk of the material and silanol groups on the external surface. Xie and Zhao (2014) performed a number of experiments to study the transesterification of soybean oil with $CaO\text{-}MoO_3$-SBA-15 catalyst prepared by impregnation . The best yield of 83% conversion to biodiesel was observed with reaction conditions of 50 : 1 methanol to oil molar ratio, 6 wt% catalyst loading, 65 °C reaction temperature, and 50 hours reaction time. The catalyst was reused after calcination at 550 °C for five hours. The conversion gradually decreased, achieving 77% after the fifth run. However, the calcium and molybdenum contents of the biodiesel produced were

21 and 13 ppm, respectively, above the permissible limit set by the EN14214 standard of 5 ppm, necessitating further purification of the biodiesel to eliminate residual metal (de Lima et al. 2016).

Several studies have been reported in the literature that use oxides as heterogeneous catalysts for the transesterification reaction, including ones involving the use of calcium oxide, magnesium oxide, and calcium oxide impregnated with potassium acetate or silica (Sousa et al. 2019). Along with commercial calcium oxide, calcium oxides obtained from the calcination of eggshells have been studied, both of which showed a high efficiency in conversion to fatty acid methyl esters (FAMEs) with a methanol to oil molar ratio of 12 : 1, a reaction temperature of 65 °C, and 3% (w/w) catalyst loading with stirring at 600 rpm for four hours. Experiments involving doping of calcium oxide with SiO_2 demonstrated a greater efficiency in conversion. This efficiency is attributed to an increase in the surface area of the catalyst caused by SiO_2 impregnation. After impregnation, there was a 20% increase in conversion rate, up to a maximum of 91%. For the ion-exchange resins, factors like density of cross-linking, degree of swelling, and porosity of the catalyst played a major part in the catalytic activity. Anion-exchange resins, meanwhile, showed more efficiency for biodiesel production at low temperatures on account of their low thermal stability. In terms of the process, high conversion efficiency at lower temperatures is an important gain that reduces the overall costs and energy demands, an attribute that can be listed as one of the advantageous feature of their usage in this process. However, numerous disadvantages are also associated with the use of these catalysts in terms of cost, intensive reaction conditions, and poor reusability, which ultimately increases the production cost of the biodiesel (Alhassan et al. 2015).

De Lima et al. (2016) reviewed zeolite-based heterogeneous catalysts for biodiesel production. Zeolites are cation-exchange materials that can be deployed to introduce many different cations by simple ion-exchange procedures. Approximately 150 different types of zeolite structures are available, with varying physical and chemical properties. Some of the most commercially used include A, X, USY, Beta, Mordenite, USY, and ZSM-5. Zeolites are crystalline aluminosilicates composed of connected SiO_4 and AlO_4^- tetrahedral. Different zeolite cation combinations have been studied by various researchers, including NaX zeolites synthesized from fly ash (FA/NaX) or ion-exchanged with K^+(FA/KX), various CaO catalysts supported on NaY, KL, and NaZSM-5 zeolites, and a CeO_2/HUSY zeolite that gave biodiesel yields in the range of 56%, 83.5%, 95%, and 99.8% on sunflower and soybean oil, respectively. However, zeolites have restricted use as basic catalysts for biodiesel production, due largely to their pore structure, which impairs diffusion of the triglyceride and thus requires calcination at high temperatures. Moreover, the conversions are modest and leaching of the supported active phase is a cause for concern.

An inexpensive support material made from biomass has received attention in recent years for its role in the production of heterogeneous catalyst. In this setting, bio-based or biomass-based carbon or activated carbon (AC) is attractive due to its varied properties (including desirable physiochemical properties), abundance, accessibility, and low cost (Umer et al. 2019). Umer et al. (2019) studied a highly active sulfonated tea waste (STW) catalyst prepared from tea waste material using wet impregnation with sulfuric acid as active agent as a method for the conversion of palm fatty acid distillate (PFAD) to FFA through esterification. Their work demonstrates that at 97% FFA conversion, the catalyst worked with high efficiency at a relatively low temperature of 65 °C, with 4 wt% catalyst and a PFAD to methanol ratio of 9 : 1, over 90 minutes of reaction time. The characterization of STW catalyst revealed that it has external thermal stability, very high acid density, and a porous nature. Recyclability of STW catalyst was high, and the catalyst was shown to be viable commercially because of its varied suitable properties.

Interestingly, pyrogenic black carbon, also called biochar, has recently been utilized as a heterogeneous catalyst. Biochar, obtained from oxygen-limited thermal degradation of lignocellulosic biomass, can act as an efficient support for active sites essentially due to its large surface area, low ash content, and high porosity (Balajii and Niju 2019). Notably, pyrolysis is one of the most promising methods for the conversion of plant material into biochar and its subsequent activation by sulfonation. Biodiesel generated using sulfonated biochar meets with ASTM and EN norms (US and EU Biodiesel Specifications), and biomass-derived biochar will act as an efficient heterogeneous catalyst in biodiesel production due to its yearlong availability, low cost, sustainability, and carbon-rich nature. Pablo et al. (2019) studied sulfonated heterogeneous catalyst using activated carbon produced from corncobs as a precursor to perform microwave-assisted transesterification of soybean oil with ethanol. The combination of a biomass-derived heterogeneous catalyst and a microwave-assisted process was revealed to be a promising, less energy-intensive means of carrying out transesterification of oils when compared to contemporary processes (Pablo et al. 2019).

Generally, in the synthesis of biofuels, the catalysts routinely employed are expensive or have disadvantages such as difficulty of removal from the product, instability, or low selectivity. Nanocatalysts represent a promising alternative, combining the high activity of homogeneous catalysts with the easy recoverability of heterogeneous ones. Those used in the production of biodiesel are base nanocatalysts, acid nanocatalysts, and bifunctional nanocatalysts (Zuliani et al. 2018). High specific surface and degeneration activity are the hallmarks of heterogeneous nanocatalysts (Tahvildari et al. 2015). High specific surface and large porosity are favorable for catalyst bonding with the sublayer, which augments the effectiveness of the transesterification reaction. Tahvildari et al. (2015) used a coupled CaO and MgO heterogenic nanocatalyst in the production of biodiesel from recycled cooking oil. They had observed that nano-MgO alone was not capable of catalyzing the transesterification reaction, but when used together with nano-CaO it proved capable of achieving a significant enhancement in the yield. The combined catalyst had a high contact area and its reusability was better than that of nano-CaO alone. The authors also observed that a higher ratio of nano-CaO to nano-MgO led to greater biodiesel production. Jadhav and Tandale (2017) studied the use of nano-MgO, nano-ZnO, and nano-SiO_2 in the production of biodiesel from mango (*Mangifera indica*). Optimal values for the different parameters tested were 15 : 1 methanol to oil molar ratio, 0.5 wt% nano-SiO_2 catalyst loading, 63.85 °C reaction temperature, 300 rpm stirrer speed, and 90 minutes reaction time. A high yield with the use of nano-SiO_2 catalyst was attributed to its super-acidity, which had a positive influence on the catalytic reaction. Bet-Moushoul et al. (2016) studied the use of Ag/bauxite nanocomposites in biodiesel production. They concluded that an increased catalytic activity was a consequence of the large porous structure of the nanocomposite, which contributed to a greater surface area for contact between the alcohol and the oil, leading to a higher efficiency of the transesterification reaction.

To conclude, broadly heterogeneous catalysts can provide an efficient and effective conversion of feedstock to biodiesel, with a beneficial attribute of recyclability (Sharma et al. 2018). The synthesis of nanocomposites, comprising both acid and base sites, with a larger pore distribution and greater surface to volume ratio, can contribute immensely to the commercialization of biodiesel production (Sze et al. 2019). Figure 7.1 shows a schematic illustration of different types of heterogeneous catalysts used in biodiesel production via the transesterification reaction.

7.4 Carbon-Based Heterogeneous Nanocatalysts

As already mentioned, conventional heterogeneous catalysts have several limitations, including fast deactivation, low efficiency, mass-transfer resistance, time inefficiency, and low industrial applicability (Pandya et al. 2019). Therefore, several research groups have attempted to develop highly efficient and novel catalysts for the production of biodiesel by various routes. In this context, development of nanocatalysts for biodiesel production is highly desirable and anticipated (Mandotra et al. 2018; Srivastava et al. 2020). Nanocatalysts with favorable features such as a high specific surface area and high catalytic activity have the ability to resolve the most common problems of heterogeneous catalysts (Akia et al. 2014; Ingle et al. 2020).

Catalytic processes are commonly employed in biodiesel production, and nanotechnology can contribute effectively in this regard (Palaniappan 2017; Nizami and Rehan 2018). They are preferred over conventional catalysts because of their role in product quality improvement and the achievement of optimal operating conditions (Akia et al. 2014; Akubude et al. 2019). Further, they exhibit several other notable advantages, such as high reactivity, good selectivity, and high yield (Velan et al. 2014; Roldan-Cuenya and Behafarid 2015). Additional benefits associated with nanocatalyst-mediated esterification or transesterification mainly include an enhanced rate of mixing with reactants, reduced reaction times, and fast and easy separation from the reaction mixture (Polshettiwar and Varma 2010).

"Nanomaterial" is a broad term used for a variety of materials including nanoparticles, nanotubes, nanosheets, nanocomposites, nanocrystalline materials, metal-based nanomaterials, and carbon-based nanomaterials (Palaniappan 2017). Various physical, chemical, and biological methods have been proposed by different groups for the synthesis of a variety of such materials (Gardy et al. 2019). Nanomaterials with features such as small size, high surface area, insolubility, recyclability, and stability have been tested for the production of biodiesel. The results reveal that such nanocatalysts could lead to more efficient transesterification reactions (Xu et al. 2019; De and Boxi 2020). Given these facts, several studies have looked at the use of various nanocatalysts in biodiesel production, such as metal-based nanocatalysts prepared from metal oxides (e.g. zinc oxide, Borah et al. 2019, Massaro et al. 2020; calcium oxide, Foroutan et al. 2020, Krishnamurthy et al. 2020; titanium dioxide, De and Boxi 2020), supported metal oxides (Montero et al. 2010), zeolites/nanozeolites (de Vasconcellos et al. 2018; Fattahi et al. 2019), and magnetic nanocatalysts (Ambat et al. 2019; Alaei et al. 2020; Dantas et al. 2020). Several reports further indicate the use of various carbon-based materials in catalysis as heterogeneous catalysts or support materials

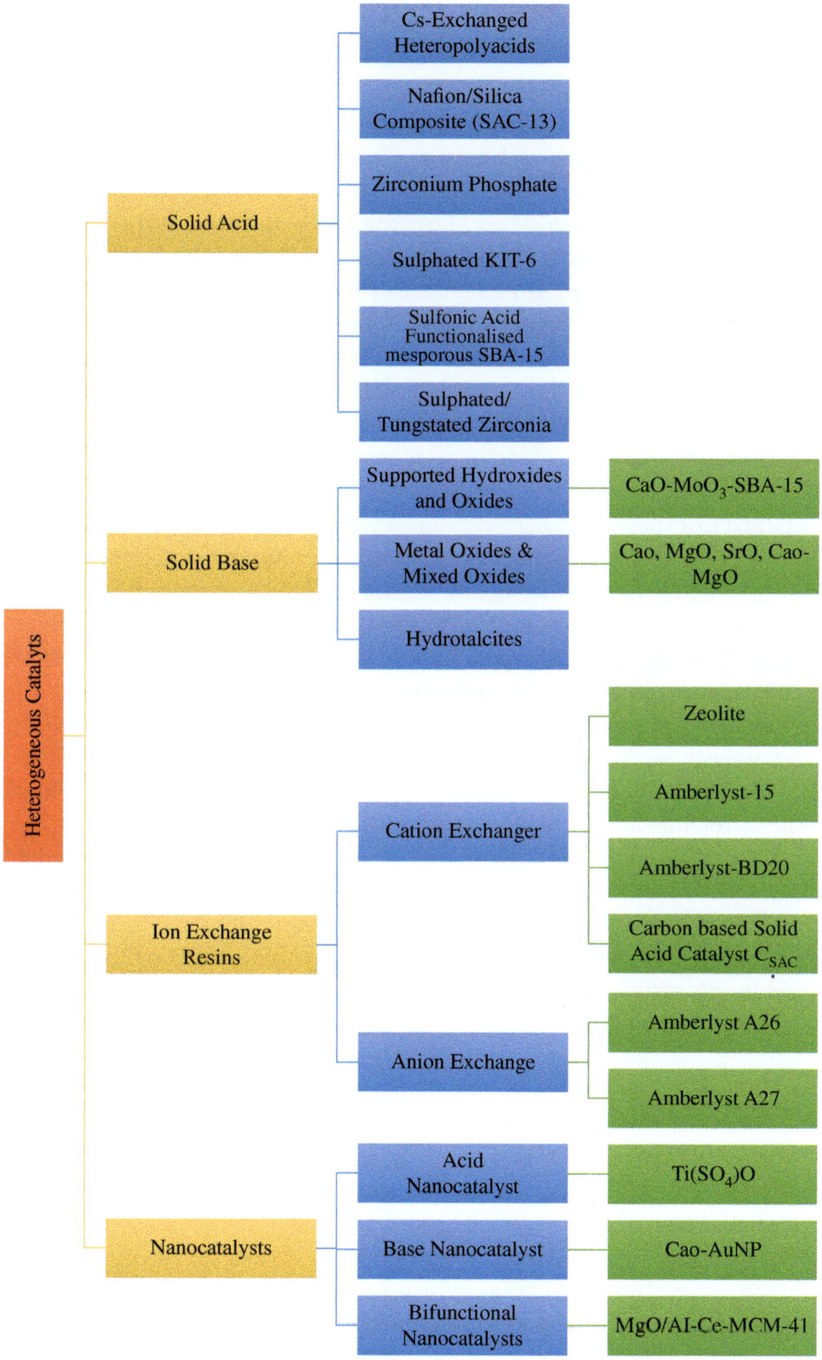

Figure 7.1 Different types of heterogeneous catalysts used in transesterification for biodiesel production.

(Abdulkareem-Alsultan et al. 2019; Ibrahim et al. 2020). However, it was proposed that the application of nanosized carbon-based catalysts is more promising as far as biodiesel production is concerned. They show low cost, wide availability, good thermal stability, high stability in acidic and basic conditions, tunable hydrophobicity and polarity, and easy recovery from the reaction mixture (Bahuguna et al. 2019; Ingle et al. 2020).

Carbon can exist in numerous forms and anisotropic variants and acts as a versatile material that can be chemically adapted via functionalization and doping. Various carbaceous materials like amorphous carbon (Hara 2009; Nakajima and Hara 2012), biochar (Dehkhoda et al. 2010; Xiong et al. 2017; Gholizadeh et al. 2019), sugar (Wang et al. 2015; Janaun et al. 2017), waste (Shi and Liang 2019; Fonseca et al. 2019), activated carbons (AC), graphite, diamond, graphene oxide (Zhang et al. 2018; Borah et al. 2018; Mohammadi et al. 2017), and carbon nanoforms including graphene oxide (GO), graphene, fullerene, carbon nanotubes (CNTs) (Shuit and Tan 2014a; Guan et al. 2017; Ibrahim et al. 2020), carbon nanofibers (CNFs) (Stellwagen et al. 2013), and carbon nanohorns (CNHs) have been used in various catalytic reactions, including biodiesel production (Collavini and Delgado 2016; Prinsen and Luque 2019).

The findings obtained from several research groups reveal that carbon-based nanomaterials, including acidic- and basic-activated carbon, sulfonated CNTs, and supported carbon materials, represent promising heterogeneous catalysts for use in biodiesel production (Liu et al. 2008; Macina et al. 2019; Ibrahim et al. 2020). Some studies propose they show other advantages over other heterogeneous catalysts due to their simple and inexpensive synthetic routes, high thermal stability, and metal-free composition (Konwar et al. 2014; Srivastava et al. 2020). However, Mahto et al. (2016) report certain limitations associated with carbon-based nanocatalysts, mainly involving deactivation, an excessive methanol to oil ratio, side reactions of active sites, high temperature requirements, and long reaction times. In this context, the development of sulfonated carbon-based nanocatalysts through the functionalization of normal carbon-based nanomaterials by the addition of SO_3H groups is preferred because of their lower cost and improved stability (Konwar et al. 2019). Moreover, sulfonated acid-functionalized solid acid catalysts have been prepared by modifying carbon supports with sulfuric acid (Zhang et al. 2020). Various carbon-based nanocatalysts are briefly discussed in this section.

7.4.1 Carbon Nanotubes

Recently, much attention has been given to the synthesis of novel nanomaterials like CNTs that have numerous applications in diverse fields. Iijima (1991) was the first to demonstrate the synthesis of CNTs using an arc evaporation approach. CNTs are carbon allotropes with a cylinder-like nanostructure. They are classified as either single-walled carbon nanotubes (SWCNTs), which have a single graphene sheet rolled into a tube, or multiwalled carbon nanotubes (MWCNTs), which consist of a number of concentric SWCNTs with different diameters (Iijima 2002; Kumar et al. 2018). CNTs exhibit extraordinary features such as a desirable structure, high surface area, high chemical stability, and low toxicity, and they can be employed as both catalysts and supports (Peng et al. 2005). Several research groups have noted the importance of CNTs in several catalytic processes, including biofuel production (Ibrahim et al. 2020). One demonstrated the synthesis of Al^{3+}-SO_4^{2-}/MWCNTs and SO_4^{2-}/MWCNTs by the impregnation of MWCNTs with concentrated sulfuric acid and Al^{3+}. Additionally, Al^{3+}-SO_4^{2-}/SWCNTs and Al^{3+}-SO_4^{2-}/C catalysts were prepared by impregnated concentrated sulfuric acid and Al^{3+} on SWCNTs and activated carbon (C), respectively. Thus synthesized SO_4^{2-}/MWCNTs, Al^{3+}-SO_4^{2-}/MWCNTs, Al^{3+}-SO_4^{2-}/SWCNTs, and Al^{3+}-SO_4^{2-}/C (CNT-based nanocatalysts) were tested for catalytic efficiency in biodiesel production via esterification of oleic acid and methanol, with Al^{3+}-SO_4^{2-}/MWCNTs showing the highest conversion ratio of oleic acid at 95% at 65 °C with 0.9 wt% catalyst, 12 : 1 methanol to oleic acid molar ratio, and seven hours reaction time. About 80% conversion was reported for the other tested catalysts (i.e. Al^{3+}-SO_4^{2-}/SWCNTs and Al^{3+}-SO_4^{2-}/C). Further, it was noted that the reusability of Al^{3+}-SO_4^{2-}/MWCNT catalysts was about 81.8% after eight cycles, indicating a high catalytic stability (Shu et al. 2017).

In another study, Deep et al. (2015) demonstrated the preparation of lipase-conjugated MWCNT catalyst via chemical reaction. They examined the catalyst produced for the production of biodiesel via methanolysis of crude Jatropha oil, and compared the further catalytic activities of immobilized and non-immobilized lipase catalyst. The results revealed that FAME production (reaction time one hour) increased linearly with increasing fraction of the catalyst support, and the maximum yield of the desired product was achieved at 15% of the immobilized catalyst. Moreover, it was also reported that MWCNTs could be effectively used as supports for the immobilization of lipase, and thus developed carbon-based nanocatalysts could be reused up to 10 times without any loss of enzyme activity. A maximum 90% FAME yield was recorded for lipase-conjugated MWCNT catalyst, compared to only 10–15% for non-immobilized lipase catalyst (Deep et al. 2015). Similarly, Ibrahim et al. (2020) recently investigated the synthesis of CNTs with impregnation of sodium oxide (Na_2O) and evaluated their potential as a heterogeneous catalyst for the transesterification of WCO. The results revealed that Na_2O (20 wt%)/CNTs catalysts showed FAME yields >97% under optimum reaction conditions (e.g. methanol to oil molar ratio of 20 : 1, catalyst loading of 3 wt%, reaction temperature of 65 °C, reation time of three hours). Moreover, the catalyst could be reused for up to three cycles without any notable loss in activity. Hence, it can be concluded that a thus prepared Lewis acidic carbon-based nanocatalyst could be used as efficient heterogeneous catalyst for the production of biodiesel from WCOs.

7.4.2 Sulfonated Carbon Nanotubes

The common problem associated with normal CNTs is their deactivation upon loss of catalyst surface area and agglomeration. This can be solved by through sulfonation. In this context, sulfonated multiwalled carbon nanotubes (S-MWCNTs) have been developed as highly effective heterogeneous acid catalysts for use in transesterification reactions (Guan et al. 2017). Given the advantages of functionalized carbon materials, numerous studies have been conducted using sulfonic acid-functionalized CNTs for biodiesel production from diverse feedstocks, including cottonseed oil, bio-oil, PFAD, and trilaurin (Gardy et al. 2019). Guan et al. (2017) reported the synthesis of a high-performing S-MWCNT exhibiting a Brunauer–Emmett–Teller (BET) surface area of 198.9 m^2/g and pore sizes ranging from 5 to 35 nm. This solid acid catalyst was further examined for its catalytic efficacy in the transesterification of trilaurin in ethanol. It showed a biodiesel yield of 97.8% for triglycerides under optimum reaction conditions of 20 : 1 ethanol to trilaurin mass ratio, 3.7 wt% catalyst loading, 150 °C reaction temperature, and one hour reaction time. It further demonstrated a higher catalytic efficiency for the transesterification of triglycerides than other catalysts like hydrothermal carbonization-synthesized sulfonated carbon and metal oxide catalyst WO_3/ZrO_2 (Guan et al. 2017). Moreover, Shuit et al. (2015) demonstrated the preparation of S-MWCNTs by adoption of a facile and acid-free approach to sulfonation via thermal decomposition of ammonium sulfate ($(NH_4)_2SO_4$) with the aid of ultrasonication. The catalytic activity of the resultant S-MWCNTs was examined through esterification of PFAD with methanol and shown to give a biodiesel yield of 84.9%.

Shuit and Tan (2014b) investigated the effects of various sulfonation methods on the MWCNTs used for the esterification of PFAD in the production of biodiesel. They noted that S-MWCNT obtained through *in situ* polymerization of 4-styrene sulfonate resulted in a FAME yield of 93.4% under reaction conditions of 20 : 1methanol to oil molar ratio, 2 wt% catalyst loading, 170 °C reaction temperature, and three hours reaction time, whereas that prepared via thermal treatment with concentrated sulfuric acid yielded 78.1% FAMEs. Moreover, the *in situ* polymerization of acetic anhydride and sulfuric acid gave an 85.8% yield, while the thermal decomposition of ammonium sulfate gave 88.0%. In another report, Shu et al. (2009a) studied the sulfonation of MWCNTs using concentrated sulfuric acid at 210 °C. They reported an acid site density 3.09 mmol/g and an average pore size 7.48 nm for the S-MWCNT catalyst. They also evaluated the production of biodiesel from feedstocks with FFAs using S-MWCNT, showing a maximum conversion of oleic acid (95.46%) at 135 °C, assisted by 0.2 wt% catalyst loading, 5.8 : 1 methanol to oleic acid molar ratio, and 1.5 hours reaction time. Similarly, in another study by the same authors, synthesis of S-MWCNTs with high acid site $-SO_3H$ density (2.04 mmol/g), 41.27 m^2/g surface area, and 7.48 nm average pore size was demonstrated and a maximum biodiesel yield of 89.9% was recorded from cottonseed oil under optimized reaction conditions (Shu et al. 2009b).

7.4.3 Graphene/Graphene Oxide-Based Nanocatalysts

Current research interests are remarkably focused on the preparation of nanocomposites from graphene. Graphene is considered an atomic layer of graphite. It acts as a basic unit of fullerenes and CNTs and has optical, mechanical, thermal, electronic, physical, and chemical properties. Structural information reveals that it is a 2D molecule with a single atomic layer of sp^2-hybridized carbon atoms closely packed into a honeycomb lattice by σ bonding (Kumar et al. 2018). Graphene oxide (GO) is increasingly popular within the research community due to its unique properties, including mechanical strength and chemically stable structure (Zhu and Xu 2010; Zaaba et al. 2017). It can be obtained from graphite and graphite oxide on a large scale and can be synthesized from waste materials like waste plastics, biomass wastes, wood, leaf, and animal bones (Kumar et al. 2018).

Mahto et al. (2016) reported the synthesis of sulfonated GO and tested it for esterification of oleic acid. This catalyst was obtained by single-step sulfonation of GO with concentrated sulfuric acid, and showed a high surface area and S_3OH groups. The results recorded for esterification indicate that this carbon-based nanocatalyst has the ability to convert 97.6% oleic acid at 100 °C after four hours with 0.1972 mol of methanol in 2 ml of oil and 5 wt%

of catalyst loading. Based on these findings, it was concluded that the sulfonated groups ($-SO_3H$) were responsible for the efficient esterification (Mahto et al. 2016).

Nongbe et al. (2017) demonstrated a simple and efficient method for the preparation of sulfonated graphene catalyst from graphite, subsequently functionalized with benzene-sulfonic acid. They then used this catalyst for the transesterification of palm oil with methanol into FAMEs and reported a 98% yield under optimum reaction conditions of 20 : 1 methanol to oil molar ratio, 10 wt% catalyst loading, 100 °C reaction temperature, and 10 hours reaction time. The authors further suggested that thus developed sulfonic acid-supported graphene catalysts may be able to overcome the problems of limited robustness and extensive leaching in the aqueous phase due to their partial solubility in methanol.

Jun et al. (2017) examined the catalytic activities of graphene and its derivatives, such as GO, sulfonated graphene, sulfonated GO, and sulfonated active carbon, for the conversion of lipids from wet microalgae into biodiesel. All the tested catalysts showed different conversion rates and efficiencies for microwave-assisted lipid transesterification. It was reported that sulfonated GO showed the best catalytic efficacy and hence the highest conversion efficiency of wet microalgae lipid at 84.6%. The conversion efficiency for sulfonated graphene was found to be 48.6% and that for GO was 73.1%. In another study, the same authors reported a yield of about 96% biodiesel from the transesterification of microalgal lipids with methanol at 90 °C and 5% catalyst loading. Hence, they concluded that wet microalgae lipids can be efficiently converted into FAMEs using a solid GO catalyst (Jun et al. 2016). However, *in situ* decorated titanium oxide (TiO_2) on reduced graphene oxide (RGO) nanocatalyst (in-situ TiO_2/RGO nanocomposite) obtained via the hydrothermal route showed efficient transesterification of WCO with biodiesel conversion of 98% (Borah et al. 2018). *Candida rugosa* lipase conjugated on to Fe_3O_4-GO nanocomposite yielded 92.8% biodiesel at 40 °C through a three-step addition of methanol (Xie and Huang 2018). Loy et al. (2019) proposed an approach for the development of bifunctional heterogeneous catalysts (i.e. silicon carbide/sodium hydroxide-GO, SiC/NaOH-GO) and evaluated their potential in microwave-assisted transesterification and esterification of rapeseed oil and oleic acid (binary model) containing 20% FFAs. Their results indicated that a FAME yield of 81% could be achieved from FFA of microalgae lipid at 85 °C, with a methanol to lipid molar ratio of 48 : 1, 4 wt% catalyst loading, and five minutes reaction time. Moreover, another report demonstrated the synthesis of a novel sulfonated graphitic carbon nitride (Sg-CN) as an organosulfonated heterogeneous acid catalyst via simple sulfonation and evaluated its catalytic performance in the efficient production of biodiesel (Nasir Baig et al. 2016). Thus synthesized Sg-CN catalyst was found to have potential catalytic activity, as confirmed by a high conversion of oleic acid (>99%) to the corresponding methyl esters at room temperature over four hours of reaction time. It was also noted that there was no significant loss in activity for up to five cycles (Nasir Baig et al. 2016).

7.4.4 Carbon Nanofibers and Carbon Dots

In addition to the various carbon-based heterogeneous nanocatalysts already discussed, there are some other new classes of carbon-based nanocatalysts that are also used in biodiesel production, such as CNFs and carbon dots. Stellwagen et al. (2013) demonstrated a simple and efficient approach for the preparation of functionalized CNF catalyst with aryl sulfonic acid groups using a selective diazonium coupling reaction. Thus developed functionalized CNF (CNF-Ar-SO_3H) showed a total acid site (sulfonic acid group) load of 0.62 mmol/g with high reactant molecule accessibility to the -Ar-SO_3H sites. This nanocatalyst was examined for the transesterification of triolein and showed a 72% yield of methyl oleate at 120 °C with 1 : 10 triolein to methanol molar ratio, 0.75 g catalyst loading, and four hours reaction time. Moreover, it was observed that the same nanocatalyst could be recovered and reused for up to four cycles of transesterification following simple treatments.

Recently, Macina et al. (2019) developed a novel heterogeneous carbon dot catalyst via hydrothermal reaction of glycine and citric acid and evaluated its catalytic potential for the transesterification of canola oil toward biodiesel production. The authors achieved biodiesel conversions of 97% at 150 °C and 1 wt% catalyst loading. Moreover, it has been found that this nanocatalyst exhibits high stability and reusability with unchanged catalytic efficiency for at least five reaction cycles.

Although there is limited literature available on these nanocatalysts, the findings recorded in previous studies clearly indicate the catalytic potential of such carbon-based nanocatalysts in the transesterification of a variety of feedstocks. It is essential to perform extensive studies on these aspects so as to investigate the catalytic efficacy of these nanocatalysts for different feedstocks and different reaction conditions.

7.4.5 Carbon Nanohorns

The search for novel and vastly effective solutions toward heterogeneous catalysts led to the introduction of CNH-based catalysts for esterification and transesterification. Recently, the scientific community has focused on the applicability of these carbon nanostructures to various reactions. CNHs are conical-shaped carbon nanostructures that exhibit remarkable properties as compared to other carbon nanostructures (Karousis et al. 2016). They have a comparatively larger diameter (2–5 nm) than the typical CNTs (1.4 nm), and a length of about 40–50 nm. They are also called carbon nanocones and are grouped into single-walled carbon nanohorns (SWCNHs) and multiwalled carbon nanohorns (MWCNHs) (Karfa et al. 2018). SWCNHs are considered aggregates of SWCNTs with closed ends (Iijima et al. 1999). Tsang et al. (1993) derived these carbon nanostructures from the waste product of shoot formed during fullerene production via an arc discharge process. They are considered superior alternatives to other CNTs, having higher chemical stability, more purity, greater specific surface area, higher yield, lower toxicity, exceptional catalytic properties, greater porosity, and good conductivity (Zhu and Xu 2010; Karfa et al. 2018).

Several reports claim the suitability of CNH-based catalysts for biodiesel production via esterification and transesterification. Sano et al. (2017) demonstrated the preparation of CNHs dispersed with a calcium ferrite $Ca_2Fe_2O_5$ ($Ca_2Fe_2O_5$-CNH) as a new catalyst and studied its catalytic performance in the conversion of triglyceride to FAMEs via transesterification. They found that a maximum yield of about 95% could be achieved using this catalyst in two hours of reaction time, increasing to 100% in three hours, and that it could be reused for up to three cycles of subsequent reactions without any loss of activity; however, a reduction in biodiesel yield to 86.6% was recorded after five cycles.

Similarly, Poonjarernsilp et al. (2014a) studied the catalytic performance of iron nanoparticles (Fe/Fe_2O_3) dispersed in sulfonated SWCNHs (SO_3H/SWCNHs) for the esterification of palmitic acid with methanol, as well as the effect of Fe loading. The synthesis of SWCNHs dispersed with iron nanoparticles (Fe-SWCNHs) was performed using a gas-injected arc-in-water (GI-AIW) method. In the case of transesterification, a yield of 92% was reported at the highest Fe loading of 13 wt% after one hour. These results indicate that higher Fe loading leads to high yields of methyl palmitate due to an increase in the adsorption of palmitic acid molecules by Fe/Fe_2O_3 on the surface of the Fe_2O_3 nanoparticles. Moreover, it has been found that the catalytic activity and magnetic susceptibility of SO_3H/Fe-SWCNHs catalyst can last for a few cycles of repeated use, unless Fe loading is high.

In another study, the same authors examined the synthesis of carbon-based solid acid catalysts from four different carbon materials – SWCNHs, oxidized single-walled carbon nanohorns (ox-SWCNHs), activated carbon (AC), and carbon black (CB) – using hydrothermal sulfonation at 200 °C in an autoclave reactor, labeling the results SO_3H-SWCNH, SO_3H-ox-SWCNH, SO_3-AC, and SO_3H-CB, respectively (Poonjarernsilp et al. 2014b). All four carbon-based nanocatalysts were evaluated for their catalytic activity in biodiesel production, and it was observed that sulfonated single-walled carbon nanohorns (SO_3H-SWCNHs) exhibited the highest acid density among the four. SO_3H-SWCNHs with a BET surface area of 418 m^2/g and an acid density of 504 ($\mu mol\ g^{-1}$-cat) gave the highest yield of 93% under conditions of 1 : 33 palmitic acid to methanol molar ratio, 3 wt% catalyst loading, 64 °C reaction temperature, and five hours reaction time. However, sulfonated SWCNH catalysts dispersed with Fe/Fe_2O_3 nanoparticles (SO_3H-Fe/Fe_2O_3-SWCNHs) showed a greater than 90% yield of methyl palmitate with three to seven hours of reaction (Poonjarernsilp et al. 2015).

7.4.6 Other Carbon-Based Nanocatalysts

Nanocatalysts can be exploited as catalysts in various forms, including as supported nanocatalysts, mixed nanocatalysts, impregnated nanocatalysts, and doped nanocatalysts. They are effective and show higher conversion and yield as compared to other conventional heterogeneous catalysts under optimum conditions (Pandya et al. 2019). Therefore, nanocomposites with high surface to volume ratio and large pore distribution may represent a good solution for bulk-scale biodiesel production.

Several research groups have studied the applicability of amorphous carbon as a solid acid catalyst for various chemical reactions, including transesterification (Hara 2009; Nakajima and Hara 2012). Hara (2009) looked at the efficiency and stability of amorphous carbon bearing various acid groups (-SO_3H, -COOH, and -OH) for the esterification of oleic acid. They reported high acidity of H0 −8 to −11 and low surface area (2 m^2/g) for such catalysts. Moreover, they found high catalytic activity in the conversion of oleic acid via esterification (99.5%) (Hara 2009). Similarly, amorphous carbon material functionalized with sulfonic acid groups showed improved catalytic performance in the esterification of higher fatty acids and transesterification of triglycerides as compared to other conventional solid acid catalysts (Hara 2010). Nakajima and Hara (2012) provided a detailed review of the application of amorphous carbon catalysts in catalytic reactions. They stated that amorphous carbon bearing SO_3H groups can act as an insoluble Brønsted acid catalyst with higher catalytic activity for various acid-catalyzed reactions.

Ballotin et al. (2019) investigated sulfonated carbon nanostructures embedded in an amorphous carbon matrix produced from bio-oil and used them for the esterification of oleic acid with methanol. These materials bear sulfonic surface groups and exhibit high acidity (up to 0.2 mmol/g). Results indicated the conversion of 95% oleic acid via esterification with 1 : 10 oleic acid to methanol molar ratio, 2.5 wt% catalyst loading, and 100 °C reaction temperature.

Dehkhoda et al. (2010) prepared two novel carbon-based catalysts in the form of "biochar" by sulfonating pyrolysis with concentrated sulfuric acid and fuming sulfuric acid. Following synthesis, they examined both catalysts for their efficacy in the transesterification of vegetable oils and the esterification of FFAs. The results revealed that the catalyst sulfonated with the concentrated sulfuric acid showed significant conversion of the feedstock in FFA through esterification, while at the same time indicating limited transesterification activity. However, the catalyst sulfonated with fuming sulfuric acid showed considerably higher transesterification activity under selected optimum reaction conditions.

Kastner et al. (2012) reported more than 90 wt% of fatty acids with sulfonated biochar catalysts. González et al. (2017) evaluated the effect of adding a sulfonic group to biochar catalysts and found more SO_3H groups on the biochar surface and a higher FAME yield of up to 90%. Hence, they proposed that sulfonation significantly improves the acidity of catalysts and thus the amount of conversion in the esterification reaction.

Moreover, a few reports demonstrate the catalytic potential of carbon-based catalyst obtained from sugars for biodiesel production. Fu et al. (2015) studied the preparation of solid acid catalyst from β-cyclodextrins through incomplete hydrothermal carbonization followed by sulfonation and further used for the simultaneous esterification and transesterification of triolein containing 50.2 wt% of oleic acid. Thus synthesized sulfonated solid acid catalyst showed a FAME yield of 80% after 12 hours reaction time with a 30 : 1 methanol to oil molar ratio, 5 wt% catalyst loading, and 85 °C reaction temperature. Moreover, the activity of this catalyst was found to be unchanged even after a sixth cycle of esterification and transesterification. Janaun et al. (2017) reported the synthesis of sulfonated carbon catalyst via incomplete carbonization of D-glucose and sulfonation with concentrated sulfuric acid. They achieved an oleic acid conversion of 52% using a sulfonated granular sugar catalyst.

In another study, Fu et al. (2013) demonstrated the synthesis of a cheap and highly stable sulfonated solid catalyst exhibiting high density of active sites (4.25 mmol/g) and a surface area of 1 m^2/g using microalgae residue via *in situ* hydrothermal carbonization with sulfuric acid. The performance of the thus synthesized catalyst was evaluated for the esterification of oleic acid and the transesterification of triolein. It was found that said catalyst had great stability and reusability for up to six cycles of subsequent reactions without any loss of activity. However, Shu et al. (2010) evaluated the synthesis of low-cost carbon-based solid acid through the sulfonation of vegetable oil asphalt and tested it for the simultaneous esterification and transesterification of cottonseed oil containing 50% oleic acid. They found that conversions of FFAs and triglyceride were 80.5 and 94.8 wt%, respectively, after 4.5 hours reaction time with a 16.8 : 1 methanol to oil molar ratio, 0.2% catalyst loading, and 220 °C reaction temperature.

Activated carbon is another type of amorphous carbon exhibiting a high degree of porosity. It can be obtained economically from carbonaceous material like coal, wood, and coconut shell (Islam and Rouf 2012; Konwar et al. 2014). Many research articles report the potential of activated carbon catalyst derived from biomass and coated with

Table 7.1 Some important carbon-based nanocatalysts used in biodiesel production.

Catalyst	Feedstock	Methanol to oil molar ratio	Catalyst loading (wt%)	Temperature (°C)	Reaction time (h)	Yield/ conversion (%)	Reference
Lipase-MWCNT	Jatropha oil	20 : 1	15	60	1	90	Deep et al. (2015)
$Al^{3+}SO_4^{2-}$/MWCNT	Oleic acid	12 : 1	0.9	65	7	95	Shu et al. (2017)
Sulfonated multiwalled carbon nanotube	Triglycerides	10 : 1[a]	3.7	60	2.5	97.8	Guan et al. (2017)
S-MWCNTs	palm fatty acid distillate	20 : 1	2	170	3	93.4	Shuit and Tan (2014b)
S-MWCNTs	PFAD	20 : 1	2	170	3	84.9	Shuit et al. (2015)
S-MWCNTs	Oleic acid	5.8 : 1	0.2	135	1.5	95.46	Shu et al. (2009a)
S-MWCNTs	Cottonseed oil	20 : 1	1.5	240	2 240 rpm (mixing rate)	89.4	Shu et al. (2009b)
CNF-Ar-SO_3H	Triolein	10 : 1	0.75	120	4	72	Stellwagen et al. (2013)
SO_3H-SWCNHs	Palmitic acid	33 : 1	3	64	5	93	Poonjarernsilp et al. (2014a)
SO_3H-Fe/Fe_2O_3-SWCNH	Oil Mixture Consisting of Palmitic Acid and Tributyrin	20 : 1	NR	60	3	90	Poonjarernsilp et al. (2015)
($Ca_2Fe_2O_5$–CNH)		NR	NR	NR	3	>100	Sano et al. (2017)
Carbon dots (GlyCDs)	Canola oil	20 : 1	1	150	3	≥97	Macina et al. (2019)
Sulfonated graphene oxide	Oleic acid	1 : 0.97	5	100	4	97.6	Mahto et al. (2016)
Sulfonated graphene oxide	Palm oil	20 : 1	10	100	10	98	Nongbe et al. (2017)
Sulfonated graphene (TiO_2/RGO)	WCO	20 : 1	1.5	65	3	98	Borah et al. (2018)
(SiC/NaOH-GO)	Rapeseed oil and oleic acid	48 : 1	4	85	5	81	Loy et al. (2019)
Sulfonated graphitic carbon nitride (Sg-CN)	Oleic acid	5 : 1	1	RT	4	>99	Nasir Baig et al. (2016)
(K_2CO_3/KLC)	Rapeseed oil and methanol	15 : 1	3	65	2	99.6	Li et al. (2013)

[a] Ethanol to oil molar ratio. NR, not reported; RT, room temperature.

certain active metals that enhance its catalysis. For instance, Li et al. (2013) investigated the synthesis of potassium carbonate (K_2CO_3) catalyst-supported Kraft Lignin activated carbon (K_2CO_3/KLC) via an *in situ* route involving mixing of K_2CO_3 with Kraft lignin followed by activation at 800 °C for two hours under nitrogen (N_2) flow. They evaluated this catalyst for the transesterification of rapeseed oil with methanol and obtained a biodiesel yield of 99.6% with a 0.6 K_2CO_3/KL mass ratio and activation at 800 °C under reaction conditions of 15 : 1 methanol to rapeseed oil molar ratio, 3.0 wt.% catalyst loading, 65 °C reaction temperature, and two hours reaction time.

From these various studies, it is clear that carbon-based heterogeneous nanocatalysts have several novel and promising features in comparison with conventional heterogeneous catalysts and normal carbon-based heterogeneous catalysts. Table 7.1 summarizes some important carbon-based nanocatalysts used in biodiesel production. They have attracted a great deal of attention from the global scientific community for their utilization in various catalytic reactions, particularly in biodiesel production. There are a few reports claiming certain limitations among the nanocatalysts, but the majority point to high catalytic performance for the carbon-based ones. We strongly believe that further extensive studies are needed in this area in order to address all the concerns raised and draw some concrete conclusions.

7.5 Conclusions

Human society is well and truly entrenched in consumerism. Increasing consumption must be met by ramping up production and related activity, causing a continuous surge in demand for energy, which is currently largely met by the burning of fossil fuel. Consequently, the global carbon footprint is rapidly spiraling out of control, leading to environmental damage and climate change. We have tipped the balance, the situation is untenable, and the only way to mitigate things is to shift to alternative fuels that cause low emissions.

Efficient biodiesel production through catalyst-based transesterification of bio-oils looks promising in this regard. Conventional homogeneous and heterogeneous catalysts are capable of delivering biodiesel but are inefficient and have not yet led to any full-scale production. They are not commercially viable: either they are too expensive, cause problems in the separation of the product, have low specificity, or are unstable. Carbon nanocatalysts, on the other hand, combine desirable features of both the conventional types of catalysts and are able to deliver high-density biofuel. They have to be tested on different feeds to ascertain their reliability before they can be positioned as viable alternatives, and a few issues have still to be ironed out by further extensive research initiatives.

References

Abdulkareem-Alsultan, G., Asikin-Mijan, N., Lee, H.V. et al. (2019). A review on thermal conversion of plant oil (edible and inedible) into green fuel using carbon-based nanocatalyst. *Catalysts* 9 (4): 350.

Abishek, M.P., Patel, J., and Rajan, A.P. (2014). Algae oil: a sustainable renewable fuel of future. *Biotechnology Research International* 2014: 272814.

Adewale, P., Dumont, M.J., and Ngadi, M. (2015). Recent trends of biodiesel production from animal fat wastes and associated production techniques. *Renewable and Sustainable Energy Reviews* 45: 574–588.

Akia, M., Yazdani, F., Motaee, E. et al. (2014). A review on conversion of biomass to biofuel by nanocatalysts. *Biofuel Research Journal* 1 (1): 16–25.

Akubude, V.C., Nwaigwe, K.N., and Dintwa, E. (2019). Production of biodiesel from microalgae via nanocatalyzed transesterification process: a review. *Materials Science for Energy Technologies* 2 (2): 216–225.

Alaei, S., Haghighi, M., Vahid, B.R. et al. (2020). Conventional vs. hybrid methods for dispersion of MgO over magnetic Mg-Fe mixed oxides nanocatalyst in biofuel production from vegetable oil. *Renewable Energy* 154: 1188–1203.

Alhassan, F.H., Rashid, U., and Taufiq-Yap, Y.H. (2015). Synthesis of waste cooking oil-based biodiesel via effectual recyclable bi-functional $Fe_2O_3MnOSO_4^{2-}/ZrO_2$ nanoparticle solid catalyst. *Fuel* 142: 38–45.

Ambat, I., Srivastava, V., Haapaniemi, E., and Sillanpää, M. (2019). Nano-magnetic potassium impregnated ceria as catalyst for the biodiesel production. *Renewable Energy* 139: 1428–1436.

Babak, S., Iman, H., and Abdullah, A.Z. (2013). Alkaline earth metal oxide catalysts for biodiesel production from palm oil: elucidation of process behaviors and modeling using response surface methodology. *Iranian Journal of Chemistry and Chemical Engineering* 32: 113–126.

Bahuguna, A., Kumar, A., and Krishnan, V. (2019). Carbon support based heterogeneous nanocatalysts: synthesis and application toward organic transformation reactions. *Asian Journal of Organic Chemistry* 8 (8): 1263–1305.

Balajii, M. and Niju, S. (2019). Biochar derived heterogeneous catalysts for biodiesel production. *Environmental Chemistry Letters* 17: 1447–1469.

Ballotin, F.C., Perdigão, L.T., Rezende, M.V.B. et al. (2019). Bio-oil: a versatile precursor to produce carbon nanostructures in liquid phase under mild conditions. *New Journal of Chemistry* 43: 2430–2433.

Banković-Ilić, I.B., Stojković, I.J., Stamenković, O.S. et al. (2014). Waste animal fats as feedstocks for biodiesel production. *Renewable and Sustainable Energy Reviews* 32: 238–254.

Barot, S., Gadhiya, N., Nawab, M., and Bandyopadhyay, R. (2017). Sulphated zirconia, an efficient solid acid catalyst for transesterification of triglycerides. *Advanced Materials Proceedings* 2 (12): 819–824.

Bet-Moushoul, E., Khalil, F., Yaghoub, M. et al. (2016). Development of novel Ag/bauxite nanocomposite as a heterogeneous. *Renewable Energy* 92: 12–21.

Borah, M.J., Devi, A., Saikia, R.A., and Deka, D. (2018). Biodiesel production from waste cooking oil catalyzed by in-situ decorated TiO_2 on reduced graphene oxide nanocomposite. *Energy* 158: 881–889.

Borah, M.J., Devi, A., Borah, R., and Deka, D. (2019). Synthesis and application of Co doped ZnO as heterogeneous nanocatalyst for biodiesel production from non-edible oil. *Renewable Energy* 133: 512–519.

Borugadda, V.B. and Goud, V.V. (2012). Biodiesel production from renewable feedstocks: status and opportunities. *Renewable and Sustainable Energy Reviews* 16 (7): 4763–4784.

Brennan, L. and Owende, P. (2010). Biofuels from microalgae – a review of technologies for production, processing, and extractions of biofuels and co-products. *Renewable and Sustainable Energy Reviews* 14 (2): 557–577.

Canakci, M. (2007). The potential of restaurant waste lipids as biodiesel feedstocks. *Bioresource Technology* 98: 183–190.

Chandra, R., Vishal, G., Sánchez, C.E.G., and Uribe, J.A.G. (2020). Bioreactor for algae cultivation and biodiesel production. *Bioreactors*: 289–307.

Collavini, S. and Delgado, J.L. (2016). Carbon nanoforms in perovskite-based solar cells. *Advanced Energy Materials* 7 (10): 160100.

Dantas, J., Leal, E., Cornejo, D.R., and Costa, A.C.F.M. (2020). Biodiesel production evaluating the use and reuse of magnetic nanocatalysts $Ni^{0.5}Zn^{0.5}Fe_2O_4$. *Arabian Journal of Chemistry* 13 (1): 3026–3042.

De, A. and Boxi, S.S. (2020). Application of Cu impregnated TiO_2 as a heterogeneous nanocatalyst for the production of biodiesel from palm oil. *Fuel* 265: 117019.

de Lima, A.L., Ronconi, C.M., and Mota, C.J.A. (2016). Heterogeneous basic catalysts for biodiesel production. *Catalysis Science & Technology* 6: 2877–2891.

Deep, A., Sharma, A.L., and Kumar, P. (2015). Lipase immobilized carbon nanotubes for conversion of *Jatropha* oil to fatty acid methyl esters. *Biomass and Bioenergy* 81: 83–87.

Dehkhoda, A.M., West, A.H., and Ellis, N. (2010). Biochar based solid acid catalyst for biodiesel production. *Applied Catalysis A: General* 382 (2): 197–204.

de Vasconcellos, A., Miller, A.H., Aranda, D.A.G., and Nery, J.G. (2018). Biocatalysts based on nanozeolite-enzyme complexes: effects of alkoxysilane surface functionalization and biofuel production using microalgae lipids feedstock. *Colloids and Surfaces B: Biointerfaces* 165: 150–157.

European Union. (2018). EU Directive 2018/2001 on the Promotion of the Use of Energy from Renewable Sources. Brussels, Belgium.

Fattahi, N., Triantafyllidis, K., Luque, R., and Ramazani, A. (2019). Zeolite-based catalysts: a valuable approach toward ester bond formation. *Catalysts* 9 (9): 758.

Fazal, M.A., Haseeb, A.S.M.A., and Masjuki, H.H. (2010). Comparative corrosive characteristics of petroleum diesel and palm biodiesel for automotive materials. *Fuel Processing Technology* 91 (10): 1308–1315.

Fazal, M.A., Rubaiee, S., and Al-Zahrani, A. (2019). Overview of the interactions between automotive materials and biodiesel obtained from different feedstocks. *Fuel Processing Technology* 196: 106178.

Fonseca, J.M., Teleken, J.G., de Cinque Almeida, V., and da Silva, C. (2019). Biodiesel from waste frying oils: methods of production and purification. *Energy Conversion and Management* 184: 205–218.

Foroutan, R., Mohammadi, R., Esmaeili, H. et al. (2020). Transesterification of waste edible oils to biodiesel using calcium oxide@magnesium oxide nanocatalyst. *Waste Management* 105: 373–383.

Fu, X., Li, D., Chen, J. et al. (2013). A microalgae residue based carbon solid acid catalyst for biodiesel production. *Bioresource Technology* 146: 767–770.

Fu, X., Chen, J., Song, X. et al. (2015). Biodiesel production using a carbon solid acid catalyst derived from β-cyclodextrin. *Journal of the American Oil Chemists' Society* 92 (4): 495–502.

Gardy, J., Rehan, M., Hassanpour, A. et al. (2019). Advances in nano-catalysts based biodiesel production from non-food feedstocks. *Journal of Environmental Management* 249: 109316.

Gholizadeh, M., Hu, X., and Liu, Q. (2019). Progress of using biochar as a catalyst in thermal conversion of biomass. *Reviews in Chemical Engineering* 35: 1–35.

Godri Pollitt, K.J., Chhan, D., Rais, K. et al. (2019). Biodiesel fuels: a greener diesel? A review from a health perspective. *Science of the Total Environment* 688: 1036–1055.

González, M., Cea, M., Reyes, D. et al. (2017). Functionalization of biochar derived from lignocellulosic biomass using microwave technology for catalytic application in biodiesel production. *Energy Conversion and Management* 137: 165–173.

Guan, Q., Li, Y., Chen, Y. et al. (2017). Sulfonated multi-walled carbon nanotubes for biodiesel production through triglycerides transesterification. *RSC Advances* 7 (12): 7250–7258.

Hara, M. (2009). Environmentally benign production of biodiesel using heterogeneous catalysts. *ChemSusChem* 2 (2): 129–135.

Hara, M. (2010). Biomass conversion by a solid acid catalyst. *Energy & Environmental Science* 3: 601–607.

Hosseinzadeh-Bandba, H., Tabatabaei, M., Aghbashlo, M. et al. (2018). A comprehensive review on the environmental impacts of diesel/biodiesel additives. *Energy Conversion and Management* 174: 579–614.

Ibrahim, M.L., Nik Abdul Khalil, N.N.A., Islam, A. et al. (2020). Preparation of Na_2O supported CNTs nanocatalyst for efficient biodiesel production from waste-oil. *Energy Conversion and Management* 205: 112445.

Iijima, S. (1991). Helical microtubules of graphitic carbon. *Nature* 354: 56–58.

Iijima, S. (2002). Carbon nanotubes: past, present, and future. *Physica B* 323: 1–5.

Iijima, S., Yudasaka, M., Yamada, R. et al. (1999). Nano-aggregates of single-walled graphitic carbon nano-horns. *Chemical Physics Letters* 309: 165–170.

Ingle, A.P., Chandel, A.K., Philippini, R. et al. (2020). Advances in nanocatalysts mediated biodiesel production: a critical appraisal. *Symmetry* 12 (2): 256.

Islam, M.S. and Rouf, M.A. (2012). Waste biomass as sources for activated carbon production: a review. *Bangladesh Journal of Scientific and Industrial Research* 47: 347–364.

Jadhav, S.D. and Tandale, M.S. (2017). Optimization of transesterification process using homogeneous and nano-heterogeneous catalysts for biodiesel production from *Mangifera indica* oil. *Environmental Progress & Sustainable Energy* 37 (1): 533–545.

Janaun, J., Lim, P.M., Balan, W.S. et al. (2017). Synthesis of biodiesel in batch and packed-bed reactors using powdered and granular sugar catalyst. *IOP Conference Series: Materials Science and Engineering* 206: 012015.

Jayathilakan, K., Sultana, K., Radhakrishna, K., and Bawa, A.S. (2011). Utilization of byproducts and waste materials from meat, poultry and fish processing industries: a review. *Journal of Food Science and Technology* 49 (3): 278–293.

Jun, C., Yi, Q., Rui, H. et al. (2016). Biodiesel production from wet microalgae by using graphene oxide as solid acid catalyst. *Bioresource Technology* 221: 344–349.

Jun, C., Yi, Q., Jie, Z. et al. (2017). Conversion of lipids from wet microalgae into biodiesel using sulfonated graphene oxide catalysts. *Bioresource Technology* 244: 569–574.

Karfa, P., De, S., Majhi, K.C. et al. (2018). Functionalization of carbon nanostructures. *Reference Module in Materials Science and Materials Engineering* 2: 123–144.

Karousis, N., Suarez-Martinez, I., Ewels, C.P., and Tagmatarchis, N. (2016). Structure, properties, functionalization, and applications of carbon nanohorns. *Chemical Reviews* 116 (8): 4850–4883.

Kastner, J.R., Miller, J., Geller, D.P. et al. (2012). Catalytic esterification of fatty acids using solid acid catalysts generated from biochar and activated carbon. *Catalysis Today* 190: 122–132.

Knothe, G. and Steidley, K.R. (2009). A comparison of used cooking oils: a very heterogeneous feedstock for biodiesel. *Bioresource Technology* 100 (23): 5796–5801.

Konwar, L.J., Boro, J., and Deka, D. (2014). Review on latest developments in biodiesel production using carbon-based catalysts. *Renewable and Sustainable Energy Reviews* 29: 546–564.

Konwar, L.J., Mäki-Arvela, P., and Mikkola, J.P. (2019). SO_3H containing functional carbon materials: synthesis, structure and acid catalysis. *Chemical Reviews* 119 (22): 11576–11630.

Krishnamurthy, K.N., Sridhara, S.N., and Ananda-Kumar, C.S. (2020). Optimization and kinetic study of biodiesel production from *Hydnocarpus wightiana* oil and dairy waste scum using snail shell CaO nano catalyst. *Renewable Energy* 146: 280–296.

Kumar, D., Sharma, S., Srivastava, N. et al. (2018). Advancement in the utilization of nanocatalyst for transesterification of triglycerides. *Journal of Nanoscience and Technology* 4 (3): 374–379.

Li, Y. and Khanal, S.K. (2016). *Bioenergy Principles and Applications*. Hoboken, NJ: Wiley.

Li, X.F., Zuo, Y., Zhang, Y. et al. (2013). In situ preparation of K_2CO_3 supported Kraft lignin activated carbon as solid base catalyst for biodiesel production. *Fuel* 113: 435–442.

Liu, R., Wang, X., Zhao, X., and Feng, P. (2008). Sulfonated ordered mesoporous carbon for catalytic preparation of biodiesel. *Carbon* 46 (13): 1664–1669.

Loy, A.C.M., Quitain, A.T., Lam, M.K. et al. (2019). Development of high microwave-absorptive bifunctional grapheme oxide-based catalyst for biodiesel production. *Energy Conversion and Management* 180: 1013–1025.

Macina, A., De Medeiros, T.V., and Naccache, R. (2019). A carbon dot-catalyzed transesterification reaction for the production of biodiesel. *Journal of Materials Chemistry A* 7: 23794–23802.

Mahto, T.K., Jain, R., Chandra, S. et al. (2016). Single step synthesis of sulfonic group bearing graphene oxide: a promising carbo-nano material for biodiesel production. *Journal of Environmental Chemical Engineering* 4 (3): 2933–2940.

Mandotra, S.K., Kumar, R., Upadhyay, S.K., and Ramteke, P.W. (2018). Nanotechnology: a new tool for biofuel production. In: *Green Nanotechnology for Biofuel Production, Biofuel and Biorefinery Technologies 5* (eds. N. Srivastava, M. Srivastava, H. Pandey, et al.). Berlin: Springer.

Massaro, M., Casiello, M., D'Accolti, L. et al. (2020). One-pot synthesis of ZnO nanoparticles supported on halloysite nanotubes for catalytic applications. *Applied Clay Science* 189: 105527.

Mofijur, M., Rasul, M.G., Hyde, J. et al. (2016). Role of biofuel and their binary (diesel–biodiesel) and ternary (ethanol–biodiesel–diesel) blends on internal combustion engines emission reduction. *Renewable and Sustainable Energy Reviews* 53: 265–278.

Mohammadi, O., Golestanzadeh, M., and Abdouss, M. (2017). Recent advances in organic reactions catalyzed by graphene oxide and sulfonated graphene as heterogeneous nanocatalysts: a review. *New Journal of Chemistry* 41 (20): 11471–11497.

Montero, J.M., Wilson, K., and Lee, A.F. (2010). Cs promoted triglyceride transesterification over MgO nanocatalysts. *Topics in Catalysis* 53 (11–12): 737–745.

Muniyappan, V.K., Tamilmani, E., Desikan, R., and Ranganathan, U. (2019). Influence of groundnut seed viability on biodiesel feedstock quality. *Industrial Crops and Products* 140: 111697.

Nakajima, K. and Hara, M. (2012). Amorphous carbon with SO_3H groups as a solid Brønsted acid catalyst. *ACS Catalyst* 2 (7): 1293–1307.

Nasir Baig, R.B., Verma, S., Nadagouda, M.N., and Varma, R.S. (2016). Room temperature synthesis of biodiesel using sulfonated graphitic carbon nitride. *Scientific Reports* 6: 39387.

Ndiaye, M., Arhaliass, A., Legrand, J. et al. (2020). Reuse of waste animal fat in biodiesel: biorefining heavily-degraded contaminant-rich waste animal fat and formulation as diesel fuel additive. *Renewable Energy* 145: 1073–1079.

Nematian, T., Salehi, Z., and Shakeri, A. (2020). Conversion of bio-oil extracted from *Chlorella vulgaris* micro algae to biodiesel via modified superparamagnetic nano-biocatalyst. *Renewable Energy* 146: 1796–1804.

Nizami, A.S. and Rehan, M. (2018). Towards nanotechnology-based biofuel industry. *Biofuel Research Journal* 5 (2): 798–799.

Nongbe, M.C., Ekou, T., Ekou, L. et al. (2017). Biodiesel production from palm oil using sulfonated graphene catalyst. *Renewable Energy* 106: 135–141.

Pablo, D.R., Leandro, S.O., and Adriana, S.F. (2019). Sulfonated activated carbon from corn cobs as heterogeneous catalysts for biodiesel production using microwave-assisted transesterification. *Renewable Energy* 143: 1710–1716.

Palaniappan, K. (2017). An overview of applications of nanotechnology in biofuel production. *World Applied Sciences Journal* 35 (8): 1305–1311.

Pandya, H.N., Parikh, S.P., and Shah, M. (2019). Comprehensive review on application of various nanoparticles for the production of biodiesel. *Energy Sources, Part A: Recovery, Utilization, and Environmental Effects* https://doi.org/10.1080/15567036.2019.1648599.

Peng, F., Zhang, L., Wang, H. et al. (2005). Sulfonated carbon nanotubes as a strong protonic acid catalyst. *Carbon* 43 (11): 2405–2408.

Pimentel, D. and Patzek, T.W. (2005). Ethanol production using corn, switchgrass, and wood; 'biodiesel production using soybean and sunflower'. *Natural Resources Research* 14 (1): 65–76.

Polshettiwar, V. and Varma, R.S. (2010). Green chemistry by nano-catalysis. *Green Chemistry* 12: 743–754.

Poonjarernsilp, C., Sano, N., Sawangpanich, N. et al. (2014a). Effect of Fe/Fe_2O_3 loading on the catalytic activity of sulfonated single-walled carbon nanohorns for the esterification of palmitic acid. *Green Chemistry* 16: 4936–4943.

Poonjarernsilp, C., Sano, N., and Tamon, H. (2014b). Hydrothermally sulfonated single-walled carbon nanohorns for use as solid catalysts in biodiesel production by esterification of palmitic acid. *Applied Catalysis B: Environmental* 147: 726–732.

Poonjarernsilp, C., Sano, N., and Tamon, H. (2015). Simultaneous esterification and transesterification for biodiesel synthesis by a catalyst consisting of sulfonated single-walled carbon nanohorn dispersed with Fe/Fe_2O_3 nanoparticles. *Applied Catalysis A: General* 497: 145–152.

Prinsen, P. and Luque, R. (2019). Introduction to nanocatalysts. In: *Nanoparticle Design and Characterization for Catalytic Applications in Sustainable Chemistry*, Catalysis Series (eds. R. Luque and P. Prinsen), 1–36. London: Royal Society of Chemistry.

Roldan-Cuenya, B. and Behafarid, F. (2015). Nanocatalysis: size- and shape-dependent chemisorption and catalytic reactivity. *Surface Science Reports* 70 (2): 135–187.

Sano, N., Yamada, K., Tsunauchi, S., and Tamon, H. (2017). A novel solid base catalyst for transesterification of triglycerides toward biodiesel production: carbon nanohorn dispersed with calcium ferrite. *Chemical Engineering Journal* 307: 135–142.

Sharma, S., Saxena, V., Baranwal, A. et al. (2018). Engineered nanoporous materials mediated heterogeneous catalysts and their implications in biodiesel production. *Materials Science for Energy Technologies* 1: 11–21.

Shi, Y. and Liang, X. (2019). Novel carbon micro-tube based solid acid from pampas grass stick for biodiesel synthesis from waste oils. *Journal of Saudi Chemical Society* 23 (5): 515–524.

Shu, Q., Zhang, Q., Xu, G., and Wang, J. (2009a). Preparation of biodiesel using s-MWCNT catalysts and the coupling of reaction and separation. *Food and Bioproducts Processing* 87 (3): 164–170.

Shu, Q., Zhang, Q., Xu, G. et al. (2009b). Synthesis of biodiesel from cottonseed oil and methanol using a carbon-based solid acid catalyst. *Fuel Processing Technology* 90: 1002–1008.

Shu, Q., Gao, J., Nawaz, Z. et al. (2010). Synthesis of biodiesel from waste vegetable oil with large amounts of free fatty acids using a carbon-based solid acid catalyst. *Applied Energy* 87 (8): 2589–2596.

Shu, Q., Tang, G., Liu, F. et al. (2017). Study on the preparation, characterization of a novel solid Lewis acid Al^{3+}-SO_4^{2-}/MWCNTs catalyst and its catalytic performance for the synthesis of biodiesel via esterification reaction of oleic acid and methanol. *Fuel* 209: 290–298.

Shuit, S.H. and Tan, S.H. (2014a). Feasibility study of various sulphonation methods for transforming carbon nanotubes into catalysts for the esterification of palm fatty acid distillate. *Energy Conversion and Management* 88: 1283–1289.

Shuit, S.H. and Tan, S.H. (2014b). Biodiesel production via esterification of palm fatty acid distillate using sulphonated multi-walled carbon nanotubes as a solid acid catalyst: process study, catalyst reusability and kinetic study. *Bioenergy Research* 8 (2): 605–617.

Shuit, S.H., Ng, E.P., and Tan, S.H. (2015). A facile and acid-free approach towards the preparation of sulphonated multi-walled carbon nanotubes as a strong protonic acid catalyst for biodiesel production. *Journal of the Taiwan Institute of Chemical Engineers* 52: 100–108.

Singh, D., Sharma, D., Soni, S.L. et al. (2020). A review on feedstocks, production processes, and yield for different generations of biodiesel. *Fuel* 262: 116553.

Sousa, R., Ferreira, B., dos Passos, R.M. et al. (2019). Heterogeneous catalysts for biodiesel production: a review. *Food and Public Health* 9 (4): 125–137.

Srivastava, M., Srivastava, N., Mishra, P.K., and Gupta, V.K. (eds.) (2020). *Nanomaterials in Biofuels Research*, Clean Energy Production Technologies. Singapore: Springer Nature.

Stellwagen, D.R., van der Klis, F., van Es, D.S. et al. (2013). Functionalized carbon nanofibers as solid-acid catalysts for transesterification. *ChemSusChem* 6 (9): 1668–1672.

Sze, Y.L., Revathy, S., Kit, W.C. et al. (2019). Waste to bioenergy: a review on the recent conversion technologies. *BMC Energy* 1: 4.

Tahvildari, K., Anaraki, Y.N., Fazaeli, R. et al. (2015). The study of CaO and MgO heterogenic nano-catalyst coupling on transesterification reaction efficacy in the production of biodiesel from recycled cooking oil. *Journal of Environmental Health Science & Engineering* 13: 73.

Tsang, S.C., Harris, P.J.F., Claridge, J.B., and Green, M.L.H. (1993). A microporous carbon produced by arc-evaporation. *Journal of the Chemical Society, Chemical Communications* 19: 1519.

Umer, R., Junaid, A., Mohd, L.I. et al. (2019). Single-pot synthesis of biodiesel using efficient sulfonated-derived tea waste-heterogeneous catalyst. *Materials* 12: 2293.

Vahid, B.R., Haghighi, M., Alaei, S., and Toghiani, J. (2017). Reusability enhancement of combustion synthesized $MgO/MgAl_2O_4$ nanocatalyst in biodiesel production by glow discharge plasma treatment. *Energy Conversion and Management* 143: 23–32.

Velan, P., Mookan, R., Krishnasamy, A., and Sundaresan, M. (2014). Biodiesel production from Pongamiapinnata oil using synthesized iron nanocatalyst. *International Journal of ChemTech Research* 6 (10): 4511–4516.

Wang, C., Yuan, F., Liu, L. et al. (2015). Transesterification of tributyrin and dehydration of fructose over a carbon-based solid acid prepared by carbonization and sulfonation of glucose. *ChemPlusChem* 80 (11): 1657–1665.

Xie, W. and Huang, M. (2018). Immobilization of *Candida rugosa* lipase onto graphene oxide Fe_3O_4 nanocomposite: characterization and application for biodiesel production. *Energy Conversion and Management* 159: 42–53.

Xie, W. and Zhao, L. (2014). Heterogeneous CaO-MoO_3-SBA-15 catalysts for biodiesel production from soybean oil. *Energy Conversion and Management* 79: 34–42.

Xiong, X., Yu, I.K.M., Cao, L. et al. (2017). A review of biochar-based catalysts for chemical synthesis, biofuel production and pollution control. *Bioresource Technology* 246: 254–270.

Xu, C., Nasrollahzadeh, M., Sajjadi, M. et al. (2019). Benign-by-design nature-inspired nanosystems in biofuels production and catalytic applications. *Renewable and Sustainable Energy Reviews* 112: 195–252.

Zaaba, N.I., Foo, K.L., Hashima, U. et al. (2017). Synthesis of graphene oxide using modified hummers method: solvent influence. *Procedia Engineering* 184: 469–477.

Zhang, X., Fan, Q., and Yang, H. (2018). Green synthesis of functionalized graphene and their use as solid acid catalysts. *Journal of Materials Research* 33 (23): 3946–3952.

Zhang, Q., Zhang, Y., Deng, T. et al. (2020). Sustainable production of biodiesel over heterogeneous acid catalysts. In: *Biomass, Biofuels, Biochemicals: Recent Advances in Development of Platform Chemicals* (eds. S. Saravanamurugan, H. Li, A. Riisager and A. Pandey), 407–432. Amsterdam: Elsevier.

Zhu, S. and Xu, G. (2010). Single-walled carbon nanohorns and their applications. *Nanoscale* 2 (12): 2538.

Zuliani, A., Ivars, F., and Luque, R. (2018). Advances in nanocatalysts design for biofuels production. *ChemCatChem* 10 (9): 1968–1981.

8

Functionalized Magnetic Nanocatalysts in Biodiesel Production

Kalyani Rajkumari[1,2] *and Lalthazuala Rokhum*[1,3]

[1] *Department of Chemistry, National Institute of Technology Silchar, Assam, India*
[2] *Department of Chemistry, Duliajan College, Duliajan, Dibrugarh, Assam, India*
[3] *Department of Chemistry, University of Cambridge, Cambridge, UK*

8.1 Introduction

Nanotechnology has brought a revolutionary hurricane of changes to such diverse domains as medicine, electronics, catalysis (Kalidindi and Jagirdar 2012), food, communication, fine chemicals (Goodman et al. 2005), energy, and environmental science (Jiang et al. 2013). Harnessing nanotechnology to catalytic processes is one of the most explored key areas in contemporary research (Grunes et al. 2003). Nanostructured catalysts are the foothold of a very dynamic research field due to the interplay of many surface and interfacial effects taking place at the nanoscale level (Batlle et al. 2011). Due to their tiny size, the relative exposed surface area of nanocatalysts is massive, giving them excellent activity.

However, despite their tremendous reactivity and accessibility, isolation of these nano-sized particles from the reaction system can prove troublesome (Polshettiwar et al. 2011). As metal nanoparticles (NPs) form highly stabilized solutions, conventional techniques like filtration are not effective in separating them, while the introduction of sophisticated techniques like cross-flow filtration and high-speed centrifugation hampers the sustainability and economy of the catalytic protocol (Ko and Jang 2006; Zhu et al. 2010). Magnetic nanocatalysts have emerged as a viable solution to overcome this shortcoming. The magnetic nature of these materials can be exploited to easily retrieve them from the reaction mixture using an external magnet. Thus, magnetically retrievable nanocatalysts not only offer attractive scope for great accessibility and activity but also mitigate the challenge of recoverability of nanocatalysts (Polshettiwar and Varma 2010; Shylesh et al. 2010; Rossi et al. 2013). This marks them as a "bridge between homogeneous and heterogeneous catalysts" (Figure 8.1) (Wu et al. 2008). Because of this highly advantageous nature, magnetic nanocatalysts have recently been vigorously studied as "green catalysts" for a wide range of organic reactions to facilitate easy and waste-minimizing purification processes.

Early reports (Polshettiwar and Varma 2010) on the applications of magnetic nanomaterials in the field of catalysis included exploitation of the intrinsic magnetic

Nano- and Biocatalysts for Biodiesel Production, First Edition. Edited by Avinash P. Ingle.

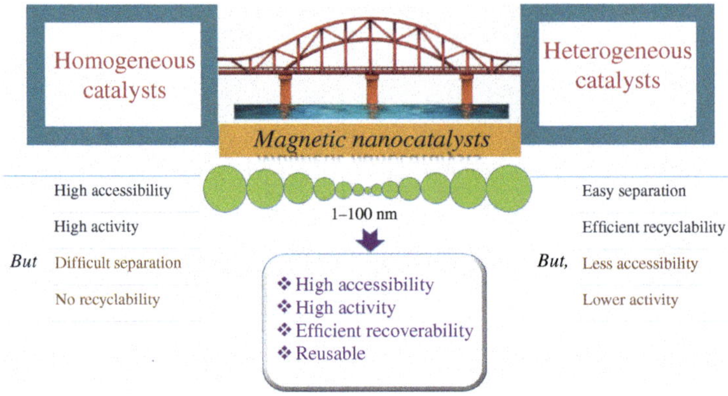

Figure 8.1 Schematic diagram of the advantages of magnetic nanocatalysis over homogeneous and heterogeneous catalysts.

properties of various metals (e.g. Fe, Co, Ni, and their oxides), which are still relevant to recent studies (Yoo 1998; Shi et al. 2007). Iron oxide NPs, especially ferrite (Fe_3O_4) and maghemite (γ-Fe_2O_3), are the most prominently investigated class of magnetic nanocatalyst due to their attractive magnetism (Ko and Jang 2006), easy synthesis, biocompatibility (Polshettiwar and Varma 2010), and wide scope of chemical modifications (Pathak et al. 2019). In order to overcome several shortcomings associated with bare NPs (Jin et al. 2013; Cheng et al. 2017; Sudarsanam et al. 2018) and to facilitate required functionality as per demand, functionalized NPs were introduced into catalytic research. To date, functionalized magnetic nanoparticles (MNPs) have been utilized as catalysts for a wide range of organic transformations, including hydrogenation (Yi et al. 2006), oxidation (El-Sheikh et al. 2009), cross-coupling reactions (Ko and Jang 2006; Yinghuai et al. 2007), multicomponent reactions (Safari and Zarnegar 2013; Dam et al. 2016), and etherification (Rostami et al. 2016; Rajkumari et al. 2017; Rostami et al. 2017).

In this chapter, we will discuss the applications of functionalized magnetic nanocatalysts in the field of biodiesel production. Biodiesel has emerged as a promising alternative to fossil fuels owing to its renewability (Abdullah et al. 2017), low CO_2 emissions (Dehkhoda et al. 2010), lack of sulfur content, nontoxicity, and eco-friendly nature (Hanna and Fangrui 1999). It can be prepared from renewable biological sources such as vegetable oils, animal fats, and algal oils based on the chemical conversion of triglyceride molecules via small alcohols (mostly methanol) using an appropriate catalyst to afford a mixture of fatty acid methyl esters (FAME) with glycerol as byproduct, in a reaction known as transesterification (Scheme 8.1). This reduces the viscosity of the oil, making it suitable for use in conventional engines (Ahmad et al. 2011). Another type of reaction employed in biodiesel production is esterification. In order to mitigate the high production cost of biodiesel and make it competitive with conventional fossil fuels, use of less-expensive feedstocks such as non-edible vegetable oils, waste oils, and animal fats has been extensively examined. Since these low-cost feedstocks contain a high amount of free fatty acids (FFAs), which are difficult to directly convert into biodiesel, a pre-treatment process involving esterification is often required before transesterification can be carried out (Di Serio et al. 2008; Borges and Díaz 2012).

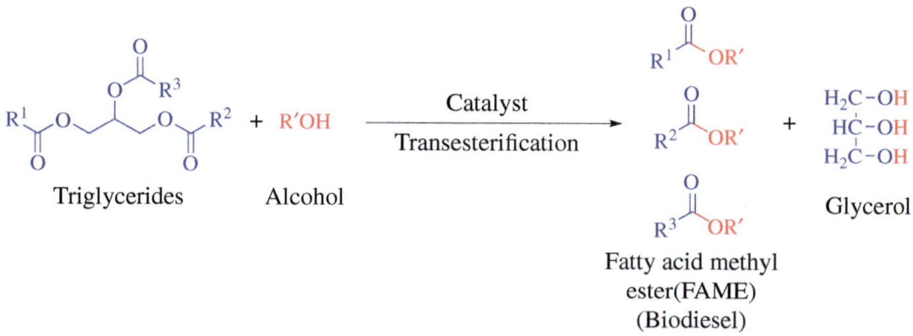

Scheme 8.1 Transesterification reaction for biodiesel production.

8.2 Relevance of Heterogeneous Catalysis in Biodiesel Production

Earlier homogeneous alkali solutions such as NaOH and KOH solutions were used as catalysts for transesterification reactions due to their miscibility and high activity. However, when reacting with triglyceride molecules, these aqueous alkali solutions often lead to saponification as a side-reaction. The soap formed in this reaction not only decreases the FAME yield but also inhibits the separation of glycerol from the product mixture (Laskar et al. 2018; Pathak et al. 2018; Rajkumari and Rokhum 2020). In this context, heterogeneous catalysts have been a preferred catalyst for biodiesel production in recent decades and have been applied at the industrial level (Changmai et al. 2019). Since catalysts play a pivotal role in the processing of biodiesel and substantially contribute to the production cost, a thesaurus of heterogeneous catalysts is being investigated for an efficient, economic, and environmentally friendly transesterification reaction. Both acid and base catalysts have been found effective in catalyzing transesterification, although base catalysts are far more active (4000 times!) than the acid ones (Konwar et al. 2014). A diverse range of heterogeneous catalysts, including metal oxides, carbonates, zeolites, polymers, and double-layered halides (LDH), have found application in the transesterification of biodiesel production (Navajas et al. 2010; Wen et al. 2010; Refaat 2011; Xie and Fan 2014). On the other hand, basic catalysts react with the FFAs present in low-cost feedstock oils, leading to deactivation of the catalyst (Jiménez-Morales et al. 2010). Earlier homogeneous acidic catalysts such as sulfuric acid, hydrochloric acid, and p-toluene sulfonic acid were used to catalyze the esterification of FFAs in these oils (Carmo et al. 2009). However, due to their highly corrosive nature and environmental impact, heterogeneous acid catalysts have become the most investigated and preferred catalyst for esterification reactions in recent times (Do Nascimento et al. 2011).

Employing functionalized magnetic nanocatalysts as heterogeneous catalysts for biodiesel production reactions (i.e. transesterification and esterification) is a lucrative idea from both an environmental and an economic perspective. The easy magnetic separation of catalyst will undoubtedly reduce the post-processing steps of biodiesel synthesis reactions, as well as minimizing the use of solvents and other chemicals and reducing the amount of waste generated by the tedious isolation of biodiesel.

8.3 Surface Modification and Functionalization of NPs

The use of bare metal oxide NPs (especially iron oxide NPs) has several downsides. Their self-interaction encourages particle aggregation, affecting their size and properties (Ko and Jang 2006). Rapid exothermic reaction with oxygen and high air sensitivity further limit their applicability (Jin et al. 2013; Cheng et al. 2017). Several surface-modification strategies have been developed recently to stabilize the NPs and prevents them from agglomerating, ensuring preservation of magnetic properties (Gardy et al. 2017). Of the various strategies used for surface modification of inorganic NPs, including chemical treatment, polymer grafting, ligand exchange, surfactant capping, and encapsulating (Kango et al. 2013), coating with different materials is the most investigated. Silica is one of the most common coating materials, especially for iron oxide MNPs. The inert coating of silica on the surface of MNPs prevents aggregation, stabilizes the NPs by shielding dipole interactions with silica particles, and enhances Coulomb repulsions due to the presence of negatively charged silica particles (Laurent et al. 2008). Most importantly, the surface silanol groups present in the silica coating layer tend to react with various coupling agents, enabling covalent attachment of various ligands and functional groups to the MNP shell (Ulman 1996). Such a structure is commonly known as a core–shell MNP. Apart from silica, other polymer materials are also used for encapsulation of the magnetic core based on their different hydrophobic properties (Sudarsanam et al. 2018).The coating of the magnetic core nanostructure with a suitable inert layer introduces required surface functionality to the MNPs and thereby greatly boosts their catalytic activity. The catalytic properties can be tailored by varying the surface functionality, widening their applicability in diverse reactions.

8.4 Applications of Functionalized Magnetic Nanocatalysts in Biodiesel Production

8.4.1 Acid-Functionalized Magnetic Nanocatalysts

Solid acid catalysts are a major focus of catalytic research in biodiesel production as they can simultaneously carry out both esterification and transesterification reactions and thereby facilitate the use of low-cost feedstock with high fatty acid content (Chouhan and Sarma 2011; Sani et al. 2014). For the same reason, acid-functionalized magnetic nanocatalysts are of immense significance as they are easily magnetically separable. In recent years, both Lewis-type acids (e.g. mixed and sulfated oxides) and Brønsted acids (e.g. sulfonic acid, sulfamic acid, etc.) have been functionalized over magnetic core nanostructures in order to utilize their catalytic properties by combining their heterogeneous nature and mineral acid properties (Lotero et al. 2005).

Duan et al. (2013) synthesized heteropolyacid-functionalized silica-coated Fe_3O_4 MNPs in which the catalyst surface was modified by organic groups to prevent its deactivation by water. After coating the magnetite core with silica shell, they first functionalized it with the silanizing agent aminopropyl trimethoxy silane (APTES) and then protonated the ammine group with $H_3PW_{12}O_{40}$ (HPA). A hydrophobic hybrid of this MNP catalyst was designed by the additional use of octyltrimethoxysilane (OTMS, apart from using APTES) in the silanization process (Figure 8.2) (Duan et al. 2013). The catalyst was successfully utilized

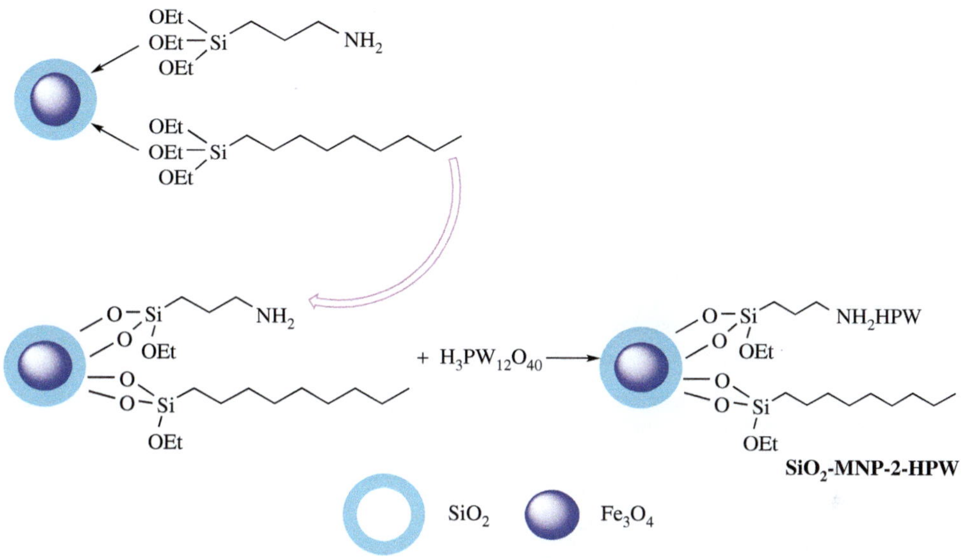

Figure 8.2 Preparation strategy of a hydrophobic hybrid of HPA-functionalized silica-coated Fe_3O_4 MNPs. Source: Reproduced with permission (Duan et al. 2013) Copyright, 2013, RSC.

in biodiesel synthesis through the esterification of palmitic acid with methanol, resulting in 90.4% FFA conversion within two hours.

A core–shell magnetic nanostructure with Fe_3O_4 MNP as core and poly(glycidyl methacrylate) (PGMA) as shell, functionalized with sulfonic acid, was designed by Zillillah et al. (2012), as depicted in Figure 8.3. The chemically active epoxide groups on the PGMA polymer chain facilitate easy covalent attachment of the sulfonic acid group after reaction with Na_2SO_4, resulting in an efficient magnetic solid acid recyclable catalyst for esterification of waste grease in a two-step biodiesel production method. The same authors also tried using polystyrene (PS) and silica as coating material, but the PGMA-coated catalyst performed better in terms of acid loading, stability, and magnetic properties. In another attempt, they synthesized a novel magnetite MNP-based nanocatalyst coated with PGMA and functionalized with phosphotungstic acid (HPW), which effectively catalyzed the esterification of waste grease (FFA 21.3%) with methanol (Zillillah et al. 2014).

Sulfonic acid- and sulfamic acid-functionalized silica-coated Fe/Fe_3O_4 crystalline nanocatalysts were developed by Wang et al. (2015). Both exhibited excellent activity against esterification of oleic ester and transesterification of glyceryl trioleate, although the sulfamic acid-functionalized catalyst outperformed the sulfonic acid-functionalized one. The same authors later fabricated a bifunctional magnetic carbonaceous acid catalyst containing both Brønsted and Lewis acid sites on its surface (Wang et al. 2017). The catalyst was based on a core–shell structure of Ferrite NPs coated with carboxymethylcellulose (CMC) and loaded with Brønsted sulfonic (-SO_3H) acid and Lewis (CMC-COOZr or CMC-COOFe) acid sites in a four-step synthetic method. Esterification of oleic acid (>90% yield), transesterification of soybean oil (> 95% FAME yield), and pretreatment with jatropha oil were efficiently carried out by this novel carbonaceous magnetic catalyst (Zr-CMC-SO_3H@3Fe-C_{400}), with a recyclability of more than 10 cycles (Figure 8.4) (Wang et al. 2017).

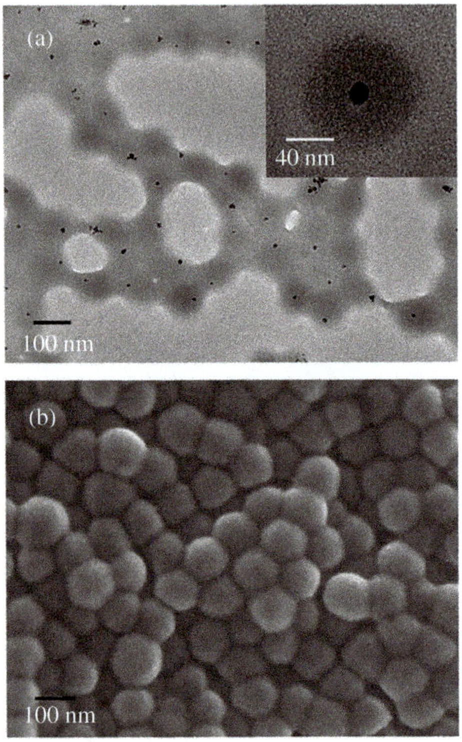

Figure 8.3 (a) TEM image and (b) FESEM images of SO_3H-PGMA-MNPs. Source: Reproduced with permission (Zillillah et al. 2012). Copyright, 2012, RSC.

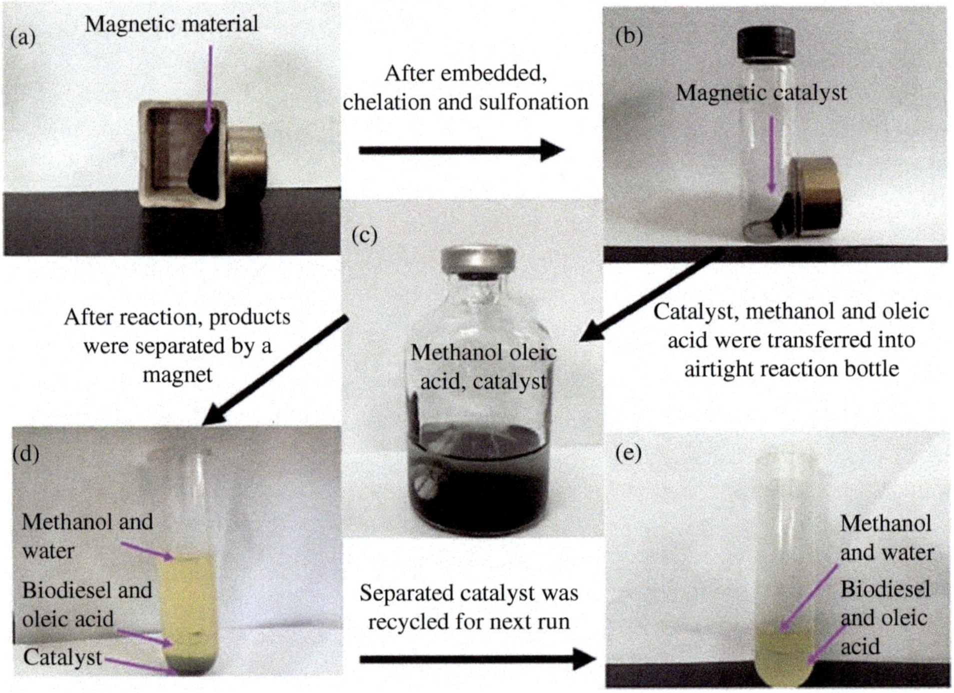

Figure 8.4 Biodiesel production by oleic acid esterification using a Zr-CMC-SO3H@3Fe-C400 catalyst. Source: Reproduced with permission (Wang et al. 2017). Copyright, 2017, Elsevier.

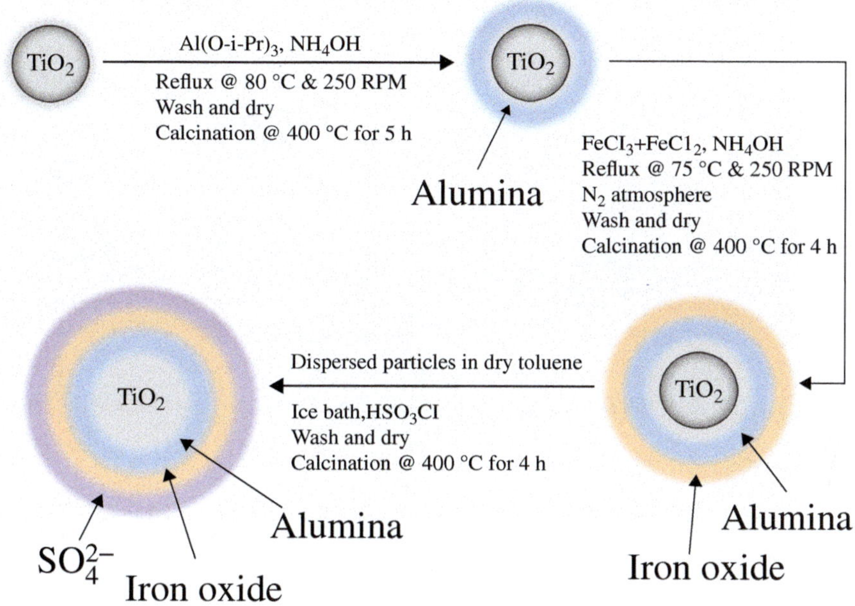

Figure 8.5 Synthesis of a super-magnetic SO_4/Fe-Al-TiO_2 catalyst. Source: Reproduced with permission (Gardy et al. 2018). Copyright, 2018, Elsevier.

In 2018, Gardy et al. (2018), reported a novel super-magnetic SO_4/Fe-Al-TiO_2 solid acid nanocatalyst that was successfully utilized in the synthesis of FAME (96% yield) by simultaneous esterification and transesterification of waste cooking oil (WCO) containing up to 20% FFA. The nanocatalyst was synthesized by impregnation of TiO_2 NPs (commercial rutile/anatase mixed phase) in the alumina layer followed by hematite functionalization through the co-precipitation of ferrous and ferric ion precursors in order to impart magnetic properties to the nanocomposite (Figure 8.5) (Gardy et al. 2018). Sulfation of the Fe-Al-TiO_2 nanostructure produced Brønsted acidity to its surface, which was later established by diffuse reflectance infrared spectroscopy (DRIFTS).

8.4.2 Base-Functionalized Magnetic Nanocatalysts

In the context of transesterification reactions, heterogeneous base catalysts are the most efficient option, and have been widely investigated. Especially for oils with low fatty acid contents, solid base catalysts are preferred (Laskar et al. 2018; Pathak et al. 2018; Rajkumari and Rokhum 2020). Homogeneous alkali solutions were once used as base catalysts for industrial-level biodiesel production, but they had several drawbacks, including saponification of fatty acid molecules, difficult product separation, and lack of recoverability (Konwar et al. 2014). Impregnation of basic sites over a magnetic solid support can simplify the product separation process and boost catalyst recyclability. Several attempts have been made toward this end in recent studies.

A triazabicyclodecene (TBD)-functionalized silica-coated Fe_3O_4-based magnetic solid base catalyst was developed in 2015 (Chiang et al. 2015). The strong base TBD was attached

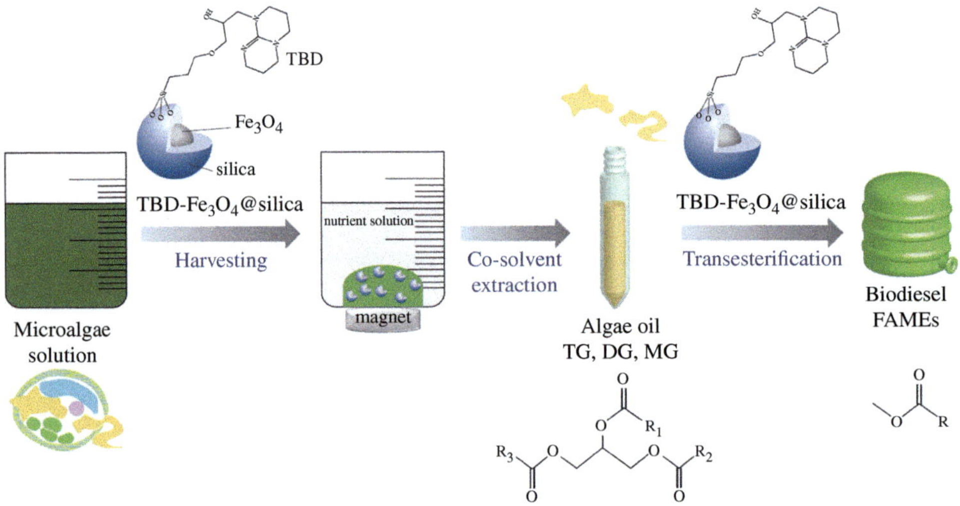

Figure 8.6 Synthesis of silica-coated TBD-functionalized ferrite MNPs. Source: Reproduced with permission (Chiang et al. 2015). Copyright, 2014, Willey-VSH.

through covalent bonding to the surface silanol group of the silica shell in two steps: silanation of the Fe_3O_4-SiO_2 MNP surface with trimethoxysilylpropoxymethyloxirane (an epoxy containing silane) followed by impregnation of 1,5,7-triazabicyclo[4.4.0]dec-5-ene (TBD). This highly basic magnetic catalyst was successfully utilized in one-pot synthesis of FAME from microalgae with 97.1% yield (Figure 8.6) (Chiang et al. 2015).

Santos et al. (2015) presented three different catalysts having ferrite MNPs as core and functionalized with the organic super bases TBD and 1,1,3,3-tetramethylguanidine (TMG) on a silica-coated or uncoated magnetic core (Figure 8.7) (Santos et al. 2015). The resulting magnetic nanostructures were MNP-TBD, MNP@SiO_2-TBD, and MNP-TMS, of which MNP-TBD showed the best performance in the transesterification of soybean oil exhibiting 96% biodiesel yield and a TON value of 480. However, the reusability of the silica-coated MNP@SiO_2-TBD MNPs was found to be better than that of the uncoated MNP-TBD particles, possibly because the silica coating prevented dissolution of iron oxide NPs instigated by FFAs of soybean oil.

A magnetic γ-Fe_2O_3-based catalyst was designed by encapsulation with hydroxyapatite (HAp) using the co-precipitation method (Xie et al. 2017). Later, the HAp-γ-Fe_2O_3 particles were anchored via biguanide functionalization, first by silanation with 3-chloropropyltrimethoxysilane (CPTS), then by immobilization with a biguanide (metformin hydrochloride). The basic strength of the catalyst was found to be moderate, and it successfully converted soybean oil into FAME by transesterification using methanol in 1 : 25 M ratio and 3 wt% catalyst.

8.4.3 Magnetic Nanocatalysts Functionalized with Waste Materials

In the quest for cost-effective methods of biodiesel production, a variety of low-cost feedstocks and catalytic systems have been tested in recent decades. In fact, catalysts play a

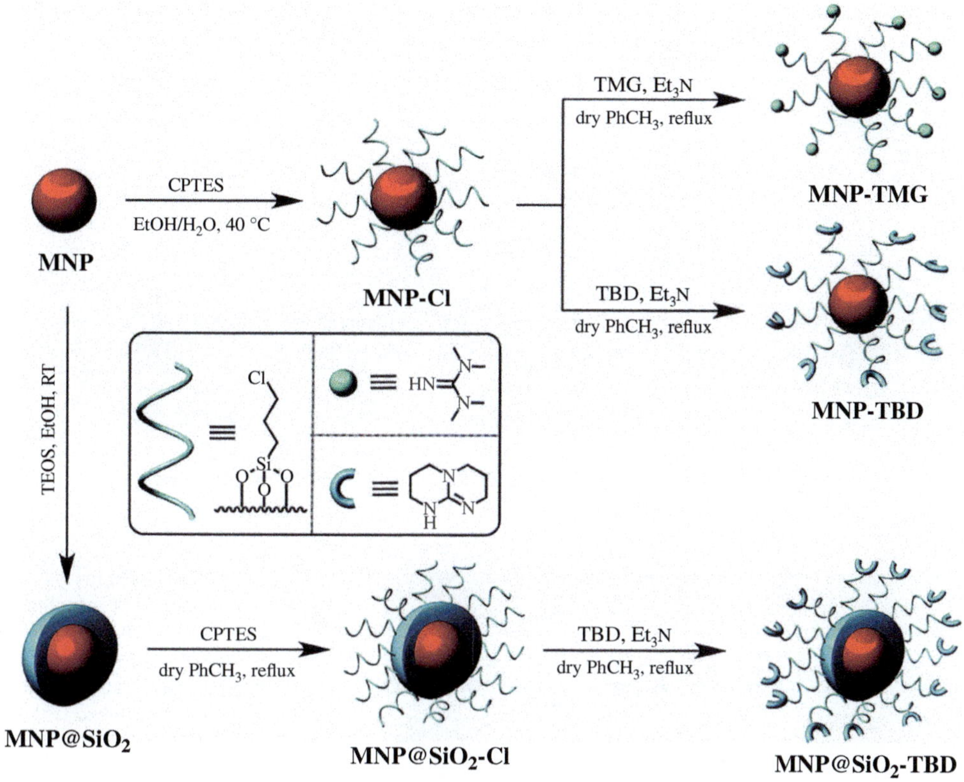

Figure 8.7 Synthesis of TBD- and TMG-functionalized magnetic silica-coated and uncoated Fe_3O_4 MNPs. Source: Reproduced with permission (Santos et al. 2015). Copyright, 2015, RSC.

major role in deciding the production cost of a synthesis protocol, as most involve precious chemicals or functional materials in their preparation. Use of waste materials as functionalizing agents therefore represents a promising means of mitigating the cost issue, as well as alleviating current environmental problems. Several studies can be found in the literature presenting magnetic nanocatalysts functionalized by waste-derived materials or chemicals.

A double shell-structured mesoporous magnetic solid acid catalyst was developed in which the Fe_3O_4 MNPs were sequentially coated with chitosan (an alkaline polysaccharide derived from chitin shell of shellfish) and silica twice over, before finally being sulfonated with p-toluenesulfonic acid (PTSA) (Wang et al. 2018). The resulting catalyst, Fe_3O_4@Chitosan-Hollow-Chitosan-SO_3H (FCHC-SO_3H), showed excellent activity (96.7% biodiesel yield at 1 : 15 oil to methanol ratio and 80 °C reaction temperature) in the esterification of oleic acid with methanol. Its catalytic performance was superior to its counterpart without a chitosan layer, probably due to the presence of additional NH_3^+ sites (apart from $-SO_3H$ sites) on the catalyst surface, derived from ammine groups of the chitosan (Figure 8.8) (Wang et al. 2018).

Teo et al. (2019) developed a novel calcium sulfate-functionalized core–shell nanostructured magnetic catalyst using Fe_2O_3 as the core and silica as the coating layer. Loading of bioderived $CaSO_4$ on to the silica-embedded Fe_2O_3 MNPs was carried out by treating the

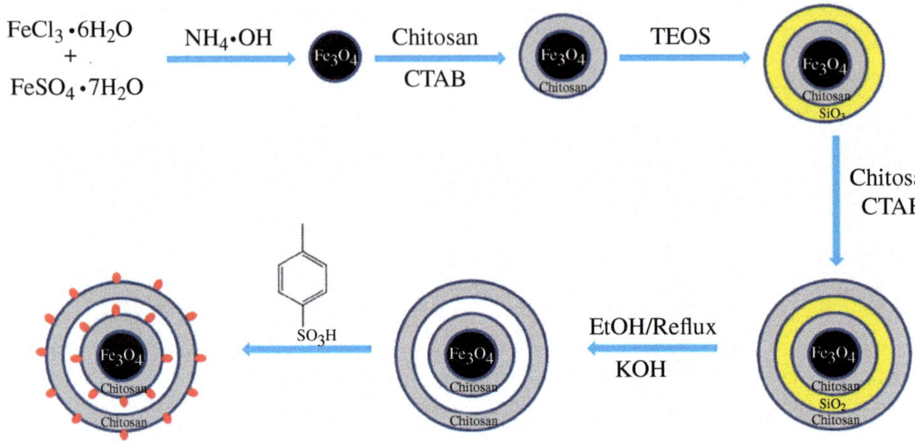

Figure 8.8 Synthesis of FCHC-SO_3H MNPs. Source: Reproduced with permission (Wang et al. 2018). Copyright, 2018, Elsevier.

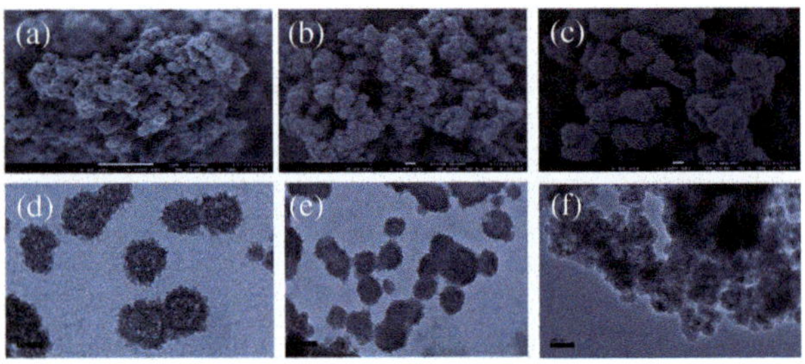

Figure 8.9 (a–c) FESEM images and (d–f) TEM images of recycled catalyst after the fourth, seventh, and ninth catalytic cycles, respectively. Source: Reproduced with permission (Teo et al. 2019). Copyright, 2019, Elsevier.

MNPs with sulfuric acid and CaO produced by the calcination of cockle shells. The resulting catalyst was observed to have a well-defined hierarchically porous structure, which successfully converted crude *Jatropha curcas* oil to biodiesel with potential activity (94% yield, 1 : 9 oil to methanol ratio, 120 °C optimum temperature). It showed significant recyclability up to its ninth catalytic cycle with little drop in activity; this could be anticipated from the blocking of active sites as revealed by transmission electron microscopy (TEM) and scanning electron microscopy (SEM) (Figure 8.9) (Teo et al. 2019).

8.4.4 Ionic Liquid-Immobilized Magnetic Nanocatalysts

Ionic liquids (ILs) are a promising new class of green catalysts and solvents. They have the advantages of a wide liquid range, negligible vapor pressure, thermal stability, flexible chemical and physical properties, and good catalytic activity. For these reasons, they have been investigated as catalysts in a variety of organic reactions (Kirchner and Clare 2009). However, their widespread use is limited by several issues, including homogeneity, difficult product separation, lack of recoverability, and unendurable viscosity (Shahabi Nejad and Sheibani 2018). Lately, heterogenization of ILs over magnetic nanocatalysts has been attempted by a number of researchers (Valkenberg et al. 2002). This allows rapid magnetic

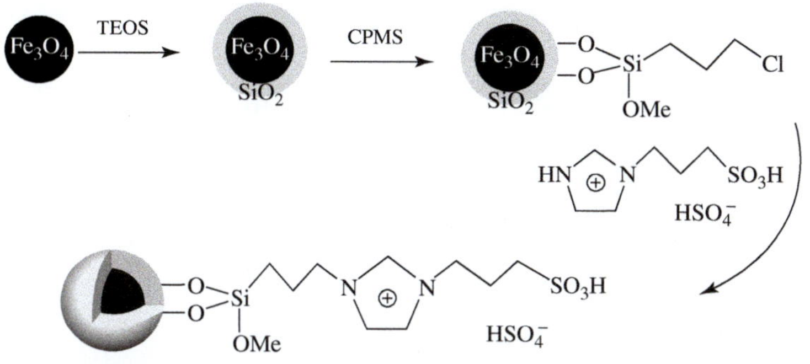

Figure 8.10 Synthesis of IL-SiO_2-Fe_3O_4 MNPs. Source: Reproduced with permission (Wu et al. 2014). Copyright, 2014, American Chemical Society.

separation of catalyst from crude product, while retaining the advantageous properties of the IL.

A silica-coated Fe_3O_4 MNP-based magnetic solid acid catalyst was developed by immobilizing an acidic IL precursor prepared by treatment with APTES (Li and Liang 2017). The acidic IL was attached to the magnetic core through Si–O–Si bonds of silica shell. The catalyst effectively converted both FFAs and triglycerides in the one-pot synthesis of biodiesel from waste oil, exhibiting a high yield of 98.9% at 70 °C.

Wu et al. (2014) synthesized a silica-coated magnetite core–shell-structured MNP catalyst functionalized by dual Brønsted acidic ILs [SO_3H-$(CH_2)_3$-HIM][HSO_4], which they successfully employed in both the esterification of oleic acid and the transesterification of soybean oil (Figure 8.10) (Wu et al. 2014).

Recently, a hydrophobic magnetic catalyst containing poly-ionic liquid (PIL) functionalized on to the silica shell of magnetite MNPs was developed by Zhang et al. (2018) PILs are ionic polymers comprising a polymeric backbone and IL head. Consequently, they offer the twin properties of both polymer and IL. The catalyst expressed as FnmS-PIL showed good physicochemical properties, high acidity, strong hydrophobicity, and, most importantly, excellent catalytic activity in the esterification of oleic acid and transesterification of crude *Euphorbia lathyris L.* seed oil (CELO). The reusability of the catalyst was satisfactorily established.

Wu et al. (2016) reported a magnetic composite of Fe_3O_4 in which amino-functionalized metal organic frameworks (MOF) NH_2-MIL-88B(Fe) acted as a carrier to support the immobilization of IL. To this magnetic composite, a Brønsted acidic IL 1,4-butanediyl-3,3′-bis-(3-sulfopropyl) imidazolium dihydrogensulfate (DAIL) was anchored (confined in the MOF) by strategic interaction between the sulfonic acid group of the DAIL and the ammine group of the MOF carrier (Figure 8.11) (Wu et al. 2016). The catalyst, DAIL-Fe_3O_4@NH_2-MIL-88B(Fe), having multi-acidic sites, was exploited in the esterification of oleic acid with good catalytic performance (93.2% oleic acid conversion).

A core–shell-structured magnetic nanocatalyst was developed using Fe_3O_4 particles as magnetic core and HKUST-1-type MOFs as coating, immobilized with a basic IL (ABIL-Im) (Xie and Wan 2018). Transesterification of soybean oil with methanol was carried out using this catalyst with 92.3% oil conversion. It showed almost no change even after being used

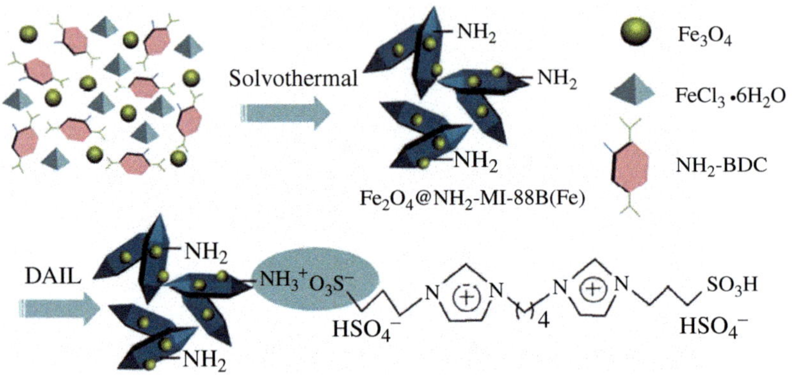

Figure 8.11 Synthesis of DAIL-Fe_3O_4@NH_2-MIL-88B(Fe) MNPs. Source: Reproduced with permission (Wu et al. 2016). Copyright, 2016, American Chemical Society.

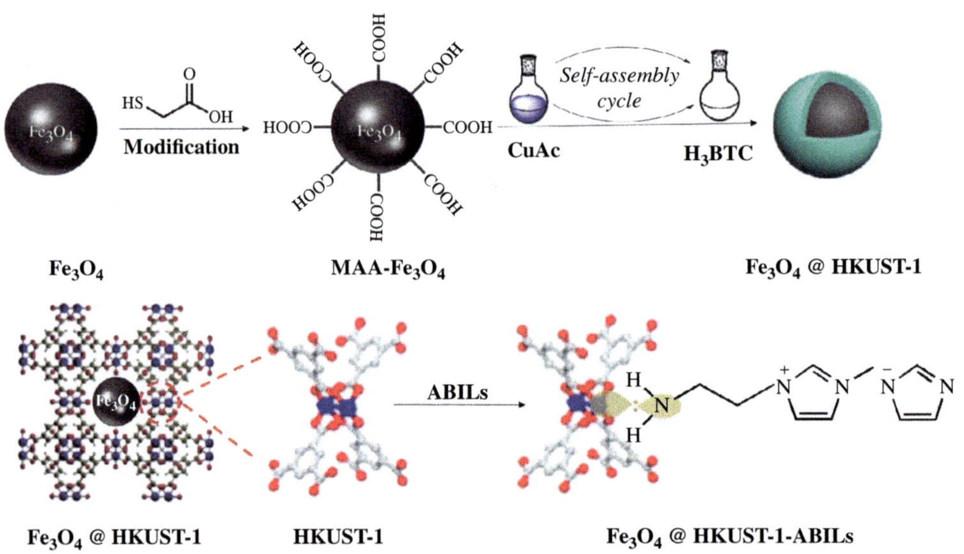

Figure 8.12 Synthesis of core–shell-structured MOF-coated basic IL-functionalized MNPs. Source: Reproduced with permission (Xie and Wan 2018). Copyright, 2018, Elsevier.

for five catalytic cycles, and excellent heterogeneity was observed in a hot leaching test experiment (Figure 8.12) (Xie and Wan 2018).

8.5 Conclusions

The climate and pollution issues associated with fossil fuels, together with their diminishing supply, represent a serious threat to the ballooning demand for fuel in the modern world. In this regard, finding eco-friendly and efficient alternative fuel options and developing sustainable methodologies are the needs of the hour. Biodiesel produced by transesterification (and esterification) of triglycerides is emerging as a viable solution, despite struggling for worldwide market exposure mainly due to its high production cost and low efficiency compared with fossil fuels. The transition from homogeneous catalysis (which still persists at the industrial level to a major extent) to heterogeneous catalysis in biodiesel production represented a revolutionary step, helping reduce post-processing in biodiesel synthesis. The development of efficient and recoverable heterogeneous catalysts is therefore a significant area of research today.

Functionalized magnetic nanocatalysts can facilitate major benefits over conventional solid catalysts in terms of ease of recoverability. The separation of conventional heterogeneous catalysts through filtration or centrifugation generally entails considerable loss, impeding their practical application. Imparting magnetic properties to catalysts makes the separation process remarkably simple, utilizing an external magnetic field in place of the tedious solvent-consuming purification steps. This boosts the recoverability of the catalytic system, significantly lowering production costs. Magnetic separation is simple, fast, clean, and easy to scale up, and avoids the need for additional solvents. On top of that, magnetic catalysis enables maximum recovery of catalysts, an important plus. The coating of magnetic core nanostructures prevents aggregation of NPs and allows immobilization of different functionalities on their surfaces, facilitating wide applicability. Hence, catalytic technology using functionalized magnetic nanocatalysts stands as a powerful and promising option in the field of biodiesel production.

References

Abdullah, S.H.Y.S., Hanapi, N.H.M., Azid, A. et al. (2017). A review of biomass-derived heterogeneous catalyst for a sustainable biodiesel production. *Renewable and Sustainable Energy Reviews* 70: 1040–1051.

Ahmad, A.L., Yasin, N.H.M., Derek, C.J.C., and Lim, J.K. (2011). Microalgae as a sustainable energy source for biodiesel production: a review. *Renewable and Sustainable Energy Reviews* 15: 584–593.

Batlle, X., Pérez, N., Guardia, P. et al. (2011). Magnetite nanoparticles with almost bulk magnetic properties: the role of the surfactant. *Journal of Applied Physics* 109: 07B524-1-6.

Borges, M.E. and Díaz, L. (2012). Recent developments on heterogeneous catalysts for biodiesel production by oil esterification and transesterification reactions: a review. *Renewable and Sustainable Energy Reviews* 16: 2839–2849.

Carmo, A.C., de Souza, L.K.C., da Costa, C.E.F. et al. (2009). Production of biodiesel by esterification of palmitic acid over mesoporous aluminosilicate Al-MCM-41. *Fuel* 88: 461–468.

Changmai, B., Sudarsanam, P., and Rokhum, L. (2019). Biodiesel production using a renewable mesoporous solid catalyst. *Industrial Crops and Products* 145: 111911.

Cheng, J., Qiu, Y., Zhang, J. et al. (2017). Conversion of lipids from wet microalgae into biodiesel using sulfonated graphene oxide catalysts. *Bioresource Technology* 244: 569–574.

Chiang, Y.D., Dutta, S., Chen, C.T. et al. (2015). Functionalized Fe3O4@silica core-shell nanoparticles as microalgae harvester and catalyst for biodiesel production. *ChemSusChem* 8: 789–794.

Chouhan, A.P.S. and Sarma, A.K. (2011). Modern heterogeneous catalysts for biodiesel production: a comprehensive review. *Renewable and Sustainable Energy Reviews* 15: 4378–4399.

Dam, B., Pal, A.K., and Gupta, A. (2016). Nano-Fe_3O_4@silica sulfuric acid as a reusable and magnetically separable potent solid acid catalyst in Biginelli-type reaction for the one-pot multicomponent synthesis of fused dihydropyrimidine derivatives: a greener NOSE and SFRC approach. *Synthetic Communications* 46: 275–286.

Dehkhoda, A.M., West, A.H., and Ellis, N. (2010). Biochar based solid acid catalyst for biodiesel production. *Applied Catalysis A: General* 382: 197–204.

Di Serio, M., Tesser, R., Pengmei, L., and Santacesaria, E. (2008). Heterogeneous catalysts for biodiesel production. *Energy and Fuels* 22: 207–217.

Do Nascimento, L.A.S., Angélica, R.S., da Costa, C.E.F. et al. (2011). Comparative study between catalysts for esterification prepared from kaolins. *Applied Clay Science* 51: 267–273.

Duan, X., Zhao, J., Shi, W. et al. (2013). Water-tolerant heteropolyacid on magnetic nanoparticles as efficient catalysts for esterification of free fatty acid. *RSC Advances* 3: 13748–13755.

El-Sheikh, S.M., Harraz, F.A., and Abdel-Halim, K.S. (2009). Catalytic performance of nanostructured iron oxides synthesized by thermal decomposition technique. *Journal of Alloys and Compounds* 487: 716–723.

Gardy, J., Hassanpour, A., Lai, X. et al. (2017). Biodiesel production from used cooking oil using a novel surface functionalised TiO_2 nano-catalyst. *Applied Catalysis B: Environmental* 207: 297–210.

Gardy, J., Osatiashtiani, A., Céspedes, O. et al. (2018). A magnetically separable SO_4/Fe-Al-TiO_2 solid acid catalyst for biodiesel production from waste cooking oil. *Applied Catalysis B: Environmental* 234: 268–278.

Goodman, R.P., Schaap, I.A.T., Tardin, C.F. et al. (2005). Rapid chiral assembly of rigid DNA building blocks for molecular nanofabrication. *Science* 310: 1661–1665.

Grunes, J., Somorjai, G.A., Sciences, M. et al. (2003). Catalysis and nanoscience. *Chemical Communications* 8: 2257–2260.

Hanna, M. and Fangrui, M. (1999). Biodiesel production: a review. *Bioresource Technology* 70: 1–15.

Jiang, Y., Wang, P., Zang, X. et al. (2013). Uniformly embedded metal oxide nanoparticles in vertically aligned carbon nanotube forests as pseudocapacitor electrodes for enhanced energy storage. *Nano Letters* 13: 3524–3530.

Jiménez-Morales, I., Santamaría-González, J., Maireles-Torres, P., and Jiménez-López, A. (2010). Zirconium doped MCM-41 supported WO_3 solid acid catalysts for the esterification of oleic acid with methanol. *Applied Catalysis A: General* 379: 61–68.

Jin, X., Dang, L., Lohrman, J. et al. (2013). Lattice-matched bimetallic CuPd-graphene nanocatalysts for facile conversion of biomass-derived polyols to chemicals. *ACS Nano* 7: 1309–1316.

Kalidindi, S.B. and Jagirdar, B.R. (2012). Nanocatalysis and prospects of green chemistry. *ChemSusChem* 5: 65–75.

Kango, S., Kalia, S., Celli, A. et al. (2013). Surface modification of inorganic nanoparticles for development of organic-inorganic nanocomposites – a review. *Progress in Polymer Science* 38: 1232–1261.

Kirchner, B. and Clare, B. (2009). *Ionic Liquids*, 1e. Heidelberg: Springer.

Ko, S. and Jang, J. (2006). A highly efficient palladium nanocatalyst anchored on a magnetically functionalized polymer-nanotube support. *Angewandte Chemie, International Edition* 45: 7564–7567.

Konwar, L.J., Boro, J., and Deka, D. (2014). Review on latest developments in biodiesel production using carbon-based catalysts. *Renewable and Sustainable Energy Reviews* 29: 546–564.

Laskar, I.B., Rajkumari, K., Gupta, R. et al. (2018). Waste snail shell derived heterogeneous catalyst for biodiesel production by the transesterification of soybean oil. *RSC Advances* 8: 20131–20142.

Laurent, S., Forge, D., Port, M. et al. (2008). Magnetic iron oxide nanoparticles: synthesis, stabilization, vectorization, physicochemical characterizations, and biological applications. *Chemical Reviews* 110: 2064–2110.

Li, J. and Liang, X. (2017). Magnetic solid acid catalyst for biodiesel synthesis from waste oil. *Energy Conversion and Management* 141: 126–132.

Lotero, E., Liu, Y., Lopez, D.E. et al. (2005). Synthesis of biodiesel via acid catalysis. *Industrial and Engineering Chemistry Research* 44: 5353–5363.

Navajas, A., Campo, I., Arzamendi, G. et al. (2010). Synthesis of biodiesel from the methanolysis of sunflower oil using PURAL® Mg–Al hydrotalcites as catalyst precursors. *Applied Catalysis B: Environmental* 100: 299–309.

Pathak, G., Das, D., Rajkumari, K., and Rokhum, L. (2018). Exploiting waste: towards a sustainable production of biodiesel using: *Musa acuminata* peel ash as a heterogeneous catalyst. *Green Chemistry* 20: 2365–2373.

Pathak, G., Rajkumari, K., and Rokhum, L. (2019). Wealth from waste: *M. acuminata* peel waste-derived magnetic nanoparticles as a solid catalyst for the Henry reaction. *Nanoscale Advances* 1: 1013–1020.

Polshettiwar, V. and Varma, R.S. (2010). Green chemistry by nano-catalysis. *Green Chemistry* 12: 743–775.

Polshettiwar, V., Luque, R., Fihri, A. et al. (2011). Magnetically recoverable nanocatalysts. *Chemical Reviews* 111: 3036–3075.

Rajkumari, K. and Rokhum, L. (2020). A sustainable protocol for production of biodiesel by transesterification of soybean oil using banana trunk ash as a heterogeneous catalyst. *Biomass Conversion and Biorefinery* 10: 839–848.

Rajkumari, K., Kalita, J., Das, D., and Rokhum, L. (2017). Magnetic Fe_3O_4@silica sulfuric acid nanoparticles promoted regioselective protection/deprotection of alcohols with dihydropyran under solvent-free conditions. *RSC Advances* 7: 56559–56565.

Refaat, A.A. (2011). Biodiesel production using solid metal oxide catalysts. *International Journal of Environmental Science and Technology* 8: 203–221.

Rossi, L.M., Costa, N.J.S., Silva, F.P., and Gonçalves, R.V. (2013). Magnetic nanocatalysts : supported metal nanoparticles for catalytic applications. 2: 597–514.

Rostami, A., Tahmasbi, B., and Abedi, F. (2016). Supported N-propylsulfamic acid onto Fe_3O_4 magnetic nanoparticles as a reusable and efficient nanocatalyst for the protection/deprotection of hydroxyl groups and protection of aldehydes. *Research on Chemical Intermediates* 42: 3689–3701.

Rostami, A., Pourshiani, O., Navasi, Y. et al. (2017). Magnetic nanoparticle-supported DABCO tribromide: a versatile nanocatalyst for the synthesis of quinazolinones and benzimidazoles and protection/deprotection of hydroxyl groups. *New Journal of Chemistry* 41: 9033–9040.

Safari, J. and Zarnegar, Z. (2013). A magnetic nanoparticle-supported sulfuric acid as a highly efficient and reusable catalyst for rapid synthesis of amidoalkyl naphthols. *Journal of Molecular Catalysis A: Chemical* 379: 269–276.

Sani, Y.M., Daud, W.M.A.W., and Abdul, A.A.R. (2014). Activity of solid acid catalysts for biodiesel production: a critical review. *Applied Catalysis A: General* 470: 140–161.

Santos, E.C.S., Santos, T.C., Guimarães, R.B. et al. (2015). Guanidine-functionalized Fe3O4 magnetic nanoparticles as basic recyclable catalysts for biodiesel production. *RSC Advances* 5: 48031–48038.

Shahabi Nejad, M. and Sheibani, H. (2018). Multi-layer functionalized poly(2-vinylpyridinium) ionic liquid immobilized on magnetic nanoparticles: highly recoverable and magnetically separable Brønsted acid catalyst. *Catalysis Letters* 148: 125–133.

Shi, F., Tse, M.K., Pohl, M.M. et al. (2007). Tuning catalytic activity between homogeneous and heterogeneous catalysis: improved activity and selectivity of free nano-Fe2O3 in selective oxidations. *Angewandte Chemie, International Edition* 46: 8866–8868.

Shylesh, S., Schünemann, V., and Thiel, W.R. (2010). Magnetically separable nanocatalysts: bridges between homogeneous and heterogeneous catalysis. *Angewandte Chemie, International Edition* 49: 3428–3459.

Sudarsanam, P., Zhong, R., Bosch, S.V. et al. (2018). Functionalised heterogeneous catalysts for sustainable biomass valorisation. *Chemical Society Reviews* 47: 8349–8402.

Teo, S.H., Islam, A., Chan, E.S. et al. (2019). Efficient biodiesel production from *Jatropha curcus* using CaSO4/Fe2O3-SiO2 core-shell magnetic nanoparticles. *Journal of Cleaner Production* 208: 816–826.

Ulman, A. (1996). Formation and structure of self-assembled monolayers. *Chemical Reviews* 96: 1533–1554.

Valkenberg, M.H., de Castro, C., and Hölderich, W.F. (2002). Immobilisation of ionic liquids on solid supports. *Green Chemistry* 4: 88–93.

Wang, H., Covarrubias, J., Prock, H. et al. (2015). Acid-functionalized magnetic nanoparticle as heterogeneous catalysts for biodiesel synthesis. *Journal of Physical Chemistry C* 119: 26020–26028.

Wang, Y.T., Fang, Z., and Yang, X.X. (2017). Biodiesel production from high acid value oils with a highly active and stable bifunctional magnetic acid. *Applied Energy* 204: 702–714.

Wang, A., Li, H., Pan, H. et al. (2018). Efficient and green production of biodiesel catalyzed by recyclable biomass-derived magnetic acids. *Fuel Processing Technology* 181: 259–267.

Wen, Z., Yu, X., Tu, S.T. et al. (2010). Biodiesel production from waste cooking oil catalyzed by TiO2-MgO mixed oxides. *Bioresource Technology* 101: 9570–9576.

Wu, W., He, Æ.Q., and Jiang, Æ.C. (2008). Magnetic iron oxide nanoparticles : synthesis and surface functionalization strategies. *Nanoscale Research Letters* 3: 397–415.

Wu, Z., Li, Z., Wu, G. et al. (2014). Brønsted acidic ionic liquid modified magnetic nanoparticle: an efficient and green catalyst for biodiesel production. *Industrial and Engineering Chemistry Research* 53: 3040–3046.

Wu, Z., Chen, C., Wan, H. et al. (2016). Fabrication of magnetic NH2-MIL-88B (Fe) confined Brønsted ionic liquid as an efficient catalyst in biodiesel synthesis. *Energy and Fuels* 30: 10739–10746.

Xie, W. and Fan, M. (2014). Biodiesel production by transesterification using tetraalkylammonium hydroxides immobilized onto SBA-15 as a solid catalyst. *Chemical Engineering Journal* 239: 60–67.

Xie, W. and Wan, F. (2018). Basic ionic liquid functionalized magnetically responsive Fe_3O_4@HKUST-1 composites used for biodiesel production. *Fuel* 220: 248–256.

Xie, W., Han, Y., and Tai, S. (2017). Biodiesel production using biguanide-functionalized hydroxyapatite-encapsulated-γ-Fe2O3 nanoparticles. *Fuel* 210: 83–90.

Yi, D.K., Lee, S.S., and Ying, J.Y. (2006). Synthesis and applications of magnetic nanocomposite catalysts. *Chemistry of Materials* 18: 2459–2461.

Yinghuai, Z., Peng, S.C., Emi, A. et al. (2007). Supported ultra small palladium on magnetic nanoparticles used as catalysts for Suzuki cross-coupling and heck reactions. *Advanced Synthesis and Catalysis* 349: 1917–1922.

Yoo, J.S. (1998). Metal recovery and rejuvenation of metal-loaded spent catalysts. *Catalysis Today* 44: 27–46.

Zhang, H., Li, H., Pan, H. et al. (2018). Magnetically recyclable acidic polymeric ionic liquids decorated with hydrophobic regulators as highly efficient and stable catalysts for biodiesel production. *Applied Energy* 223: 416–429.

Zhu, Y., Stubbs, P., Ho, F. et al. (2010). Magnetic nanocomposites: a new perspective in catalysis. *ChemCatChem* 2: 365–374.

Zillillah, Ngu, T.A., and Li, Z. (2012). Highly active, stable, and recyclable magnetic nano-size solid acid catalysts: efficient esterification of free fatty acid in grease to produce biodiesel. *Green Chemistry* 14: 3077–3086.

Zillillah, Tana, G., and Li, Z. (2014). Phosphotungstic acid-functionalized magnetic nanoparticles as an efficient and recyclable catalyst for the one-pot production of biodiesel from grease via esterification and transesterification. *Green Chemistry* 16: 1202–1210.

9

Bio-Based Catalysts in Biodiesel Production

Umer Rashid[1], Shehu-Ibrahim Akinfalabi[1], Naeemah A. Ibrahim[1] and Chawalit Ngamcharussrivichai[2,3]

[1] *Institute of Advanced Technology, University Putra Malaysia, Serdang, Malaysia*
[2] *Center of Excellence in Catalysis for Bioenergy and Renewable Chemicals (CBRC), Faculty of Science, Chulalongkorn University, Bangkok, Thailand*
[3] *Center of Excellence on Petrochemical and Materials Technology (PETROMAT), Chulalongkorn University, Bangkok, Thailand*

9.1 Introduction

The term "biomass" usually refers to all organic matter and compounds produced from crops, forests, and marine life (McKendry 2002). Municipal solid and sewage wastes are also categorized as biomass. The basic composition of biomass compounds includes oxygen, nitrogen, carbon, and hydrogen, with some also containing sulfur and inorganic substances.

A huge amount of biomass is generated from palm oil production in Malaysia, the second largest producer and exporter of palm oil. The main agricultural commodities grown there are oil palm, rubber, rice, cocoa, and coconut. A whopping 77% of the total production of biomass is from oil palm production (17 Mt) alone, with the rest shared among rubber, rice, and other varieties of crops (Sulaiman et al. 2011; Hamzah et al. 2019). Palm tree biomass holds a huge economic significance if properly exploited. Virtually every part of a palm tree (palm mesocarp fiber, palm trunk, palm fronds, spikelets, palm kernels, empty fruit bunches, etc.) can be harnessed and converted to a steady flow of income (Hassan et al. 1991; Aziz et al. 2011). Spikelet and palm fronds have been shown to contribute about 80% of the total emptied fruit bunch. Of the total biomass produced, only 10% is utilized to produce oil, the remaining 90% being discarded as wastes in palm oil mill (Abdullah et al. 2013). Hence, there is an urgent need to convert this remaining biomass waste to wealth.

Generally, fibers and shells are the main sources of oil palm biomass, so palm oil mills used these materials as a fuel for power production (Awalludin et al. 2015). The biomass wastes of the oil palm industry are usually categorized into two parts: one includes the waste of harvested trunks and pruned fronds from oil palm plantations (about 75%), the other comes from palm oil extraction mills and includes empty fruit bunch (EFB), mesocarp fibers (MF), and palm kernel shell (PKS) (about 25%), as shown in Figure 9.1. Furthermore, oil palm fronds and trunks are used as sources of pulp, paper, and fiberboard by the wood industry. In recent years, an increasing number of studies have reported on the potential use

Nano- and Biocatalysts for Biodiesel Production, First Edition. Edited by Avinash P. Ingle.

Figure 9.1 Biomass wastes from palm oil mills: (a) empty fruit bunches (EFB); (b) mesocarp fiber; and (c) palm kernel shells (PKS).

of oil palm biomass in catalysts for biodiesel production. Table 9.1 shows the compositions of different types of oil palm biomass; the major components include cellulose, lignin, inert ash, and hemicellulose (Viana et al. 2018).

The continuous exploitation of fossil resources has led to severe environmental degradation to the ecosystem. The combustion and flaring of gas to produce petroleum products as fuel generates carbon dioxide (CO_2), which is deposited into the atmosphere and so contributes to increasing global temperature. As a result, greenhouse gases have dramatically increased, leading to drastic climate changes (Fauzi and Amin 2013). Thus, there is an imperative need for alternative fuel sources. In recent years, biofuels have gained traction as a viable alternative to fossil resources as they are derived from different renewable sources. Biofuels, by their nature, are environmentally benign, economically feasible, and readily available (Avhad and Marchetti 2016).

In the last two decades, biofuels such as biodiesel have attracted a lot of interest due to their environmental nontoxicity, ability to degrade easily, and potential for renewability (Li et al. 2010; Marchetti 2013). Additionally, fuel produced from biomass resources (biodiesel) does not add to the total atmospheric CO_2, chlorofluorocarbons (CFCs), and methane. Most

Table 9.1 Compositions of different types of oil palm biomass.

Material	Ash	Lignin	Cellulose	Hemicellulose	Glucose	References
EFB	7.54	29.20	23.70	21.60	62.9	Onoja et al. (2019)
PKS	2.10	44.00	27.70	21.60	–	Abnisa et al. (2011)
OPF	0.24	29.50	40.01	30.78	49.8	Kristiani et al. (2013)
OPL	12.32	26.00	32.49	22.97	–	Abnisa et al. (2011)
OPFL	4.26	29.20	23.70	21.60	–	Abnisa et al. (2011)
OPT	5.27	35.89	34.44	23.94	37.14	Abnisa et al. (2011)

CO_2 emitted during biodiesel combustion can be recycled simply by photosynthesis of the plant from which the biodiesel is derived. This shows that emissions from biodiesel processes are greatly minimized (Peterson and Hustrulid 1998). The most popular methods used in biodiesel production are transesterification and esterification, where feedstocks with high free fatty acid (FFA) contents are converted to fatty acid methyl ester (FAME) with the aid of catalysts in the presence of excess alcohol. Biodiesel contains about 10% oxygen, and so speeds up combustion rate and lowers environmental pollutants such as polycyclic aromatic compounds (PACs), particulate matter (PM), and carbon monoxide (CO) (Agarwal 2007). Lower sulfur content, better cetane number, higher flash point, and increased lubrication are some of the other advantages of using biodiesel in place of petroleum diesel (Knothe 2009; Marchetti 2012). The two forms, biodiesel and petro-diesel, are currently being used in blends in diesel engines around the world.

Availability and geographical location play an important role in the selection of feedstock for the production of biodiesel, particularly when production is large-scale. In countries like the United States, where soybean oil is mostly used (about 9431 million pounds in 2015), and in Europe, where rapeseed oil is abundant (about 6.17 million tons, making up 55% of total production in 2014), the choice of feedstock is straightforward. In Asia, of the total 59 million tons of palm oil produced globally in 2014, Malaysia produced 19.4 million, or approximately 33% (MPOB 2016; Akinfalabi et al. 2017; Kapor et al. 2017). Therefore, it makes sense to produce biodiesel from palm waste, along with other biomass sources. Non-edible fats/oils, refined vegetable oils, and some crude wastes are other sources of biodiesel production globally. The fuel properties of biodiesel are largely dependent on the type of feedstock used for its production. These include viscosity, oxidation stability, cetane number, heating value, density, and low temperature properties (Ferrero et al. 2016). Some of the edible vegetable oils used for the synthesis of biodiesel are oil palm, coconut, canola, and sunflower oil (Demirbas 2007). However, the food versus fuel debate will not allow these oils to be used, as they are major sources of food and their use would lead to high prices (Choudhury et al. 2013; Talebian-Kiakalaieh et al. 2013a).

A viable solution would be to opt for non-edible or waste oils such as palm fatty acid distillate (PFAD), waste cooking oils (WCOs), Jatropha, and rubber seed. Among these, it has been reported that PFAD represents the best alternative to serve as a feedstock for biodiesel production. (Chongkhong et al. 2007; Yujaroen et al. 2009; Cho et al. 2012; Lokman et al. 2016; Al-Jaberi et al. 2017). PFAD is a byproduct derived from oil palm refinery during stripping and deodorization. It contains a large amount of FFAs and has a production capacity of about 649 459 tons. Of each ton of crude palm oil (CPO) produced in Malaysia, PFAD makes up about 3.25% (Kapor et al. 2017).

The aim of this chapter is to explore the synthesis of bio-based catalysts that could be extracted from different biomass sources and their utilization for the production of biodiesel from PFAD. To this end, detailed descriptions are given of homogeneous and heterogeneous catalysts used in biodiesel production, heterogeneous bio-based acidic catalysts from different sources, and bio-based magnetic catalysts. Following a detailed survey of bio-based heterogeneous catalysts, a brief overview is presented on the use of bio-based catalysts for biodiesel production and their methods of characterization.

9.2 Biodiesel: A Potential Source of Renewable Energy

As already mentioned, biofuel is a promising replacement for renewable sources of energy, being environmentally friendly due to its biodegradability and nontoxic nature, all well as free of sulfur and having low CO emissions (Nur et al. 2014). It is a FAME, usually produced by esterification or transesterification from animal fats or vegetable oils with methanol or ethanol in the presence or absence of catalyst (García-Martínez et al. 2017; Ambat et al. 2018; Abdulkareem-Alsultan et al. 2019). It has similar fuel properties to petroleum diesel (i.e. cetane number, viscosity, phase changes, and energy content) (Atadashi et al. 2010; Lim and Teong 2010), and is nonflammable, nontoxic, biodegradable, renewable, and environmentally benign (Ribeiro et al. 2007; Hoekman et al. 2012; Shahabuddin et al. 2012). It can be blended with petroleum diesel in any proportion to create a stable fuel that can be utilized in a diesel engine without modification (Jain and Sharma 2011).

As a result of these qualities, biodiesel has emerged as one of the most common biofuels in the world. Vegetable oils such as peanut oil, sunflower oil, and rapeseed oil have been used in diesel engines since as far back as August 10, 1893 (Jayed et al. 2011), when Rudolf Diesel utilized peanut oil as a fuel in his prime diesel engine model, which ran on its own power. The exploitation of vegetable sources continued until the 1920s, when petroleum diesel became more famous. Due to their high availability and government intervention in the form of subsidies, petroleum products became cheap and readily abundant. As a result, existing diesel engines were modified to allow the use of petroleum diesel fuel (Agarwal 2007).

9.2.1 Progress in Biodiesel Development

In the 1970s, falling supplies and questions over petroleum diesel fuel security globally encouraged the development of alternative fuel from renewable sources such as vegetable oils. However, key fuel properties of vegetable oils such as high volatility and high viscosity made it almost impossible for biodiesel from these sources to be used in diesel engines. Notable among the problems was very poor fuel atomization in the fuel spray, which resulted in the cooking and deposition of the injectors and combustion and valve chambers. In order to improve the quality of fuels obtained from vegetable oils, processes such as pyrolysis and micro-emulsification have been proposed to enhance their viscosity

(Srivastava and Prasad 2000) and blending (Ma and Hanna 1999; Pramanik 2003). Even with these processes, however, problems with contamination and carbon deposition still persist (Van Gerpen 2005; Sarin et al. 2007).

In 1853, years before the invention of the first functional diesel engine, E. Duffly and J. Patrick, Belgian scientists, successfully converted vegetable oil into methyl esters by the transesterification reaction. They filed a patent on a "procedure for the transformation of vegetable oils for their uses as fuels," which was accepted (Knothe and Dunn 2001). As time went by, biodiesel was found to have similar properties to petroleum diesel and was widely accepted as into existing diesel engines without any modifications or alterations. In 1977, a Brazilian scientist, E. Parente, successfully applied for the first patent for an "individual process for biodiesel." Also in 1977, in South Africa, research on producing and refining biodiesel from sunflower was initiated. By 1983, engine-tested biodiesel had become globally accepted. In 1987, Gaskoks, an Austrian company, acquired the technology from a South African Agricultural Engineer and produced the first ever pilot plant, as well as an industrial-scale plant in 1989. Around the late 1990s, because of increasing concerns over environmental sustainability and the need to reduce cost, a huge number of commercial productions of biodiesel were proposed (Lim and Teong 2010). At this time, plants of varying qualities were established in European countries. In France, for example, biodiesel produced from rapeseed oil is known as "diester." Table 9.2 shows the key milestones in the growth of the biodiesel industry.

9.2.2 Development of Biodiesel in Malaysia

The Malaysian government originally started development of biodiesel and biofuel in the early 1980s. As one of the largest producers and exporters of palm oil globally, it was important that Malaysia took the initiative in becoming a pioneer in the production of biodiesel and biofuel from palm sources. Thus, plans were developed to produce palm biodiesel at the laboratory level by the Malaysian Palm Oil Board (MPOB) in 1982. About two years later, in 1984, following encouraging results from the laboratory research, construction was commenced on a pilot palm oil biodiesel plant, with full support and collaboration of Petronas, a local petroleum fuel company (Lim and Teong 2010). The commissioning of this plant saw an annual production of palm oil biodiesel of 3000 metric tons (MT).

As the years rolled by, biodiesel from palm oil was assessed to be a viable replacement for petroleum diesel. It was tested successfully in different types of vehicles powered by diesel engines. Blending of biodiesel from palm oil with petroleum diesel was also tested, and the results showed positive outcomes in comparison to the performance of petroleum diesel alone. The technology required to produce winter-grade biodiesel was successfully achieved in 1992, based on research on palm oil biodiesel, solving the gelling problems that arose when biodiesel was used in regions with temperatures close to freezing degrees (Lim and Teong 2010).

Malaysia's biodiesel industry faced setbacks over the years, but from 2001 to 2005 the Fifth Fuel Policy was announced, which recognized energy from renewable source as the fifth fuel in the energy supply mix in the country. The main aim of this announcement was to ensure that the petroleum-derived fuel import bill was lowered in order to raise demand for palm oil-derived biodiesel. This led to a steady upward trend in the growth of

Table 9.2 Key milestones in the industrial development of biodiesel (Lin et al. 2009; Mofijur et al. 2013).

Date	Event
August 10, 1893	Rudolf Diesel's prime diesel engine model, fueled by peanut oil, run for the first time in Augsburg, Germany
1900	Rudolf Diesel's engine shown at the World Exposition in Paris, running on 100% peanut oil
August 31, 1937	Patent granted for a "procedure for the transformation of vegetable oils for their uses as fuels," by G. Chavanne, a Belgian scientist. For the first time, the idea of what is today known as "biodiesel" is proposed
1977	First patent for an industrial process for biodiesel production applied for by Expedito Parente, a Brazilian scientist
1979	South Africa initiates research into the utilization of transesterified sunflower oil, and refines it to diesel fuel standards
1983	Production process for fuel-quality, engine-tested biodiesel completed and published internationally
November 1987	First biodiesel pilot plant established by Gaskoks, an Austrian company
April 1989	First industrial-scale plant established by Gaskoks
1991	First biodiesel standard issued by Austria
1997	German standard DIN 51606 formalized
2002	ASTM D 6751 published
October 2003	DIN EN 14214, a new Europe-wide biodiesel standard, published
September 2005	Minnesota becomes the first US state to authorize that all diesel fuel sold in the state should contain part biodiesel, demanding a content of at least 2%
October 2008	New Biodiesel Blend Specifications Standards published by ASTM
November 2008	European Standard EN 14214 current version published, superseding EN 14214:2003

biodiesel in Malaysia, and by 2006, with the introduction of Envo Diesel (a 5% blend of refined palm oil and 95% petroleum diesel), the position of biodiesel production from palm oil was cemented. According to the Malaysian Biodiesel Association, there were 10 standard operative biodiesel plants in the country in 2008, with a cumulative annual installed capacity of 1.2 million tons of biodiesel (Lim and Teong 2010).

9.2.3 Biodiesel Feedstocks

In biodiesel production, the primary source for fatty acids is usually renewable natural sources such as vegetable oils and fats (Singh and Singh 2010). Globally, about 95% of biodiesel production comes from edible vegetable oils (Gui et al. 2008). Several crops bearing oil, like sunflower, rapeseed, and Jatropha, are used in parts of Europe and Asia as feedstock, while soybean is commonly used in the United States. Palm oil and coconut are common sources in Malaysia and Indonesia (Marchetti et al. 2007; Demirbas 2009).

Because of the high cost of vegetable oils compared with petroleum diesel, due to their use as foods, non-edible waste vegetable oil is preferred as a potential source for low-cost feedstock. Several types of non-edible vegetable oils have been researched, such as rubber seed tree (*Hervia brasiliensis*) (Ramadhas et al. 2005), Karanja (*Pongamia pinnata*) (Naik et al. 2008; Agarwal and Rajamanoharan 2009), microalgae (Gao et al. 2010; Huang et al. 2010), Jatropha oil (Tiwari et al. 2007; Ganapathy et al. 2009), rice bran (Sinha et al. 2008; Lin et al. 2009), neem (*Azadirachta indica*) (Agarwal et al. 2008), sunflower oil (Rashid et al. 2008), and *Moringa oleifera* oil (Rashid et al. 2011).

Although animal fats have been mentioned regularly, they have not been as rigorously investigated as feedstocks for biodiesel production due to their other natural properties. An ideal feedstock that can serve as a primary precursor for the production of biodiesel should not raise questions of cost and the food versus fuel issue; in other words, a viable feedstock must meet the necessary requirements of low-cost production and high production scale (Singh and Singh 2010). As a result, the use of cheap and abundant waste oils from animal and plant fats and vegetable oils is currently receiving great attention.

In Malaysia, one of the byproducts of palm oil factories is PFAD, a waste obtained from the palm refining, bleaching, and deodorizing processes. Appropriate use of this byproduct would lead to a drastic reduction in the cost of production of biodiesel and help in waste management in the palm oil industry.

9.2.3.1 PFAD as a Biodiesel Feedstock

PFAD is a second-generation low-cost feedstock, derived from purified palm oil (Lee and Ofori-Boateng 2013). In the process of refining CPO, PFAD is generated during the fatty acid stripping and deodorization stages. It is commonly used as a fatty acid source for non-edible application in biodiesel production. It is also used as a greener alternative fuel in engines and boilers (Lee and Ofori-Boateng 2013). The use of PFAD as a potential alternative to fatty acids in the production of biodiesel has reduced issues surrounding the food versus fuel debate. However, a major drawback is that it is expensive compared to crude petroleum diesel (Fry 2010). PFAD is commonly used in the animal feed, cosmetics, pharmaceuticals, and soap industries (Mielke 2010).

In 2009, Malaysia and Indonesia were the largest producers of CPO, with a capacity of 17.5 and 20.9 million MT, respectively (Mielke 2010). It is estimated that about 700 000 MT of PFAD is generated annually from the production of palm oil in Malaysia alone (Lokman et al. 2015).

Reports suggest the availability of low-cost, non-edible PFAD is abundant, making it attractive for biodiesel producers. Moreover, compared to other refined oil feedstocks, PFAD costs much less (Metre and Nath 2015). Figure 9.2 shows the cost of PFAD as compared to CPO and refined, bleached, and deodorized (RBD) palm oil (Kapor et al. 2017). It is estimated that the cost of PFAD was RM 1905.50 (USD 439.40) per ton in 2015, while that of CPO was RM 2153.50 and that of RBD palm oil was RM 2279.50 (Figure 9.2).

At room temperature, PFAD is a light brown semi-solid fatty acid material, which melts and becomes a brown liquid on heating. It primarily consists of high FFA (>85 wt%) (most of which are palmitic and oleic acids), triglycerides (>10 wt%), glycerides (>3 wt%), and other small unsaponifiable sterols, vitamin E, and squalene (Ping and Yusof 2009; Metre and Nath 2015). Table 9.3 describes the general properties of PFAD according to two different reports.

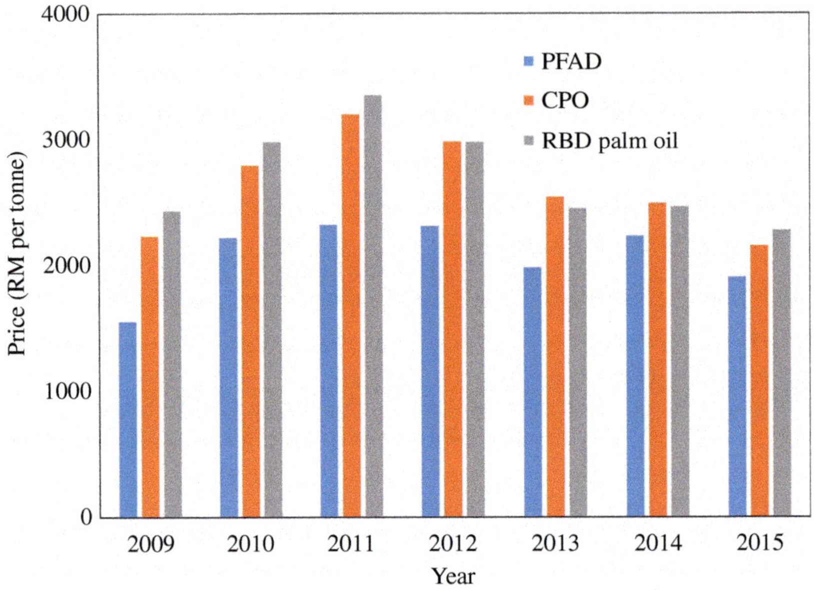

Figure 9.2 Prices of RBD, CPO, and PFAD oil (Kapor et al. 2017).

In a typical esterification reaction, fatty acid and methanol are measured in a 1 : 1 M ratio, stoichiometrically. However, excess alcohol is usually used to drive the reaction forward. Most researchers have qualified methyl ester based on yield and conversion, but a few findings make use of the reduction of FFA as a means of showing the progress of esterification. Due to the high content of FFA (80–90% wt), esterification easily converts PFAD into biodiesel in the presence of either heterogeneous or homogeneous catalyst (Kapor et al. 2017). Generally, before the commencement of esterification, the PFAD, which is usually solid at room temperature, is subjected to heat (50–80 °C) to melt it for a period of 10–15 minutes. Transesterification of PFAD with a base catalyst is impossible, since the catalyst will interact with the acidic FFA content, resulting in its deactivation by way of saponification (soap formation). In addition, homogeneous (liquid) acid esterification of PFAD requires numerous purification steps in order to recover the product. To ameliorate these concerns, application of heterogeneous solid acid catalysts has been preferred (Nakpong and Wootthikanokkhan 2010). Recently, sugar (glucose) catalyst, known also as carbohydrate-derived acid catalyst, has been found to have unique properties and the potential to esterify feedstocks with high FFA content because of its acidic properties and the stability of its polycyclic carbon structure (Zong et al. 2007; Lou et al. 2008).

9.2.4 Common Methods Used for Biodiesel Reaction

Low sulfur content, biodegradability, ready availability, low aromatic content, and renewability are some of the great advantages of vegetable oil as a diesel source (Goering et al. 1982). However, drawbacks include low volatility, high viscosity, and the chain behavior of polyunsaturated hydrocarbons. In particular, the viscosity of all vegetable oils ranges from

Table 9.3 General properties of PFAD feedstocks.

Properties	Report 1 Ping and Yusof (2009)	Report 2 Lokman et al. (2015)
FFA content (%)	72.7–92.6	86.3 ± 1.75
Iodine value, Wijs/g·(100 g)$^{-1}$	46.3–57.6	
Conventional mass per volume at 50 °C (kg·l^{-1})		0.8634
Titer (°C)	46.0–48.3	
Saponification value (mg KOH)·(g sample)$^{-1}$	200.3–215.4	149.74 ± 4.72
Moisture content (%)	0.03–0.24	0.089 ± 0.004
Unsaponifiable matter (%)	1.0–2.5	
Fatty acid composition		
C14:0 myristic acid (wt%)	0.9–1.6	1.9 ± 0.05
C16:0 palmitic acid (wt%)	43.0–49.1	45.7 ± 1.32
C18:0 stearic acid (wt%)	4.0–4.5	4.3 ± 0.09
C18:1 oleic acid (wt%)	34.7–37.2	40.2 ± 1.21
C18:1 linoleic acid (wt%)	8.5–9.7	7.9 ± 0.21

10 to 20 times greater than that of petroleum diesel (Keera et al. 2011). Furthermore, problems related to carbon deposits, gelling of lubricant oil in the engine, coking of injectors, thickening of fuel, and oil ring sticking have been reported (Darnoko and Cheryan 2000; Komers et al. 2001; Demirbaş 2003).

Several methods have been suggested to avoid these issues, especially the problems around viscosity, including dilution or blending, pyrolysis, esterification, transesterification, and micro-emulsification. However, micro-emulsification and blending have been reported to cause fouling of the engine, which results in technical issues such as sticking and gunning of piston rings and engine coking (Ramadhas et al. 2004; Di Serio et al. 2007) and incomplete combustion (Sahoo and Das 2009). The methods that are most promising for reducing the viscosity of biodiesels from vegetable oil and animal fat sources are thus esterification and transesterification (Leung et al. 2010).

9.2.4.1 Esterification

The reaction of carboxylic acids (fatty acids) with a short-chain alcohol (methanol) in the presence of an acid catalyst, leading to the synthesis of alkyl esters and water as byproduct, is termed esterification (Moser 2009). A typical esterification reaction is presented in Figure 9.3a. This reaction is reversible, which means excess alcohol is used to push it toward the product. This process is more pronounced and relevant in biodiesel production from low-cost materials with high FFA contents (Lotero et al. 2005; Mbaraka and Shanks 2005; Liu et al. 2006).

A low-grade byproduct from the oil refining industry, soap stock, has been reported for the production of biodiesel using an acid catalyst (Haas et al. 2003). This low-grade material is a mixture of water, soaps, and oil. However, before it can be used, this material must be

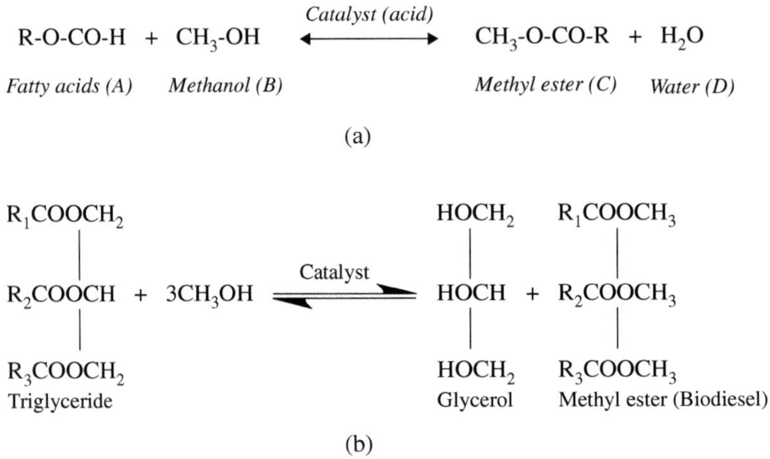

Figure 9.3 (a) General esterification reaction of fatty acids. (b) General transesterification reactions of triglycerides with methanol (Dehkhoda et al. 2010).

dried, saponified, and esterified with a short-chain alcohol such as methanol while using an organic catalyst to produce biodiesel.

The utility of this process depends largely on the recovery cost of the excess alcohol used. Biodiesel production from feedstocks with high FFA contents is usually a two-step catalytic process. The first step involves esterification with an acid catalyst like sulfuric acid (pretreatment of the feedstock) in order to reduce the FFA content. The second step is to transesterify the triglyceride using an alkaline catalyst for the biodiesel synthesis (Canakci and Gerpen 2001; Hayyan et al. 2010; Charoenchaitrakool and Thienmethangkoon 2011). Nevertheless, there are studies that have reported a one-step esterification of high-FFA feedstocks in biodiesel production (Borges and Díaz 2012). Here, it is important to mention that esterification of carboxylic acids has found application in specialty chemical and fine production or manufacture of intermediates for fabrics, fragrances, polymers, and lubricants (Hoydonckx et al. 2004).

9.2.4.2 Transesterification

Transesterification, or alcoholysis, is defined as a chemical reaction that involves triglycerides of oils or fats and alcohol in the production of esters and glycerol with or without a catalyst (Singh and Singh 2010). This process has shown the most promise for resolving the problems of viscosity around biodiesel produced using pure vegetable oils. Typically, it is more convenient and easier to convert from an existing alkyl group of esters to another alkyl group. This reaction can be catalyzed by either an acid, an alkali, or an enzyme. However, the homogeneous alkali-catalyzed transesterification reaction is the most popular in commercial industry (Balat and Balat 2010).

Alkaline catalysts such as sodium hydroxide (NaOH) and potassium hydroxide (KOH), alkoxide, and carbonates of potassium and sodium are most frequently used. The transesterification of triglycerides and alcohol is presented in Figure 9.3b. This reaction is highly reversible; hence, a huge amount of alcohol is usually needed to ensure that more methyl esters are produced by forcing the equilibrium to the product side. Short-chain alcohols such as methanol, ethanol, butanol, and propanol are generally utilized in the transesterification process (Speight 2008). Ethanol, due to its green advantages, is preferred over the other alcohols since it is derived from agricultural products, is renewable, and is environmentally benign. However, methanol is more popularly used because it is cheap, readily available, and has physicochemical advantages such as a short chain and polarity. In addition, methanol and triglycerides can react quickly, and the alkaline catalyst is readily soluble in it (Ma and Hanna 1999). It is a well-known fact that the activities of catalysts are strongly dependent on the nature of the alcohol and hence alcohols with a linear chain positively affect the reaction when the chain is short but negatively affect it when the chain length is increased (Macedo et al. 2006).

Transesterification generally depends on key factors such as the reaction temperature, reaction time, molar ratio of methanol to oil/fat, catalyst concentration, and type of catalyst. Conventionally, the reaction temperature is around 25–65 °C, where the biodiesel yield increases with temperature (Vicente et al. 2004; Bambase et al. 2007). For instance, biodiesel yields of approximately 25% at 25 °C, 84% at 40 °C, and 92% at 60 °C have been reported for the transesterification of sunflower oil with 6 : 1 methanol to sunflower oil molar ratio, 0.5 wt% catalyst concentratixon (NaOH), and 20 minutes reaction time (Bambase et al. 2007). Transesterification of vegetable oil via a supercritical fluid process to produce biodiesel without catalyst has also been reported (Kusdiana and Saka 2001; Glisic and Skala 2009; Lee and Ofori-Boateng 2013).

At the commercial scale, the production of biodiesel from the transesterification of vegetable oils through supercritical alcohol and enzyme-catalyzed processes have not yet been exploited due to the high operation costs. Mostly, supercritical alcoholysis of oils is not cost-effective due to the need for high pressures (~8 MPa) and temperatures (~350 °C) (Kusdiana and Saka 2001). Additionally, enzyme-catalyzed reaction for the production of biodiesel cannot be effectively utilized for any meaningful industrial application because of drawbacks such as the high cost of enzymes, slow rates of reaction, and deactivation of enzymes (Lam et al. 2010). Hence, the use of homogeneous or heterogeneous catalysts has been the main focus of most recent researches.

9.3 Homogeneous Catalysis in Biodiesel Production

Homogeneous catalysts are highly advantageous over heterogeneous catalysts due to the fact that they have low mass transfer resistance, which is a key component in the conversion of FFA to FAME. Those most commonly used for alcoholysis reactions are either acidic or basic by nature. Commercially, in the modern-day industry, most catalysts used for the production of biodiesel are homogeneous basic catalysts due to the fact the homogeneous base catalysts are faster than the homogenous acid ones in a transesterification reaction. Moreover, they provide a better biodiesel yield under mild reaction conditions.

The choice of catalyst rests mainly on the moisture content and FFA composition of the feedstock. Some of the most commonly used homogeneous base catalysts that are soluble in polar reactants are hydroxides and alkoxides of alkali metals, such as KOH, NaOH, sodium methoxide (CH_3ONa), sodium ethoxide (C_2H_5ONa), and potassium methoxide (KCH_3O) (Dias et al. 2008; Uzun et al. 2012). At the industrial level, NaOH and KOH are commonly

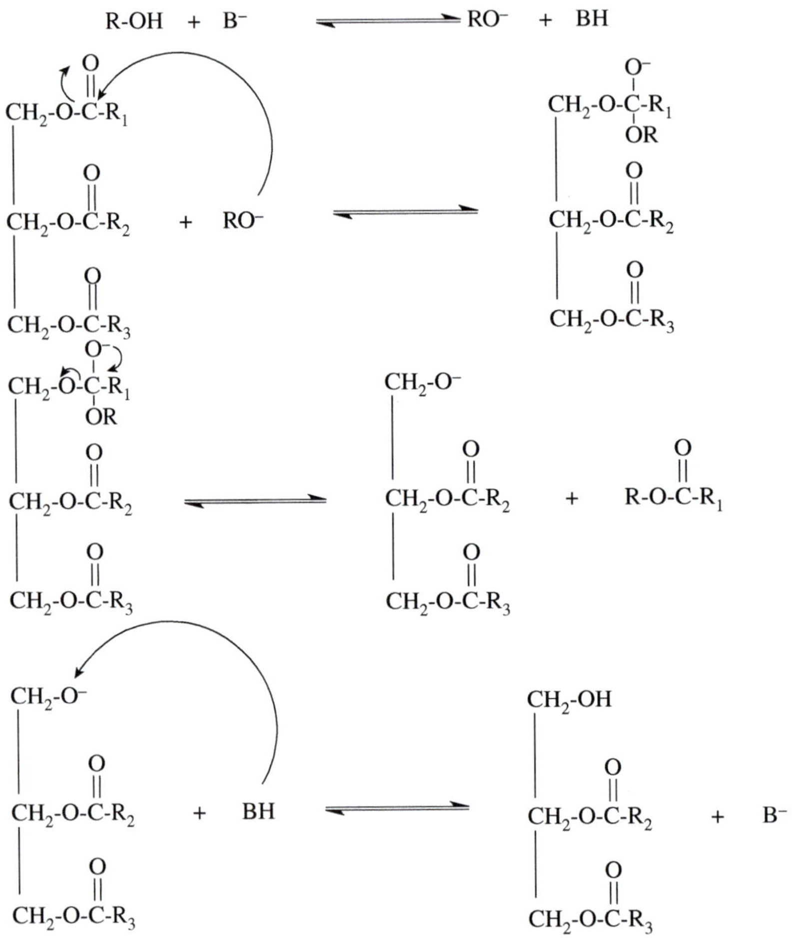

Figure 9.4 Reaction mechanism for a homogeneous base-catalyzed transesterification reaction (Avhad and Marchetti 2016).

used for biodiesel production due to their availability, low cost, short reaction time, and superior biodiesel yield. The mechanism of the homogeneous base-catalyzed transesterification reaction is still somewhat unclear. However, in the process of alcoholysis, it is expected that the reaction will occur in two phases: as a biphasic reaction medium consisting of an alcohol and oil phase where the catalyst dissolves, and in a single-phase medium that is dependent mainly on the reaction operating parameters, such as molar ratio of alcohol to oil/fat, concentration of catalyst, and reaction temperature. Figure 9.4 shows a feasible pathway for the mechanism of the homogeneous base-catalyzed transesterification reaction. First, the basic homogeneous catalyst is shown to dissolve in the alcohol, which leads to the formation of an alkoxide and a protonated catalyst. Next, the nucleophile attacks the carbonyl group of the triglycerides, leading to a tetrahedral intermediate. Finally, the intermediates rearrange to form alkyl esters and the corresponding anion of diglycerides. This process occurs twice, where the transformation of the diglycerides and monoglycerides is seen to form a mixture of alkyl esters of fatty acids and glycerol, then the post-reaction ensures the mixture is separated into two liquid phases of fatty acid alkyl ester (FAAE) and crude glycerol (Brunschwig et al. 2012; Issariyakul and Dalai 2014).

It is worthy of note that the homogeneous basic catalysts have many technical barriers that limit their utilization in biodiesel production. A negative influence on the catalytic performance of the basic homogeneous catalysts was observed when low-cost feedstocks with high moisture and FFA contents were used. This led to an unavoidable saponification reaction between the FFA and the basic catalyst. Hence, the catalyst was wasted, and more cost was incurred in separation and purification in order to achieve the final biodiesel. Saponification, otherwise called soap formation, also prevents the separation of methyl ester from glycerol; consequently, it reduces the biodiesel yield.

Homogeneous acid catalysts could serve as a means of rectifying some of the disadvantages involved in the use of low-grade feedstocks in biodiesel production. Homogeneous acid-catalyzed transesterification has been reported to be about 4000 times slower than the reaction using homogeneous base catalyst. However, homogeneous acid catalysts are insensitive to high FFA contents and can simultaneously carry out the esterification of FFA and the transesterification of triglycerides. The major problem in the industrial production of biodiesel by upscaling of the output of the biodiesel produced while using a homogeneous acid catalyst is the harsh operative reaction involved; that is, prolonged reaction time and temperature and the use of large amounts of alcohol. Expensive and complicated neutralization steps and equipment corrosion are also seen. Sulfuric acid (H_2SO_4), followed by hydrochloric acid (HCl), is the most widely used homogeneous acid catalyst in alcoholysis to produce biodiesel. Some other acids have been tested too, such as aluminum chloride ($AlCl_3$), nitric acid (HNO_3), phosphoric acid (H_3PO_4), and other organic sulfonic acids. A schematic representation showing the reaction mechanism of homogeneous acid-catalyzed transesterification is presented in Figure 9.5.

A viable alternative process that is more tolerant to FFAs can be used for acid catalysts (Aksoy et al. 1988; Liu 1994). This involves protonation of the carbonyl group, including the triglycerides, quickly followed by a nucleophilic attack of the alcohol, producing a tetrahedral intermediate (carbonation). The intermediate then goes through a migration of protons followed by production of esters after its breakdown. The process is repeated twice in order to convert the glycerides (di- and mono-) into three molecules of FAAE. At the end of the reaction process, glycerol is produced as a byproduct and the unused catalyst is regenerated. The most important reason for the differences in how the catalysts function may be that the base catalysts forms more nucleophilic species while the acid catalysts form more electrophilic ones (Demirbas 2009; Lam et al. 2010).

9.4 Heterogeneous Catalysis in Biodiesel Production

Heterogeneous catalysts have been the subject of much interest among both the scientific community and industry due to their economic and environmental advantages (Yusuf et al. 2012; Avhad and Marchetti 2016). Synthesized heterogeneous catalysts are expected to ameliorate the prominent issues arising from the use of homogeneous catalysts for the production of biodiesel, including reducing the amount of wastewater generated by the

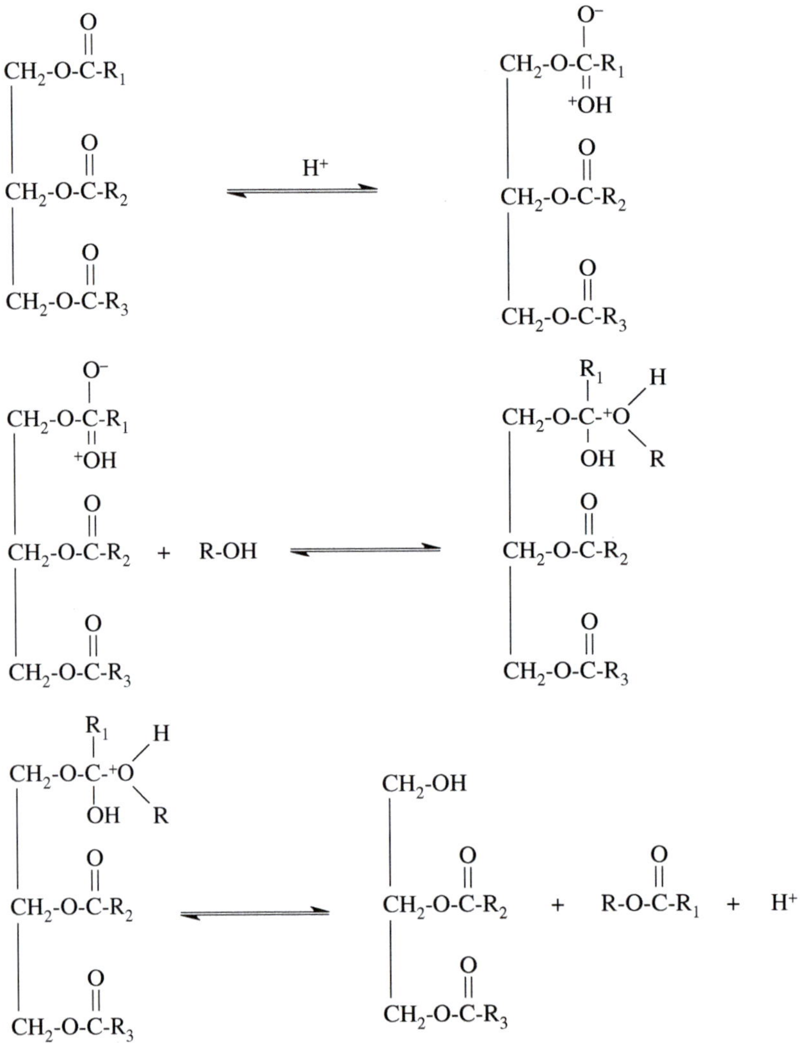

Figure 9.5 Reaction mechanism for acid-catalyzed homogeneous transesterification (Avhad and Marchetti 2016).

purification and separation process, thereby reducing the post-treatment of wastewater and other residual contaminants. This greatly reduces the total cost of the process, since the synthesized catalyst is neither consumed nor dissolved during the reaction. After the reaction, the residual catalysts can be recovered by filtration and centrifugation and reused. This process will ensure that there is no need for acid post-treatment of the wastewater, which directly enables a simple and efficient catalyst synthesis (Puna et al. 2010; Tonetto and Marchetti 2010; Sánchez-Cantú et al. 2013). Nonetheless, one of the outstanding disadvantage of heterogeneous (solid) catalysts is the limitation on their activity, which necessitates severe reaction conditions in order to achieve a similar fatty acid conversion of oils/fats to that obtained in thomogeneous catalyzed reactions (Kazemian et al. 2013).

Recently, biodiesel production by transesterification with the aid of heterogeneous base catalysts has been widely researched in the hope of obtaining a simplified reaction process and reducing waste.

This process requires that the fatty acid source (i.e. feedstock; a vegetable oil or animal fat) has a low FFA content as the solid base catalyst will react with FFA to form soaps and the presence of water will inadvertently hydrolyze the triglycerides to form more FFAs (Leung et al. 2010). A huge number of basic materials have been synthesized and utilized as heterogeneous (solid) catalysts for the alcoholysis reaction, including metal oxides, single alkaline earth metal oxides, mixed metal oxides, alkaline earth metal oxides, supported alkali metals, anionic ion-exchange resins, and hydrotalcites. Mostly, homogeneous and heterogeneous base catalysts follow similar transesterification catalytic mechanism pathways.

It is hoped that advances in research will eventually provide an excellent development of heterogeneous acid catalysts for biodiesel synthesis. The successful production of heterogeneous solid acid catalysts is expected to yield alkyl esters of high quality. These catalysts, once successfully synthesized, will not cause equipment corrosion or generation of unwanted wastewater. Heterogeneous acid catalysts have the added advantage of functioning as two catalysts in one (bifunctional): heterogeneous base catalysts and mineral catalysts (Lewis acid, containing mixed and sulfated oxides, and Brønsted acid, containing sulfonic agents) (Lotero et al. 2005). In addition, heterogeneous acid catalysts can simultaneously speed up the transesterification and the esterification of FFAs in the alcoholysis reaction. Generally, the esterification reaction is normally accelerated by the Brønsted acid component while the transesterification reaction is accelerated by the Lewis component.

The hydrophobic surfaces present in heterogeneous acid catalysts prevent water and glycerol (polar byproduct) from deactivating the active sites, thereby enhancing the selective adsorption of oily hydrophobic molecules to the catalyst surfaces.

Heterogeneous and homogeneous acid catalysts are reported to have similar pathways with regard to the alcoholysis reaction mechanism, the only difference being that alcoholysis accelerated by heterogeneous acid catalysts is controlled by the catalyst surface, relying heavily on the interconnectivity of the large pores, surface-active acid sites, and hydrophobicity. The pathway for the alcoholysis of acid-catalyzed esterification of carboxylic acids mainly includes protonation of the organic acid carbonyl oxygen by the catalyst (Brønsted acid), which leads to the formation of carbocation. Deprotonation of the oxygen then follows from the methanol, which results in the production of a nucleophile, which attacks the carbocation to form an intermediate tetrahedral. The intermediate tetrahedral then removes hydrogen from the molecules of water to form alkyl esters. This reaction is reversible because nucleophilic attack by the water of the protonated alkyl ester regenerates the carboxylic acid and alcohol. Figure 9.6 shows a breakdown of the reaction mechanism of acid-catalyzed esterification.

9.5 Catalyst Supports

The catalyst support is regarded as highly effective in the process of reducing mass transfer limitations for heterogeneous catalysts in liquid phase reactions. Such a support could be

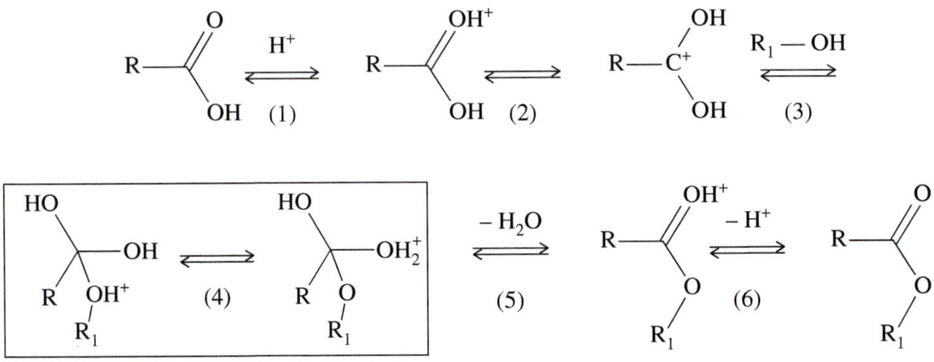

Figure 9.6 Reaction mechanism for acid-catalyzed esterification (Liu et al. 2006).

very useful in providing higher surface areas by showcasing existing pores in the surface where metal particles could be anchored. Supporting catalysts with potential in biodiesel production include alumina, silicate, zirconium oxide, and activated carbon.

9.5.1 Alumina

Different types of aluminum oxide exist as supports in the catalysis process. Among them, γ-alumina and β-alumina forms are known as porous alumina, whereas α-alumina is known as nonporous crystalline alumina. These forms are considered highly effective in supporting catalysis processes through polymerization, reforming, steam reforming, hydrogenation, and dehydration. Alumina is mostly used due to its high thermal strength and mechanical stability in the process of supporting catalysis, as well as its large pore size, highly specific surface area, and high pore volume.

Delfort et al. (2005) described the catalytic activity of alumina in the transesterification of colza oil with methanol, which provided necessary support to titanium oxide-zinc and titanium oxide-bismuth oxide. According to the authors, the catalysts were prepared by co-mixing followed by calcination at 600 °C for three hours. In the supporting process, the specific surface areas were presented specifically in the form of alumina titanium oxide-zinc oxide (ATZ) and alumina titanium oxide-bismuth oxide (ATB) at 62 and 54 g, respectively. Furthermore, both the catalysts (i.e. ATZ and ATB) were found to be stable and achieved up to 94% yield of methyl esters under reaction conditions of 200 °C reaction temperature, 50 bars reaction pressure, and eight hours reaction time. Both catalysts were reusable and showed similar activity in the second cycle.

9.5.2 Silicate

Various types of mesoporous silica provide necessary support to improve the catalytic properties of calcium oxide (CaO). This is shown by X-ray diffraction (XRD) characterization and highlighted in the impregnation of the CaO. The hexagonal structure of SBA-15 shows stability while the MCM-41 structure and fumed silica are collapsed, indicating that SBA-15 could remain stable with CaO species on its surface (Zhang et al. 2016).

Siddiqui et al. (2019) presented the process by which porous silica materials like SBA-15, MCM-41, and fumed silica supported with CaO can be used in the transesterification of vegetable oil using ethyl butyrate in combination with methanol under different reaction parameters. Furthermore, they reported that the same combination of catalysts and methanol can show about 95% of biodiesel conversion under optimal reaction conditions in the case of sunflower oil, as opposed to about 65% in the case of castor oil.

9.5.3 Zirconium Oxide

Baskar and Aiswarya (2015) specifically described zirconia oxide as a supporting catalyst that provides support to two types of acid, heteropoly and isopoly tungsten acids, as used in the synthesis of biodiesel from sunflower oil. The results indicated that isopoly tungsten was more active in comparison to heteropoly tungsten because it was associated with more Brønsted acid sites. A maximum 97% oil conversion was recorded under reaction conditions of 15 : 1 alcohol to oil molar ratio, 15 wt% catalyst content, and 200 °C reaction temperature.

9.5.4 Activated Carbon

Both the chemical and the physical methods of biodiesel production show that activated carbon is very supportive in the absorption of gases, liquids, and solids because it provides high surface areas. Activated carbon is usually prepared through physical and chemical methods, the physical method involving two steps and the chemical method involving heating a mixture of the organic materials and activation agents (e.g. NaOH, KOH, and $ZnCl_2$), rinsing to remove the activation agents, and drying to obtain the carbon. Erdem et al. (2017) demonstrated that an increase in temperature (700 °C) and impregnation ratio (40 to 30%) can help to achieve maximum biodiesel conversion. The carbon-based catalysts can be effectively used in the production of biodiesel fuels because of their high stability, low toxicity, and convenient facile preparation.

9.6 Heterogeneous Bio-Based Acid Catalysts

Bio-based acid catalysts have recently been exploited from sugarcane bagasse (Chin et al. 2012), hardwood biochar (Dehkhoda et al. 2010), rice husk (Li et al. 2014), and numerous palm parts (palm frond, palm spikelets, palm EFB, and palm trunk) (Akinfalabi et al. 2017; Sani et al. 2015). About 7.3 metric tons of palm waste biochar (PWB) were processed in Malaysia in 2016, showing the abundance of palm waste there (Index Mundi 2017).

The use of heterogeneous acid catalysts for the transesterification of oils/fats has been widely researched and reported. Some of the most used catalysts include sulfated tin oxide ($SO4_2$-/SnO_2) (Furuta et al. 2004), heteropoly acids (HPWs) (Zhang et al. 2009), sulfated zirconium oxide ($SO_4^{2-}ZrO_2$) (Furuta et al. 2004; Park et al. 2008), and sulfonated carbonized acid catalyst (Lou et al. 2008; Zong et al. 2007). Ideally, a heterogeneous carbon-based solid acid catalyst is expected to have increased pores, thermal stability, strong acid sites, nontoxicity, no corrosion effect, a low cost of separation and preparation, a high surface area, a hydrophobic surface, and reusability (Konwar et al. 2014a; Lotero et al. 2005). Similarly, there are reports showing some disadvantages of carbon-based heterogeneous catalysts such as a higher molar ratio of oil to methanol, leaching of active acid sites, and slow reaction rates.

Nakajima and Hara (2012) have demonstrated the preparation of amorphous carbon with an attached sulfonic (SO_3H) group in the form of Brønsted solid acid catalyst. Thus synthesized acid carbon catalyst with a small polycyclic aromatic structure attached with a $-SO_3H$ group ensures successful esterification of feedstocks with high FFA contents. Furthermore, the catalysts can be recovered and reused for about 50 consecutive cycles, maintaining catalytic activity. The carbohydrate carbon-based solid acid catalyst (sugar catalyst) is a typical example of a carbon-based solid catalyst with polycyclic aromatic carbon rings providing a strong and stable structure (Alves et al. 2013).

9.7 Synthesis of Bio-Based Solid Acid Catalysts

Bio-based solid acid catalysts can be synthesized by the functionalization of bio-based materials with sulfonic agents or acid groups ($-SO_3H$ and COOH) that serve as strong catalytic active sites. As far as production of biodiesel goes, most of the functionalized bio-based acid catalysts reported are sulfonated acid catalysts. They are mostly employed to catalyze the esterification reaction of fatty acids in oils/fats or to catalyze transesterification and esterification reactions simultaneously to produce biodiesel (Marchetti and Errazu 2008).

There are several reports available stating different methods for the preparation of bio-based catalysts and other carbon-based catalysts, but two methods are widely accepted for the attachment of SO_3H groups (Nakajima and Hara 2012). One involves the sulfonation of incomplete carbonization of organic materials like biomass, glucose, cellulose, starch, or sucrose (Toda et al. 2005; Okamura et al. 2006), and the other the incomplete carbonization of sulfoaromatic compounds (Hara et al. 2004). Most of the recent work on catalysts synthesis applies the first method because it is largely convenient and safe (Konwar et al. 2014b).

Various parameters, such as sulfonation agent, sulfonation time, quality of carbon precursor, and carbonization time of catalyst synthesis, have been extensively studied. For instance, Guldhe et al. (2017) synthesized chromium aluminum mixed oxide (CRAL) heterogeneous catalyst and utilized it for the esterification of PFAD. They achieved a FAME conversion of 98.28% under optimum reaction conditions of 15 : 1 methanol to oil molar ratio, 15 wt% catalyst concentration, and 80 °C reaction temperature. In a related study, rice husk-derived sodium silicate was utilized for the transesterification of oils under optimal reaction conditions of 12 : 1 methanol to oil molar ratio, 2.5 wt% catalyst loading, 65 °C reaction temperature, and 30 minutes reaction time, giving a FAME yield of 97% (Roschat et al. 2016). Similarly, the methanolysis of RBD palm oil was demonstrated using CaO catalyst under optimal conditions of 12 : 1 methanol to oil ratio, 9 wt% catalyst loading, 60 °C reaction temperature, and two hours reaction time, giving a conversion yield of 90.11% (Uprety et al. 2016).

Typically, synthesis of these bio-based catalysts starts with the physical preparation of the precursor, which involves drying the waste biochar material, grinding, and sieving to a fine powder (<2 μm). Further, 4 g of the fine powder is soaked in orthophosphoric acid for 24 hours (impregnation period) to chemically activate the bio-based material. It is then pyrolyzed with a crucible tube furnace under nitrogen flow for two hours at 400 °C. The biochar produced is washed with warm de-ionized water (80 °C) until a pH ~7 is achieved, then placed in an oven set at 80 °C to dry for 24 hours, in order to obtain the soaked palm waste biochar (PWB-soaked) material (Akinfalabi et al. 2017). After drying, a mortar and pestle is used to powder the PWB materials. Attachment of a sulfonic group (–SO_3H) is performed by sulfonating the PWB (4 g) with 100 ml of concentrated H_2SO_4 at 150 °C under a nitrogen flow in a reflux system for 12 hours. At the end of heating, the sample is allowed to cool to room temperature and then slowly diluted with distilled water. The black precipitate obtained is later collected. The sulfonated palm waste biochar (SPWB-SO_3H) catalyst is washed with hot (>80 °C) distilled water in order to remove any possible remnants of the sulfate ions and unwanted impurities. The excess moisture is removed by drying the SPWB-SO_3H catalyst in an oven for 24 hours at 80 °C (Akinfalabi et al. 2019a).

In catalyst synthesis, the temperature used for the carbonization is essential, since the optimal temperature ensures that the polyaromatic carbon ring is formed and that carbonization of the sulfo-aromatic compound does not occur (Nakajima and Hara 2012). It has been reported that the optimum temperature at which incomplete carbonization can be performed is 400 °C for organic carbohydrate materials. The use of nitrogen (inert environment) is also important to ensure the precursor does not burn to ash (Zong et al. 2007).

9.7.1 Palm Tree Fronds and Spikelets

Biomass-based feedstock offers a reliable alternative for use as both biodiesel and catalyst because it is sustainable and environmentally benign. Sulfonated palm frond spikelet has been used as a catalyst to produce biodiesel (Sani et al. 2015). Bio-based catalysts have been synthesized from palm frond and spikelet with concentrated sulfuric acid as the sulfonating agent. Sani et al. (2015) investigated the application of the synthesized catalyst for the esterification of FFA into FAME. Palm tree fronds and spikelet were air-dried for one week in sunlight, then oven-dried for 24 hours. Partially carbonized materials could be obtained after 24 hours' pyrolysis in a tubular furnace, which helped achieve the formation of a polycyclic aromatic carbon ring. The sulfonic group (SO_3H) was then introduced on to the partially carbonized material via sulfonation with concentrated H_2SO_4. Afterward, the material was washed with distilled water to ensure radical ions were washed off. Succinic acid was used to homogenize the new synthesized catalyst in order to achieve a complete sulfonation.

9.7.2 *Jatropha curcas*

Jatropha curcas has been exploited for the synthesis of bio-based catalyst in recent years. Solid bases, when used in catalytic reactions, require favorable reaction conditions and are more active than acids (Bozell and Petersen 2010). A major reason why they are no longer popular is that they have thin pore sizes and low surface areas (FitzPatrick et al. 2010). Activated carbon, on the other hand has been established to have a large surface area, large pore size, and high stability, and it can be produced from many different raw materials (Buasri et al. 2012). In a number of Asian countries, *J. curcas* fruit seeds have been used to produce biodiesel because of their very rich oil content. After the oil has been pressed out, what is left is generally regarded as a waste, which can be effectively converted to an activated carbon and used as a catalyst for the production of biodiesel. In one study, *J. curcas* fruit shell was utilized as a base for the attachment of KOH as a catalyst for biodiesel production from waste frying oil via transesterification in the presence of methanol. This fruit shell was converted to activated carbon via thermal conversion (Buasri et al. 2012).

Activated carbon from waste *J. curcas* was prepared by first drying it at 100 °C for 12 hours, then grinding and sieving to obtain a particle size of 1.8–2.8 mm. An impregnation period followed, when the sample was soaked in phosphoric acid for 24 hours at ambient temperature with 1 : 7 w/w ratio (Gratuito et al. 2008). The sample was then further dried for 12 hours at 100 °C and put in a tubular furnace for pyrolysis under the flow of nitrogen for 24 hours at 500 °C (Tongpoothorn et al. 2011). Then, it was thoroughly washed until the pH reach 7, dried at 110 °C for 12 hours, and stored in desiccators. The final stage of preparation involved the addition of the activated carbon into a concentrated solution of KOH followed by proper stirring at ambient temperature for 24 hours (Baroutian et al. 2011).

9.7.3 Coconut Shells

The search for greener and environmentally benign sources for catalysts has led to the exploitation of various biomass sources such as coconut and palm shells. Biomass sources have proven to be very promising and are readily available in abundance (Bozell and Petersen 2010). Activated carbon should preferentially be used as an anchorage for the catalyst because of its previously discussed advantages, such as its high surface area which presents many active sites that can undergo catalytic activity. Biomass sources are cost-effective, drastically reduce waste, and promote a greener environment and economy.

A coconut shell-derived catalyst was prepared by cleaning it with water to remove fine sands and dirt and dried at 100 °C for 12 hours. The resultant sample was grinded and sieved to a particle size of 1.8–2.8 mm. It was then soaked in phosphoric acid (H_3PO_4) solution with an impregnation ratio of 1.7 at ambient temperature for 24 hours (Gratuito et al. 2008). It was later dried for 12 hours in an oven set at 100 °C. Next, it was further activated in a horizontal automatic tubular furnace at 500 °C for 24 hours under nitrogen atmosphere (Tongpoothorn et al. 2011). After cooling, the activated carbon was washed successively several times with distilled water until the pH became neutral. The washed sample was then dried at 110 °C for 12 hours to obtain activated carbon of coconut shell as the final product, which was cooled in a desiccator and stored in glass containers.

9.7.4 Rice Husks

Rice husk is a major byproduct of the rice milling industry. It has been used to prepare SiO_2, SiC, and activated carbon (Zhao et al. 2018).

The development of a bio-based catalyst from rice husk char has recently been demonstrated (Zhao et al. 2018). Rice husk biochar supported with CaO was synthesized as a bio-based catalyst for the production of biodiesel. The silica containing the hybrid material was prepared by pyrolysis of the husk. After synthesis, the bio-based catalyst was characterized using different analytical tools, such as TGA EDX, SEM, BET, FT-IR, and XRD. The catalyst was prepared by washing it with distilled water to remove impurities and drying it overnight at 90 °C. The resultant rise husk biochar was pyrolyzed at 700 °C and had a retention time of three hours in the tube furnace with the flow of nitrogen. Afterward, it was pounded to a fine powder then chemically activated with the treatment of KOH solution. It was later washed with distilled water several times to remove extra organic matter and potassium until the filtrate attained a pH of 7. It was then dried in an oven for five hours at 110 °C. The incorporation of CaO into the rice husk biochar was achieved simply by adding 2 g of CaO to 100 ml of de-ionized water and mixing properly, then adding 8 g of dried rice husk biochar and stirring vigorously. The stirring continued for four hours at atmospheric temperature and the resultant mixture was kept in an oven at 90 °C overnight. Finally, the mixture calcinated under a flow of nitrogen at 600–800 °C for 4 hours to obtain a rice husk biochar/CaO catalyst (Zhao et al. 2018).

9.7.5 Bamboo

Bamboo is one of the few plants that both grows and propagates quickly and has high productivity. As a result, it has attracted a lot of interest from researchers looking to find a use for it in the bioenergy field, particularly as a catalyst for the production of biodiesel (Shi et al. 2011). Indeed, after carbonization, bamboo charcoal possesses potential as a catalyst support due to its high specific surface area and large pore volume (Wang et al. 2013).

Zhou et al. (2016) recently reported on the preparation of a carbon-based heterogeneous acid catalyst for use in biodiesel production. The catalyst preparation process went through successive phases of carbonization, sulfonation, and washing. The bio-based acid catalyst was prepared through carbonization, where the freshly acquired bamboo was dried at 105 °C. It was later heated under nitrogen flow with a ramp of 5 °C/min until it reached the desired carbonization temperature of between 300 and 500 °C. Afterward, the material was kept within the carbonization temperature at a retention time of 0.5–4 hours. The partially carbonized bamboo was treated with concentrated sulfuric acid (>98%) with vigorous stirring for one to five hours. Finally, the carbonized and sulfonated bamboo was washed with hot distilled water (>80 °C) until neutral pH was achieved and dried overnight at 105 °C to obtain bamboo-derived sulfonated bio-based catalyst (Zhou et al. 2016).

9.7.6 Cocoa Pod Husks

Cocoa pod husks (CPHs), a major source of potash, are a group of agricultural wastes found in most parts of the world where cocoa is produced. Potash from inorganic or mineral sources has been used as a catalyst for biodiesel production and has proven to possess a high catalytic activity (Baroi et al. 2009; Liang et al. 2009; Lee and Ofori-Boateng 2013). Carbonate contents of 40–60% have been reported in the ashed CPHs, and 3–9% in the dried pod husks (Afrane 1992; Lee and Ofori-Boateng 2013). CPHs have been investigated to check the plausibility of using the potash derived from them for the transesterification of soybean oil into biodiesel. Green heterogeneous catalysts were derived from CPHs, CPH/MgO, and CPH ash.

Catalyst preparation started with washing the CPHs with tapwater, cutting them into small cubes, and drying them under the sun for 14 days. Milling of the dried husks achieved a fine powder, which was calcinated in a furnace for four hours at 650 °C. 250 g of the calcinated sample was vigorously stirred in a mixture with the addition of 1 l of water in a blending vessel. Leaching of the potash in the mixture followed, with the help of a filter. The filtrate was evaporated for three hours at 140 °C to achieve an impure brown crystal solution, which was further refined by washing with distilled water and evaporated to obtain the pure colorless form of potash, which was used as an unsupported catalyst (named CPH) for the transesterification of soybean oil. Supported catalysts were derived by loading the K_2CO_3 potash on magnesium oxide (CPH/MgO) at a catalyst loading ratio of 0.6, with calcination at 550 °C for three hours. Both catalysts were used for the transesterification of soybean oil. Under optimum conditions of 1 : 6 oil to methanol ratio of, 1 wt% catalyst loading of MgO-doped CPH ash, 60 °C reaction temperature, and 60 minutes reaction time, CPH/MgO catalyst showed a biodiesel yield of 98.7%. With conditions of 1 : 6 oil to methanol ratio of, 1 wt% catalyst loading of CPH ash, 60 °C reaction temperature, and 120 minutes reaction time, a 91.4% yield was recorded (Lee and Ofori-Boateng 2013).

9.7.7 Hardwoods

Pyrolysis, a thermochemical process usually involving the conversion of biomass into char via heating in a furnace tube at approximately 500 °C in the absence of oxygen, is used to synthesize biobased catalysts from hardwood, with a yield typically between 10 and 25 wt% of feedstock (Dehkhoda et al. 2010). Biochar is also an excellent candidate for a source of low-price carbonaceous catalyst support. Dehkhoda et al. (2010) used biochar as a support for a heterogeneous acid catalyst for use in the production of biodiesel. Catalyst preparation was achieved with the aid of sulfuric acid as sulfonating agent applied on the pyrolyzed samples. The sulfonating method was similar to one reported by Toda et al. (2005), with 200 ml of concentrated sulfuric acid added to 20 g of biochar in a 500 ml round-bottom flask. The mixture was heated to 150 °C with a heating mantle and monitored with a temperature controller (corrosion-resistant Type-J thermocouple) for 24 hours. After heating, the resultant slurry was placed in cool distilled water and filtered. The biochar catalyst was washed with distilled water at 80 °C until the wash water became neutral and free from sulfate ions. Following filtration, the biochar was dried in an oven at 70 °C for approximately one hour (Dehkhoda et al. 2010).

9.7.8 Peanut Hulls

It is believed that temperatures between 400 and 600 °C can be used for the slow pyrolysis of biomass to generate a highly multiringed, cross-linked, aromatic structure attached to lignin capable of being easily functionalized with catalytically active acidic groups (Kastner et al. 2012; Lee et al. 2017). Based on an analysis of peanut hull (biochar) generated by pyrolysis of pure biomass components (cellulose, hemicelluloses, and lignin) and whole biomass, it is theorized that lignin would undergo partial decomposition and that hemicellulose and cellulose would undergo a series of thermal homolysis, hydrolysis, dehydration, and molecular rearrangement to form a polymerized aromatic structure (Kastner et al. 2012).

Reusable bio-based catalysts were synthesized from peanut hulls and pyrolyzed at between 400 and 500 °C to generate a biochar capable of serving as a backbone or anchorage for the attachment of a sulfonic functional group after sulfonation. Pelletized peanut hulls (typically 15 mm in length, 8 mm in diameter, 1–5 mm in thickness) and pine logging residues were used to generate biochar (Kastner et al. 2012). Sulfonation was achieved using concentrated sulfuric acid at varying temperatures of 100, 150, and 200 °C over 12 hours with gaseous SO_3 at 23 °C. The synthesized sulfonated catalysts were used for the esterification of FFA in mixtures with vegetable oils and animal fats at between 5 and 15 wt% in the presence of methanol. Esterification of the fatty acids was typically complete (~90–100% conversion) within 30–60 minutes at 55–60 °C (large methanol excess), but this decreased when using the biochar catalysts (e.g. 70% conversion, 20 : 1 methanol to oil ratio, six hours reaction time) (Kastner et al. 2012).

9.7.9 Wood Mixtures

Mixtures of woods have been exploited as a source of biochar for the synthesis of catalysts for biodiesel production. Yu et al. (2010) generated biochar from the fast pyrolysis of woody biomass, a mixture of wood waste, white wood, bark, and shavings. Preparation of the catalysts started with slight grinding of the biochar followed by sieving and collection of the desired particle sizes. For each sample, approximately 24 g of the finely filtered biochar was dried in an oven at 110 °C for two hours. Thereafter, 150 ml of 7 M KOH solution was mixed with the dried biochar for two hours, as reported by Azargohar and Dalai (2008), before again being dried in an oven at 110 °C overnight after filtration (Yu et al. 2010). Next, the samples were weighed,d loaded on to ceramic boats, and placed in a tube furnace for pyrolysis, with temperatures ranging from 450 to 675 to 875 °C at a rate of 3 °C/min over two hours' residence time under a nitrogen flow (Yu et al. 2010). The activated char samples were then washed with 500 ml increments of distilled water at 15-minute intervals until neutral pH was achieved. The washed char samples were placed in 0.1 M HCl for one hour, then washed again with 500 ml increments of distilled water as before until neutral pH was achieved, before being dried in an oven at 110 °C overnight (Azargohar and Dalai 2008). Fuming sulfuric acid containing 20 wt% free SO_3 was used to incorporate sulfonic groups on to the carbon matrix of the chemically prepared activated char samples. The samples were sulfonated using 16.5 ml of fuming H_2SO_4/g char at 150 °C for 15 hours under a nitrogen flow rate of 50 ml/min (Okamura et al. 2006; Zong et al. 2007). After sulfonation, the char samples were washed with hot distilled water until a neutral pH was achieved, and finally dried at 110 °C overnight to get the desired catalyst.

9.7.10 Palm Kernel Shells

PKSs are derived from the processing of palm fruits to produce palm oil. The oil palm tree originated in West Africa but is now grown in many parts of the world as a hybrid. The price for palm oil is relatively low, and it has found uses in a variety of areas, including as a precursor for catalysts and feedstocks for the production of biofuel. Because of the large amounts of solid wastes generated from oil palm mills, there is an urgent need for their recovery, reuse, and recycling. Thus generated wastes are either disposed of in landfills or pyrolyzed to produce energy. It is known that PKS takes a long time to decompose, and substantial furnace modifications are usually required if it is to be used in boilers in place of firewood or fuel oil (Prasertsan and Prasertsan 1996; Bazargan et al. 2015). It has been reported that when burned in its raw form, PKS produces too much smoke and not enough energy (Kuti 2007).

The preparation of PKSB was initiated with calcination under atmospheric pressure for two hours at 800 °C, immediately before it was due to be used. The calcination temperature was chosen based on the results of thermogravimetric analysis (TGA) and XRD analysis of the PKSB. The catalyst was used after cooling in a desiccator containing calcium chloride and potassium hydroxide pellets. For comparative purposes, CaO was calcined at the optimum temperature of 550 °C for two hours, as suggested elsewhere (Veljković et al. 2009), and used immediately after cooling in the desiccator (Bazargan et al. 2015).

9.8 Magnetic Bio-Based Catalysts for Biodiesel Production

Heterogeneous catalysts are gaining increasing attention from researchers due to their ease of separation from reaction products. In the context of recovering solid catalyst through filtration or centrifugation, conventional homogeneous (liquid) catalysts are invariably difficult and time-consuming to recover (Zhang et al. 2016). Recently, magnetic solid catalysts have been studied extensively due to their promising magnetic properties. The magnetic nature of these catalysts facilitates their easy separation and recovery by applying an external magnetic field even in cases of high-viscosity reaction mixtures (Liu et al. 2010; Guo et al. 2012). It has been reported that magnetic porous carbon possesses a much higher surface area (391.7 m^2/g) than nonmagnetic carbon (14.0 m^2/g). The presence of iron (Fe) in magnetic catalysts helps catalyze the release of volatile materials during the dehydration and decomposition of $FeCl_3$.

A typical approach to the synthesis of a magnetic catalyst involves impregnation of a ferrite ion-containing reagent, followed by carbonization and functionalization of the magnetic support derived from biomass. The synthesis pathway starts from hydrolysis of ferric chloride into iron (III) oxide-hydroxide, before it further transforms into its anhydrous form, FeO(OH). Subsequently, precursors impregnated with $FeCl_3$ undergo carbonization at elevated temperature. Carbonization induces the formation of hydrogen, carbon, and carbon monoxide, which act as reducing agents for the reduction of FeO(OH) into iron (II, III) oxide, Fe_3O_4. The product is known as the magnetite that provides the magnetic properties to the catalyst (Liu et al. 2013).

The attachment of iron not only plays an important role in providing magnetism to the catalyst, but also improves pore formation on the synthesized porous carbon during carbonization due to the decomposition and dehydration of $FeCl_3$ (Liu et al. 2013). Further, Fe-containing compounds such as Fe_3O and FeO(OH) also contribute to the pore formation, acting as *in situ* catalysts during carbonization (Atkinson et al. 2011). The biomass source of the catalyst, optimal transesterification and esterification reaction conditions, and the product yield are summarized in Table 9.4.

A study performed by Liu et al. (2013) demonstrated the conversion of sawdust into a magnetic porous carbonaceous (MPC) solid acid catalyst. The catalyst produced (MPC-SO_3H) was used for three acid-catalyzed reactions: esterification of acetic acid and benzyl alcohol to produce biodiesel, dehydration of xylose to form furfural, and hydrolysis of sucrose into glucose and fructose. The synthesis method consisted in a sequence of processes: (i) biosorption of sawdust with $FeCl_3$ solution; (ii) fast pyrolysis of Fe-loaded sawdust to stabilize and restructure the magnetic particles on carbon; and (iii) sulfonation

Table 9.4 Carbonization and activation conditions of different types of biomass.

Raw material	Particle size	Carbonization conditions	Activation conditions	Chemical treatment	References
Palm shell	1–2 mm	N_2, 800 °C/2 h, 15 l/min, 10 °C/min	NA	Chemical (K_2CO_3)	Adinata et al. (2007)
Palm shell	2.0–2.8 mm	N_2, 200 °C–600 °C/2 h, 150 ml/min	NA	Chemical (H_3PO_4)	Guo and Lua (2003)
Peach stone	2.0–2.8 mm	N_2, 850 °C/2 h, 80 ml/min	8 h	CO_2	Zhang et al. (2005)
Peach stone	1.6–0.5 mm	N_2, 450 °C, 100 ml/min	NA	Chemical (H_3PO_4)	Muñoz-González et al. (2009)
Wood (*Salix viminalis*)	10 mm	Ar, 700 °C–800 °C /1 h	NA	NA	Łukaszewicz and Wesołowski (2008)
Eucalyptus wood	NA	N_2, 450 °C, 500 ml/min	825 °C	N_2/CO_2, 50 ml/min	Amaya et al. (2007), Łukaszewicz and Wesołowski (2008)
Eucalyptus wood	50 mm	N_2, 830 °C–870 °C/2 h, 300 ml/min, 4 °C/min	NA	NA	Amaya et al. (2007)
Coconut shell	<0.25 mm	N_2, 775 °C–900 °C /8 h, 2 °C–10 °C/min	650 °C–900 °C	CO_2, steam, O_2	Farris et al. (1992)
Walnut shell	1–2 mm	N_2, 800 °C/1 h	NA	Chemical (KOH)	Hu and Vansant (1995)
Spruce wood	Cube (10 mm^3)	N_2, 700 °C/1 h	NA	Chemical (HNO_3)	Onyestyák and Bóta (2009)
Spruce wood	Cube (10 mm^3)	N_2, 700 °C/1 h	900 °C/15 min	Steam	Onyestyák and Bóta (2009)
Spruce, beech, and oak wood	Cube (10 mm^3)	N_2, 700 °C/1 h	900 °C	Steam	Onyestyák and Bóta (2009)

of MPC material with concentrated H_2SO_4 to obtain MPC-SO_3H. During pyrolysis at high temperature, $FeCl_3$ dehydrated the carbohydrate structure of the sawdust, altering the decomposition of lignocellulosic biomass. This phenomenon inhibited the formation of tars blocking the pore structures, improving the open pore formation in the carbon matrix. Overall, MPC-SO_3H catalyst had a high surface area of 296.4 m^2/g and a high total acid density of 2.57 mmol/g. The catalyst was experimentally proven to be functional in biodiesel production, with a yield of 93% at a 1 : 5 methanol to oil ratio, 35 wt% catalyst loading, and 88 °C reaction temperature. The magnetism and surface morphology of MPC-SO_3H remained unchanged even after being reused five times, and the average biodiesel yield recorded was above 90%.

A novel method of double hydrothermal precipitation, pyrolysis, and sulfonation was developed by Zhang et al. (2015) to synthesize a new magnetic carbonaceous acid catalyst from glucose and iron chloride (AC- 600-SO_3H @Fe/C). First, a mixture of $FeCl_3$ solution, glucose, and solid urea was subjected to hydrothermal reaction at 180 °C for 14 hours to obtain Fe_2O_3 solid. Then, the solid sample was heated (pyrolysis) at 700 °C for 90 minutes to produce a magnetic carbon-based core (Fe/C). The Fe/C core was recoated with glucose as the catalyst showed weak magnetism after sulfonation, causing the dissolution of Fe_3O_4/Fe. The recoated Fe/C was hydrothermally recoated again with a carbonaceous layer to prevent Fe_3O_4/Fe from leaching during sulfonation, and was then sent to pyrolysis (600 °C) to form stabile carbonized magnetic particles. Finally, AC-600- SO_3H @Fe/C was formed after sulfonating the carbonized magnetic particles with concentrated H_2SO_4. It was further was observed that Fe was formed at 600 °C during the second pyrolysis, which had stronger magnetism than either Fe_3O_4 or Fe_3C (formed at 400 and 800 °C, respectively). Thus synthesized magnetic acid catalyst showed strong magnetism (14.4 emu/g), high reusability (biodiesel yield >90% after three cycles), and high recovery (96.3%). The catalyst also had a high total acid density of 2.79 mmol/g, mainly due to recoating of the carbon layer, which provided more functional groups for –SO_3H to bond to. AC-600-SO_3H@Fe/C was used to catalyze the esterification and transesterification of Jatropha oil, and the highest yield of 91.8% was achieved under optimal conditions of 24 : 1 methanol to oil ratio, 10 wt% catalyst loading, 200 °C reaction temperature, and 10 hours reaction time.

Contrarily, a new magnetic base catalyst supported with nickel, sodium silicate, and carbon derived from bamboo powder was successfully synthesized, known as Na_2SiO_3@Ni/C. The reported catalyst was used for the production of biodiesel, and subsequently utilized the byproduct (crude glycerol) and deactivated catalyst to produce H_2 gas through hydrothermal gasification. The Ni element in the catalyst provided strong magnetism (15.7 emu/g) and was found to be an excellent catalyst for the conversion of glycerol to hydrogen. Initially, nickel nitrate solution, solid urea, and bamboo powder were mixed to obtain $Ni(OH)_2$, the magnetic precursor of the catalyst, before calcinating it at high temperature to form Ni/C particles. The gel form of the Ni/C particle and sodium silicate solution mixture was obtained through reflux and further calcinated at 400 °C for two hours to obtain the final catalyst product. Zhang et al. (2016) recorded the highest biodiesel yield (98.1%) and catalyst reusability (>93% after four usages) among all biomass-origin magnetic catalysts. The optimum conditions for biodiesel production were 9 : 1 methanol to oil ratio, 7 wt% catalyst loading, 65 °C reaction temperature, and 100 minutes reaction time. However, the study highlighted that the catalyst activity dropped after the fifth cycle due to leaching of Na_2SiO_3 into methanol, which was later used for H_2 production.

In another study, Zhang et al. (2017) developed carbon-based magnetic solid acid (C-SO_3H@Fe/JHC) and base (Na_2SiO_3@Ni/JRC) catalysts from Jatropha hull hydrolysate and hydrolysis residue, which they used in the esterification and transesterification of Jatropha oil to produce biodiesel (JHC: Jatropha hydrolysate carbon, JRC: Jatropha residue carbon). The synthesis of C-SO_3H@Fe/JHC was performed through several steps, starting from the hydrothermal precipitation of Fe nanoparticles and Jatropha-hull hydrolysate, followed by pyrolysis to form the magnetic core (Fe/JHC). Then, the magnetic core was recoated with a layer of carbon (derived from the Jatropha hull hydrolysate) through incubation and pyrolysis to avoid dissolution of Fe during sulfonation. Finally, the carbon-coated magnetic catalyst was sulfonated by concentrated H_2SO_4 to form C-SO_3H@Fe/JHC. For the magnetic solid base catalyst, a Jatropha hydrolysis residue and nickel nitrate mixture was incubated in an oil bath to form Ni/JRC, before being calcinated at very high temperature to synthesize the Ni/JRC supporter, which was reacted with $Na_2SiO_3.9H_2O$ to add basicity to the catalyst. Both catalysts displayed a great magnetism (11.2 and 15.0 emu/g), with high total acid/base sites (2.96 and 3.24 mmol/g). Both possessed an absolute carbonaceous supporting structure with abundant functional groups that gave rise to a high specific surface area and pore volume. Both were proven effective when tested in a two-step biodiesel production (esterification using C-SO_3H@Fe/JHC, transesterification using Na_2SiO_3@Ni/JRC). The highest yield achieved was 96.7%, with an average recovery yield of C-SO_3H@Fe/JHC and Na_2SiO_3@Ni/JRC after five cycles of 90.3% and 86.7%, respectively.

Han et al. (2016) presented a novel cotton-regenerated magnetic cellulose microsphere (MCM) supported with HPW catalyst. They regenerated cotton into cellulose microsphere (CM), which they modified through grafting of triethylenetetramine (TETA, a kind of micromolecular amine) and subsequent *in situ* co-precipitation of Fe_3O_4 magnetic nanoparticles on to modified CM to obtain MCM. They then immobilized HPW with MCM to form the final product, MCM-HPW. The grafting or surface modification of CM enhanced the attachment of chemical functional groups including $-SO_3H$, –COOH, –CHO, and $-NH_2$, thus boosting the capacity and reusability of the cellulose materials. Likewise, HPW was selected over conventional acids such as H_2SO_4 due to its strong Brønsted acidity, low corrosiveness, and distinct Keggin structure. This structure was well protected by MCM, and provided extra proton mobility that induced all possible mobile protons to involve in the catalytic reaction on both the surface and the bulk of HPW's crystalline structure. Apart from this, the purpose of the HPW immobilization on cellulose was to improve the reusability, thermal instability, and surface area. MCM-HPW demonstrated high catalytic activity and reusability when it was used to catalyze the microwave-assisted transesterification of *Pistacia chinensis* seed oil. A biodiesel yield of 93.1% was achieved under optimum conditions of 10 : 1 methanol to oil ratio, 15 wt% catalyst loading, 60 °C reaction temperature, and 80 minutes reaction time. The study reported a conversion yield >85% even after the third cycle of reuse, but it dropped to 80.7% in the fourth run.

Liu et al. (2018) demonstrated the synthesis of a bifunctional magnetic solid base catalyst from bamboo charcoal (K/BC-Fe_2O_3) and its use in biodiesel production. Thus developed K/BC-Fe_2O_3 catalyst showed a mesoporous structure with a high specific surface area of 28.7 m^2/g. Moreover, the catalyst demonstrated strong magnetism (35.4 emu/g) due to stable impregnation of γ-Fe_2O_3 particles on to the carbon skeleton of the bamboo charcoal. Transesterification of soybean oil and methanol catalyzed by K/BC-Fe_2O_3 was carried out, and a 98.0% yield of biodiesel was recorded under conditions of 8 : 1 methanol to oil ratio, 2.5 wt% catalyst loading, 60 °C reaction temperature, and one hour reaction time, with a recovery yield exceeding 94% through four cycles.

Another novel magnetic mesoporous acid catalyst, Fe_3O_4@Chitosan-Hollow-Chitosan-SO_3H (FCHC-SO_3H), was developed using chitosan as the catalyst carrier. The special property of this catalyst is its double-shell hollow structure, at the core of which is Fe_3O_4, while the outer part consists of a chitosan shell, with hollow space in between. This specialized structure imparts several important properties to the catalyst: (i) it prevents the magnetic material in the core shell from being etched; (ii) it increases the specific surface area and active sites; and (iii) it improves the mass transfer of catalyst during reaction. The authors further compared the catalytic performances of different catalysts to produce biodiesel through esterification of oleic acid. They found that catalyst containing both chitosan and $-SO_3H$ groups performed excellently in esterification because the $-SO_3H$ groups grafted and reacted well with $-NH_2$ groups in the chitosan. Furthermore, $-NH_2$ and –OH groups in the chitosan were protonated under strong acidic conditions, giving the catalyst more active sites. Nonetheless, the hollow structure of FCHC-SO_3H produced the highest biodiesel yield (96.7%) as compared to nonhollow Fe_3O_4@Chitosan-SO_3H (75.1%) under optimum conditions of 15 : 1 methanol to oil ratio, 4 wt% catalyst loading, 80 °C reaction temperature, and three hours reaction times. Besides this, FCHC-SO_3H catalyst displayed strong magnetism of 18.9 emu/g and total acid density of 1.20 mmol/g, as well as a good reusability with 84.1% biodiesel yield after five cycles (Wang et al. 2018).

Zhang et al. (2015) further reported the synthesis of magnetic acid catalyst from glucose for biodiesel production. The method involved a series of stages: (i) synthesis of the magnetic core (Fe/C) by hydrothermal precipitation; (ii) hydrothermal recoating of glucose to prevent leaching of Fe_3O_4/Fe during sulfonation; (iii) pyrolysis at high temperature to stabilize and restructure the magnetic carbon; and (iv) sulfonation of the magnetic carbon by direct sulfonation with concentrated H_2SO_4. Thus obtained magnetic acid catalyst showed strong magnetism (14.4 Am^2/kg), high total acid density (2.79 mmol/g), and high catalytic activity.

9.9 Characterization of Bio-Based Catalysts

Bio-based catalysts are characterized after synthesis to confirm their morphological, structural, and textural properties and physicochemical and thermal characteristics using a variety of characterization processes and analyses. Field emission scanning electron microscopy (FESEM) is used to ascertain their morphology and textural quality, Fourier transform infrared (FT-IR) is used to analyze their functional groups, XRD is used to confirm their amorphous nature, TGA is used to check their thermal stability, temperature programmed desorption – ammonia (TPD-NH_3) is used to check their acidity, and Brunauer–Emmett–Teller (BET) analysis is used to assess the quality of their surface area .

9.9.1 Field Emission Scanning Electron Microscopy (FESEM)

The bio-based materials and microstructural behaviors of all samples are normally investigated using FESEM, which reveals their morphology and shape.

In the synthesis of a bamboo-based catalyst for the esterification of oleic acid with methanol, the morphological imaging and elemental analysis of the catalyst is performed using a field emission scanning electron microscope (FESEM NovaSem 230) fitted with an energy-dispersive X-ray microanalyzer (EDX). The FESEM produces a very high-resolution image of the sample with very high magnification. During sample preparation, the powder catalyst is dispersed on a carbon double-sided tape on top of an aluminum holder. The samples are subsequently coated with a gold thin layer using a sputter coater to avoid electron charging during scanning. The images are recorded at different magnifications. Quantitative elemental analysis of the samples is subsequently performed by the EDX (Akinfalabi et al. 2019a; Zhou et al. 2016).

9.9.2 Fourier Transform Infrared (FT-IR)

In the synthesis of bio-based catalysts, the FT-IR Research Instrument Thermo Nicolet 6700 is mainly used to determine the functional group present on the polycyclic carbon structure of the optimized sulfonated catalyst (Akinfalabi et al. 2019b; Kefas et al. 2018). A small quantity of the catalyst sample is mixed with 200 mg of KBr powder and ground to achieve homogeneity. The mixture is then pressed on a sample holder and scanned at wavelengths between 400 and 4000 cm^{-1}. A FT-IR spectrum is generated and recorded by the FT-IR software (OMNIC) (Kapor et al. 2017).

Functional groups are present on biochar whether it is treated with chemicals or without. Various adsorption peaks and regions are identified as specific for each functional group in different biochars, as shown in Table 9.5. Biochar and sulfonated biochar exhibit aromatic groups (e.g. phenol), C–O–C, and –COOH. Through hydrogen bonding, hydroxyl groups such as (–OH), C–O, and C=O interact with phenol groups (Liu et al. 2011). The sulfonic groups (–SO_3H and S=O) are mostly found on sulfonated biochar. Importantly, the desired temperature pre- and post-treatment greatly affects the number of functional groups. For biochar that is sulfonated, heat treatment at 450 °C produces more aromatic functional groups than treatment at between 650 and 850 °C (Joylene et al. 2011). As a result, biochar activation also affects functionality. KOH-activated biochar develops more C–O and C=O functional groups relative to hydrothermal activation (Cha et al. 2010). Therefore, it is concluded that physical properties of biochar such as pore volume, surface area, and pore diameter can be influenced and modified to match desired target values. Chemical properties like functionality meanwhile can be modified by carbonization temperature, chemicals, and the gas environment.

9.9.3 X-Ray Diffraction (XRD)

XRD is commonly used to determine the composition and bulk structure of solid catalysts having crystalline and amorphous structures. Most carbon-based catalysts have the form of polycrystalline powders, while bio-based catalysts are usually amorphous in nature. XRD analysis is mainly restricted to specific lattice planes that produces sharp or broad peaks at their respective angular position of 2θ, as determined by Bragg's law: $2d \sin \theta = n\lambda$. XRD is quite useful in spite of some limitations in providing information about amorphous bio-based carbon sheets. It can provide accurate information about the identification of bulk crystalline components solid catalysts, for example.

XRD analysis usually requires samples to be in powder form. The crystalline size of a material may be obtained from XRD spectra by applying the Paul Scherer formula, which

Table 9.5 Functional group characterization of biochar using FT-IR.

Functionality	Wavenumber (cm^{-1})	References
Sulfonated biochar $-SO_3H$	1706–1717	Joylene et al. (2011), Dehkhoda et al. (2010), Li et al. (2014)
Aromatic C–H bending	731–903	Joylene et al. (2011)
Aromatic C=C bending	1554–1580	Joylene et al. (2011), Dehkhoda and Ellis (2013)
C–O–C stretching	1130–1150	Joylene et al. (2011)
C=O stretching	1700–1706	Joylene et al. (2011), Dehkhoda and Ellis (2013), Li et al. (2014)
C–O stretching	1215	Joylene et al. (2011)
S=O stretching	1032–1100	Joylene et al. (2011), Kastner et al. (2012)

gives the relation between the size of submicrometer particles, or crystal size, in a solid material and the broadening of a peak in a diffraction pattern:

$$D = K\lambda/(\beta \cos\theta) \tag{9.1}$$

where D is the mean crystalline size of the ordered domains, which is less than or equal to the grain size; K is a constant referred to as dimensionless shape factor with a given value close to unity, typically 0.9; λ is the X-ray wavelength; and β is the broadening line at half maximum intensity (FWHM), calculated in radians.

The crystalline phase of the bamboo-derived catalyst is analyzed by a Shimadzu XRD 6000 diffractometer utilizing Cu-Ka radiation electrons generated by a Philips diffraction X-ray tube (Akinfalabi et al. 2019a; Zhou et al. 2016). The powder is first leveled flat on an aluminum sample holder and sample scanning is performed at a range of theta programmed from 2° to 80° at a scanning rate of 4°/min, with electron radiation accelerated by a broad-focus 2.7 kW type on the catalyst. Data from the Joint Committee on Powder Diffraction Standards (JCPDS) are then compared with diffractograms obtained from the analysis to identify existing information about the sample.

9.9.4 Thermogravimetric Analysis (TGA)

TGA determines the quantity and rate at which modification in the mass of a certain sample relates to temperature and time in a precise atmosphere. This analysis can be used to examine the composition of a sample and to postulate its thermal stability at certain temperatures up to about 1000 °C. The analysis is capable of characterizing materials that reveal a weight gain or loss due to decomposition, dehydration, or oxidation. The TGA machine weighs a specimen continuously as heat is applied to it up to about 2000 °C; many components of the sample give up and decompose as the temperature increases, and the fraction mass loss of every subsequent change is recorded. The results are obtainable by plotting a graph of temperature against weight loss. The result may be adjusted by curve smoothing. The first derivatives are usually plotted in order to investigate points of variation.

Thermal stability analysis of bio-based catalysts is in most cases performed using a Mettler Toledo Thermogravimetric Analyzer (TGA) equipped with a differential scanning calorimeter (TGA/DSC1 HT). The weight of the bio-based catalyst sample should be between 10 and 20 mg. It is weighed and placed in a crucible and put inside the furnace. Heating is set at 25 to 1000 °C with a rate of 10 °C/min under nitrogen gas flow of 50 mL/min. The weight loss of the derivative is calculated automatically by TGA software (STARe software) and recorded and plotted against temperature (Akinfalabi et al. 2019a; Zhou et al. 2016).

9.9.5 Temperature-Programmed Desorption – Ammonia (TPD-NH_3)

TPD is usually used to analyze specific acid or base sites in solid catalysts. The temperature at which desorption occurs indicates the strength of adsorption, while either the amount of gas consumed in the uptake or the amount of desorption upon heating indicates the concentration of the surface sites (Ma and Zaera 2006). The most common molecules used in TPD are NH_3 and CO_2, which probe acidic and basic sites, respectively, but experiments with pyridine, O_2, H_2, CO, H_2O, and other molecules are often performed as well.

A Thermo Finnigan TPDRO 1100 TPD-ammonia analyzer is used to analyze the acid density, strength, and acid active sites of a synthesized bio-based catalyst. About 0.05 g of the catalyst sample is weighed and placed in an inner tube of the reactor equipped with a thermal conductivity detector. The catalyst sample is first pretreated to remove moisture and unwanted substances by passing N_2 gas through the sample at 150 °C for 45 minutes (Akinfalabi et al. 2020a; Lokman et al. 2016). Next, it is subjected to ammonia adsorption by passing ammonia gas through it for one hour at ambient temperature. It is then pretreated a second time by passing nitrogen gas (20 ml/min) through it for 25 minutes at ambient temperature to remove excess ammonia in the gas phase. Ammonia desorption analysis is performed by heating the sample from 50 to 950 °C at a rate of 10 °C/min, with helium gas flowing at 20 ml/min as carrier gas. The desorbed ammonia gas from the catalyst sample is detected by the thermal conductivity detector (thermocouple) and a graph of intensity versus temperature is plotted (Akinfalabi et al. 2017).

9.9.6 Brunauer–Emmett–Teller (BET) Analysis

The surface area, pore size, pore volume, and surface functionality are extremely important variables in terms of biomass origin and thermolysis conditions (e.g. gaseous composition, temperature, heating rate, and time). For example, an increment in the surface area of biochar produced from poplar wood via gasification under 10% CO_2/90% N_2 atmospheric conditions increased from 435 to 687 $m^2\ g^{-1}$ with a temperature rise from 750 to 920 °C (Klinghoffer et al. 2012). While this biochar developed more micropores, one produced via gasification in 90% H_2O/10% N_2 did not (Klinghoffer et al. 2012). An increment was also noticed in the pore volume from 0.18 to 0.3 cm^3/g with a steady rise in temperature from 750 to 920 °C (Klinghoffer et al. 2012). Similarly, biochar was generated from the pyrolysis of lignocellulose and saccharide by treating with H_2SO_4 for 48 hours at three different temperatures (180, 280, and 380 °C) (Liu et al. 2017). The resulting products had pore volume and surface area proportional to the pyrolytic temperature.

The surface area and pore analysis of bio-based catalyst samples are determined by a BET Micromeritics 3-flex surface characterizer. Prior to beginning the analysis, the samples are degassed at 150 °C for 12 hours. An N_2 adsorption–desorption isotherm is performed to evaluate the samples at 77 K temperature. The surface area is obtained from the Langmuir and BET results (Kapor et al. 2017).

9.10 Reaction Parameters Affecting Biodiesel Production

9.10.1 Reaction Time

The time required to complete an esterification reaction is very important as far as yield and conversion of biodiesel are concerned. The longer the reaction time, the better the yield, reaching up to 99%. However, this is largely dependent on the reactant variables in the mixture. If the reaction variables are not agitated or properly adjusted, the reaction may be reversed, resulting eventually in a decrease in yield (Wu et al. 2007; Refaat 2010). A time range of 7–48 hours is required for a lipase-catalyzed reaction (Ahmad et al. 2009). It is also reported that reaction time directly affects the cost of product (Kapilakarn and Peugtong 2007). A biodiesel yield of 96.1% was obtained when the reaction time was set for one hour; when it was set to less than three hours, no substantial increase was recorded (96.35%) (Refaat et al. 2008). This study proved the importance of reaction time.

9.10.2 Catalyst Concentration

Catalyst concentration in an esterification reaction plays an important role in supporting and accelerating the reaction rate. The yield of biodiesel increases with increasing concentration as it helps enhance the reaction rate. However, if the concentration of the catalyst is excessively high, it can reduce the yield due to viscosity of the reaction mixture (Talebian-Kiakalaieh et al. 2013b). Various reports have attempted to optimize the concentration of catalyst based on the yield of product and the amount of catalyst recovered.

Heterogeneous catalysts consisting of CaO and ZrO_2 mixed oxides were studied with different molar ratios of Ca and Zr (Dehkordi and Ghasemi 2012). It was observed that when the molar ratios were increased, the biodiesel yield also increased, even though the catalyst was not stable (Dehkordi and Ghasemi 2012). Hence, it was confirmed that the optimal concentration of catalyst may vary based on the type of catalyst used and the catalyst precursor. It has also been reported that the reaction rate increases with the optimum catalyst concentration, although it was shown that the type of catalyst chosen will depend on the feedstock composition (Nakpong and Wootthikanokkhan 2010). For example, reactions with feedstocks with higher FFA and alkaline heterogeneous or homogeneous catalysts may lead to the formation of soap, because the process for recovering the catalyst from the final product requires difficult steps. Therefore, most researchers have chosen to use heterogeneous acid catalysts because they are not sensitive to high-FFA feedstocks and high moisture contents (Li et al. 2009).

9.10.3 Methanol to Fat/Oil Molar Ratio

The effect of the molar ratio of fat/oil to alcohol is an important variable that determines the rate of esterification and transesterification. The most commonly used alcohol in both reactions is methanol, because it can be easily recovered from the final product (Mathiyazhagan et al. 2011). Moreover, the viscosity of biodiesel remains low when methanol is used as compared to other alcohols. However, while methanol is cheaper, ethanol is less toxic.

In esterification, 1 mol of methanol is expected to convert 1 mol of fatty acids to produce 1 mol of methyl ester, whereas 3 mol of methanol and 1 mol of triglycerides are expected to produce 3 mol of alkyl esters. This shows a linear relationship between the alcohol and the fat/oil, since it is always positive with respect to biofuel/biodiesel conversion (Noshadi et al. 2012).

Le Chatelier's principle supports the transesterification reaction, because it is reversible. The principle states that the "rate of formation of products increases as concentration of reactant increases." The most popular molar ratio of fat/oil to alcohol is 1 : 6, and some studies have shown that this produces essentially a complete conversion (Freedman et al. 1984). In spite of the important role of alcohol in the conversion of fatty acids to alkyl esters, it must be noted that an unnecessary increment of the alcohol does not increase the biodiesel yield significantly; rather, it adds an unusual cost to the production (Banerjee and Chakraborty 2009). Nonetheless, the molar ratio of alcohol to fat/oil can be up to 15 : 1 provided the feedstock has a high FFA and the catalyst is pretreated with acid (Leung et al. 2010; Highina et al. 2011). While using acid catalysts, some researchers have reported higher yields of biodiesel when they applied a high amount of alcohol relative to the oil (Zheng et al. 2006).

The use of excess amounts of alcohol (methanol) sometimes creates problems in product separation due to the formation of emulsion as a result of the polar and emulsification nature of the methanol. Furthermore, it can cause the active acid sites of the catalyst to be flooded, which will reduce or inhibit the protonation of TAGs present in them (Leung and Guo 2006). An unnecessary increase in the amount of alcohol and catalyst beyond the optimized reaction conditions is not significant for either esterification or transesterification (Tomasevic and Siler-Marinkovic 2003). Hence, it is imperative to optimize the molar ratio of alcohol to fat/oil to make sure reaction equilibrium is attained.

9.10.4 Reaction Temperature

The reaction temperature is one of the important variables for the success of esterification and transesterification. With increasing temperature, the reaction rate and yield of conversion will also increase (Jiang et al. 2010; Highina et al. 2011). Do Nascimento et al. (2011) showed that the application of high temperature helped reduce the intermolecular association of methanol in both transesterification and esterification. This aided in dispersing the adsorption, preventing the clustering of alcohols around the Brønsted acid on the catalyst due to the bonding of hydrogen ions. Moreover, Akinfalabi et al. (2020b) showed that the reaction temperature should not exceed the boiling point of the alcohol as this prevents the alcohol's vaporization. In the case of methanol use, the expected reaction temperature should thus not exceed 65 °C (the boiling point of methanol). If the reaction temperature is maintained below 50 °C, the viscosity of the biodiesel will increase (Kapilakarn and Peugtong 2007).

The influence of temperature on the esterification and transesterification of FFA conversion of PFAD to biodiesel was studied by Hidayat et al. (2015). Their findings showed an increment of biodiesel conversion from 64% to 82% when the temperature increased from

Table 9.6 Bio-based catalysts used for biodiesel production.

Catalyst	Feedstock/FFA (wt%)	Reactor	Optimum reaction conditions	Conversion/yield (%)	References
Hardwood biochar	Waste vegetable oil	Batch reflux	60 °C; EtOH/oil (28 : 1); 3 h	C = 88	Dehkhoda et al. (2010)
Sulfonated glucose-derived solid acid	PFAD (86.3)	Batch reflux	T = 75 °C, t = 2 h Methanol/oil (10 : 1) Catalyst amount = 2.5 wt%	Y = 92.4	Lokman et al. (2015)
Hardwood biochar	Canola oil		65 °C; MeOH/oil (15 : 1); 24 h	C = 85.1	Dehkhoda et al. (2010)
Sugar cane bagasse	PFAD	Stainless steel batch	T = 170 °C, t = 0.5 h Methanol/oil (20 : 1) Catalyst amount = 11.5 wt%	C = 80	Chin et al. (2012)
Wood mixture biochar	Canola oil		150 °C; 1.52 MPa N_2; MeOH/oil (15 : 1); 3 h	C = 44	Yu et al. (2010)
Peanut hull biochar	Mixture of palmitic and stearic acids and soybean oil		60 °C; MeOH/oil (20 : 1); 4 wt% catalyst; 6 h	C = 70	Kastner et al. (2012)
Commercial biochar	Mixture of oleic acid and canola oil		150 °C; 1.52 MPa N_2; MeOH/oil (10 : 1), 3 h	C = 48	Dehkhoda and Ellis (2013)
Sulfonated ZnAl2O4	PFAD (approx. 90)	Autoclave	T = 100 °C, t = 1 h Methanol/oil (9 : 1) Catalyst amount = 1 wt%	Y = 94.65	Soltani et al. (2016)
KOH and NaOCH3	*Jatropha curcas* (0.5)	Pulsed loop	T = 60 °C, t = 10 min Methanol/oil (6 : 1) Catalyst amount = 1 wt% Mixing rate = 6 Hz	Y = 99.7 and 98.8	Syam et al. (2012)
Rice husk biochar	Waste cooking oil		110 °C; MeOH/oil (20 : 1); 15 h	Y = 88	Li et al. (2014)
ZnO	Refined *Jatropha curcas* oil	Oscillation baffled flow	T = 65 °C, t = 5 min Methanol/oil (12 : 1) Catalyst amount = 15 wt% Mixing rate = 5 Hz	Y = 96	Highina et al. (2011)
Carbon-supported copper oxide (CuO/C)	Rubber seed oil (18)	Batch reflux	T = 65 °C, t = 2 h Methanol/oil (10 : 1) Catalyst amount = 8 wt%	C = 95	Long et al. (2014)

T = temperature, t = time, C = conversion, Y = yield.

40 °C to 60 °C. A maximum conversion of 91% biodiesel was achieved while using a 10 wt% catalyst derived from coconut shell at a 12 : 1 alcohol to PFAD molar ratio, 60 °C reaction temperature, and two hours reaction time. Furthermore, the authors explained that higher temperatures increased the reaction rate and the mass transfer between the reactants and the catalysts.

9.10.5 Mixing Rate

Canakci and Gerpen (2003) showed the influence of mixing rate on biodiesel production, as achieved by the stirrer attached to the reflux machine. They posited that an increase in temperature would also increase the yield of product. A proper agitation increases the collision of the particles, allowing the diffusion of one reactant into another and proper mixing of reactants with catalyst. Increasing the speed of the stirrer will reduce the reaction time (Jiang et al. 2010) and increase conversion (Adeyemi et al. 2011). At a certain stirring speed, there is a significant increase in yield. Hence, the optimization of stirring speed for different materials and precursors is necessary for the esterification of biodiesel.

Kumari et al. (2009) studied the efficacy of reactions using a mixing rate from 100 to 200 rpm, finding that each specific mixing rate significantly influenced the rate of reaction. However, when the stirring speed was increased to 250 rpm, no significant increase in the conversion was reported, as a result of the shearing in the molecules of the enzyme.

Table 9.6 summaries different catalysts, feedstocks, reactors, and operating conditions of esterification and transesterification used for biodiesel production.

9.11 Conclusions

This chapter has shown the development and utilization of bio-based catalysts in the biodiesel industry. It discussed the extensive use and exploitation of petroleum fuels and their sources, which pose serious environmental and socioeconomic challenges. These concerns collectively create the need to search for novel ecofriendly, sustainable, and alternative renewable energies. Among biofuels, biodiesel has attracted considerable attention as a transport fuel due to its various merits. As a result of issues like competition between fuel and food, edible vegetable oils are not commonly used for biodiesel production. Hence, use of non-edible sources of fats/oils and biomass feedstocks becomes the chief choice for the production of FAME. In Malaysia, for example, biodiesel is usually produced from PFAD (a byproduct obtained during palm oil processing) because it is abundantly available.

The important role catalyst synthesis plays in biodiesel production has brought about a need to appraise which catalysts will be best suitable for this work. In the mainstream biodiesel factories, we have shown how homogeneous acidic and basic catalysts are predominantly used for the esterification of fats and oils. However, they are not currently sustainable as they have lots of drawbacks, such as wastewater generation, reactor corrosion, low recyclability, and environmental pollution. Therefore, it is important to synthesize catalysts that are effective, environmentally friendly, recyclable, abundant, and cheap. The sulfonated bio-based catalysts demonstrate a promising and effective application for biodiesel synthesis, especially for feedstocks with high FFA contents. Furthermore, the performance of nanomagnetic catalysts with activated carbon sourced from biomass shows high biodiesel conversion and excellent recovery.

With the rate of success and challenges, some recommendations can be made for future research work in the area of bio-based catalyst development for biodiesel production:

1. To bring up to speed the industrialization of palm fatty acid distillate methyl ester (PFADME), use of bio-based materials, especially non-edible wastes, as catalysts and feedstocks in biodiesel production should be seriously encouraged.
2. The regeneration of bio-based catalysts should also be encouraged, so that they can be resulfonated and applied as catalysts, since bio-based material retains its key morphological components even after several reaction runs.
3. In order to increase the porosity of catalysts, the use of phosphoric acid and other agents that aid porosity should be encouraged as a pretreatment procedure in catalysis.
4. The mixed metal oxide nanomagnetic catalysts should be prepared using other methods such as sol-gel and compared with wet pore impregnation in order to develop the best possible affordable method.

References

Abdulkareem-Alsultan, G., Asikin-Mijan, N., Lee, H.V. et al. (2019). A review on thermal conversion of plant oil (edible and inedible) into green fuel using carbon-based nano-catalyst. *Catalysts* 9 (4): 350.

Abdullah, N., Sulaiman, F., and Aliasak, Z. (2013). A case study of pyrolysis of oil palm wastes in Malaysia. *AIP Conference Proceedings* 1528: 331–336.

Abnisa, F., Daud, W.M.A.W., Husin, W.N.W., and Sahu, J.N. (2011). Utilization possibilities of palm shell as a source of biomass energy in Malaysia by producing bio-oil in pyrolysis process. *Biomass and Bioenergy* 35 (5): 1863–1872.

Adeyemi, N.A., Mohiuddin, A., and Jameel, T. (2011). Waste cooking oil transesterification: influence of impeller type, temperature, speed and bottom clearance on FAME yield. *African Journal of Biotechnology* 10 (44): 8914–8929.

Adinata, D., Wan Daud, W.M., and Aroua, M.K. (2007). Production of carbon molecular sieves from palm shell based activated carbon by pore sizes modification with benzene for methane selective separation. *Fuel Processing Technology* 88: 599–605.

Afrane, G. (1992). Leaching of caustic potash from cocoa husk ash. *Bioresource Technology* 41 (2): 101–104.

Agarwal, A.K. (2007). Biofuels (alcohols and biodiesel) applications as fuels for internal combustion engines. *Progress in Energy and Combustion Science* 33 (3): 233–271.

Agarwal, A.K. and Rajamanoharan, K. (2009). Experimental investigations of performance and emissions of Karanja oil and its blends in a single cylinder agricultural diesel engine. *Applied Energy* 86 (1): 106–112.

Agarwal, D., Kumar, L., and Agarwal, A.K. (2008). Performance evaluation of a vegetable oil fuelled compression ignition engine. *Renewable Energy* 33 (6): 1147–1156.

Ahmad, M., Rashid, S., Khan, M.A. et al. (2009). Optimization of base catalyzed transesterification of peanut oil biodiesel. *African Journal of Biotechnology* 8 (3): 441–446.

Akinfalabi, S.I., Rashid, U., Yunus, R., and Taufiq-Yap, Y.H. (2017). Synthesis of biodiesel from palm fatty acid distillate using sulfonated palm seed cake catalyst. *Renewable Energy* 111: 611–619.

Akinfalabi, S.I., Rashid, U., Yunus, R., and Taufiq-Yap, Y.H. (2019a). Appraisal of sulphonation processes to synthesize palm waste biochar catalysts for the esterification of palm fatty acid distillate. *Catalysts* 9 (2): 184.

Akinfalabi, S.I., Rashid, U., Shean, T.Y.C. et al. (2019b). Esterification of palm fatty acid distillate for biodiesel production catalyzed by synthesized Kenaf seed cake-based sulfonated catalyst. *Catalysts* 9 (5): 482.

Akinfalabi, S.I., Rashid, U., Ngamcharussrivichai, C., and Nehdi, I.A. (2020a). Synthesis of reusable biobased nano-catalyst from waste sugarcane bagasse for biodiesel production. *Environmental Technology and Innovation* 18: 100788.

Akinfalabi, S.I., Rashid, U., Nehdi, I.A. et al. (2020b). Optimization and blends study of heterogeneous acid catalyst-assisted esterification of palm oil industry by-product for biodiesel production. *Royal Society Open Science* 7 (1): 191592.

Aksoy, H., Kahraman, I., Karaosmanoglu, F., and Civelekoglu, H. (1988). Evaluation of Turkish sulphur olive oil as an alternative diesel fuel. *Journal of the American Oil Chemists' Society* 65 (6): 936–938.

Al-Jaberi, S.H., Rashid, U., Al-Doghachi, F.A. et al. (2017). Synthesis of MnO-NiO-SO_4^{-2}/ZrO_2 solid acid catalyst for methyl ester production from palm fatty acid distillate. *Energy Conversion and Management* 139: 166–174.

Alves, C., Oliveira, A., Carneiro, S. et al. (2013). Transesterification of waste frying oil using a zinc aluminate catalyst. *Fuel Processing Technology* 106: 102–107.

Amaya, A., Píriz, J., Tancredi, N., and Cordero, T. (2007). Activated carbon pellets from eucalyptus char and tar TG studies. *Journal of Thermal Analysis and Calorimetry* 89 (3): 987–991.

Ambat, I., Srivastava, V., and Sillanpää, M. (2018). Recent advancement in biodiesel production methodologies using various feedstock: a review. *Renewable and Sustainable Energy Reviews* 90: 356–369.

Atadashi, I., Aroua, M., and Aziz, A.A. (2010). High quality biodiesel and its diesel engine application: a review. *Renewable and Sustainable Energy Reviews* 14 (7): 1999–2008.

Atkinson, J.D., Fortunato, M.E., Dastgheib, S.A. et al. (2011). Synthesis and characterization of iron-impregnated porous carbon spheres prepared by ultrasonic spray pyrolysis. *Carbon* 49 (2): 587–598.

Avhad, M.R. and Marchetti, J.M. (2016). Innovation in solid heterogeneous catalysis for the generation of economically viable and ecofriendly biodiesel: a review. *Catalysis Reviews* 58 (2): 157–208.

Awalludin, M.F., Sulaiman, O., Hashim, R. et al. (2015). An overview of the oil palm industry in Malaysia and its waste utilization through thermochemical conversion, specifically via liquefaction. *Renewable and Sustainable Energy Reviews* 50: 1469–1484.

Azargohar, R. and Dalai, A.K. (2008). Steam and KOH activation of biochar: experimental and modeling studies. *Microporous and Mesoporous Materials* 110 (2–3): 413–421.

Aziz, M.K.A., Morad, N.A., Wambeck, N., and Shah, M.H. (2011). Optimizing palm biomass energy though size reduction. Fourth International Conference on Modeling, Simulation and Applied Optimization, Kuala Lumpur, Malaysia.

Balat, M. and Balat, H. (2010). Progress in biodiesel processing. *Applied Energy* 87 (6): 1815–1835.

Bambase, M.E., Nakamura, N., Tanaka, J., and Matsumura, M. (2007). Kinetics of hydroxide-catalyzed methanolysis of crude sunflower oil for the production of fuel-grade methyl esters. *Journal of Chemical Technology and Biotechnology* 82 (3): 273–280.

Banerjee, A. and Chakraborty, R. (2009). Parametric sensitivity in transesterification of waste cooking oil for biodiesel production – a review. *Resources, Conservation and Recycling* 53 (9): 490–497.

Baroi, C., Yanful, E.K., and Bergougnou, M.A. (2009). Biodiesel production from *Jatropha curcas* oil using potassium carbonate as an unsupported catalyst. *International Journal of Chemical Reactor Engineering* 7 (1).

Baroutian, S., Aroua, M.K., Raman, A.A.A., and Sulaiman, N.M.N. (2011). A packed bed membrane reactor for production of biodiesel using activated carbon supported catalyst. *Bioresource Technology*, 102 (2), 1095–1102.

Baskar, G. and Aiswarya, R. (2015). Biodiesel production from waste cooking oil using copper doped zinc oxide nanocomposite as heterogeneous catalyst. *Bioresource Technology* 188: 124–127.

Bazargan, A., Kostić, M.D., Stamenković, O.S. et al. (2015). A calcium oxide-based catalyst derived from palm kernel shell gasification residues for biodiesel production. *Fuel* 150: 519–525.

Borges, M. and Díaz, L. (2012). Recent developments on heterogeneous catalysts for biodiesel production by oil esterification and transesterification reactions: a review. *Renewable and Sustainable Energy Reviews* 16 (5): 2839–2849.

Bozell, J.J. and Petersen, G.R. (2010). Technology development for the production of biobased products from biorefinery carbohydrates – the US Department of Energy's "top 10" revisited. *Green Chemistry* 12 (4): 539–554.

Brunschwig, C., Moussavou, W., and Blin, J. (2012). Use of bioethanol for biodiesel production. *Progress in Energy and Combustion Science* 38 (2): 283–301.

Buasri, A., Chaiyut, N., Loryuenyong, V. et al. (2012). Continuous process for biodiesel production in packed bed reactor from waste frying oil using potassium hydroxide supported on *Jatropha curcas* fruit shell as solid catalyst. *Applied Sciences* 2 (3): 641–653.

Canakci, M. and Gerpen, J.V. (2001). Biodiesel production from oils and fats with high free fatty acids. *Transactions of the American Society of Agricultural Engineers* 44 (6): 1429.

Canakci, M. and Gerpen, J.V. (2003). A pilot plant to produce biodiesel from high free fatty acid feedstocks. *Transactions of the American Society of Agricultural Engineers* 46 (4): 945–954.

Cha, J.S., Choi, J.C., Ko, J.H. et al. (2010). The low-temperature SCR of NO over rice straw and sewage sludge derived char. *Chemical Engineering Journal* 156 (2): 321–327.

Charoenchaitrakool, M. and Thienmethangkoon, J. (2011). Statistical optimization for biodiesel production from waste frying oil through two-step catalyzed process. *Fuel Processing Technology* 92 (1): 112–118.

Chin, L.H., Abdullah, A.Z., and Hameed, B.H. (2012). Sugar cane bagasse as solid catalyst for synthesis of methyl esters from palm fatty acid distillate. *Chemical Engineering Journal* 183: 104–107.

Cho, H.J., Kim, J.-K., Hong, S.W., and Yeo, Y.-K. (2012). Development of a novel process for biodiesel production from palm fatty acid distillate (PFAD). *Fuel Processing Technology* 104: 271–280.

Chongkhong, S., Tongurai, C., Chetpattananondh, P., and Bunyakan, C. (2007). Biodiesel production by esterification of palm fatty acid distillate. *Biomass and Bioenergy* 31 (8): 563–568.

Choudhury, H., Malani, R., and Moholkar, V.S. (2013). Acid catalyzed biodiesel synthesis from Jatropha oil: mechanistic aspects of ultrasonic intensification. *Chemical Engineering Journal* 231: 262–272.

Darnoko, D. and Cheryan, M. (2000). Kinetics of palm oil transesterification in a batch reactor. *Journal of the American Oil Chemists' Society* 77 (12): 1263–1267.

Dehkhoda, A.M. and Ellis, N. (2013). Biochar-based catalyst for simultaneous reactions of esterification and transesterification. *Catalysis Today* 207: 86–92.

Dehkhoda, A.M., West, A.H., and Ellis, N. (2010). Biochar based solid acid catalyst for biodiesel production. *Applied Catalysis A: General* 382 (2): 197–204.

Dehkordi, A.M. and Ghasemi, M. (2012). Transesterification of waste cooking oil to biodiesel using Ca and Zr mixed oxides as heterogeneous base catalysts. *Fuel Processing Technology* 97: 45–51.

Delfort, B., Le Pennec, D., and Lendresse, C. (2005). Process for transesterification of vegetable oils or animal oils by means of heterogeneous catalysts based on zinc or bismuth, titanium and aluminum. US patent 7151187.

Demirbaş, A. (2003). Biodiesel fuels from vegetable oils via catalytic and noncatalytic supercritical alcohol transesterifications and other methods: a survey. *Energy Conversion and Management* 44 (13): 2093–2109.

Demirbas, A. (2007). Importance of biodiesel as transportation fuel. *Energy Policy* 35 (9): 4661–4670.

Demirbas, A. (2009). Progress and recent trends in biodiesel fuels. *Energy Conversion and Management* 50 (1): 14–34.

Dias, J.M., Alvim-Ferraz, M.C., and Almeida, M.F. (2008). Comparison of the performance of different homogeneous alkali catalysts during transesterification of waste and virgin oils and evaluation of biodiesel quality. *Fuel* 87 (17–18): 3572–3578.

Di Serio, M., Tesser, R., Pengmei, L., and Santacesaria, E. (2007). Heterogeneous catalysts for biodiesel production. *Energy & Fuels* 22 (1): 207–217.

Do Nascimento, L.A.S., Angélica, R.S., Da Costa, C.E. et al. (2011). Conversion of waste produced by the deodorization of palm oil as feedstock for the production of biodiesel using a catalyst prepared from waste material. *Bioresource Technology* 102 (17): 8314–8317.

Erdem, S., Öksüzoğlu, R.M., Avşar, S.B., and Erdem, B. (2017). Magnetic mesoporous silica nanocomposite for biodiesel production. *Acta Physica Polonica A* 132 (3): 763–766.

Farris, T., Coe, C., and Armor, J.S.J. (1992). High capacity coconut shell char for carbon molecular sieves. US patent 5164355A.

Fauzi, A.H.M. and Amin, N.A.S. (2013). Catalysis in biodiesel synthesis: challenges and future perspectives. In: *Advances in Biofuels* (eds. R. Pogaku and R.H. Sarbatly), 127, Springer–152. Boston, MA.

Ferrero, G.O., Rojas, H.J., Argaraña, C.E., and Eimer, G.A. (2016). Towards sustainable biofuel production: design of a new biocatalyst to biodiesel synthesis from waste oil and commercial ethanol. *Journal of Cleaner Production* 139: 495–503.

FitzPatrick, M., Champagne, P., Cunningham, M.F., and Whitney, R.A. (2010). A biorefinery processing perspective: treatment of lignocellulosic materials for the production of value-added products. *Bioresource Technology* 101 (23): 8915–8922.

Freedman, B., Pryde, E., and Mounts, T. (1984). Variables affecting the yields of fatty esters from transesterified vegetable oils. *Journal of the American Oil Chemists Society* 61 (10): 1638–1643.

Fry, J. (2010). Is it smooth sailing or choppy waters ahead? Palm and Lauric Oil Conference and Exhibition Price Outlook (POC), Kuala Lumpur, Malaysia, March 8–10.

Furuta, S., Matsuhashi, H., and Arata, K. (2004). Biodiesel fuel production with solid superacid catalysis in fixed bed reactor under atmospheric pressure. *Catalysis Communications* 5 (12): 721–723.

Ganapathy, T., Murugesan, K.A., and Gakkhar, R. (2009). Performance optimization of Jatropha biodiesel engine model using Taguchi approach. *Applied Energy* 86 (11): 2476–2486.

Gao, C., Zhai, Y., Ding, Y., and Wu, Q. (2010). Application of sweet sorghum for biodiesel production by heterotrophic microalga *Chlorella protothecoides*. *Applied Energy* 87 (3): 756–761.

García-Martínez, N., Andreo-Martínez, P., Quesada-Medina, J. et al. (2017). Optimization of non-catalytic transesterification of tobacco (*Nicotiana tabacum*) seed oil using supercritical methanol to biodiesel production. *Energy Conversion and Management* 131: 99–108.

Glisic, S. and Skala, D. (2009). The problems in design and detailed analyses of energy consumption for biodiesel synthesis at supercritical conditions. *Journal of Supercritical Fluids* 49 (2): 293–301.

Goering, C., Schwab, A., Daugherty, M. et al. (1982). Fuel properties of eleven vegetable oils. *Transactions of ASAE* 25 (6): 1472–1477.

Gratuito, M.K.B., Panyathanmaporn, T., Chumnanklang, R.A. et al. (2008). Production of activated carbon from coconut shell: optimization using response surface methodology. *Bioresource Technology* 99 (11): 4887–4895.

Gui, M.M., Lee, K., and Bhatia, S. (2008). Feasibility of edible oil vs. non-edible oil vs. waste edible oil as biodiesel feedstock. *Energy* 33 (11): 1646–1653.

Guldhe, A., Moura, C.V., Singh, P. et al. (2017). Conversion of microalgal lipids to biodiesel using chromium–aluminum mixed oxide as a heterogeneous solid acid catalyst. *Renewable Energy* 105: 175–182.

Guo, P. and Lua, A.C. (2003). Textural and chemical properties of adsorbent prepared from palm shell by phosphoric acid activation. *Materials Chemistry and Physics* 80 (1): 114–119.

Guo, P., Huang, F., Zheng, M. et al. (2012). Magnetic solid base catalysts for the production of biodiesel. *Journal of the American Oil Chemists' Society* 89: 925–933.

Haas, M.J., Michalski, P.J., Runyon, S. et al. (2003). Production of FAME from acid oil, a by-product of vegetable oil refining. *Journal of the American Oil Chemists' Society* 80 (1): 97–102.

Hamzah, N., Tokimatsu, K., and Yoshikawa, K. (2019). Solid fuel from oil palm biomass residues and municipal solid waste by hydrothermal treatment for electrical power generation in Malaysia: a review. *Sustainability* 11: 1060.

Han, Y.-Z., Hong, L., Wang, X.-Q. et al. (2016). Biodiesel production from *Pistacia chinensis* seed oil via transesterification using recyclable magnetic cellulose-based catalyst. *Industrial Crops and Products* 89: 332–338.

Hara, M., Yoshida, T., Takagaki, A. et al. (2004). A carbon material as a strong protonic acid. *Angewandte Chemie International Edition* 43: 2955–2958.

Hassan, K., Hassan, A.H., Ramli, R., and Jusoh, M.Z. (1991). Improvement on the brightness of unbleached chemithermomechanical pulps from oil palm frond. Proceedings of National Seminar on Oil Palm Trunk and Other Palmwood Utilisation, Kuala Lumpur, Malaysia, FRIM.

Hayyan, A., Alam, M.Z., Mirghani, M.E. et al. (2010). Sludge palm oil as a renewable raw material for biodiesel production by two-step processes. *Bioresource Technology* 101 (20): 7804–7811.

Hidayat, A., Wijaya, K., and Budiman, A. (2015). Esterification of free fatty acid on palm fatty acid distillate using activated carbon catalysts generated from coconut shell. *Procedia Chemistry* 16: 365–371.

Highina, B., Bugaje, I., and Umar, B. (2011). Biodiesel production from Jatropha caucus oil in a batch reactor using zinc oxide as catalyst. *Journal of Petroleum Technology and Alternative Fuels* 2 (9): 146–149.

Hoekman, S.K., Broch, A., Robbins, C. et al. (2012). Review of biodiesel composition, properties, and specifications. *Renewable and Sustainable Energy Reviews* 16 (1): 143–169.

Hoydonckx, H., De Vos, D., Chavan, S., and Jacobs, P. (2004). Esterification and transesterification of renewable chemicals. *Topics in Catalysis* 27 (1–4): 83–96.

Hu, Z. and Vansant, E.F. (1995). Carbon molecular sieves produced from walnut shell. *Carbon* 33 (5): 561–567.

Huang, G., Chen, F., Wei, D. et al. (2010). Biodiesel production by microalgal biotechnology. *Applied Energy* 87 (1): 38–46.

Index Mundi. (2017). Malaysia palm oil production by year. Available from https://www.indexmundi.com/agriculture/?country=my&commodity=palm-oil&graph=production (accessed December 19, 2020).

Issariyakul, T. and Dalai, A.K. (2014). Biodiesel from vegetable oils. *Renewable and Sustainable Energy Reviews* 31: 446–471.

Jain, S. and Sharma, M. (2011). Oxidation stability of blends of Jatropha biodiesel with diesel. *Fuel* 90 (10): 3014–3020.

Jayed, M., Masjuki, H., Kalam, M. et al. (2011). Prospects of dedicated biodiesel engine vehicles in Malaysia and Indonesia. *Renewable and Sustainable Energy Reviews* 15 (1): 220–235.

Jiang, S., Zhang, F., and Pan, L. (2010). Sodium phosphate as a solid catalyst for biodiesel preparation. *Brazilian Journal of Chemical Engineering* 27 (1): 137–144.

Joylene, T.Y., Dehkhoda, A.M., and Ellis, N. (2011). Development of biochar-based catalyst for transesterification of canola oil. *Energy & Fuels* 1 (25): 337–344.

Kapilakarn, K. and Peugtong, A. (2007). A comparison of costs of biodiesel production from transesterication. *International Energy Journal* 8: 1–6.

Kapor, N.Z.A., Maniam, G.P., Rahim, M.H.A., and Yusoff, M.M. (2017). Palm fatty acid distillate as a potential source for biodiesel production-a review. *Journal of Cleaner Production* 143: 1–9.

Kastner, J.R., Miller, J., Geller, D.P. et al. (2012). Catalytic esterification of fatty acids using solid acid catalysts generated from biochar and activated carbon. *Catalysis Today* 190 (1): 122–132.

Kazemian, H., Turowec, B., Siddiquee, M.N., and Rohani, S. (2013). Biodiesel production using cesium modified mesoporous ordered silica as heterogeneous base catalyst. *Fuel* 103: 719–724.

Keera, S., El Sabagh, S., and Taman, A. (2011). Transesterification of vegetable oil to biodiesel fuel using alkaline catalyst. *Fuel* 90 (1): 42–47.

Kefas, H.M., Yunus, R., Rashid, U., and Taufiq-Yap, Y.H. (2018). Modified sulfonation method for converting carbonized glucose into solid acid catalyst for the esterification of palm fatty acid distillate. *Fuel* 229: 68–78.

Klinghoffer, N.B., Castaldi, M.J., and Nzihou, A. (2012). Catalyst properties and catalytic performance of char from biomass gasification. *Industrial & Engineering Chemistry Research* 51 (40): 13113–13122.

Knothe, G. (2009). Improving biodiesel fuel properties by modifying fatty ester composition. *Energy & Environmental Science* 2 (7): 759–766.

Knothe, G. and Dunn, R.O. (2001). Biofuels derived from vegetable oils and fats. In: *Oleochemical Manufacture and Applications* (eds. F.D. Gunstone and R.J. Hamilton), 106–163. Sheffield: Sheffield Academic.

Komers, K., Stloukal, R., Machek, J., and Skopal, F. (2001). Biodiesel from rapeseed oil, methanol and KOH. 3. Analysis of composition of actual reaction mixture. *European Journal of Lipid Science and Technology* 103 (6): 363–371.

Konwar, L.J., Boro, J., and Deka, D. (2014a). Review on latest developments in biodiesel production using carbon-based catalysts. *Renewable and Sustainable Energy Reviews* 29: 546–564.

Konwar, L.J., Das, R., Thakur, A.J. et al. (2014b). Biodiesel production from acid oils using sulfonated carbon catalyst derived from oil-cake waste. *Journal of Molecular Catalysis A: Chemical* 388: 167–176.

Kristiani, A., Abimanyu, H., Setiawan, A.H. et al. (2013). Effect of pretreatment process by using diluted acid to characteristic of oil palm's frond. *Energy Procedia* 32: 183–189.

Kumari, A., Mahapatra, P., Garlapati, V.K., and Banerjee, R. (2009). Enzymatic transesterification of Jatropha oil. *Biotechnology for Biofuels* 2 (1): 1.

Kusdiana, D. and Saka, S. (2001). Kinetics of transesterification in rapeseed oil to biodiesel fuel as treated in supercritical methanol. *Fuel* 80 (5): 693–698.

Kuti, O.A. (2007). Impact of charred palm kernel shell on the calorific value of composite sawdust briquette. *Journal of Engineering and Applied Sciences* 2 (1): 62–65.

Lam, M.K., Lee, K.T., and Mohamed, A.R. (2010). Homogeneous, heterogeneous and enzymatic catalysis for transesterification of high free fatty acid oil (waste cooking oil) to biodiesel: a review. *Biotechnology Advances* 28 (4): 500–518.

Lee, K.T. and Ofori-Boateng, C. (2013). Environmental sustainability assessment of biofuel production from oil palm biomass. In: *Sustainability of Biofuel Production from Oil Palm Biomass* (eds. K.T. Lee and C. Ofori-Boateng), 149–187. Singapore: Springer.

Lee, J., Kim, K.H., and Kwon, E.E. (2017). Biochar as a catalyst. *Renewable and Sustainable Energy Reviews* 77: 70–79.

Leung, D. and Guo, Y. (2006). Transesterification of neat and used frying oil: optimization for biodiesel production. *Fuel Processing Technology* 87 (10): 883–890.

Leung, D.Y., Wu, X., and Leung, M. (2010). A review on biodiesel production using catalyzed transesterification. *Applied Energy* 87 (4): 1083–1095.

Li, S., Wang, Y., Dong, S. et al. (2009). Biodiesel production from Eruca Sativa Gars vegetable oil and motor, emissions properties. *Renewable Energy* 34 (7): 1871–1876.

Li, X., Zhu, Z., De Marco, R. et al. (2010). Modification of coal as a fuel for the direct carbon fuel cell. *Journal of Physical Chemistry A* 114 (11): 3855–3862.

Li, M., Zheng, Y., Chen, Y., and Zhu, X. (2014). Biodiesel production from waste cooking oil using a heterogeneous catalyst from pyrolyzed rice husk. *Bioresource Technology* 154: 345–348.

Liang, X., Gao, S., Wu, H., and Yang, J. (2009). Highly efficient procedure for the synthesis of biodiesel from soybean oil. *Fuel Processing Technology* 90 (5): 701–704.

Lim, S. and Teong, L.K. (2010). Recent trends, opportunities and challenges of biodiesel in Malaysia: an overview. *Renewable and Sustainable Energy Reviews* 14 (3): 938–954.

Lin, L., Ying, D., Chaitep, S., and Vittayapadung, S. (2009). Biodiesel production from crude rice bran oil and properties as fuel. *Applied Energy* 86 (5): 681–688.

Liu, K.-S. (1994). Preparation of fatty acid methyl esters for gas-chromatographic analysis of lipids in biological materials. *Journal of the American Oil Chemists' Society* 71 (11): 1179–1187.

Liu, Y., Lotero, E., and Goodwin, J.G. Jr. (2006). Effect of water on sulfuric acid catalyzed esterification. *Journal of Molecular Catalysis A: Chemical* 245: 132–140.

Liu, C., Lv, P., Yuan, Z. et al. (2010). The nanometer magnetic solid base catalyst for production of biodiesel. *Renewable Energy* 35 (7): 1531–1536.

Liu, W.J., Zeng, F.X., Jiang, H., and Zhang, X.S. (2011). Preparation of high adsorption capacity bio-chars from waste biomass. *Bioresource Technology* 102 (17): 8247–8252.

Liu, W.J., Tian, K., Jiang, H., and Yu, H.Q. (2013). Facile synthesis of highly efficient and recyclable magnetic solid acid from biomass waste. *Scientific Reports* 3: 1–7.

Liu, Q.Y., Yang, F., Sun, X.F. et al. (2017). Preparation of biochar catalyst with saccharide and lignocellulose residues of corncob degradation for corncob hydrolysis into furfural. *Journal of Material Cycles and Waste Management* 19 (1): 134–143.

Liu, K., Wang, R., and Yu, M. (2018). An efficient, recoverable solid base catalyst of magnetic bamboo charcoal: preparation, characterization, and performance in biodiesel production. *Renewable Energy* 127: 531–538.

Lokman, I.M., Rashid, U., and Taufiq-Yap, Y.H. (2015). Production of biodiesel from palm fatty acid distillate using sulfonated-glucose solid acid catalyst: characterization and optimization. *Chinese Journal of Chemical Engineering* 11: 1857–1864.

Lokman, I.M., Goto, M., Rashid, U., and Taufiq-Yap, Y.H. (2016). Sub- and supercritical esterification of palm fatty acid distillate with carbohydrate-derived solid acid catalyst. *Chemical Engineering Journal* 284: 872–878.

Long, Y.D., Fang, Z., Su, T.-C., and Yang, Q. (2014). Co-production of biodiesel and hydrogen from rapeseed and Jatropha oils with sodium silicate and Ni catalysts. *Applied Energy* 113: 1819–1825.

Lotero, E., Liu, Y., Lopez, D.E. et al. (2005). Synthesis of biodiesel via acid catalysis. *Industrial & Engineering Chemistry Research* 44 (14): 5353–5363.

Lou, W.-Y., Zong, M.-H., and Duan, Z.-Q. (2008). Efficient production of biodiesel from high free fatty acid-containing waste oils using various carbohydrate-derived solid acid catalysts. *Bioresource Technology* 99 (18): 8752–8758.

Łukaszewicz, J. and Wesołowski, R. (2008). Fabrication of molecular-sieve-type carbons from *Salix viminalis*. *Microporous and Mesoporous Materials* 116: 723–726.

Ma, F. and Hanna, M.A. (1999). Biodiesel production: a review. *Bioresource Technology* 70 (1): 1–15.

Ma, Z. and Zaera, F. (2006). Characterization of heterogeneous catalysts. In: *Surface and Nanomolecular Catalysis* (ed. R. Richards), 1–37. Abingdon: Routledge.

Macedo, C., Abreu, F.R., Tavares, A.P. et al. (2006). New heterogeneous metal-oxides based catalyst for vegetable oil trans-esterification. *Journal of the Brazilian Chemical Society* 17 (7): 1291–1296.

Malaysian Palm Oil Board (MPOB). (2016). Annual Report. Available from www.mpob.gov.my (accessed December 19, 2020)

Marchetti, J.M. (2012). A summary of the available technologies for biodiesel production based on a comparison of different feedstock's properties. *Process Safety and Environmental Protection* 90 (3): 157–163.

Marchetti, J.M. (2013). Influence of economical variables on a supercritical biodiesel production process. *Energy Conversion and Management* 75: 658–663.

Marchetti, J. and Errazu, A. (2008). Esterification of free fatty acids using sulfuric acid as catalyst in the presence of triglycerides. *Biomass and Bioenergy* 32 (9): 892–895.

Marchetti, J., Miguel, V., and Errazu, A. (2007). Possible methods for biodiesel production. *Renewable and Sustainable Energy Reviews* 11 (6): 1300–1311.

Mathiyazhagan, M., Ganapathi, A., Jaganath, B. et al. (2011). Production of biodiesel from non-edible plant oils having high FFA content. *International Journal of Chemical and Environmental Engineering* 2 (2): 119–122.

Mbaraka, I.K. and Shanks, B.H. (2005). Design of multifunctionalized mesoporous silicas for esterification of fatty acid. *Journal of Catalysis* 229 (2): 365–373.

McKendry, P. (2002). Energy production from biomass (part 1): overview of biomass. *Bioresource Technology* 83: 37–46.

Metre, A.V. and Nath, K. (2015). Super phosphoric acid catalyzed esterification of palm fatty acid distillate for biodiesel production: physicochemical parameters and kinetics. *Polish Journal of Chemical Technology* 17 (1): 88–96.

Mielke, T. (2010). The price outlook of palm and lauric oils and impacts from the global vegetable oil markets: a fundamental approach. Palm and Lauric Oils Conference & Exhibition Price Outlook (POC), Kuala Lumpur, Malaysia.

Mofijur, M., Masjuki, H., Kalam, M., and Atabani, A. (2013). Evaluation of biodiesel blending, engine performance and emissions characteristics of Jatropha curcas methyl ester: Malaysian perspective. *Energy* 55: 879–887.

Moser, B.R. (2009). Biodiesel production, properties, and feedstocks. *Vitro Cellular & Developmental Biology-Plant* 45 (3): 229–266.

Muñoz-González, Y., Arriagada-Acuña, R., Soto-Garrido, G., and García-Lovera, R. (2009). Activated carbons from peach stones and pine sawdust by phosphoric acid activation used in clarification and decolorization processes. *J. Chem. Technol. Biotechnol.*, 84: 39–47.

Naik, M., Meher, L., Naik, S., and Das, L. (2008). Production of biodiesel from high free fatty acid Karanja (*Pongamia pinnata*) oil. *Biomass and Bioenergy* 32 (4): 354–357.

Nakajima, K. and Hara, M. (2012). Amorphous carbon with SO3H groups as a solid Brønsted acid catalyst. *ACS Catalysis* 2 (7): 1296–1304.

Nakpong, P. and Wootthikanokkhan, S. (2010). High free fatty acid coconut oil as a potential feedstock for biodiesel production in Thailand. *Renewable Energy* 35 (8): 682–1687.

Noshadi, I., Amin, N., and Parnas, R.S. (2012). Continuous production of biodiesel from waste cooking oil in a reactive distillation column catalyzed by solid heteropolyacid: optimization using response surface methodology (RSM). *Fuel* 94: 156–164.

Nur, Z.A.S., Nizah, M.F.R., Teo, S.H. et al. (2014). Production of biodiesel from palm oil using modified Malaysian natural dolomites. *Energy Conversion and Management* 78: 738–744.

Okamura, M., Takagaki, A., Toda, M. et al. (2006). Acid-catalyzed reactions on flexible polycyclic aromatic carbon in amorphous carbon. *Chemistry of Materials* 18 (13): 3039–3045.

Onoja, E., Chandren, S., Abdul Razak, F.I. Mahat, N. A., & Wahab, R. A. (2019). Oil palm (*Elaeis guineensis*) Biomass in Malaysia: The Present and Future Prospects.. *Waste and Biomass Valorization* 10: 2099–2117.

Onyestyák, G. and Bóta, A. (2009). Sorption dynamics of N_2 and O_2 in carbon monoliths from spruce, beech and oak affected by activation. *Microporous and Mesoporous Materials* 120 (1–2): 84–90.

Park, Y.-M., Lee, D.-W., Kim, D.-K. et al. (2008). The heterogeneous catalyst system for the continuous conversion of free fatty acids in used vegetable oils for the production of biodiesel. *Catalysis Today* 131 (1–4): 238–243.

Peterson, C.L. and Hustrulid, T. (1998). Carbon cycle for rapeseed oil biodiesel fuels. *Biomass and Bioenergy* 14 (2): 91–101.

Ping, B.T.Y. and Yusof, M. (2009). Characteristics and properties of fatty acid distillates from palm oil. *Oil Palm Bulletin* 59: 5–11.

Pramanik, K. (2003). Properties and use of *Jatropha curcas* oil and diesel fuel blends in compression ignition engine. *Renewable Energy* 28 (2): 239–248.

Prasertsan, S. and Prasertsan, P. (1996). Biomass residues from palm oil mills in Thailand: an overview on quantity and potential usage. *Biomass and Bioenergy* 11 (5): 387–395.

Puna, J., Gomes, J., Correia, M.J.N. et al. (2010). Advances on the development of novel heterogeneous catalysts for transesterification of triglycerides in biodiesel. *Fuel* 89 (11): 3602–3606.

Ramadhas, A., Jayaraj, S., and Muraleedharan, C. (2004). Use of vegetable oils as IC engine fuels – a review. *Renewable Energy* 29 (5): 727–742.

Ramadhas, A.S., Jayaraj, S., and Muraleedharan, C. (2005). Biodiesel production from high FFA rubber seed oil. *Fuel* 84 (4): 335–340.

Rashid, U., Anwar, F., Moser, B.R., and Ashraf, S. (2008). Production of sunflower oil methyl esters by optimized alkali-catalyzed methanolysis. *Biomass and Bioenergy* 32 (12): 1202–1205.

Rashid, U., Anwar, F., Ashraf, M. et al. (2011). Application of response surface methodology for optimizing transesterification of *Moringa oleifera* oil: biodiesel production. *Energy Conversion and Management* 52 (8–9): 3034–3042.

Refaat, A. (2010). Different techniques for the production of biodiesel from waste vegetable oil. *International Journal of Environmental Science and Technology* 7 (1): 183–213.

Refaat, A., Attia, N., Sibak, H.A. et al. (2008). Production optimization and quality assessment of biodiesel from waste vegetable oil. *International Journal of Environmental Science and Technology* 5 (1): 75–82.

Ribeiro, N.M., Pinto, A.C., Quintella, C.M. et al. (2007). The role of additives for diesel and diesel blended (ethanol or biodiesel) fuels: a review. *Energy & Fuels* 21 (4): 2433–2445.

Roschat, W., Siritanon, T., Yoosuk, B., and Promarak, V. (2016). Rice husk-derived sodium silicate as a highly efficient and low-cost basic heterogeneous catalyst for biodiesel production. *Energy Conversion and Management* 119: 453–462.

Sahoo, P. and Das, L. (2009). Process optimization for biodiesel production from Jatropha, Karanja and Polanga oils. *Fuel* 88 (9): 1588–1594.

Sánchez-Cantú, M., Pérez-Díaz, L.M., Pala-Rosas, I. et al. (2013). Hydrated lime as an effective heterogeneous catalyst for the transesterification of castor oil and methanol. *Fuel* 110: 54–62.

Sani, Y.M., Raji-Yahya, A.O., Alaba, P.A. et al. (2015). Palm frond and spikelet as environmentally benign alternative solid acid catalysts for biodiesel production. *BioResources* 10 (2): 3393–3408.

Sarin, R., Sharma, M., Sinharay, S., and Malhotra, R.K. (2007). Jatropha–palm biodiesel blends: an optimum mix for Asia. *Fuel* 86 (10–11): 1365–1371.

Shahabuddin, M., Masjuki, H., Kalam, M.A. et al. (2012). Effect of additive on performance of CI engine fueled with bio diesel. *Energy Procedia* 14: 1624–1629.

Shi, Z.J., Xiao, L.P., Deng, J. et al. (2011). Isolation and characterization of soluble polysaccharides of *Dendrocalamus brandisii*. *BioResources* 6 (4): 5151–5166.

Siddiqui, M.T.H., Nizamuddin, S., Baloch, H.A. et al. (2019). Fabrication of advance magnetic carbon nano-materials and their potential applications: a review. *Journal of Environmental Chemical Engineering* 7 (1): 102812.

Singh, S. and Singh, D. (2010). Biodiesel production through the use of different sources and characterization of oils and their esters as the substitute of diesel: a review. *Renewable and Sustainable Energy Reviews* 14 (1): 200–216.

Sinha, S., Agarwal, A.K., and Garg, S. (2008). Biodiesel development from rice bran oil: transesterification process optimization and fuel characterization. *Energy Conversion and Management* 49 (5): 1248–1257.

Soltani, S., Rashid, U., Yunus, R., and Taufiq-Yap, Y.H. (2016). Biodiesel production in the presence of sulfonated mesoporous ZnAl2O4 catalyst via esterification of palm fatty acid distillate (PFAD). *Fuel* 178: 253–262.

Speight, J.G. (2008). *Synthetic Fuels Handbook: Properties, Process and Performance*. New York: McGraw-Hill.

Srivastava, A. and Prasad, R. (2000). Triglycerides-based diesel fuels. *Renewable and Sustainable Energy Reviews* 4 (2): 111–133.

Sulaiman, O.O., Saharuddin, A.H., and Wan Nik, W.B. (2011). Potential of waste-based biomass cogeneration for Malaysia energy sector. *Arabian Journal of Business and Management Review* 1 (4): 77–101.

Syam, A.M., Yunus, R., Ghazi, T.I.M., and Yaw, T.C.S. (2012). Synthesis of *Jatropha curcas*-based methyl ester and ethyl ester as biodiesel feedstocks. *Pertanika Journal of Science and Technology* 20 (1): 165–173.

Talebian-Kiakalaieh, A., Amin, N.A.S., and Mazaheri, H. (2013a). A review on novel processes of biodiesel production from waste cooking oil. *Applied Energy* 104: 683–710.

Talebian-Kiakalaieh, A., Amin, N.A.S., Zarei, A., and Noshadi, I. (2013b). Transesterification of waste cooking oil by heteropoly acid (HPA) catalyst: optimization and kinetic model. *Applied Energy* 102: 283–292.

Tiwari, A.K., Kumar, A., and Raheman, H. (2007). Biodiesel production from jatropha oil (*Jatropha curcas*) with high free fatty acids: an optimized process. *Biomass and Bioenergy* 31 (8): 569–575.

Toda, M., Takagaki, A., Okamura, M. et al. (2005). Green chemistry: biodiesel made with sugar catalyst. *Nature* 438 (7065): 178.

Tomasevic, A. and Siler-Marinkovic, S. (2003). Methanolysis of used frying oil. *Fuel Processing Technology* 81 (1): 1–6.

Tonetto, G.M. and Marchetti, J.M. (2010). Transesterification of soybean oil over me/Al2O3 (Me = Na, Ba, Ca, and K) catalysts and monolith K/Al2O3-cordierite. *Topics in Catalysis* 53 (11–12): 755–762.

Tongpoothorn, W., Sriuttha, M., Homchan, P. et al. (2011). Preparation of activated carbon derived from *Jatropha curcas* fruit shell by simple thermo-chemical activation and characterization of their physico-chemical properties. *Chemical Engineering Research and Design* 89 (3): 335–340.

Uprety, B.K., Chaiwong, W., Ewelike, C., and Rakshit, S.K. (2016). Biodiesel production using heterogeneous catalysts including wood ash and the importance of enhancing byproduct glycerol purity. *Energy Conversion and Management* 115: 191–199.

Uzun, B.B., Kılıç, M., Özbay, N. et al. (2012). Biodiesel production from waste frying oils: optimization of reaction parameters and determination of fuel properties. *Energy* 44 (1): 347–351.

Van Gerpen, J. (2005). Biodiesel processing and production. *Fuel Processing Technology* 86 (10): 1097–1107.

Veljković, V.B., Stamenković, O.S., Todorović, Z.B. et al. (2009). Kinetics of sunflower oil methanolysis catalyzed by calcium oxide. *Fuel* 88 (9): 1554–1562.

Viana, H.F., dos, S., Rodrigues, A.M. et al. (2018). Evaluation of the physical, chemical and thermal properties of Portuguese maritime pine biomass. *Sustainability* 10 (8): 1–15.

Vicente, G., Martınez, M., and Aracil, J. (2004). Integrated biodiesel production: a comparison of different homogeneous catalysts systems. *Bioresource Technology* 92 (3): 297–305.

Wang, R., Amano, Y., and Machida, M. (2013). Surface properties and water vapor adsorption–desorption characteristics of bamboo-based activated carbon. *Journal of Analytical and Applied Pyrolysis* 104: 667–674.

Wang, Y.T., Fang, Z., Yang, X.X. et al. (2018). One-step production of biodiesel from Jatropha oils with high acid value at low temperature by magnetic acid-base amphoteric nanoparticles. *Chemical Engineering Journal* 348: 929–939.

Wu, Y.P.G., Lin, Y.F., and Chang, C.T. (2007). Combustion characteristics of fatty acid methyl esters derived from recycled cooking oil. *Fuel* 86 (17–18): 2810–2816.

Yu, J.T., Dehkhoda, A.M., and Ellis, N. (2010). Development of biochar-based catalyst for transesterification of canola oil. *Energy & Fuels* 25 (1): 337–344.

Yujaroen, D., Goto, M., Sasaki, M., and Shotipruk, A. (2009). Esterification of palm fatty acid distillate (PFAD) in supercritical methanol: effect of hydrolysis on reaction reactivity. *Fuel* 88 (10): 2011–2016.

Yusuf, N.N., Kamarudin, S.K., and Yaakob, Z. (2012). Overview on the production of biodiesel from *Jatropha curcas* L. by using heterogenous catalysts. *Biofuels, Bioproducts and Biorefining* 6 (3): 319–334.

Zhang, T., Walawender, W.P., and Fan, L.T. (2005). Preparation of carbon molecular sieves by carbon deposition from methane. *Bioresource Technology* 96 (17): 1929–1935.

Zhang, X., Li, J., Chen, Y. et al. (2009). Heteropolyacid nanoreactor with double acid sites as a highly efficient and reusable catalyst for the transesterification of waste cooking oil. *Energy & Fuels* 23 (9): 4640–4646.

Zhang, F., Fang, Z., and Wang, Y.T.T. (2015). Biodiesel production direct from high acid value oil with a novel magnetic carbonaceous acid. *Applied Energy* 155: 637–647.

Zhang, F., Wu, X.H., Yao, M. et al. (2016). Production of biodiesel and hydrogen from plant oil catalyzed by magnetic carbon-supported nickel and sodium silicate. *Green Chemistry* 18 (11): 3302–3314.

Zhang, F., Tian, X.F., Fang, Z. et al. (2017). Catalytic production of Jatropha biodiesel and hydrogen with magnetic carbonaceous acid and base synthesized from Jatropha hulls. *Energy Conversion and Management* 142: 107–116.

Zhao, C., Yang, L., Xing, S., Luo, W., Wang, Z., & Lv, P. (2018). Biodiesel production by a highly effective renewable catalyst from pyrolytic rice husk. *Journal of cleaner production* 199: 772–780.

Zheng, S., Kates, M., Dubé, M., and McLean, D. (2006). Acid-catalyzed production of biodiesel from waste frying oil. *Biomass and Bioenergy* 30 (3): 267–272.

Zhou, Y., Niu, S., and Li, J. (2016). Activity of the carbon-based heterogeneous acid catalyst derived from bamboo in esterification of oleic acid with ethanol. *Energy Conversion and Management* 114: 188–196.

Zong, M.-H., Duan, Z.-Q., Lou, W.-Y. et al. (2007). Preparation of a sugar catalyst and its use for highly efficient production of biodiesel. *Green Chemistry* 9 (5): 434–437.

10

Heterogeneous Nanocatalytic Conversion of Waste to Biodiesel

Nilutpal Bhuyan[1,2], *Manash J. Borah*[1], *Neelam Bora*[1], *Dipanka Saikia*[1], *Dhanapati Deka*[1] *and Rupam Kataki*[1]

[1] *Department of Energy, Tezpur University, Tezpur, Assam, India*
[2] *Department of Chemistry, D.C.B. Girls College, Jorhat, Assam, India*

10.1 Introduction

The increasing world population is leading to ever increasing energy consumption worldwide. The current rate of consumption could lead all available oil reserves to be completely depleted by 2050 (Kim et al. 2018). The increasing cost of crude oil and environmental problems such as global warming caused by its combustion are of serious concern (Akia et al. 2014). In the next three decades, the world's total energy requirement is expected to be increase by half over the existing consumption rate due to increasing industrial and transportation growth. The continued consumption of fossil fuel will lead to rises in emissions of greenhouse gases (GHGs) such as carbon dioxide (CO_2), sulfur dioxide (SO_2), nitrogen oxide (NO_x), and carbon monoxide (CO), which in turn will cause increases in the global temperature, leading to the deaths of millions of humans and other species (Atabani et al. 2012). It is estimated that by the year 2035, CO_2 emissions from global energy will have doubled compared to present rates, representing an increase of 1.6% a year. Worldwide CO_2 emissions in 2035 are forecasted to reach above 140 billion metric tons, as compared with 110 billion in 2010 (BP 2014).

Given the issue of fossil fuel depletion and its impact on the environment, the world scientific community has long been investigating the production of clean liquid fuels from renewable resources (i.e. biofuels) (Armaroli and Balzani 2007). Of the various alternative energy sources, biomass is gaining increasing attention because of its abundance and environmental friendly nature (leading to low CO_2 and sulfur emissions). In the next three decades, the world's energy demands are expected increasingly to be met by renewables, with the use of renewable energy increasing at an average rate of 2.6% a year from 2012 to 2040. Over the last few decades, extensive investigations have been made into the feasibility of replacing fossil-based fuels with biofuels. By 2050, it is expected that biomass will contribute 50% of the world's total energy demands (Demirbas 2008), compared with about 10–15% at present. Biomass feedstocks used for the production of biofuels include wood, crops, grass, algae, wheat straw, rice straw, corn, miscanthus, non-edible oils, green and wood landfill waste, animal fats, waste frying oils, agricultural residues, municipal wastes,

Nano- and Biocatalysts for Biodiesel Production, First Edition. Edited by Avinash P. Ingle.

forest product wastes, paper, and food waste (Demirbas 2008; Luo and Zhou 2012). Different biofuels such as biodiesel, bioethanol, biohydrogen, and biomethane can be prepared from these biomass feedstocks and are considered as potential renewable alternatives.

In 1893, Rudolf Diesel attempted to use peanut oil instead of petro-diesel in a model diesel engine, although he did not succeed due to the high viscosity, low volatility, and cold flow properties of vegetable oil (Bezergianni and Dimitriadis 2013). Since then, various researchers have tried to modify these properties in order to generate a biofuel comparable to petroleum oil. Biodiesel is an eco-friendly, nontoxic, and biodegradable fuel produced from biomass by transesterification. In the transesterification reaction, the long chains of triglycerides are converted into fatty acid methyl esters (FAMEs). Biodiesel provides low emissions of CO, NO_x, sulfur, and particulate matter during its combustion (Al-Zuhair 2007). Moreover, in comparison to petro-diesel, biodiesel offers a higher combustion efficiency, flash point, cetane number, and lubricant efficiency. Usually, the raw materials used in the preparation of biodiesel are classified into three different types: edible oils, non-edible oils, and waste edible oils (WEOs). Because of their high cost and the food versus fuel controversy, edible oils such as vegetable oil are not very suitable for use as raw materials in biodiesel production (Gharat and Rathod 2013). Thus, many investigations have been made into the application of non-edible oils, waste oils, animal fats, and microalgae. Various sources of non-edible oil exist, including jatropha, karanja, mahua, polanga, rubber, and castor oil, which are of low cost in comparison to edible oil sources (Hama and Kondo 2013). Likewise, there is an abundance of waste oil originating from sources such as waste cooking oil (WCO), grease, and soap stock, with WCO alone accounting for about 16.6 million tons per year (Azócar et al. 2010). Another non-edible feedstock is microalgae, which are environment friendly because of their high carbon dioxide fixation ability. They also have the advantage of offering a high growth rate in comparison to terrestrial plants. Microalgae can generate 25 times more oil than other plant sources, and some may have oil contents up to 50% by weight of dry biomass (Ahmad et al. 2011; Lee et al. 2015).

10.2 Role of Catalysts in Biodiesel Production

Transesterification is the major process involved in the chemical conversion of various feedstocks to biodiesel. In transesterification, triglyceride reacts with a mono-alcohol to produce a mixture of FAME and glycerol in the presence of catalyst (Razzak et al. 2013). Basically, there are two different ways to perform transesterification: the conventional method (two-stage method) and direct transesterification (single-stage method, *in situ* transesterification). In the conventional method, mechanical or chemical means are first used for lipid extraction from dried biomass, followed by transesterification and purification steps. In direct transesterification, lipid extraction and transesterification take place simultaneously (Johnson and Wen 2009; Zeng et al. 2011; Velasquez-Orta et al. 2012). Another process involved in the chemical conversion of biomass to biodiesel is esterification, which is a reversible reaction process. In esterification, free fatty acids (FFAs) are converted to alkyl ester using acid catalyst, which reduces the formation of foam and improves conversion of biodiesel (Razzak et al. 2013).

The role of the catalyst in transesterification and esterification is to alter the rate of reaction without consuming. An advantage of the application of catalyst in these reactions is the achievement of a lower activation energy compared to noncatalytic processes. Generally, two types of catalyst are available for application in chemical conversion of biomass to biodiesel: homogenous and heterogeneous. In homogenous catalysis, both the catalyst and the reactant function in the same phase, whereas in heterogeneous catalysis, they are in different phases. Extensive investigations have been performed in the last few years into the various classifications of catalyst for biodiesel production, the effects of factors such as particle size, shape, porosity, type, and process conditions on catalyst preparations, and the reaction kinetics, mass transfer, scale-up, and optimization of esterification and transesterification processes (Gardy et al. 2019a,b). The disadvantages of homogeneous catalysts are their corrosive nature and difficulty in recovery, which leads to costly fuel purification processes (Balat and Balat 2010; Osatiashtiani et al. 2016). Heterogeneous catalysts have thus begun to gain attention as alternative catalysts for good-quality biodiesel production from non-edible feedstocks (Gardy et al. 2019a,b).

Investigation of nanocatalysts is a recent trend in the generation of catalysts with high activity, selectivity, and stability that provide a greater yield of reaction product. Nanocatalysts can be generated by altering the surface functionality, elemental composition, or number of atoms in the particle. These properties can help a heterogeneous catalytic system perform reactions at the same rate as a homogeneous one (Thangaraj et al. 2019).

10.3 Feedstocks for Biodiesel Production

The choice among feedstocks for biodiesel production depends on the availability, geographical locations, regional climates, agricultural practices, and local soil conditions in a region or country. Numerous renewable feedstocks, including vegetable oils, animal fats, greases, WCOs, and algal oils, have been trialed in this regard (Christophe et al. 2012; Duarte et al. 2013; Eloka-Eboka et al. 2017; Gaurav et al. 2019). Generally, industrial production utilizes oil crops: mainly soybean, sunflower, and rapeseed oils (Santori et al. 2012). The choice of oil feedstock influences the biodiesel composition and hence fuel properties (Singh and Singh 2010; Karmakar et al. 2010), notably acid value, oxidation stability, cloud point, cetane number, and cold filter plugging point. Also, the selection of the cheapest feedstock is a crucial factor to ensure low production costs, because the cost of raw materials accounts for about 60–80% of the total cost of biodiesel production (Bhuyan et al. 2019). Biodiesel feedstocks can be classified into four main categories as shown in Table 10.1.

10.3.1 First-Generation Feedstocks or Edible Oils

Depending on availability, different edible oils are utilized as feedstocks for biodiesel production by different countries. First-generation feedstocks include rapeseed, soybean, peanut, sunflower, palm, and coconut oil. Among them, palm and coconut oil are commonly used in Malaysia and Indonesia (Sani et al. 2013), soybean oil is used in the United States and Brazil, and rapeseed oil is mostly used in Europe (Narasimharao et al. 2007; Sani et al. 2013). The major problems associated with these feedstocks are the food versus fuel controversy and their high cost.

Table 10.1 Different types of feedstock used for biodiesel production.

Type of feedstock	Examples
First-generation feedstock or edible oil	Rapeseed, apricot, soybean, sesame, coffee, peanut, grapeseed, sunflower, palm, linseed, coconut, olive, sesame, safflower, rice bran oil, etc.
Second-generation feedstock or non-edible oil	Jatropha, karanja, mahua, nahar, karabi, pongamia, rubber seed, babassu, jojoba, castor, etc.
Third-generation feedstock or algae	*Scenedesmus* sp., *Chlorella* sp., *Tetraselmis* sp., *Chaetoceros* sp., *Thalassiosira* sp., *Nannochloropsis* sp., etc.
Others	Animal fats (poultry fat, pork lard, beef tallow, etc.), fish oil, WCO, yellow grease, etc.

10.3.2 Second-Generation Feedstocks or Non-Edible Oils

In order to reduce production costs and the reliance on edible oil feedstocks, inedible oils are used as major sources for biodiesel production. These oils are not suitable for human consumption due to their toxic components. Non-edible oil plants can be grown in barren earth or wasteland, and their cost of cultivation is much lower than that of edible oil plants as intensive care is not required to maintain a reasonably high yield (Banković-Ilić et al. 2012; Atabani et al. 2013). Globally, more than 400 oil-bearing tree species are recognized as potential feedstocks for the production of biodiesel. The Government of India's (2018) biofuel policy, for example, envisages biodiesel production from non-edible oil seeds. Among plant species, *Jatropha curcas* and *Pongamia pinnata* (Karanja) have been found to be the most suitable for biodiesel production (Azam et al. 2005). During World War II, oil from jatropha seeds was used in blends with and substituted for petro-diesel (Sani et al. 2013). It has been reported that biodiesel produced from palm and jatropha oils has physical properties in the right balance, conferring adequate oxidation stability and cold performance (Sarin et al. 2007). In India and South East Asia, a number of tree-borne non-edible oil seeds are used as feedstocks for biodiesel production (Atabani et al. 2012).

10.3.3 Third-Generation Feedstocks or Algae

Recently, algae have gained traction as a potential feedstock for biodiesel production due to their high lipid content, which can be converted to biodiesel through transesterification (Sani et al. 2013). Due to their fast-growing nature, lack of competition for land, and lack of food versus fuel conflicts, algae are considered a promising feedstock for commercialization. They are capable of reducing the GHG effects via CO_2 sequestration (Chisti 2007). Compared with oilseeds, the harvesting and transportation costs of microalgae are relatively low. Algae can provide sufficient oil for global consumption: 1 acre of microalgae can produce 5000 gal of biodiesel annually, compared to only 70 gal from an equivalent area of soybean (Chisti 2007). Further, algae can flourish on land unusable for plant cultivation and lacking in fresh water. *Nannochloropsis*, a members of the marine green algae, is considered the most suitable candidate for biodiesel production. This strains has shown a high lipid content and biomass productivity. Algal oil yields vary by species, nutrient supply, and harvest time (Tran et al. 2010). However, the properties of the resulting FAMEs are not superior to those derived from plant oils, and further research into algal oils rich in saturated long-chain fatty acids is required in order to improve the quality of the final biodiesel (Knothe 2011).

10.3.4 Other Feedstocks

Other feedstocks include animal fats, fish oil, waste vegetable oil, and yellow grease. Their low cost and high availability make them suitable for reducing production costs. Biodiesel derived from WCOs has been gaining prominence due to its low cost compared to that from fresh vegetable oil. Because it has been found that feedstock alone contributes 60–80% of the overall biodiesel production cost (Borugadda and Goud 2012; Atadashi et al. 2012a,2012b), recent studies suggest that the cost of biodiesel could be halved through use of WCOs in comparison with virgin oils (Escobar et al. 2009). In 2018, technologies involving animal fat and waste oil provided about 6–8% of all biofuel output worldwide (Ingle et al. 2020). In the United Kingdom, recycled cooking oil is currently the main feedstock of choice for biodiesel production due to its low purchase price.

Among the advantages of using waste feedstocks for biodiesel production are that they help resolve waste management issues and reduce the environmental impacts associated with dumping in landfills. Around 15 million tons of waste cooking/frying oils are disposed of worldwide each year (Lee et al. 2014). However, due to their less predictable supply, high melting point, and viscosity, WCOs have been less extensively investigated than vegetable oils for use in biodiesel production (Lee et al. 2014). Also, some feedstocks such as beef tallow and pork lard may entail additional processing costs in order to produce an acceptable biodiesel (Kinast 2003).

10.4 Biodiesel Production Process

Transesterification is a chemical route to the production of methyl ester (i.e. biodiesel) from fatty acid found in vegetable oils or animal fats (Pathak 2015). It consists essentially in oxygenation reactions whereby triglycerides derived from animal fats and vegetable oils with alcohol in the presence of a homogeneous or heterogeneous catalyst are converted to FAME, a well-known biodegradable and renewable energy source (Jousseaume et al. 2003). This process has its widest application in the production of biodiesel. Originally, it was used to separate glycerine from fatty acids to produce soap (Tan et al. 2013).

Fatty acid chain composition influences NO_x emissions, with biodiesel containing esters of saturated fatty acids emitting less NO_x than petroleum diesel, and emissions increasing with the degree of unsaturation but decreasing with fatty acid chain length (Lee et al. 2014). Oxidation stability also depends upon the degree of unsaturation of fatty acid chains within the oil feedstock, since double bonds are prone to oxidation. Biodiesel produced from feedstocks containing linoleic and linolenic acid, with one or two bis-allylic positions, are highly susceptible to oxidation (Lee et al. 2014). The viscosity of biodiesel also increases with chain length and saturation of fatty acids within the feedstock (Allen et al. 1999), influencing fuel lubricity and flow properties. Low-viscosity biodiesel can be obtained from low-molecular-weight triglycerides, but such biodiesel cannot be used directly as a fuel due to its poor cold temperature flow properties. The lubricity of biodiesel increases with chain length and the presence of double bonds and alcohol groups. The feedstocks used for biodiesel production contain variable amounts of fatty acids (Table 10.2), along with variable amounts of water and FFA, affecting the activity of the catalyst due to poisoning (Fukuda et al. 2001; Lotero et al. 2005; Pathak 2015). To deal with the problem of catalyst poisoning due to water and FFA, a number of acid and enzymatic catalysts have been developed and reported, but the search is ongoing. In this section, different types of transesterification catalyst (acid, base, and other) are discussed.

10.4.1 Acid-Catalyzed Transesterification

Generally, the acid catalysts used for transesterification reactions are HCl, BF_3, H_2SO_4, and sulfonic acid (Otera 1993; Samios et al. 2009; Salvini et al. 2012). Usually, acid-catalyzed reactions are very much suitable for unrefined oil or waste oils. This process has not gained much attention commercially due to its slower rate of reaction and higher reaction temperature (above 100 °C) compared with base catalyst reactions. However, the introduction of the two-step production of biodiesel using acid catalysis has addressed this problem to a large extent (Salvini et al. 2012; Sasidharan and Kumar 2004). Acid-catalyzed transesterifications are advantageous over base-catalyzed ones in that the performance of the catalyst is not strongly affected by the presence of FFA in the feedstock. In fact, acid catalysts can simultaneously catalyze both esterification and transesterification reactions. Thus, a great advantage of acid catalysts is that they can directly produce biodiesel from low-cost lipid feedstocks, which are generally associated with high FFA concentrations (Srivastava and Prasad 2000).

Garcia et al. (2008) studied the transesterification of soybean oil and the simultaneous esterification of oleic acid with methanol and ethanol as a solvent in a high-pressure reactor using sulfonated zirconia as a solid acid catalyst. The sulfated zirconia was prepared by either the solvent-free method (S-ZrO_2) or the standard precipitation method (SZ). The conversion in alcoholysis catalyzed by S-ZrO_2 obtained under optimized conditions of 5 wt% catalyst loading (w/w), 120 °C reaction temperature, and one hour reaction time was 98.6% (methanolysis) and 92% (ethanolysis), respectively. The performance of ethanolysis was not as good as that of methanolysis due to the higher water content of ethanol (0.44%) compared to methanol (0.08%).

Tang et al. (2020) studied the cyclic utilization of biomass wastes from papaya seeds and corncob as catalyst supports for the synthesis of sulfonic acid catalyst used for biodiesel production through esterification and tranesterification of palm fatty acid distillate (PFAD) in the medium of methanol. From the BET result, the authors observed that the activated carbon (AC) support showed a porosity ranging between 639.68 and 972.66 m^2/g, decreasing after impregnation of the sulfonic acid group. The catalyst synthesized under a carbonization temperature of 600–1000 °C, sulfonation time of 0.5–2.5 hours, and sulfanilic acid to activated carbon (AC) weight ratio of 3 : 1–13 : 1 showed the highest FAME yield and a conversion of 72.09–93.49%. The optimized reaction conditions for maximum conversion and selectivity were found to be 30 : 1 alcohol to oil ratio, 5 wt% catalyst loading,

Table 10.2 Fatty acid composition profile of few feedstocks used for biodiesel production.

Vegetable oils	Fatty acid composition, wt%										FFA, wt%	References
	Lauric acid (C12 : 00)	Myristic acid (C14 : 00)	Palmitic acid (C16 : 00)	Palmitoleic acid (C16 : 01)	Stearic acid (C18 : 00)	Oleic acid (C : 18 : 01)	Linoleic acid (C18 : 02)	Linolenic acid (C18 : 03)	Arachidic acid (C20 : 00)	Erica acid (C22 : 0)		
Sunflower oil	–	–	3.7	–	3.1	83.5	9.6	0.1	–	–	10.2	Dossat et al. (2002)
Virgin olive oil	–	–	9.2	0.8	3.4	80.4	4.5	0.6	–	–	12.6	Lotero et al. (2005)
Soybean oil	–	0.1	12.5	–	5.2	23.5	47.8	10	–	–	15.5	Liu et al. (2008)
Palm oil	0.5	1.1	43.8	0.2	3.8	40.7	8.8	0.2	–	–	53.3	Kallio et al. (2001); Lotero et al. (2005)
Edible tallow	–	3.9	28.1	1.4	37.7	26.4	1.4	0.6	–	–	52.0	Kallio et al. (2001)
Cottonseed oil	–	0.7	20.1	0.1	2.6	19.2	55.2	0.6	–	–	13.5	Liang et al. (2010)
Rapeseed oil	–	–	3.3	–	1.8	27.32	11.87	6.5	10.54	39.42	4.4	Ma et al. (2008)
Safflower oil	–	–	10.78	–	3.4	59.34	16.2	2.39	0.59		7.4	Kumar et al. (2017)
Coconut oil	51.0	16.46	6.09	–	1.90	8.54	9.2	–	–	–	5.67	Benjapornkulaphong et al. (2009)
Madhuca indica (Mahua)	–	–	18.2	–	17.9	47.5	15	–	1.4	–	21.0	Bahadur et al. (2015)
White grease		–	23.3	3.5	11	47.1	11	1	–	–	37.8	Adewale et al. (2015)
Yellow grease	0.17	0.84	15.88	2.8	8.43	48.43	20.11	1.89	1.17	–	14.80	Plascencia et al. (1999)
Griddle grease	2.41	1.98	16.56	2.03	9.61	49.33	14.28	1.06	1.02	–	42.30	Plascencia et al. (1999)
Poultry fat	–	–	22.2	8.4	5.1	42.3	19.3	1	–	–	35.7	Lotero et al. (2005)
Jatropha oil	–	1.4	12.7	0.7	5.5	39.1	41.6	0.2	–	–	6.3	Singh and Singh (2010); Joshi et al. (2015)
Karanja oil	–	–	2.2	–	4.8	61.2	25.2	–	2.2	2.8	8.8	Joshi et al. (2015)
Mesua ferrea	–	1	19.1	–	13.6	49.2	13.0	3.42	–	–	0.1	Dhawane et al. (2019)
Camelina	0.4	5.2	8.7	–	3.5	20.6	5.1	30.5	19.5	–	1.58	Patil et al. (2010)
Calophyllum inophyllum	0.1	0.1	14.2	0.3	15.9	39.8	28.1	0.2	0.8	–	31.1	Ong et al. (2014)
Azadirachta indica (Neem)	–	–	18.1	–	18.1	44.5	18.3	–	–	–	–	Demirbas (2008)
Apricot	–	–	4.42	0.76	0.99	75.0	18.34	–	0.17	–	6.17	Dwivedi and Ram (2008)

100 °C reaction temperature, and four hours reaction time. Reusability analysis indicated that corncob waste-derived sulfonated AC catalyst was able to produce a relatively higher FAME yield during the first two reaction cycles. The higher catalytic efficiency showed high potential to provide a cost-effective and sustainable framework for biodiesel production.

Gardy et al. (2016) synthesized [$Ti(SO_4)O$] acid catalyst for both esterification and transesterification of FFAs in used cooking oil (UCO) to produce biodiesel. Referring to X-ray photoelectron spectroscopy (XPS) results, they claimed the bidentate sulfate coordinated to the Ti^{4+} metal in the nanocatalyst product. A novel method of metal impregnation was used to impregnate the sulfate group into TiO_2-NPs to synthesize titanium sulfate oxide or titanium oxide sulfate [$Ti(SO_4)O$] nanocatalyst, which is a super acidic catalyst that can be used for different applications due to its high activity and stability. The authors claimed a 97.1% FAME yield using the synthetized catalyst under the following reaction conditions: 9 : 1 methanol to UCO ratio, 1.5 wt% catalyst to UCO ratio, 75 °C reaction temperature, and three hours reaction time. The nanocatalyst showed appreciable catalytic activity for almost all investigated feedstock containing ≤6 wt% FFAs. The catalytic activity and reusability of the $Ti(SO_4)O$ for the esterification/transesterification of UCO were also investigated. The biodiesel produced further matched with American Society Testing and Methods (ASTM) and European Norm standards, showing satisfactory result as per the authors.

10.4.1.1 Mechanism of Acid-Catalyzed Transesterification

Although the acid catalyst is used for the transesterification of vegetable oil, an acid-catalyzed system is not a popular choice for commercial application of transesterification due to its slow rate of reaction, high temperature requirement, high alcohol to oil molar ratio, serious corrosion-related problems, and environmental harm (Fukuda et al. 2001; Tan et al. 2013). However, Lotero et al. (2005) investigated the reason behind the slow catalytic characteristics of the acid-catalyzed reaction and concluded it was a mechanism of acid-catalyzed transesterification of vegetable oil. The transesterification of triglyceride present in the oil undergoes three conjugative and reversible reactions: (i) protonation of the carbonyl group by the acid catalyst; (ii) nucleophilic attack of the alcohol, forming a tetrahedral intermediate; and (iii) proton migration and breakdown of the intermediate (Lam et al. 2010). In these reaction sequences, triglycerides are converted first to diglyceride and glyceride and then to glycerol, accomplished by the liberation of ester at every step. Figure 10.1 shows the transesterification mechanism of vegetable oil in the presence of acid catalyst.

The catalyst–substrate interaction causes protonation of carbonyl oxygen, which in turn increases the electrophilicity of the adjoining carbon atom, making the electrophilicity of the carbonyl group or carbocation more susceptible to nucleophilic attack. At the same time, the nucleophile from the methanol attacks the carbocation, forming a new tetrahedral ester intermediate, which eliminates the alcohol to form a new ester and regenerate the acid catalyst H^+ (Lotero et al. 2005; Pathak 2015).

10.4.2 Alkali- or Base-Catalyzed Transesterification

Currently, biodiesel is commonly produced from homogenous base catalysts such as sodium hydroxide or potassium hydroxide. These two catalysts are used industrially for

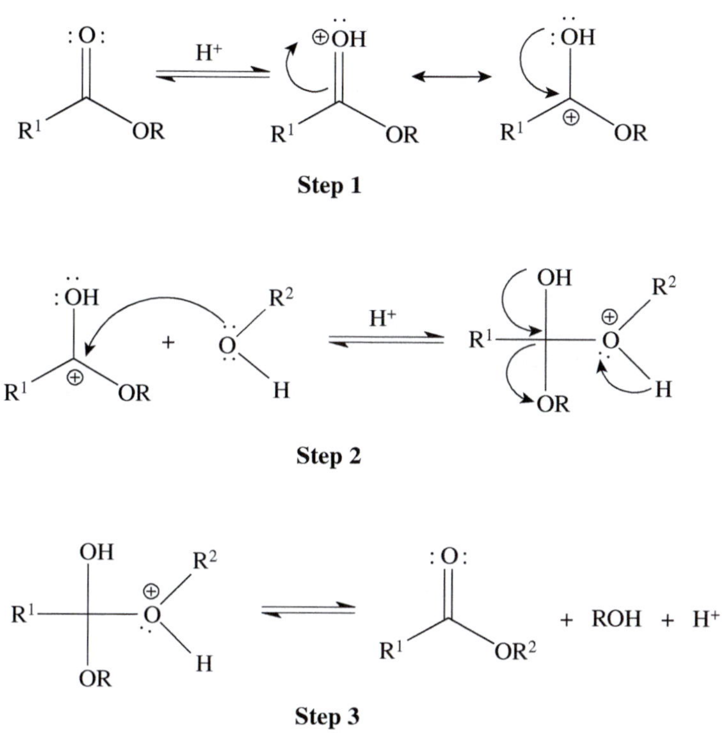

Figure 10.1 Mechanism of acid transesterification.

the following reasons: (i) ability to catalyze reactions at low reaction temperature and atmospheric pressure; (ii) ability to achieve high conversion in a minimal time; and (iii) wide availability and economic feasibility. However, the use of base catalyst is limited to refined vegetable oil only, where the FFA content ranges from 0.5% to 2% (wt/wt). This major drawback has turned the attention of researchers toward acid and other types of catalyst for use in transesterification reactions. Moreover, many prefer the use of heterogeneous rather than homogenous catalysts for biodiesel production due to the ease of catalyst separation and regeneration (Lotero et al. 2005; Lam et al. 2010).

Yadav and Sharma (2019) synthesized the heterogenous base catalyst barium aluminate ($BaAl_2O_4$) for transesterification of used vegetable oil. The catalyst showed 93.28% conversion of FAME under optimized reaction conditions of 21 : 1 methanol oil molar ratio, 4 wt% catalyst dose, 600 rpm stirring rate, 65 °C reaction temperature, and 150 minutes reaction time. Reusability analysis showed five cycles of reuue with up to 80% conversion efficiency. The biodiesel produced was matched with ASTM D 6751 in respect of important fuel properties such as acid value, density, kinematic viscosity, calorific value, flash point, fire point, cloud point, pour point, and cetane number, and was found to be satisfactory.

Du et al. (2019) developed a highly efficient and stable carbon-based magnesium oxide (MgO) solid base catalyst for the production of biodiesel from castor oil. The catalyst was prepared by the sol-gel method and calcined. Catalytic activity tests indicated excellent catalytic performance for the transesterification of castor oil with ethanol, with a FAME yield of 96.5%.

Junior et al. (2019) developed a $K_2CO_3/\gamma\text{-}Al_2O_3$ catalyst with different configurations of K_2CO_3 to produce biodiesel from sunflower oil in the medium of ethanol. To prepare the catalysts, boehmite was extruded with the aid of a binder and different percentages of K_2CO_3 active phase (15–45%) were impregnated on the supports for comparison. Transesterification reactions were carried out under conditions of 1 : 12 oil to ethanol molar ratio, 5 wt% catalyst loading, 80 °C reaction temperature, and four hours reaction time. The best result (99.3% conversion) was obtained with 35% K_2CO_3 in 65% $\gamma\text{-}Al_2O_3$. This catalyst showed a promising potential as an alternative to conventional homogeneous catalytic systems used for biodiesel production at industrial scale.

10.4.2.1 Mechanism of Alkali- or Base-Catalyzed Transesterification

The mechanism of base-catalyzed transesterification is very simple. Usually, alkali metal hydroxides such as NaOH and KOH are used as base catalysts for FAME production purposes via the transesterification route. However, the process is very sensitive to the water present in the vegetable oil or produced during the reaction as a result of reagent impurity. Water present or produced in the system reacts with alkyl esters to produce carboxylic acids, which react very rapidly with alkaline metals to form "soaps" such as $RCOO^-Na^+$. This soap formation reduces the yield of FAME and interrupts glycerol recovery during separation of biodiesel from the system (Freedman et al. 1986). Thus, a small amount of catalyst (approx. 1–2 mol%) can complete the whole reaction within a few hours of reaction time.

The steps involved in base-catalyzed transesterification of vegetable oil in the presence of solvent methanol are summarized in Figure 10.2. In the first, the base catalyst (B) reacts with the alcoholic solvent and produces alkoxide and protonated catalyst. This alkoxide group (OR^-) is primary responsible for producing biodiesel from the vegetable oil. In the second step, the triglyceride present in the vegetable oil undergoes nucleophilic attack of alkoxide OR^- . The nucleophile attacks very precisely on the carbonyl group of the triglyceride molecule and forms a tetrahedral intermediate. From this intermediate molecule, there later forms an alkyl ester and a diglyceride ion molecule, as shown in step 3. Then, the diglyceride ions react with protonated base catalyst (BH^+) and reaerate the catalyst sites. The diglyceride is then able to react with the alcohol medium and start the catalytic cycle again. Di- and monoglycerides are converted by the same mechanism to a mixture of alkyl esters and glycerol (Schuchardt et al. 1998).

10.4.3 Other Types of Transesterification

In the last decade, enzymatic transesterification, especially using lipase, has gained much attention from researchers due to the downstream processing problem posed by acid- and alkali-based transesterification. The major problems in these processes are the production of large amounts of water as byproduct and the difficulty in glycerol recovery, which eventually increases the overall biodiesel production cost. Moreover, they are not environmentally benign. In contrast, enzymatic catalysis doesn't generate any byproducts as a result of product-specific enzymatic actions, has much easier product recovery, offers mild reaction conditions, is insensitive to high-FFA oil, and offers easy catalyst regeneration (Kulkarni and Dalai 2006). These advantages have made this process a high-potential, eco-friendly, and suitable alternative to chemical processes for processing vegetable oil

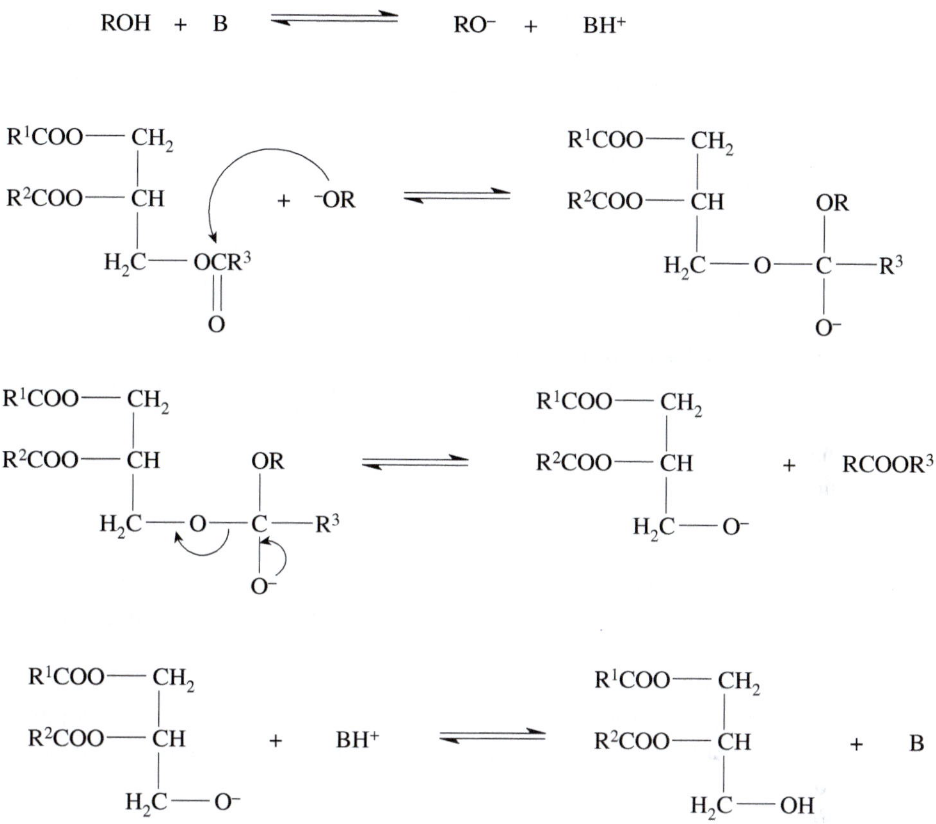

Figure 10.2 Mechanism of alkali (base)-catalyzed reaction.

to biodiesel. However, it still needs intensive research in order to be implemented at industrial scale, as well as optimization of the reaction parameters such as enzyme cost, reaction rate, and enzyme deactivation (Balaji and Chanda 1998). Various researchers have experimented on enzymes such as *Mucor miehei* (Lipozym IM 60), immobilized *Rhizopus oryzae* lipase, *Pseudomonas cepacia* (PS 30), and immobilized *Penicillium expansum* on resin D4020 and found them promising for use in the transesterification of vegetable oil to produce biodiesel through a green and efficient process (Nelson et al. 1996; Wu et al. 1999; Chen et al. 2009).

10.5 Variables Affecting Transesterification

The transesterification of vegetable oil is affected by a number of different factors, such as molar ratio, reaction temperature, reaction time, moisture and FFA content, and catalyst choice. One of the major variables affecting biodiesel yield is the molar alcohol to triglyceride ratio. The stoichiometric proportion required to complete the transesterification process is 3 mol of alcohol to 1 mol of glyceride, in order to produce 3 mol of fatty acid ester and 1 mol of glycerol. An excess amount of alcohol is needed to forward the equilibrium transesterification reaction to the right. In industrial processes, a molar ratio of 6 : 1 is commonly used for the optimum conversion to methyl ester. This ratio has no effect on the acid value or peroxide, saponification, and iodine values of the biodiesel produced. However, due to its increasing solubility, the greater molar ratio of alcohol to oil influences the separation of glycerol. With glycerol remaining in solution, it helps move the equilibrium back to the left, hence reducing the biodiesel yield (Ma and Hanna 1999; Sharma et al. 2008).

Temperature is another variable that affects the rate of reaction and the yield of esters. In the transesterification of alkali (NaOH or KOH), the temperature is held within 45–65 °C; values above that would burn the alcohol, resulting in a much lower yield, as the boiling point of methanol is 64.7 °C. Esterification, however, can occur at various temperatures, depending on the oil used (Leung and Guo 2006; Sharma et al. 2008).

If the duration of reaction increases, the conversion rate increases. Ma et al. (1998) investigated the effect of reaction time on the transesterification of beef tallow with methanol. They found that the reaction was very slow during the first minute due to the mixing and dispersal of the methanol into the tallow. However, it progressed very rapidly between one and five minutes, and the target was achieved at around 15 minutes. Freedman et al. (1984) performed transesterification of peanut, cotton, sunflower, and soybean oils under operating conditions of 6 : 1 methanol to oil molar ratio, 0.5% sodium methoxide catalyst, and 60 °C reaction temperature. They found an 80% biodiesel yield after one minute using soybean and sunflower oils. After one hour, the conversions for all of the oils were almost the same.

The FFA and moisture contents are crucial parameters for assessing the viability of the transesterification process. However, there are variations between studies recommending the selection of catalysts based on FFA: several authors recommend the use of alkaline catalyst with a feedstock content up to 5% FFA, while others suggest avoiding alkaline catalysts above 0.5% FFA. A higher FFA content of oil reduces the conversion efficiency of biodiesel. Feedstocks with a high content of FFA require two-step transesterification. In the first step, a pretreatment process catalyzed by acid transforms the FFA into FAME. In the second, an alkaline catalyst completes the reaction (Feuge and Gros 1949; Ma et al. 1998). Water content is also considered a critical variable in the transesterification process, as it causes soap formation, which consumes the catalyst and reduces its efficiency. Moreover, it contributes to an increase in viscosity and gel forming and hinders glycerol separation. Additionally, $NaOCH_3$ and NaOH react with carbon dioxide and moisture in the air, which decreases their efficacy (Canakci and Van Gerpen 1999; Demirbas 2006; Atadashi et al. 2012a,b).

10.6 Heterogeneous Nanocatalysts for Biodiesel Production

The research on development of environmentally friendly and economically feasible techniques for biomass conversion has focused on novel catalysts that are stable, easy to recycle, highly selective, and highly efficient (Chouhan and Sarma 2011). These catalysts can be categorized as either homogeneous or heterogeneous, acid or base. All have their own advantages and disadvantages. Homogeneous base catalysts offer high catalytic activity, short reaction times, and modest operating conditions, while homogeneous acid catalysts have suitability for feedstock with high FFA content such as WCOs, crude vegetable oils, and animal fats, provides insensitivity to the presence of FFA and water, and can be used in both transesterification and esterification reactions with no soap formation (Semwal et al. 2011). Although the use of homogeneous catalysts produces high yields of biodiesel under mild temperatures and short reaction times, they have shortcomings related to catalyst separation, reusability, and generation of excess wastewater. The use of heterogeneous catalysts is thus preferred, as they eliminate the cost associated with purification and separation and hence can be applied in commercial production. Wide ranges of heterogeneous catalysts, such as MgO loaded with KOH, Mg-Zn mixed-metal oxide, copper vanadium phosphate, sulfated zirconia, heteropoly acids, CaO, PbO, PbO_2, ZnO, Li-incorporated CaO, ZnO-La_2O_3, Co-doped ZnO, and hydrotalcites catalyst, have been developed and applied in biodiesel production. However, their activity is lower as compared to that of homogeneous catalyst (Borah et al. 2019a,b; Borges and Díaz 2012; Marwaha et al. 2018). Thus, the focus now is on developing novel improved catalysts that can overcome the shortcomings associated with both homogeneous and heterogeneous catalysts.

Today, research on heterogeneous catalysts is focused on developing novel solid catalysts in nanoscale range, in which mass transfer resistance is reduced due to their intrinsically large surface to volume ratio. The use of nanocatalysts appears to have gained special attention due to their high catalytic efficiency, and is capable of eliminating various problems related to homogeneous and heterogeneous catalysts. Nanocatalysts, having large surface areas, can solve various problems related to transesterification using other catalysts (Shuttleworth et al. 2014; Zuliani et al. 2018; Gardy et al. 2019a,b; Pandya et al. 2019). A number of methods can be used in their synthesis, including chemical vapor deposition, the sol-gel technique, co-precipitation, hydrothermal (conventional and microwave) methods, solvothermal methods, impregnation, high-temperature self-propagation, microwave combustion, vacuum deposition, electrochemical deposition, and gas condensation (Rizwanul Fattah et al. 2020). Compared to macroscopic catalysts, the amplified catalytic properties of nanoparticles are related to their high activity, stability, surface reactivity, high surface area, large pore size, and immunity to saponification. To provide a better understanding of the potentiality of nanocatalysis, three crucial points must be defined (Polshettiwar et al. 2011):

1. In order to initiate a chemical reaction, interactions between catalyst and reactants must take place. As the size of the particles in a nanocatalyst is small, the exposed surface area of the active component is large, which increases the contact between reactants and catalyst, enabling the heterogeneous catalytic system to achieve a reaction rate close to its homogeneous counterpart.
2. Because of the extremely small size of nanocatalysts, the ratio of surface area to volume increases during nanocatalysis, which increases catalyst activity and accessibility.
3. Materials produced at nanoscale particle size offer various novel fascinating characteristics not observed in their bulk counterparts, which are responsible for the efficacy and competency of nanocatalysts.

Nanocatalysts are proven to have great potential in the production of biofuel, allowing higher yield and selectivity and more favorable economics. As the surface area to volume ratio increases, there is an increase in low-coordinated atoms obtained on the surface, which are highly active for interaction with chemical adsorbates; hence, turnover frequency, which is a measure of the activity of the catalytic site, increases. Among nanocatalysts that have been successfully exploited and applied directly or indirectly for production of biodiesel are magnetic nanoparticles, metal nanoparticles, metal oxide nanoparticles, and acid-functionalized nanoparticles. Use of nanomaterials in biofuel production may provide a clean and cheap energy source in the near future and could lead to the establishment of a strong global industry (Jain et al. 2017; Sharma et al. 2018). Researchers have reported various nanocatalysts for production of biodiesel from various feedstocks, as summarized in Table 10.3.

10.7 Characterization of Nanoparticles Used for Biodiesel Production

The fundaments of nanotechnology lie in the fact that when the size of materials is reduced to nanometer range, their properties change dramatically. Such features make nanomaterials attractive for unique sensing applications, but at the same time cause complications in their characterization processes. Therefore, the challenge lies in choosing the right characterization techniques having optimum capabilities for studying the characteristics of nanomaterials. Two of the main parameters that are reviewed are shape and size. Moreover, the degree of aggregation, surface charge, and surface area can also be measured (Srivastava 2012; Joshi et al. 2008; Kalantar-zadeh and Fry 2008; Mourdikoudis et al. 2018). Various techniques can be used in the characterization of nanoparticles, as discussed in this section.

10.7.1 X-Ray Diffraction (XRD)

X-ray diffraction (XRD) provides a variety of information about the lattice structure of a crystalline substance, such as its crystallographic structure, unit cell dimensions, chemical composition, and bond angles. It is used for phase identification of crystalline materials, where it can reveal information on unit cell dimensions and atomic spacing.

10.7.2 Scanning Electron Microscopy (SEM)

Scanning electron microscopy (SEM) is the most utilized instrument for the characterization of nanomaterials. It can provide secondary electron images of organic and inorganic materials with nanoscale resolution, allowing topographical and morphological studies to be carried out by scanning an electron probe across a surface and monitoring the secondary electrons emitted. SEM can be used to keep track of the formation and growth of thin films and nanostructures. Nowadays, it is specifically designed for purposes ranging from routine morphological studies to cryogenic studies, high-speed compositional analyses, and the study of environmentally sensitive materials.

10.7.3 Energy Dispersive X-Ray Analysis (EDX)

Energy dispersive X-ray analysis (EDX) is an analytical technique used to analyze near-surface elements and determine their proportion at different positions, giving an overall mapping of the sample. The composition or amount of nanoparticles present near or at the surface can be determined using EDX, provided they contain some metal ions.

Table 10.3 Summary of nanocatalysts and operating conditions for biodiesel production.

Catalyst	Feedstock	Catalyst loading	Time	Temperature	Methanol to oil ratio	Conversion (%)	Yield (%)	References
Manganese-doped zinc oxide	Mahua oil	8 wt%	50 min	50 °C	7 : 1	–	97%	Baskar et al. (2017)
Iron (II)-doped zinc oxide	Castor oil	14 wt%	50 min	55 °C	12 : 1	–	91%	Baskar and Soumiya (2016)
Ag-ZnO nanoparticles	Simarouba oil	–	–	–	–	–	84.5%	Nagaraju et al. (2017)
Ni-doped ZnO	Castor oil	11 wt%	60 min	55 °C	8 : 1	–	95.20%	Baskar et al. (2018)
Co-doped ZnO	Mesua ferrea oil	2.5 wt%	3 h	60 °C	9 : 1		98.03%	Borah et al. (2019a,2019b)
$MgFe_2O_4$@CaO	Soybean oil	1.0 wt%	3 h	70 °C	12 : 1	–	98.3%	Liu et al. (2016)
Zn-doped CaO	Waste cooking oil	5 wt%	4 h	65 °C	20 : 1	–	96.74%	Borah et al. (2019b)
CaO-based/Au nanoparticles	Sunflower oil	3 wt%	3 h	65 °C	9 : 1	90 – 97%	–	Bet-Moushoul et al. (2016)
[Ti (SO_4)O]	Used cooking oil	1.5 wt%	3 h	75 °C	9 : 1	97.1%	–	Gardy et al. (2016)
TiO_2/RGO	Waste cooking oil	1.5 wt%	3 h	65 °C	12 : 1	98%	–	Borah et al. (2018)
Mesoporous $TiO_2/PrSO_3H$	Used cooking oil	4.5 wt%	9 h	60 °C	1.15 : 1	98.3%	–	Gardy et al. (2017)
$MgO/MgAl_2O_4$	Sunflower oil	3 wt%	3 h	110 ° C	12 : 1	95%	–	Vahid and Haghighi (2016)
WO_3/graphene quantum dot	Waste cooking oil	2 wt%	3.5 h	70 °C	6 : 1	–	(96.8 ± 0.16%)	Borah et al. (2020)
$ZnO/BiFeO_3$	Canola oil	4 wt%		65 °C	15 : 1	95.43%		Salimi and Hosseini (2019)
Diatomite@CaO/MgO	Waste edible oil	6 wt%	2 h	90 °C	15 : 1	–	96.47%	Rabie et al. (2019)

10.7.4 Transmission Electron Microscopy (TEM)

Transmission electron microscopy (TEM) provides information about the morphology, crystallography, particle size distribution, and elemental composition of a sample. It is capable of providing atomic-resolution lattice images and of giving chemical information at a spatial resolution of 1 nm or better. It is unique in that it can be used to focus on a single nanoparticle in a sample and directly identify and quantify its chemical and electronic structure. The most important application of TEM is in the atomic-resolution real-space imaging of nanoparticles. It is capable of creating both electron microscope images (information in real space) and diffraction patterns (information in reciprocal space) for the same region by adjusting the strength of the magnetic lenses. The electron diffraction pattern can be obtained from an area as small as several hundred to just a few nanometers in diameter by inserting a selected area aperture and using parallel incident beam illumination.

10.7.5 Atomic Force Microscopy (AFM)

Atomic force microscopy (AFM) is a microscopy technique used to create 3D images of surfaces at high magnification. It is capable of providing nanometer-scale surface roughness and of visualizing the surface nanotexture on many types of material surfaces, including polymer nanocomposites and nano-modified textiles.

10.7.6 Raman Spectroscopy

Raman spectroscopy is a spectroscopic technique used to examine rotational, vibrational, and other low-frequency modes in a system. It is based on inelastic scattering or Raman scattering of monochromatic laser light. The light interacts with the phonons or other excitations of the system, resulting in the energy of the laser photons being shifted up or down, which provides information about the phonon modes of the system.

10.7.7 Fourier Transform Infrared Spectroscopy (FT-IR)

Fourier transform infrared spectroscopy (FT-IR) is based on the measurement of the absorption of electromagnetic radiation with wavelengths in the mid-infrared region (4000–400 cm^{-1}). If a molecule absorbs IR radiation, its dipole moment is somehow modified and it becomes IR-active. A recorded spectrum gives the position of bands related to the strength and nature of the bonds and specific functional groups, thus providing information concerning molecular structures and interactions.

10.7.8 X-Ray Photoelectron Spectroscopy (XPS)

X-ray photoelectron spectroscopy (XPS) is a spectroscopic surface chemical analysis technique used to determine the empirical formula or elemental composition, chemical state, and electronic state of the elements on the surface (up to 10 nm) of a material. It can also elucidate the ligand exchange interactions and surface functionalization of nanoparticles and core–shell structures, and it operates under ultra-high vacuum conditions.

10.7.9 Thermogravimetric Analysis (TGA)

Thermogravimetric analysis (TGA) is a thermal analysis technique that measures the change in weight in a material as a function of temperature and time in a controlled environment. This is used in investigating the thermal stability of a material or in investigating its behavior in different atmospheres (e.g. inert or oxidizing). It is suitable for use with all types of solid materials, including organic and inorganic materials. Differential thermal analysis (DTA) is a calorimetric technique used to record the temperature and heat flow associated with thermal transitions in a material. This enables phase transitions to be determined (e.g. melting point, glass transition temperature, crystallization, etc.).

10.8 Influence of Nanoparticle Properties on Biodiesel Production

In addition to the operating parameters of the tranesterification of vegetable oils, different properties of catalysts have major effects on biodiesel production, including acidic and basic properties, hydrophilic and hydrophobic character, calcination temperature, size, surface area, and pore volume of nanoparticles (Mbaraka and Shanks 2006; Georgogianni et al. 2009). One of the major advantages of nanosized particles is that their physical and chemical properties can be optimized by controlling their shape and size (Saoud 2018; Bar et al. 2009).

The selectivity and activity of a catalyst depend upon its acidic or basic sites; however, it is difficult to assess the comparative significance of these two types of site in a reaction (Di Serio et al. 2006; Serio et al. 2008; Xie et al. 2007a,b). Omar and Amin (2011) found that a balanced acidity and surface basicity of Sr/ZrO_2 catalysts led to high FAME yields from WCO. Rahman et al. (2019) produced biodiesel from *Nannochloropsis* sp. lipid using several synthesized catalysts and found that both the total acidic site density and the total basic site density affected catalytic performance (Rahman et al. 2019). The hydrophilic/hydrophobic character of the catalyst surface is also an important factor in promoting a higher FAME yield. Catalyst surfaces possessing hydrophobic character are not easily deactivated as they avoid the adsorption of water, glycerol, and other polar compounds (Melero et al. 2006).

There is much literature to suggest that the calcination of a catalyst can improve its performance or restore its activity. The temperature at which calcination is performed has a vital effect on the structure and properties of the catalyst, which in turn influence its activity (Islam et al. 2013). Sistani et al. (2018) investigated the effect of calcination temperature ranging between 100 and 500 °C on the properties and performance of CaO/TiO_2–ZrO_2 nanocatalyst in an esterification reaction and found that a calcined sample at 400 °C showed the highest catalytic activity, converting 86.2% of oleic acid into ester. The calcination temperature increased the performance of the catalyst particles and provided uniform particles with a uniform scale (Sistani et al. 2018). Above the optimum condition, the conversion rate decreased, perhaps due to the adsorption of organic materials, which led to carbonization on the catalyst surface at high temperatures (Wan et al. 2009).

Catalyst loading also has an influence the biodiesel production process. Sistani et al. (2018) measured the impact of catalyst loading ranging from 1 to 5 wt% and found that conversion increased from 18.5% to 89.5% as loading increased from 1 to 3 wt%, but improved only marginally to 90.4% with further increase in loading, indicating that using more catalyst does not directly affect conversion.

MacLeod et al. (2008) transesterified rapeseed oil under the following conditions: 6 : 1 methanol to oil molar ratio, $LiNO_3/CaO$, $NaNO_3/CaO$, or KNO_3/CaO catalyst, and 60 °C reaction temperature. A yield of more than 90% was found after three hours for all three types of catalyst. However, reusability results showed some degree of catalyst leaching. The deactivation of catalysts is the main hurdle to the heterogeneous-catalyzed process. Several attempts have thus been made to develop a stable and effective catalyst for biodiesel production (Zong et al. 2007; Serio et al. 2008). However, this issue was improved to some extent when using a nanocatalyst. For example, Booramurthy et al. (2020) produced biodiesel from tannery waste using a nanosulfated zirconia catalyst (ferric manganese-doped sulfated zirconia). They found the catalyst gave a maximum yield above 90 wt% up to five cycles of reuse with a catalyst recovery of 88 wt%. A reduction in biodiesel yield was observed due to the leaching of active sites and the adsorption of debris from the catalyst surface (Booramurthy et al. 2020). However, a lower yield of FAME was observed after the fifth cycle of reusability. Similar results have been observed for other nanocatalysts (Vijaya Kumar et al. 2020; Kaur et al. 2018). Similarly, Reddy et al. (2016) used nanosized CaO catalyst in the transesterification of jatropha oil via a two-step process, finding an increase in biodiesel yield up to 98.54%. Moreover, the catalyst was reused for about nine cycles and maintained stability up to six with a yield of 95.8% (Reddy et al. 2016). In transesterification using magnetic nanocatalysts, leaching of the active species is a major challenge leading to low reusability. In magnetic nanocatalysts, leaching from the surface is associated with a partial dissolution of iron oxide nanoparticles (Ajala et al. 2020).

The transesterification of oil to biofuel can be significantly improved by improving the physical properties of the catalyst, such as its specific surface area and mean pore size (Xiao 2004; Akia et al. 2014). Based on various studies of the application of nanocatalysts in biodiesel production, it is apparent that a large porous catalytic surface increases contact between alcohol and oil, resulting in an increase in nanocatalytic efficacy (Akia et al. 2014). Compared with bulk catalysts, the high specific surface area of nanostructured materials is favorable for interaction between catalyst and substrates, effectively enhancing product yields (Akia et al. 2014). Increasing the surface area through surfactant-assisted nanoparticles results in a better FAME yield (Rahman et al. 2019). Furthermore, a catalyst with a high surface area could make reactions proceed at rates high enough to permit their commercial exploitation on a large scale. The lowest FAME yield is obtained with the lowest surface area and acid/basic site densities, and with small pore openings, which prevent voluminous triglyceride molecules from reaching the active site (Rahman et al. 2019). The existence of a small pore size restricts the diffusion of oil molecules into the catalytic sites within the porous channel and thus decreases biodiesel conversion. For smaller-pored materials, the accumulation of methanol and water around the hydrophilic sites impedes access to the catalyst by FFA ester molecules, resulting in a diffusion-controlled reaction.

10.9 Safety Issues Around the Application of Nanocatalysts in Biodiesel Production

Very little research has been done on the safety of the application of nanoparticles in the production of biodiesel. Nanoparticles can be released into the atmosphere during their application and synthesis, causing risk to human beings and the environment (Gupta et al. 2015). Due to the nano size of the particles, they are capable of penetrating and impacting the most sensitive areas of the human body or animal cells through ingestion and inhalation, and through intact and broken layers of the dermis (Tang et al. 2009; Pourmand and Abdollahi 2012), inhibiting normal functioning of the cells (Vishwakarma et al. 2010). Examples of the toxicity of some nanoparticles are mentioned in Antunes et al. (2017). Nanoparticles or nanocatalysts may also be accumulated by breathing in the lung tissues, leading to several respiratory disorders, decreased heart rate, and decreased touch responsiveness (Upadhyay et al. 2015). Although a comprehensive toxicity mechanism is not yet known, toxicity depends on particle size, shape, composition, dosage, structure, and surface cap (Antunes et al. 2017). The nanoparticles interact with the cell's lipid bilayer membrane at the cellular level, disturbing the regular functioning of the membrane, creating pits within it, and shaping the leakage of the cell material. If exposed to biological systems, nanoparticles will interact and invade the cells via the mitochondrial membrane, thereby disrupting their potential (Chen et al. 2009). Damage to mitochondria influences a cell's energy balance, disrupting normal metabolism. Nanoparticles also generate oxygen ions and other reactive oxygen species. The accumulated reactive radicals interact with proteins, especially with the enzymatic machinery. DNA damage caused by nanoparticles has also been reported to be induced by interaction with nanoparticles (Kim et al. 2009; Guan et al. 2012).

As already mentioned, the application of nanocatalysts in transesterification is an effective way of enhancing biodiesel production. In recent years, the use of nanocatalysts in this area has increased, resulting in increased risk of contamination affecting humans and the environment. Research is urgently needed on the subject of nanotoxicity, and legislation must be aimed at the use of nanomaterials in biofuel production. Safety assessment is of paramount importance in this area.

10.10 Future Perspectives

The replacement of fossil-based energy sources with clean and renewable alternatives is of grave importance. Of the various renewable energy sources, biomass can be considered the best alternative because of its environmental friendly nature. Various techniques are available for the production of biodiesel from different biomass feedstocks. The production and application of biodiesel is growing at a fast pace, in order to manage the increasing energy demands in all sectors, including residential, industrial, and transportation. Although biodiesel shows promise as an alternate fuel, there are some key challenges in biomass conversion that may pave the way for new research on the improvement of product quality and minimization of related environmental problems. It is clear that the introduction of nanotechnology in the conversion of biomass to biodiesel may play a significant role here, because of the various beneficial properties of nanoparticles, including their high specific surface area. However, there are certain problems that must be overcome before it can be applied, including the selection and preparation of nontoxic, cheap, and environmentally friendly nanoparticles and issues around their extraction and recycling. In the coming years, investigations must be made into computational, algorithmic, and optimization processes that may be useful in gathering information about favorable nanoparticles for biodiesel production and improving the quality of the biodiesel produced.

In the waste-to-biodiesel conversion process, a major challenge is to obtain a perfectly homogenous combination of catalytic nanoparticles and feedstocks. Moreover, in-depth study is required into the effect of nanoparticle size and morphology on the performance of the biodiesel preparation process. The potentiality of the application of nanoparticles in large-scale production of biodiesel from biomass, and the subsequent use of the biodiesel produced in engines, also needs to be investigated.

Another topic of analysis is the heavy cost of nanomaterials, which usually hampers their use in commercial applications. Moreover, it is vital to check the hazardous properties of nanocatalysts and related public health issues before they can be implemented in industrial sectors. Even though the biodiesel produced from biomass through application of nanocatalysts is considered environmentally friendly, a thorough evaluation is necessary in order to confirm the greenness of the process.

Furthermore, there are many challenges in the optimization of biodiesel quality and yield, with research needed on the improvement of cold flow properties, viscosity, and oxidation stability. High-quality, economic biodiesel can be synthesized by pre-blending of different feedstocks and post-blending of biodiesels from various sources. Therefore, analysis is required into the proper implementation of the type and amount of catalyst and other additives so that the performance of the biodiesel produced is comparable with that of fossil diesel.

10.11 Conclusions

This chapter has discussed available feedstocks for biodiesel production in relation to the necessities of the application of catalysts in the biodiesel process. Various types of catalytic process using homogeneous and heterogeneous acid/basic catalysts have been explored for the preparation of biodiesel from different feedstocks, but certain obstacles such as low reactivity, small pore size, low stability, long reaction times, and high reaction temperatures still need to be resolved. Nanotechnology has introduced the intermediate character of the nanocatalyst between homogeneous and heterogeneous catalysts, combining the high activity of the former with the easy recoverability of the latter. Information has been obtained on several nanocatalysts suitable for biodiesel production and the advantages of the implementation of nanotechnology in the biomass conversion process. In addition, the chapter has also described the challenges in terms of safety issues around the application of nanocatalysis, catalyst characterization methods, and future perspectives on the use of nanotechnology in biodiesel production. This will pave the way toward the investigation of new and novel nanocatalysts for the generation of green and renewable fuel from existing resources.

References

Adewale, P., Dumont, M., and Ngadi, M. (2015). Enzyme-catalyzed synthesis and kinetics of ultrasonic-assisted methanolysis of waste choice white grease for fatty acid methyl ester production. *Energy & Fuels* 29 (10): 6412–6421.

Ahmad, A.L., Yasin, N.M., Derek, C.J., and Lim, J.K. (2011). Microalgae as a sustainable energy source for biodiesel production: a review. *Renewable and Sustainable Energy Reviews* 15 (1): 584–593.

Ajala, E.O., Ajala, M.A., Ayinla, I.K. et al. (2020). Nano-synthesis of solid acid catalysts from waste-iron-filling for biodiesel production using high free fatty acid waste cooking oil. *Scientific Reports* 10 (1): 1–21.

Akia, M., Yazdani, F., Motaee, E. et al. (2014). A review on conversion of biomass to biofuel by nanocatalysts. *Biofuel Research Journal* 1 (1): 16–25.

Allen, C.A., Watts, K.C., Ackman, R.G., and Pegg, M.J. (1999). Predicting the viscosity of biodiesel fuels from their fatty acid ester composition. *Fuel* 78 (11): 1319–1326.

Al-Zuhair, S. (2007). Production of biodiesel: possibilities and challenges. *Biofuels, Bioproducts and Biorefining: Innovation for a Sustainable Economy* 1 (1): 57–66.

Antunes, F.A.F., Gaikwad, S., Ingle, A.P. et al. (2017). Bioenergy and biofuels: nanotechnological solutions for sustainable production. In: *Nanotechnology for Bioenergy and Biofuel Production* (eds. M. Rai and S.S. da Silva), 3–18. Cham: Springer.

Armaroli, N. and Balzani, V. (2007). The future of energy supply: challenges and opportunities. *Angewandte Chemie International Edition* 46 (1–2): 52–66.

Atabani, A.E., Silitonga, A.S., Badruddin, I.A. et al. (2012). Comprehensive review on biodiesel as an alternative energy resource and its characteristics. *Renewable and Sustainable Energy Reviews* 16 (4): 2070–2093.

Atabani, A.E., Silitonga, A.S., Ong, H.C. et al. (2013). Non-edible vegetable oils: a critical evaluation of oil extraction, fatty acid compositions, biodiesel production, characteristics, engine performance and emissions production. *Renewable and Sustainable Energy Reviews* 18: 211–245.

Atadashi, I.M., Aroua, M.K., Aziz, A.A., and Sulaiman, N.M.N. (2012a). Production of biodiesel using high free fatty acid feedstock. *Renewable and Sustainable Energy Reviews* 16 (5): 3275–3285.

Atadashi, I.M., Aroua, M.K., Aziz, A.A., and Sulaiman, N.M.N. (2012b). The effects of water on biodiesel production and refining technologies: a review. *Renewable and Sustainable Energy Reviews* 16 (5): 3456–3470.

Azam, M.M., Waris, A., and Nahar, N.M. (2005). Prospects and potential of fatty acid methyl esters of some non-traditional seed oils for use as biodiesel in India. *Biomass and Bioenergy* 18: 29–293.

Azócar, L., Heipieper, H.J., and Navia, R. (2010). Biotechnological processes for biodiesel production using alternative oils. *Applied Microbiology and Biotechnology* 88 (3): 621–636.

Bahadur, S., Goyal, P., Sudhakar, K., and Prakash, B.J. (2015). A comparative study of ultrasonic and conventional methods of biodiesel production from mahua oil. *Biofuels* 6 (1–2): 107–113.

Balaji, B.S. and Chanda, B.M. (1998). Simple and high yielding syntheses of β-keto esters catalysed by zeolites. *Tetrahedron* 54 (43): 13237–13252.

Balat, M. and Balat, H. (2010). Progress in biodiesel processing. *Applied Energy* 87 (6): 1815–1835.

Banković-Ilić, I.B., Stamenković, O.S., and Veljković, V.B. (2012). Biodiesel production from non-edible plant oils. *Renewable and Sustainable Energy Reviews* 16 (6): 3621–3647.

Bar, H., Bhui, D.K., Sahoo, G.P. et al. (2009). Green synthesis of silver nanoparticles using latex of *Jatropha curcas*. *Colloids and Surfaces A: Physicochemical and Engineering Aspects* 339 (1–3): 134–139.

Baskar, G. and Soumiya, S. (2016). Production of biodiesel from castor oil using iron (II) doped zinc oxide nanocatalyst. *Renewable Energy* 98: 101–107.

Baskar, G., Gurugulladevi, A., Nishanthini, T. et al. (2017). Optimization and kinetics of biodiesel production from Mahua oil using manganese doped zinc oxide nanocatalyst. *Renewable Energy* 103: 641–646.

Baskar, G., Selvakumari, I.A., and Aiswarya, R. (2018). Biodiesel production from castor oil using heterogeneous Ni doped ZnO nanocatalyst. *Bioresource Technology* 250: 793–798.

Benjapornkulaphong, S., Ngamcharussrivichai, C., and Bunyakiat, K. (2009). Al_2O_3-supported alkali and alkali earth metal oxides for transesterification of palm kernel oil and coconut oil. *Chemical Engineering Journal* 145: 468–474.

Bet-Moushoul, E., Farhadi, K., Mansourpanah, Y. et al. (2016). Application of CaO-based/Au nanoparticles as heterogeneous nanocatalysts in biodiesel production. *Fuel* 164: 119–127.

Bezergianni, S. and Dimitriadis, A. (2013). Comparison between different types of renewable diesel. *Renewable and Sustainable Energy Reviews* 21: 110–116.

Bhuyan, N., Sut, D., Gogoi, L. et al. (2019). Rural biorefinery: a viable solution for production of fuel and chemicals in rural India. *Sustainable Bioenergy* 2019: 21–47.

Booramurthy, V.K., Kasimani, R., Pandian, S., and Ragunathan, B. (2020). Nano-sulfated zirconia catalyzed biodiesel production from tannery waste sheep fat. *Environmental Science and Pollution Research* 27: 1–8.

Borah, M.J., Devi, A., Saikia, R.A., and Deka, D. (2018). Biodiesel production from waste cooking oil catalyzed by in-situ decorated TiO_2 on reduced graphene oxide nanocomposite. *Energy* 158: 881–889.

Borah, M.J., Das, A., Das, V. et al. (2019a). Transesterification of waste cooking oil for biodiesel production catalyzed by Zn substituted waste egg shell derived CaO nanocatalyst. *Fuel* 242: 345–354.

Borah, M.J., Devi, A., Borah, R., and Deka, D. (2019b). Synthesis and application of Co doped ZnO as heterogeneous nanocatalyst for biodiesel production from non-edible oil. *Renewable Energy* 133: 512–519.

Borah, M.J., Sarmah, H.J., Bhuyan, N. et al. (2020). Application of Box-Behnken design in optimization of biodiesel yield using WO_3/graphene quantum dot (GQD) system and its kinetics analysis. *Biomass Conversion and Biorefinery* 2020: 1–12.

Borges, M.E. and Díaz, L. (2012). Recent developments on heterogeneous catalysts for biodiesel production by oil esterification and transesterification reactions: a review. *Renewable and Sustainable Energy Reviews* 16 (5): 2839–2849.

Borugadda, V.B. and Goud, V.V. (2012). Biodiesel production from renewable feedstocks: status and opportunities. *Renewable and Sustainable Energy Reviews* 16: 4763–4784.

BP. (2014). BP Energy Outlook 2035. Available from https://www.bp.com/content/dam/bp/business-sites/en/global/corporate/pdfs/energy-economics/energy-outlook/bp-energy-outlook-2014.pdf (accessed December 19, 2020).

Canakci, M. and Van Gerpen, J. (1999). Biodiesel production via acid catalysis. *Transactions of ASAE* 42 (5): 1203–1210.

Chen, Y., Xiao, B., Chang, J. et al. (2009). Synthesis of biodiesel from waste cooking oil using immobilized lipase in fixed bed reactor. *Energy Conversion and Management* 50 (3): 668–673.

Chisti, Y. (2007). Biodiesel from microalgae. *Biotechnology Advances* 25 (3): 294–306.

Chouhan, A.S. and Sarma, A.K. (2011). Modern heterogeneous catalysts for biodiesel production: a comprehensive review. *Renewable and Sustainable Energy Reviews* 15 (9): 4378–4399.

Christophe, G., Kumar, V., Nouaille, R. et al. (2012). Recent developments in microbial oils production: a possible alternative to vegetable oils for biodiesel without competition with human food. *Brazilian Archives of Biology and Technology* 55: 29–46.

Demirbas, A. (2006). Biodiesel production via non-catalytic SCF method and biodiesel fuel characteristics. *Energy Conversion and Management* 47 (15–16): 2271–2282.

Demirbas, A. (2008). New liquid biofuels from vegetable oils via catalytic pyrolysis. *Energy Education Science and Technology* 21 (1–2): 1–59.

Dhawane, S.H., Chowdhury, S., and Halder, G. (2019). Lipase immobilised carbonaceous catalyst assisted enzymatic transesterification of *Mesua ferrea* oil. *Energy Conversion and Management* 184: 671–680.

Di Serio, M., Ledda, M., Cozzolino, M. et al. (2006). Transesterification of soybean oil to biodiesel by using heterogeneous basic catalysts. *Industrial & Engineering Chemistry Research* 45 (9): 3009–3014.

Dossat, V., Combes, D., and Marty, A. (2002). Lipase-catalysed transesterification of high oleic sunflower oil. *Enzyme and Microbial Technology* 30 (1): 90–94.

Du, L., Li, Z., Ding, S. et al. (2019). Synthesis and characterization of carbon-based MgO catalysts for biodiesel production from castor oil. *Fuel* 258: 116122.

Duarte, S.H., Ghiselli, G., and Maugeri, F. (2013). Influence of culture conditions on lipid production by *Candida* sp. LEB-M3 using glycerol from biodiesel synthesis. *Biocatalyst and Agricultural Biotechnology* 2: 339–343.

Dwivedi, D.H. and Ram, R.B. (2008). Chemical composition of bitter apricot kernels from Ladakh, India. *Acta Horticulture* 765: 335–338.

Eloka-Eboka, A.C., Igbum, G.O., and Inambao, F.L. (2017). Biodiesel methyl ester production and testing from selected African tropical seed oil feedstocks. *Energy Procedia* 142: 755–767.

Escobar, J.C., Lora, E.S., Venturini, O.J. et al. (2009). Biofuels: environment, technology and food security. *Renewable and Sustainable Energy Reviews* 13 (6–7): 1275–1287.

Feuge, R.O. and Gros, A.T. (1949). Modification of vegetable oils VII. Alkali catalysed interesterification of peanut oil with ethanol. *Journal of the American Oil Chemists' Society* 26 (3): 97–102.

Freedman, B.E.H.P., Pryde, E.H., and Mounts, T.L. (1984). Variables affecting the yields of fatty esters from transesterified vegetable oils. *Journal of the American Oil Chemists Society* 61 (10): 1638–1643.

Freedman, B., Butterfield, R.O., and Pryde, E.H. (1986). Transesterification kinetics of soybean oil. *Journal of the American Oil Chemists' Society* 63 (10): 1375–1380.

Fukuda, H., Kondo, A., and Noda, H. (2001). Biodiesel fuel production by transesterification of oils. *Journal of Bioscience and Bioengineering* 92 (5): 405–416.

Garcia, C.M., Teixeira, S., Marciniuk, L.L., and Schuchardt, U. (2008). Transesterification of soybean oil catalyzed by sulfated zirconia. *Bioresource Technology* 99 (14): 6608–6613.

Gardy, J., Hassanpour, A., Lai, X., and Ahmed, M.H. (2016). Synthesis of $Ti(SO_4)O$ solid acid nano-catalyst and its application for biodiesel production from used cooking oil. *Applied Catalysis A: General* 527: 81–95.

Gardy, J., Hassanpour, A., Lai, X. et al. (2017). Biodiesel production from used cooking oil using a novel surface functionalised TiO_2 nano-catalyst. *Applied Catalysis B: Environmental* 207: 297–310.

Gardy, J., Rehan, M., Hassanpour, A. et al. (2019a). Advances in nano-catalysts based biodiesel production from non-food feedstock. *Journal of Environmental Management* 249: 109316.

Gardy, J., Rehan, M., Hassanpour, A. et al. (2019b). Advances in nano-catalysts based biodiesel production from non-food feedstocks. *Journal of Environmental Management* 249: 109316.

Gaurav, A., Dumas, S., Mai, C.T.Q., and Ng, T.T. (2019). A kinetic model for a single step biodiesel production from a high free fatty acid (FFA) biodiesel feedstock over a solid heteropolyacid catalyst. *Green Energy and Environment* 4: 328–341.

Georgogianni, K.G., Katsoulidis, A.P., Pomonis, P.J., and Kontominas, M.G. (2009). Transesterification of soybean frying oil to biodiesel using heterogeneous catalysts. *Fuel Processing Technology* 90: 671–676.

Gharat, N. and Rathod, V. (2013). Ultrasound assisted enzyme catalyzed transesterification of waste cooking oil with dimethyl carbonate. *Ultrasonics Sonochemistry* 20: 900–905.

Government of India. (2018). National Policy on Biofuels. Ministry of Petroleum and Natural Gas. Available from http://petroleum.nic.in/sites/default/files/biofuelpolicy2018_1.pdf (accessed December 19, 2020).

Guan, R., Kang, T., Lu, F. et al. (2012). Cytotoxicity, oxidative stress, and genotoxicity in human hepatocyte and embryonic kidney cells exposed to ZnO nanoparticles. *Nanoscale Research Letters* 7 (1): 602.

Gupta, I.R., Anderson, A.J., and Rai, M. (2015). Toxicity of fungal-generated silver nanoparticles to soil inhabiting *Pseudomonas putida* KT2440, a rhizospheric bacterium responsible for plant protection and bioremediation. *Journal of Hazardous Materials* 286: 48–54.

Hama, S. and Kondo, A. (2013). Enzymatic biodiesel production: an overview of potential feedstocks and process development. *Bioresource Technology* 135: 386–395.

Ingle, A.P., Chandel, A.K., Philippini, R. et al. (2020). Advances in nanocatalysts mediated biodiesel production: a critical appraisal. *Symmetry* 12: 256.

Islam, A., Taufiq-Yap, Y.H., Chu, C.M. et al. (2013). Studies on design of heterogeneous catalysts for biodiesel production. *Process Safety and Environmental Protection* 91 (1–2): 131–144.

Jain, S., Pant, K., Kassaye, S., and Singh, S. (2017). Nanocatalysts: green fuel and chemical production. In: *Dekker Encyclopedia of Nanoscience and Nanotechnology* (eds. N. Srivastava, M. Srivastava, H. Pandey, et al.). Abingdon: Taylor & Francis.

Johnson, M.B. and Wen, Z. (2009). Production of biodiesel fuel from the microalga *Schizochytrium limacinum* by direct transesterification of algal biomass. *Energy & Fuels* 23 (10): 5179–5183.

Joshi, M., Bhattacharyya, A., and Ali, S.W. (2008). Characterization techniques for nanotechnology applications in textiles. *Indian Journal of Fibre and Textile Research* 33 (3): 304–317.

Joshi, G., Rawat, D.S., Lamba, B.Y. et al. (2015). Transesterification of Jatropha and Karanja oils by using waste egg shell derived calcium based mixed metal oxides. *Energy Conversion and Management* 96: 258–267.

Jousseaume, B., Laporte, C., Rascle, M.C., and Toupance, T. (2003). Dichlorodistannoxane transesterification catalysts, pure Lewis acids. *Chemical Communications* 23 (12): 1428–1429.

Junior, E.G.S., Perez, V.H., Reyero, I. et al. (2019). Biodiesel production from heterogeneous catalysts based K_2CO_3 supported on extruded γ-Al_2O_3. *Fuel* 241: 311–318.

Kalantar-zadeh, K. and Fry, B. (2008). Characterization techniques for nanomaterials. In: *Nanotechnology-Enabled Sensors* (eds. K. Kalantar-zadeh and B. Fry), 211–218. Cham: Springer.

Kallio, H., Yli-Jokipii, K., Kurvinen, J.P. et al. (2001). Regioisomerism of triacylglycerols in lard, tallow, yolk, chicken skin, palm oil, palm olein, palm stearin, and a transesterified blend of palm stearin and coconut oil analyzed by tandem mass spectrometry. *Journal of Agricultural and Food Chemistry* 49 (7): 3363–3369.

Karmakar, A., Karmakar, S., and Mukherjee, S. (2010). *Bioresource Technology* 101: 7201–7210.

Kaur, M., Malhotra, R., and Ali, A. (2018). Tungsten supported Ti/SiO_2 nano flowers as reusable heterogeneous catalyst for biodiesel production. *Renewable Energy* 116: 109–119.

Kim, Y.J., Choi, H.S., Song, M.K. et al. (2009). Genotoxicity of aluminum oxide (Al_2O_3) nanoparticle in mammalian cell lines. *Molecular & Cellular Toxicology* 5: 172–178.

Kim, K.H., Lee, O.K., and Lee, E.Y. (2018). Nano-immobilized biocatalysts for biodiesel production from renewable and sustainable resources. *Catalysts* 8: 68.

Kinast, J.A. (2003). Production of Biodiesels from Multiple Feedstocks and Properties of Biodiesels and Biodiesel/Diesel Blends: Final Report; Report 1 in a Series of 6 (No. NREL/SR-510-31460). National Renewable Energy Lab., Golden, CO.

Knothe, G. (2011). A technical evaluation of biodiesel from vegetable oils vs. algae. Will algae-derived biodiesel perform? *Green Chemistry* 13: 3048–3065.

Kulkarni, M.G. and Dalai, A.K. (2006). Waste cooking oil an economical source for biodiesel: a review. *Industrial & Engineering Chemistry Research* 45 (9): 2901–2913.

Kumar, N., Mohapatra, S.K., Ragit, S.S. et al. (2017). Optimization of safflower oil transesterification using the Taguchi approach. *Petroleum Science* 14 (4): 798–805.

Lam, M.K., Lee, K.T., and Mohamed, A.R. (2010). Homogeneous, heterogeneous and enzymatic catalysis for transesterification of high free fatty acid oil (waste cooking oil) to biodiesel: a review. *Biotechnology Advances* 28 (4): 500–518.

Lee, A.F., Bennett, J.A., Manayil, J.C., and Wilson, K. (2014). Heterogeneous catalysis for sustainable biodiesel production via esterification and transesterification. *Chemical Society Reviews* 43 (22): 7887–7916.

Lee, O.K., Seong, D.H., Lee, C.G., and Lee, E.Y. (2015). Sustainable production of liquid biofuels from renewable microalgae biomass. *Journal of Industrial and Engineering Chemistry* 29: 24–31.

Leung, D.Y.C. and Guo, Y. (2006). Transesterification of neat and used frying oil: optimization for biodiesel production. *Fuel Processing Technology* 87 (10): 883–890.

Liang, J.H., Ren, X.Q., Wang, J.T., and Li, Z.J. (2010). Preparation of biodiesel by transesterification from cottonseed oil using the basic dication ionic liquids as catalysts. *Journal of Fuel Chemistry and Technology* 38 (3): 275–280.

Liu, X., He, H., Wang, Y. et al. (2008). Transesterification of soybean oil to biodiesel using CaO as a solid base catalyst. *Fuel* 87: 216–221.

Liu, Y., Zhang, P., Fan, M., and Jiang, P. (2016). Biodiesel production from soybean oil catalyzed by magnetic nanoparticle $MgFe_2O_4$@ CaO. *Fuel* 164: 314–321.

Lotero, E., Liu, Y., Lopez, D.E. et al. (2005). Synthesis of biodiesel via acid catalysis. *Industrial & Engineering Chemistry Research* 44 (14): 5353–5363.

Luo, Z. and Zhou, J. (2012). Thermal conversion of biomass. In: *Handbook of Climate Change Mitigation* (eds. W.Y. Chen, J. Seiner, T. Suzuki and M. Lackner), 1002–1037. Berlin: Springer Science Business Media.

Ma, F. and Hanna, M.A. (1999). Biodiesel production: a review. *Bioresource Technology* 70 (1): 1–15.

Ma, F., Clements, L.D., and Hanna, M.A. (1998). The effects of catalyst, free fatty acids, and water on transesterification of beef tallow. *Transactions of the ASAE-American Society of Agricultural Engineers* 41 (5): 1261–1264.

Ma, H., Li, S., Wang, B. et al. (2008). Transesterification of rapeseed oil for synthesizing biodiesel by K/KOH/γ-Al_2O_3 as heterogeneous base catalyst. *Journal of the American Oil Chemists' Society* 85 (3): 263–270.

MacLeod, C.S., Harvey, A.P., Lee, A.F., and Wilson, K. (2008). Evaluation of the activity and stability of alkali-doped metal oxide catalysts for application to an intensified method of biodiesel production. *Chemical Engineering Journal* 135: 63–70.

Marwaha, A., Dhir, A., Mahla, S.K., and Mohapatra, S.K. (2018). An overview of solid base heterogeneous catalysts for biodiesel production. *Catalysis Reviews* 60 (4): 594–628.

Mbaraka, I.K. and Shanks, B.H. (2006). Conversion of oils and fats using advanced mesoporous heterogeneous catalysts. *Journal of the American Oil Chemists' Society* 83: 79–91.

Melero, J.A., Grieken, R.V., and Morales, G. (2006). Advances in the synthesis and catalytic applications of organosulfonic-functionalized mesostructured materials. *Chemical Reviews* 106: 3790–3812.

Mourdikoudis, S., Pallares, R.M., and Thanh, N.T. (2018). Characterization techniques for nanoparticles: comparison and complementarity upon studying nanoparticle properties. *Nanoscale* 10 (27): 12871–12934.

Nagaraju, G., Prashanth, S.A., Shastri, M. et al. (2017). Electrochemical heavy metal detection, photocatalytic, photoluminescence, biodiesel production and antibacterial activities of Ag–ZnO nanomaterial. *Materials Research Bulletin* 94: 54–63.

Narasimharao, K., Lee, A., and Wilson, K. (2007). Catalysts in production of biodiesel: a review. *Journal of Biobased Materials and Bioenergy* 1 (1): 19–30.

Nelson, L.A., Foglia, T.A., and Marmer, W.N. (1996). Lipase-catalyzed production of biodiesel. *Journal of the American Oil Chemists' Society* 73 (9): 1191–1195.

Omar, W.N. and Amin, N.A. (2011). Biodiesel production from waste cooking oil over alkaline modified zirconia catalyst. *Fuel Processing Technology* 92 (12): 2397–2405.

Ong, H.C., Masjuki, H.H., Mahlia, T.M.I. et al. (2014). Optimization of biodiesel production and engine performance from high free fatty acid *Calophyllum inophyllum* oil in CI diesel engine. *Energy Conversion and Management* 81: 30–40.

Osatiashtiani, A., Durndell, L.J., Manayil, J.C. et al. (2016). Influence of alkyl chain length on sulfated zirconia catalysed batch and continuous esterification of carboxylic acids by light alcohols. *Green Chemistry* 18 (20): 5529–5535.

Otera, J. (1993). Transesterification. *Chemical Reviews* 93 (4): 1449–1470.

Pandya, H.N., Parikh, S.P., and Shah, M. (2019). Comprehensive review on application of various nanoparticles for the production of biodiesel. *Energy Sources, Part A: Recovery, Utilization, and Environmental Effects* 2019: 1–4.

Pathak, S. (2015). Acid catalyzed transesterification. *Journal of Chemical and Pharmaceutical Research* 7 (3): 1780–1786.

Patil, P.D., Gude, V.G., Camacho, L.M., and Deng, S. (2010). Microwave-assisted catalytic transesterification of camelina sativa oil. *Energy & Fuels* 24 (2): 1298–1304.

Plascencia, A., Estrada, M., and Zinn, R.A. (1999). Influence of free fatty acid content on the feeding value of yellow grease in finishing diets for feedlot cattle. *Journal of Animal Science* 77 (10): 2603–2609.

Polshettiwar, V., Luque, R., Fihri, A. et al. (2011). Magnetically recoverable nanocatalysts. *Chemical Reviews* 111 (5): 3036–3075.

Pourmand, A. and Abdollahi, M. (2012). Current opinion on nanotoxicology. *DARU* 20 (1): 95.

Rabie, A.M., Shaban, M., Abukhadra, M.R. et al. (2019). Diatomite supported by CaO/MgO nanocomposite as heterogeneous catalyst for biodiesel production from waste cooking oil. *Journal of Molecular Liquids* 279: 224–231.

Rahman, N.J., Ramli, A., Jumbri, K., and Uemura, Y. (2019). Tailoring the surface area and the acid–base properties of ZrO_2 for biodiesel production from *Nannochloropsis* sp. *Scientific Reports* 9 (1): 1–2.

Razzak, S.A., Hossain, M.M., Lucky, R.A. et al. (2013). Integrated CO_2 capture, waste water treatment and biofuel production by microalgae culturing – a review. *Renewable and Sustainable Energy Reviews* 27: 622–653.

Reddy, A.N.R., Saleh, A.A., Islam, M.S. et al. (2016). Biodiesel production from crude Jatropha oil using a highly active heterogeneous nanocatalyst by optimizing transesterification reaction parameters. *Energy & Fuels* 30 (1): 334–343.

Rizwanul Fattah, I.M., Ong, H.C., Mahlia, T.M. et al. (2020). State of the art of catalysts for biodiesel production. *Frontiers in Energy Research* 8: 101.

Salimi, Z. and Hosseini, S.A. (2019). Study and optimization of conditions of biodiesel production from edible oils using ZnO/$BiFeO_3$ nano magnetic catalyst. *Fuel* 239: 1204–1212.

Salvini, A., Giomi, D., Cipriani, G. et al. (2012). Sustainable formation of fatty acid alkyl esters by transesterification of triglycerides with chlorotrimethylsilane. *RSC Advances* 2 (11): 4864–4868.

Samios, D., Pedrotti, F., Nicolau, A. et al. (2009). A transesterification double step process – TDSP for biodiesel preparation from fatty acids triglycerides. *Fuel Processing Technology* 90 (4): 599–605.

Sani, Y.M., Daud, W.M.A.W., and Abdul-Aziz, A.R. (2013). Biodiesel feedstock and production technologies: successes, challenges and prospects. *Biodiesel – Feedstocks, Production, and Applications* 10: 52790.

Santori, G., Nicola, G.D., Moglie, M., and Polonara, F. (2012). A review analyzing the industrial biodiesel production practice starting from vegetable oil refining. *Applied Energy* 92: 109–132.

Saoud, K. (2018). Nanocatalyst for biofuel production: a review. In: *Green Nanotechnology for Biofuel Production* (eds. N. Srivastava, M. Srivastava, H. Pandey, et al.), 39–62. Cham: Springer.

Sarin, R., Sharma, M., Sinharay, S., and Malhotra, R.K. (2007). Jatropha–palm biodiesel blends: an optimum mix for Asia. *Fuel* 86 (10–11): 1365–1371.

Sasidharan, M. and Kumar, R. (2004). Transesterification over various zeolites under liquid-phase conditions. *Journal of Molecular Catalysis A: Chemical* 210 (1–2): 93–98.

Schuchardt, U., Sercheli, R., and Vargas, R.M. (1998). Transesterification of vegetable oils: a review. *Journal of the Brazilian Chemical Society* 9 (3): 199–210.

Semwal, S., Arora, A.K., Badoni, R.P., and Tuli, D.K. (2011). Biodiesel production using heterogeneous catalysts. *Bioresource Technology* 102 (3): 2151–2161.

Serio, M.D., Tesser, R., Pengmei, L., and Santacesaria, E. (2008). Heterogeneous catalysts for biodiesel production. *Energy & Fuels* 22: 207–217.

Sharma, Y.C., Singh, B., and Upadhyay, S.N. (2008). Advancements in development and characterization of biodiesel: a review. *Fuel* 87 (12): 2355–2373.

Sharma, S., Saxena, V., Baranwal, A. et al. (2018). Engineered nanoporous materials mediated heterogeneous catalysts and their implications in biodiesel production. *Materials Science for Energy Technologies* 1 (1): 11–21.

Shuttleworth, P.S., Parker, H.L., Hunt, A.J. et al. (2014). Applications of nanoparticles in biomass conversion to chemicals and fuels. *Green Chemistry* 16 (2): 573–584.

Singh, S.P. and Singh, D. (2010). Biodiesel production through the use of different sources and characterization of oils and their esters as the substitute of diesel: a review. *Renewable and Sustainable Energy Reviews* 14 (1): 200–216.

Sistani, A., Saghatoleslami, N., and Nayebzadeh, H. (2018). Influence of calcination temperature on the activity of mesoporous CaO/TiO_2–ZrO_2 catalyst in the esterification reaction. *Journal of Nanostructure in Chemistry* 8 (3): 321–331.

Srivastava, R. (2012). Synthesis and characterization techniques of nanomaterials. *International Journal of Green Nanotechnology* 4 (1): 17–27.

Srivastava, A. and Prasad, R. (2000). Triglycerides-based diesel fuels. *Renewable and Sustainable Energy Reviews* 4 (2): 111–133.

Tan, H.W., Aziz, A.R.A., and Aroua, M.K. (2013). Glycerol production and its applications as a raw material: a review. *Renewable and Sustainable Energy Reviews* 27: 118–127.

Tang, J., Xiong, L., Wang, S. et al. (2009). Distribution, translocation and accumulation of silver nanoparticles in rats. *Journal of Nanoscience and Nanotechnology* 9 (8): 4924–4932.

Tang, Z.E., Lim, S., Pang, Y.L. et al. (2020). Utilisation of biomass wastes based activated carbon supported heterogeneous acid catalyst for biodiesel production. *Renewable Energy* 158: 91–102.

Thangaraj, B., Solomon, P.R., Muniyandi, B. et al. (2019). Catalysis in biodiesel production – a review. *Clean Energy* 3 (1): 2–23.

Tran, N.H., Bartlett, J.R., Kannangara, G.S.K. et al. (2010). Catalytic upgrading of biorefinary oil from micro-algae. *Fuel* 89: 265–274.

Upadhyay, S., Ganguly, K., and Palmberg, L. (2015). Wonders of nanotechnology in the treatment for chronic lung diseases. *Journal of Nanomedicine & Nanotechnology* 6 (6): 1.

Vahid, B.R. and Haghighi, M. (2016). Urea-nitrate combustion synthesis of $MgO/MgAl_2O_4$ nanocatalyst used in biodiesel production from sunflower oil: influence of fuel ratio on catalytic properties and performance. *Energy Conversion and Management* 126: 362–372.

Velasquez-Orta, S.B., Lee, J.G.M., and Harvey, A. (2012). Alkaline in situ transesterification of *Chlorella vulgaris*. *Fuel* 94: 544–550.

Vijaya Kumar, B., Ramesh, K., Deepalakshmi, S., and Sivakumar, P. (2020). Production of biodiesel from tannery waste using a stable and recyclable nano-catalyst: an optimization and kinetic study. *Fuel* 260: 116373.

Vishwakarma, V., Samal, S.S., and Manoharan, N. (2010). Safety and risk associated with nanoparticles – a review. *Journal of Managed Care Continuing Education* 9 (5): 455.

Wan, T., Yu, P., Wang, S., and Luo, Y. (2009). Application of sodium aluminate as a heterogeneous base catalyst for biodiesel production from soybean oil. *Energy & Fuels* 23: 1089–1092.

Wu, W.H., Foglia, T.A., Marmer, W.N., and Phillips, J.G. (1999). Optimizing production of ethyl esters of grease using 95% ethanol by response surface methodology. *Journal of the American Oil Chemists' Society* 76 (4): 517–521.

Xiao, F.S. (2004). Hydrothermally stable and catalytically active ordered mesoporous materials assembled from preformed zeolites nanoclusters. *Catalysis Surveys from Asia* 35: 151–159.

Xie, W., Huang, X., and Li, H. (2007a). Soybean oil methyl esters preparation using NaX zeolites loaded with KOH as a heterogeneous catalyst. *Bioresource Technology* 98 (4): 936–939.

Xie, W., Yang, Z., and Chun, H. (2007b). Catalytic properties of lithium-doped ZnO catalysts used for biodiesel preparations. *Industrial & Engineering Chemistry Research* 46: 7942–7949.

Yadav, M. and Sharma, Y.C. (2019). Transesterification of used vegetable oil using $BaAl_2O_4$ spinel as heterogeneous base catalyst. *Energy Conversion and Management* 198: 111795.

Zeng, X., Danquah, M.K., Chen, X.D., and Lu, Y. (2011). Microalgae bioengineering: from CO_2 fixation to biofuel production. *Renewable and Sustainable Energy Reviews* 5 (6): 3252–3260.

Zong, M.H., Duan, Z.Q., Lou, W.Y. et al. (2007). Preparation of a sugar catalyst and its use for highly efficient production of biodiesel. *Green Chemistry* 9: 434–437.

Zuliani, A., Ivars, F., and Luque, R. (2018). Advances in nanocatalyst design for biofuel production. *ChemCatChem* 10 (9): 1968–1981.

11

Application of Rare Earth Cation-Exchanged Nanozeolite as a Support for the Immobilization of Fungal Lipase and their Use in Biodiesel Production

Guilherme de Paula Guarnieri, Adriano de Vasconcellos, Fábio Rogério de Moraes and José Geraldo Nery

Physics Department, São Paulo State University – UNESP, São Paulo, Brazil

11.1 Introduction

Biodiesel is a biodegradable and renewable form of energy consisting of alkyl ester fatty acids. It can be produced through the transesterification of vegetable oils or animal fats with short-chain alcohols, such as methanol and ethanol, in the presence of a catalyst. Excellent reviews describing the chemistry of biodiesel and recent advances in biodiesel technology abound in the literature (Ma and Hanna 1999; Meher et al. 2006; Vyas et al. 2010; Abbaszaadeh et al. 2012; Hoekmana et al. 2012; Wong et al. 2019). Chemically, the transesterification reaction consists of a sequence of three consecutive reversible reactions: conversion of triglycerides to diglycerides, conversion of diglycerides to monoglycerides, and conversion of glycerides to glycerol. These reactions yield one ester molecule in each step, and the overall reaction can be summarized by the scheme displayed in Figure 11.1.

The current catalytic technologies applied to the production of biodiesel can be grouped into three categories: (i) base- and acid-catalyzed transesterification (Leung et al. 2010; Chouhan and Sarma 2011; Borges and Díaz 2012; Singh and Gaurav 2018); (ii) enzyme-catalyzed transesterification (Tan et al. 2010; Zhang et al. 2012; Guldhe et al. 2015); and (iii) supercritical alcohol transesterification (Tan and Lee 2011; Abbaszaadeh et al. 2012; Ang et al. 2014; Silva and Oliveira 2014; Marx 2016; Deshpande et al. 2017). Each of these classes of catalysts has its own advantages and disadvantages in terms of catalytic efficiency, separation and purification of the final product, catalyst turnover and reuse, and scale-up for industrial application.

Among the enzymatic catalysts, lipases are an attractive alternative for biodiesel production due to their high specificity for the transesterification of triacylglycerides to fatty acid methyl esters or fatty acid ethyl esters (FAMEs or FAEEs) in comparison with the other conventional chemical catalysts employed in industrial biodiesel production. Enzymes exhibit greater selectivity and higher catalytic activity in transesterification reactions under mild operational conditions (25–60 °C) compared to homogeneous basic catalysts. In addition, lipases can catalyze the transesterification of raw materials that contain high levels of free fatty acids (FFAs) and water, with decreased risk of soap and emulsion

Nano- and Biocatalysts for Biodiesel Production, First Edition. Edited by Avinash P. Ingle.

$$\text{Triglycerides} + CH_3CH_2OH \underset{}{\overset{\text{Catalyst}}{\rightleftharpoons}} \text{Diglycerides} + R_1COOCH_2CH_3$$

$$\text{Diglycerides} + CH_3CH_2OH \overset{\text{Catalyst}}{\rightleftharpoons} \text{Monoglycerides} + R_2COOCH_2CH_3$$

$$\text{Monoglycerides} + CH_3CH_2OH \overset{\text{Catalyst}}{\rightleftharpoons} \text{Glycerol} + R_3COOCH_2CH_3$$

Figure 11.1 Transesterification of triglycerides with alcohol. Source: Adapted from Ma and Hanna (1999).

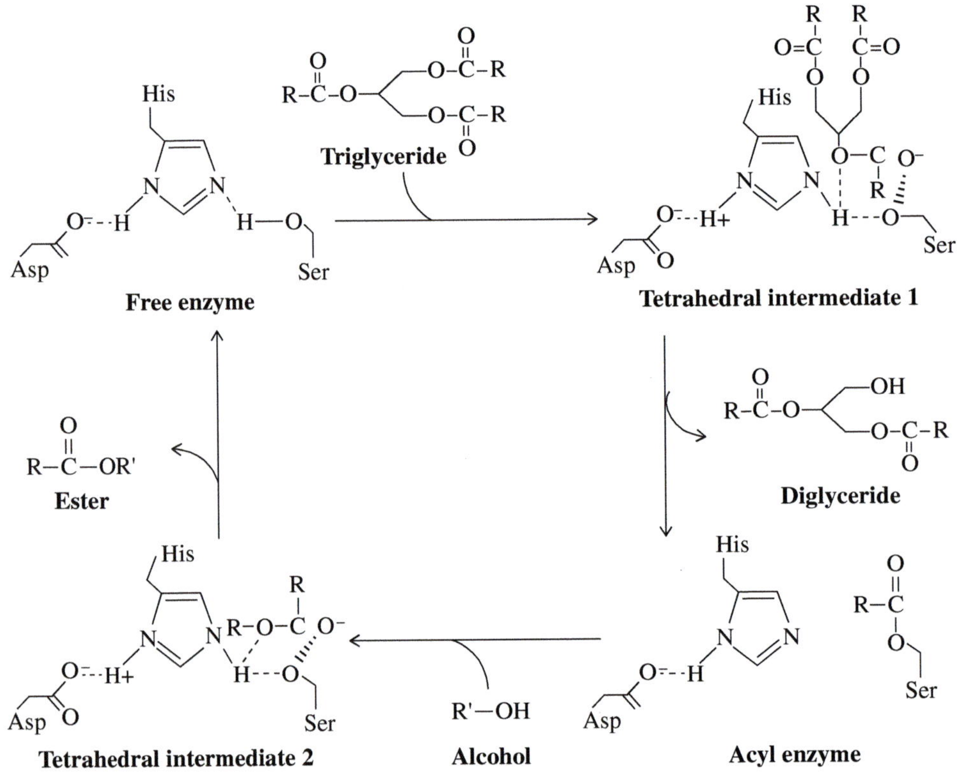

Figure 11.2 Mechanism of lipase in transesterification. Source: Jegannathan et al. (2008).

formation (Jegannathan et al. 2008). Lipases also have the ability to utilize all of the mono-, di-, and triglycerides and FFAs in the transesterification reaction, achieving high activity and yields in non-aqueous media, carrying out the reactions at low temperatures in short periods of time, and showing considerable alcohol resistance (Guldhe et al. 2015; de Vasconcellos et al. 2015, 2018). The classical mechanisms of the enzymatic transesterification of triacylglycerides into FAMEs or FAEEs are displayed in Figure 11.2.

Despite all of these advantages, some of the main stumbling blocks preventing increased use of enzymes for biodiesel production at the industrial scale are their expensive cost, difficulties in their separation and reuse after reaction, and the possible deactivation of their

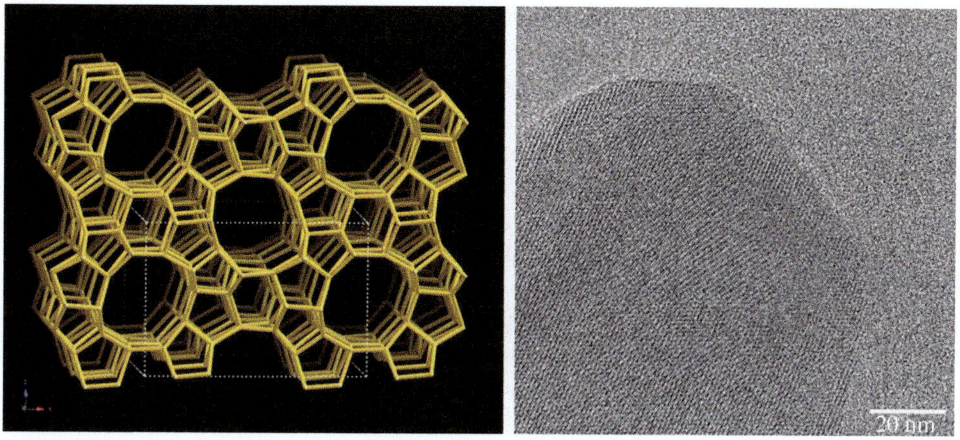

Figure 11.3 Nanozeolite MFI. Source: PhD thesis of de Vasconcellos (2015).

active sites by glycerol, which is a subproduct of the transesterification of triacylglycerides into FAEEs (Szczesna-Antczak et al. 2009; Singh and Singh 2010; Fan 2012).

One way to overcome these obstacles is to immobilize the enzymes on solid carriers (Bajaj et al. 2010; Tan et al. 2010; Rodrigues et al. 2013). Immobilization may result in improved stability of the enzymes toward harsh reaction conditions such as extreme pH, high temperature, or the presence of organic solvents, and enable their easier separation from the reaction media and subsequent reuse. However, the selection of suitable solid supports must take into account such aspects as the overall enzymatic activity of the immobilized enzyme, the cost of the immobilization procedure, identification of the best operating conditions to avoid enzyme inactivation, and the ability to regenerate and reuse the enzyme–support complex (Hanefeld et al. 2008; Kannan and Jasra 2011; Hartmann and Kostrov 2013).

Among a plethora of inorganic solid supports available for enzyme immobilization, micrometric and nanometric zeolites with different framework structures have been systematically studied and explored as enzymes carriers. The use of nanometric zeolites as solid supports for enzyme and protein immobilization has attracted the attention of researchers from a number of different fields, as the reduction of the zeolite particles' size from the micrometer to the nanometer scale can lead to substantial changes in their physicochemical properties, such as increasing their external surface area, allowing better access to and exposure of their active sites, increasing their dispersibility in both aqueous and organic solutions, and providing easy adjustment of their tunable surface properties, especially regarding surface charge and hydrophilicity/hydrophobicity (Tosheva and Valtchev 2005; de Vasconcellos et al. 2018). These properties, together with the typical high adsorption capacity and facile self-assembly character of nanozeolites, create a friendly microenvironment for enzyme and protein immobilization, thereby making them promising materials for use as solid supports for enzyme immobilization. Figure 11.3 shows an example of a nanostructured MFI zeolite.

Zeolite ions exchanged with rare earth cations have been extensively studied in the literature, especially with lanthanum- and cerium-exchanged faujasites, which are mostly used as fluid catalytic cracking catalysts (FCCs), as well as a few reports with neodymium

(Nd), samarium (Sm), gadolinium (Gd), and dysprosium (Dy) (Frising and Leflaive 2008). Nevertheless, applying such catalysts to the transesterification of triglycerides with methanol or ethanol to produce biodiesel has not been explored in the current literature (Shu et al. 2007; Chew and Bhatia 2009; Russbueldt and Hoelderich 2010; Borges et al. 2013; Vieira et al. 2013). Previous studies on the immobilization of fungal lipases such as *Thermomyces lanuginosus* lipase (TLL) and *Rhizomucor miehei* lipase (RML) on to nanozeolitic supports ion-exchanged with transition metals show a synergistic effect caused by the interaction between aspartic acid residues 102 and 158 of the TLL and the positively charged zeolitic support surface, which induced the stabilization of the opening of the enzyme lid and left its catalytic triad permanently exposed to the reaction medium (de Vasconcellos et al. 2018). In this chapter, the research work of authors on the interaction of TLL and RML immobilized on to nanozeolites ion-exchanged with rare earth cations such as lanthanum (La^{3+}), cerium (Ce^{3+}), gadolinium (Gd^{3+}), and dysprosium (Dy^{3+}), and the application of the nanozeolite–enzyme complexes as heterogeneous catalysts for the transesterification of microalgae oils to FAMEs, is presented in the form of a case study. Moreover, the potential of the new catalysts for biodiesel production is discussed.

11.2 Case Study

11.2.1 Origins of Materials and Enzymes

Flavorless high-oleic (>80%) oil extracted from the genetically modified heterotrophic algal strain *Prototheca moriformis* S2532 was obtained from Solazyme (Campinas, São Paulo, Brazil). The oil consisted of triglycerides (95%), small amounts of diglycerides (<5%), trace amounts of monoglycerides (<0.5%), and small quantities of fatty acid (1%). Ethanol (99.8%), phosphate buffer, gum arabic, inorganic salts, organic bases, and enzymes are usually purchased from commercial companies, such as Sigma-Aldrich, and used as received, without further purification. The results shown in the present work were achieved using the enzymes PALATASE 20000 l from RML and LIPOLASE 100 l from TLL, also purchased from Sigma-Aldrich. These are purified 1,3-specific lipases (EC 3.1.1.3) and are industrially produced through the submerged fermentation of a genetically modified *Aspergillus oryzae* (Russbueldt and Hoelderich 2010).

11.2.2 Preparation of Na-FAU Nanozeolites

High-aluminum-content, nanometer-sized Na-FAU zeolites were synthesized through hydrothermal crystallization in a temperature-controlled shaker following a procedure reported in the literature (de Vasconcellos et al. 2018). In a typical synthesis procedure, aluminosilicate gel is prepared by mixing together freshly prepared aluminate and silicate solutions at a molar ratio of 5.5 Na_2O : 1.0 Al_2O_3 : 4.0 SiO_2 : 190 H_2O. Aluminosilicate gel containing 5.34 g of NaOH, 2.42 g of $NaAlO_2$, 3.43 g of SiO_2, and 50.0 g of H_2O is generally used. Fumed silica is normally chosen as the silicate source and is directly mixed with a freshly prepared aluminate solution at room temperature, then immediately transferred to the temperature-controlled shaker. Hydrothermal crystallization is performed for two days at 60 °C, with a shaker rotation rate of 250 rpm. The powder product is recovered via centrifugation, washed with DI until achieving a pH <8, and dried at room temperature for 24 hours. The sample is referred to as Nano-FAU/Na nanozeolite.

11.2.3 Ion-Exchange Experiments

Prior to enzyme immobilization, the Nano-X/Na was subjected to ion-exchange experiments to replace the sodium cations with metal transition ones. These experiments took two different forms. The first was carried out according to the following conditions: 1 g of Nano-NaX was added to several Teflon-lined digestion reactors (Parr Instruments), followed by 30 ml of $La(NO_3)_3$, $Ce(NO_3)_3$, $Dy(NO_3)_3$, or $Gd(NO_3)_3$ (0.1 M) solution. Ion-exchange reactions were maintained at 60 °C for 16 hours, after which the products were centrifuged and the solids were washed with distilled water three times and dried at room temperature. These ion-exchanged derivatives were designated Nano-FAU/La/0.1 M, Nano-FAU/Ce/0.1 M, Nano-FAU/Dy/0.1 M, and Nano-FAU/Gd/0.1 M, respectively. The second procedure was similar to the first, except in the concentration of the salt solutions, which was 0.5 M, and the temperature and time of the reactions, which were 60 °C and 72 hours. After the reactions, all of the samples were again centrifuged, and the solids were washed three times with distilled water and dried at room temperature. The thus produced ion-exchanged derivatives were designated Nano-FAU/La/0.5 M, Nano-FAU/Ce/0.5 M, Nano-FAU/Gd/0.5 M, and Nano-FAU/Dy/0.5 M.

11.2.4 Enzyme Immobilization on to Nanozeolitic Supports

Lipase (PALATASE 20000 L from RML, molecular weight 31.6 KD, pI 3.8) and Lipolase 100 L (from TLL, molecular weight 30 KDa, pI 4.4) were immobilized on the nanozeolitic ion-exchanged derivatives with **La^{3+}** (Nano-FAU/La/0.1 M, Nano-FAU/La/0.5 M), **Ce^{3+}** (Nano-FAU/Ce/0.1 M, Nano-FAU/Ce/0.5 M), **Dy^{3+}** (Nano-FAU/Dy/0.1 M, Nano-FAU/Dy/0.5 M), and **Gd^{3+}** (Nano-FAU/Gd/0.1 M, Nano-FAU/Gd/0.5 M). The enzyme immobilization procedure can be found in the literature, although adaptation and optimization may be required (de Vasconcellos et al. 2015, 2018). The initial concentration of enzymes and amount of ion-exchanged nanozeolites before immobilization are typically around 2 mg/ml of enzyme per 200 mg of nanozeolite. The enzymes and nanozeolites were mixed in 5 ml of 0.3 M phosphate buffer pH 7 and stirred at 600 rpm for 16 hours at room temperature. The nanozeolite–enzyme complexes were separated via centrifugation at 13400 g, washed twice with de-ionized water, dried at 25 °C overnight, and stored at 4 °C for two days before being used in transesterification. The amount of immobilized enzymes adsorbed on the zeolitic matrixes, P_g (mg/g), was determined according to the method described by Bradford (1976), which uses bovine serum albumin as a standard, and is calculated as follows:

$$P_g = \frac{C_0 V_0 - C_f V_f}{w} \tag{11.1}$$

where C_0 is the initial protein concentration (mg/ml), C_f is the protein concentration in the filtrate (mg/ml), V_0 is the initial volume of the lipase solution (ml), V_f is the volume of the filtrate (ml), and w is the weight of the zeolitic support used (g) (Yagiz et al. 2007).

11.2.5 Physicochemical Characterization of As-Synthesized Nanozeolites and Nanozeolite–Enzyme Complexes

The prepared materials were characterized using X-ray diffraction (XRD) and scanning electron microscopy (SEM). The XRD analyses described here were performed with a MiniFlex II instrument (Rigaku, Tokyo, Japan) equipped with a rotating anode source with flat-plate Bragg–Brentano geometry and a graphite monochromator, operating at 40 kV and 40 mA, with Cu Kα radiation (wavelength = 1.5418 Å). The powder diffraction patterns were recorded in the 2θ range from 3° to 50°, with scanning at a goniometer rate of 2°/min.

Analyses of the XRD patterns (Figure 11.4) of the synthesized Nano-X/Na sample matches those previously reported in the literature (de Vasconcellos et al. 2018), but the ion-exchanged derivatives obtained with the rare earth cations under different ion-exchange conditions showed that the as-made Nano-X/Na was highly sensitive to the cation-dominant medium, ion concentration, and temperature condition parameters, due to the fact that the ion-exchanged derivatives of La^{3+} (Nano-FAU/La/0.1 M, Nano-FAU/La/0.5 M), Ce^{3+} (Nano-FAU/Ce/0.1 M, Nano-FAU/Ce/0.5 M), Dy^{3+} (Nano-FAU/Dy/0.1 M, Nano-FAU/Dy/0.5 M), and Gd^{3+} (Nano-FAU/Gd/0.1 M, Nano-FAU/Gd/0.5 M) had undergone drastic structural changes. In Figure 11.4, it can be seen that many of the Nano-FAU/Na original Bragg reflections were absent in the ion-exchanged derivative materials and that the ion-exchange experiments had caused some degree of amorphization in the samples.

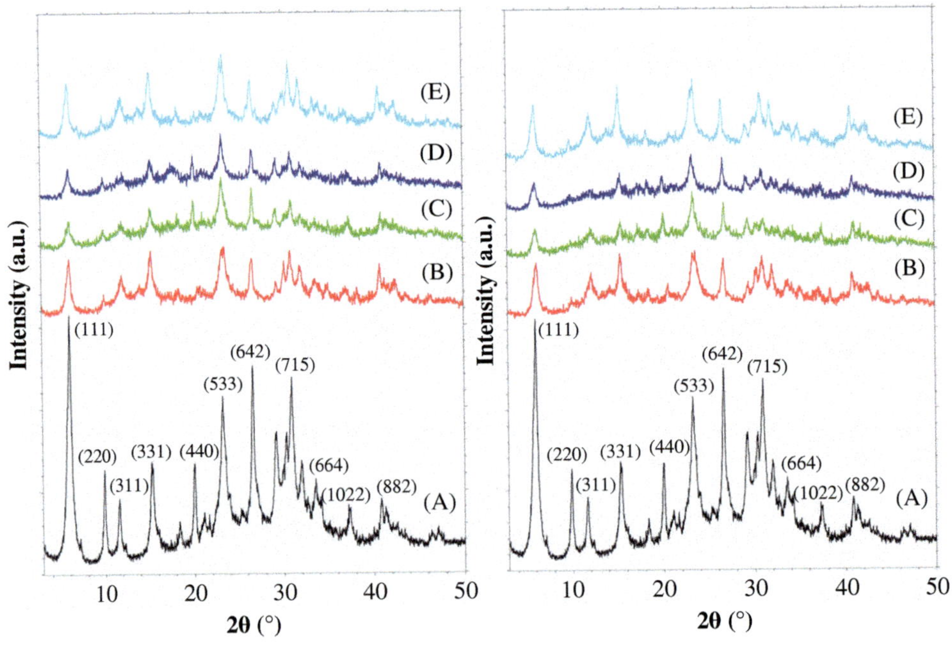

Figure 11.4 XRD patterns of as-made Nano-FAU/Na and its ion-exchanged derivatives with $\mathbf{La^{3+}}$ (Nano-FAU/La/0.1 M, Nano-FAU/La/0.5 M), $\mathbf{Ce^{3+}}$(Nano-FAU/Ce/0.1 M, Nano-FAU/Ce/0.5 M), $\mathbf{Dy^{3+}}$ (Nano-FAU/Dy/0.1 M, Nano-FAU/Dy/0.5 M), and $\mathbf{Gd^{3+}}$(Nano-FAU/Gd/0.1 M, Nano-FAU/Gd/0.5 M).

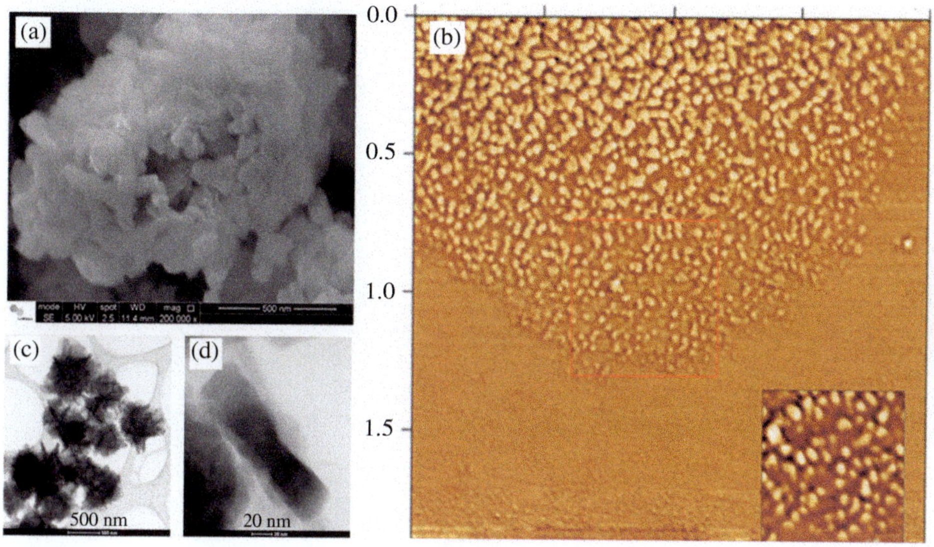

Figure 11.5 SEM, TEM, and AFM of as-made Nano-FAU/Na.

Analysis of the morphology of the nanozeolite Nano-X/Na were also performed by SEM, transmission electron microscopy (TEM), and atomic force microscopy (AFM) (Figure 11.5). AFM analyses were performed at the Brazilian National Laboratory of Nanotechnology (LNNano, Campinas), using a Dimension 3000 scanning probe microscope (SPM) equipped with a NanoScopeIIIa SMP controller (Digital Instruments Inc.). TESP tapping mode AFM images were acquired in ambient air using etched silicon probes. The particle size of the synthesized Nano-X/Na (Figure 11.5) was within the range of 20–100 nm and the AFM images of the Nano-X/Na indicated the formation of smaller nanometric aggregates of around 100–150 nm.

Fourier transform infrared microscopy (FT-IR) spectra were acquired using a PerkinElmer Frontier FT-IR spectrometer equipped with an ATR accessory. The samples were scanned 64 times between 4000 and 400 cm^{-1} at a resolution of 4 cm^{-1}. ATR-FT-IR spectra of the as-synthesized and functionalized nanozeolites are usually measured in the range 400–4000 cm^{-1} (Figure 11.6). According to the literature, infrared spectroscopy spectra for aluminosilicate zeolites can be divided into two distinct regions, corresponding to the skeletal IR spectrum (at 1600–500 cm^{-1}) and the surface hydroxyl groups (at 4000–1500 cm^{-1}) (Flanigen 1973). Analyses of the skeletal regions (1600–500 cm^{-1}) of the ATR-FT-IR spectra of the Nano-X/Na and its ion-exchange derivatives showed that the absorption bands for the as-synthesized nanozeolite structures were in agreement with the values reported in the literature. However, in rare earth ion-exchanged derivatives, the absorption bands in the skeletal region (at 425, 552, 795, 912, and 1064 cm^{-1}) were shifted from the positions expected for aluminosilicates. This is a clear indication that the ion-exchange treatment affected the original Nano-X/Na framework, corroborating the XRD results. Elemental chemical analyses of the nanozeolitic supports were performed using inductively coupled plasma atomic emission spectroscopy (ICP-AES) at the Chemical Analysis Laboratory Facility of Sao Paulo University.

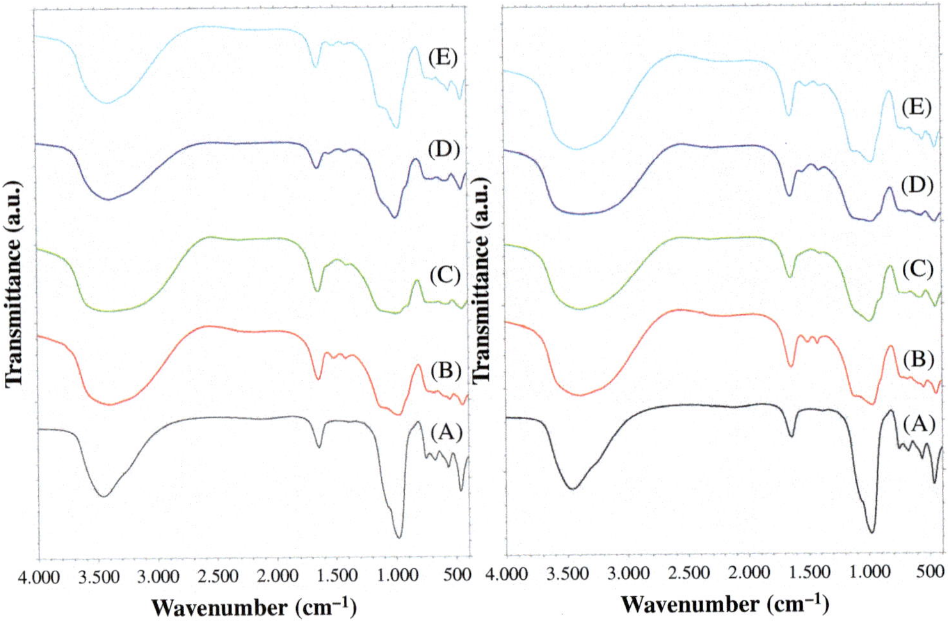

Figure 11.6 FT-IR of as-made Nano-FAU/Na and its ion-exchanged derivatives with **La^{3+}** (Nano-FAU/La/0.1 M, Nano-FAU/La/0.5 M), **Ce^{3+}**(Nano-FAU/Ce/0.1 M, Nano-FAU/Ce/0.5 M), **Dy^{3+}** (Nano-FAU/Dy/0.1 M, Nano-FAU/Dy/0.5 M), and **Gd^{3+}**(Nano-FAU/Gd/0.1 M, Nano-FAU/Gd/0.5 M).

11.2.6 Synthesis of FAAEs

Transesterification reactions of high-oleic heterotrophic microalgae oil to biodiesel were performed with all of the nanozeolite–enzyme biocatalysts produced: **La^{3+}** (Nano-FAU/**La**/0.1 M, Nano-FAU/**La**/0.5 M), **Ce^{3+}**(Nano-FAU/**Ce**/0.1 M, Nano-FAU/**Ce**/0.5 M), **Dy^{3+}** (Nano-FAU/**Dy**/0.1 M, Nano-FAU/**Dy**/0.5 M), and **Gd^{3+}**(Nano-FAU/**Gd**/0.1 M, Nano-FAU/**Gd**/0.5 M). Several experiments were performed in order to establish appropriate reaction conditions that would allow comparison among the performances. Transesterification reactions catalyzed by the different nanozeolite–enzyme complexes were performed at 40 °C using an oil to ethanol molar ratio of 1 : 5, with slow addition of ethanol in order to avoid inactivation of the catalysts. After a reaction time of 48 hours, the product (biodiesel) and byproduct (glycerol) were separated by centrifugation at 13400 g. The progress of the transesterification reactions was followed by using thin-layer chromatography (TLC), according to the procedure described by Yang et al. (2009). At predetermined time intervals, a small volume (100 μl) of the reaction mixture was collected and mixed with 500 μl of hexane for two minutes. After separation by centrifugation, 3 μl of the upper layer were applied to a silica gel plate. A solution of hexane/ethyl acetate/acetic acid (90 : 10 : 1) was used as the developing solvent and iodine was used as the color reagent. Gas chromatographic analyses were performed with a PerkinElmer Clarus580 gas chromatograph equipped with a flame ionization detector (GC-FID) and the quantification of fatty acid esters was performed according to EN 14103 (de Vasconcellos et al. 2018).

11.2.7 FAEE Yields Obtained with Nanozeolite Complexes

The amounts of TLL and RML immobilized on to the nanozeolitic supports and their respective enzymatic activities are shown in Table 11.1. Although the experimental data are quite reproducible, a straightforward correlation between immobilized enzyme amount, enzymatic activity, and FAEE yield cannot be given for the system under investigation. Analyses of the experimental data obtained for lipase from *R. miehei* immobilized on to the supports clearly indicate these nanozeolite–enzyme complexes are poor biocatalysts for the production of biodiesel. On the other hand, the enzyme from *T. lanuginosus* shows an opposite behavior, specially with regard to Nano-FAU/Dy/0.1 M, Nano-FAU/Dy/0.5 M, Nano-Fau/Gd/0.1 M, Nano-FAU/Gd/0.5 M, and Nano-FAU/Ce/0.5 M, which yield high amounts of FAEEs (Table 11.1). Attempts to correlate this behavior with the superficial area of the nanozeolitic supports ($\approx$114.61 m^2/g), microporous volume (0.310 cm^3/g), and enzymes structures have shed no light on the reasons behind the factors causing this behavior, as observed for nanozeolitic supports ion-exchanged with metal-transition cations (de Vasconcellos et al. 2018).

11.2.8 Model of Lipase Immobilization on to Zeolite Supports

Structural information about the lipase structures of a variety of organisms is available in the Protein Data Bank (PDB) (Berman et al. 2000), which contained over 3000 protein structures related to lipases as of August 2020. In particular, for lipases from *T. lanuginosus* and *R. miehei*, both open and closed conformations are solved. *T. lanuginosus* open (PDB id: 1DTE) and closed (PDB id: 1DT3) conformations were solved by Brzozowski et al. (2000), while *R. miehei* open (PDB id: 4TGL) and closed (PDB id: 3TGL) were solved by Derewenda et al. (1992) and Brzozowski et al. (1992), respectively. All structures were solved by XRD, at a resolution of 2.6 Å or better, giving insights into the interfacial activation of lipases during catalysis. According to Brzozowski et al. (2000), isomerization of the C22–C268 disulfide bond and R84 flipping are key steps in the transition from low-activity to activated form, referred to as a hinge rigid-body motion. Regarding lipase from *R. miehei*, Derewenda et al. (1992) describe how the hinge region is composed of two serine residues (83–84) and the segment over residues 91–95.

Electrostatics is known to play an important role in protein interaction with other molecules (Rocchia and Neshich 2007). Local charge distribution gives rise to an electrostatic potential surface that is important in stabilizing protein–substrate, protein–ligand, protein–protein (Rocchia and Neshich 2007), and protein–support interactions (Macario et al. 2007). To evaluate the electrostatic potential pattern on the lipase surface, the Poisson–Boltzmann equation is solved using specific software such as the Adaptive Poisson–Boltzmann Solver (APBS, version 1.3) (Baker et al. 2001). The parameters available on the APBS Web server may be tuned to simulate the ionic strength of positive and negative charges (radius 2 Å). The dielectric constants are usually set to 2.00 and 78.54 for protein and water, respectively. The default probe radius for surface assessment is 1.4 Å and the temperature may be chosen. As shown in Figure 11.7, electrostatic potential surface patterns from *T. lanuginosus* and RMLs show a region surrounding the lid and catalytic site that is rather neutral and positive, whereas the opposite surface is mainly negative. It

Table 11.1 Amounts of lipases immobilized, enzymatic activities, and FAEE yields generated by biocatalysts based on nanozeolite–enzyme complexes.

Materials	Activity per mg of complex (U/mg)	Enzyme immobilization (%)	FAEE (%)
TLL			
Nano-FAU/Na	1.3 ± 0.1	14.5 ± 1.6	Low ≥5
Nano-FAU/Ce/0.1 M	3.9 ± 0.4	14.6 ± 1.1	Low ≥5
Nano-FAU/Ce/0.5 M	20.0 ± 1.9	2.1 ± 0.8	78
Nano-FAU/Dy/0.1 M	26.7 ± 4.1	6.7 ± 1.0	75
Nano-FAU/Dy/0.5 M	15.6 ± 0.7	16.4 ± 1.2	70
Nano-FAU/Gd/0.1 M	19.2 ± 0.4	7.3 ± 1.0	65
Nano-FAU/Gd/0.5 M	27.8 ± 0.7	9.2 ± 0.8	75
Nano-FAU/La/0.1 M	4.5 ± 1.4	3.3 ± 0.9	Low ≥5
Nano-FAU/La/0.5 M	2.5 ± 0.2	6.1 ± 0.8	Low ≥5
RML			
Nano-FAU/Na	0.9 ± 0.1	12.2 ± 2.4	Low ≥2
Nano-FAU/Ce/0.1 M	3.0 ± 1.3	6.1 ± 2.6	Low ≥2
Nano-FAU/Ce/0.5 M	1.9 ± 0.1	11.0 ± 1.8	Low ≥5
Nano-FAU/Dy/0.1 M	13.0 ± 4.9	9.7 ± 1.3	20
Nano-FAU/Dy/0.5 M	4.5 ± 0.35	18.1 ± 2.1	Low ≥3
Nano-FAU/Gd/0.1 M	2.5 ± 0.2	27.2 ± 1.6	Low ≥5
Nano-FAU/Gd/0.5 M	5.3 ± 0.2	8.9 ± 1.6	Low ≥5
Nano-FAU/La/0.1 M	4.8 ± 0.3	1.7 ± 1.6	Low ≥5
Nano-FAU/La/0.5 M	3.7 ± 1.4	13.9 ± 2.5	Low ≥5

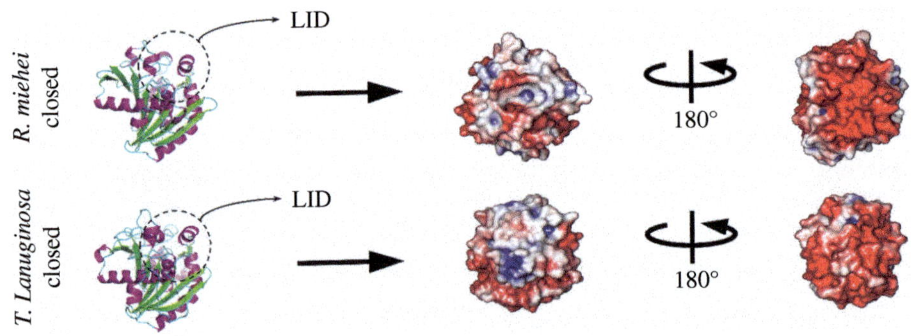

Figure 11.7 Structures from *T. lanuginosus* and *R. miehei* lipases in closed conformation. The electrostatic potential surfaces are observed in the catalytic site view and rotating 180°. The catalytic site face is mainly composed of neutral and positively charged areas, whereas the opposite face is primarily negatively charged.

is interesting to point out that the open lid reveals a hydrophobic (neutral) surface that is hidden from water in the closed conformation. The lid surface that is exposed to solvent in the closed conformation is negatively charged. The catalytic site is composed of Ser-209, Glu-341, and His-449, which display a positively charged environment surrounded by a neutral area in the open conformation state. Although protein structures feed biochemists and computational biologists with plenty of information to allow the study of interfacial activation, there is a lack of bioinformatic tools and experiments that allow a detailed description of lipase function when immobilized on zeolite supports. In order to come up with a model for this, and show how it could lead to interfacial activation, Macario et al. (2007) describe a situation where residual negative charges on the support interact with the enzyme N-terminal amino group. The interaction proposed could be either electrostatic or acid–base binding, depending on the support characteristics. Also, the authors propose that R86 will play an important role in lipase immobilization on the zeolite support. Indeed, Herrgård et al. (2000) have found that R86 residue is responsible for the stabilization of the open state conformation using a continuum electrostatics model of *R. miehei*. However, the results presented by Herrgård et al. are for the lipase simulation only, as generalized force fields for both proteins and zeolite supports are non-existent as yet. Although the model is generalized, the authors have used only lipase from *R. miehei* in their experimental approach.

On the other hand, de Vasconcellos et al. (2015) proposed a different model using bioinformatic tools and analysis of the lipase structures from *R. miehei* and *T. lanuginosus*. In order to understand the differences between immobilization and interfacial activation of both lipases, the authors used normal mode analysis (NMA) to compare low-frequency movements of the proteins. NMA is a simulation approach to protein dynamics and functional annotation that models protein atoms that are covalently bonded by a harmonic potential (Case 1994; Bahar and Rader 2005; Skjaerven et al. 2009). The force field with harmonic constants between each pair of atoms is well established and is available in protein simulation packages. By performing small displacements on each atom, the Hessian matrix, a $3\,N \times 3\,N$ matrix (N being the number of atoms), where the terms are the second derivative of the potential energy (approximated by a quadratic function), can be estimated and the eigenvalues and eigenvectors calculated. Eigenvectors give detailed information about collective movements of protein atoms that are inherent to protein structures. The first eigenvectors are related to the protein lowest-frequency movements and were evaluated for lipase interfacial activation.

The authors have described an allosteric mechanism where lid opening is connected to the region in the opposite face, comparable to the catalytic site view in Figure 11.7. Both TLLs and RMLs have a conserved negatively charged aspartic acid residue in position 158. Structurally close to this position, TLL has another aspartic acid residue in position 102, where RML has a shorter loop. The NMA model described by de Vasconcellos et al. (2015) proposes that positive Zeta potential zeolite supports will interact with lipases in this negatively charged region, stabilizing the open-state conformation. This is due to the result that the aspartic acid region moves collectively with the lid. Also, D102 residue, presented only in TLL, lies in the extension of the protein lid, potentializing the allosteric mechanism.

11.3 Conclusions

TLLs immobilized on nanozeolites ion-exchanged with rare earth cations are more efficient in the transesterification of triacylglycerides into biodiesel compared with RMLs. Modeling of the main forces leading to lipase immobilization is discussed in the literature, although there is no consensus on how lipase–nanozeolite systems are stabilized. Bioinformatics approaches have indicated that the reason is an allosteric interaction of zeolite support in the opposite region of the lipase lid, matching the support positive surface with negatively charged aspartic acids in the lipase composition. This interaction stabilizes the lid open conformation, improving FAEE yield. The model described by de Vasconcellos et al. (2015) correctly describes the results observed for rare earth cation-exchanged nanozeolites and FAEE yields for TLL and RML enzymes. The latter lack the negatively charged aspartic acid residue in position 102, which is responsible for a correct orientation during immobilization, and thus for improving enzymatic turnover rate.

References

Abbaszaadeh, A., Ghobadian, B., Omidkhah, M.R., and Najafi, G. (2012). Current biodiesel production technologies: a comparative review. *Energy Conversion and Management* 63: 138–148.

Ang, G.T., Tan, K.T., and Lee, K.T. (2014). Recent development and economic analysis of glycerol-free processes via supercritical fluid transesterification for biodiesel production. *Renewable and Sustainable Energy Reviews* 31: 61–70.

Bahar, I. and Rader, A.J. (2005). Coarse-grained normal mode analysis in structural biology. *Current Opinion in Structural Biology* 15: 586–592.

Bajaj, A., Lohan, P., Jha, P.N., and Mehrotra, R. (2010). Biodiesel production through lipase catalyzed transesterification: an overview. *Journal of Molecular Catalysis B: Enzymatic* 62: 9–14.

Baker, N.A., Sept, D., Joseph, S. et al. (2001). Electrostatics of nanosystems: application to microtubules and the ribosome. *Proceedings of the National Academy of Sciences* 98: 10037–10041.

Berman, H.M., Westbrook, J., Feng, Z. et al. (2000). The protein data bank. *Nucleic Acids Research* 28: 235–242.

Borges, M.E. and Díaz, L. (2012). Recent developments on heterogeneous catalysts for biodiesel production by oil esterification and transesterification reactions: a review. *Renewable and Sustainable Energy Reviews* 16: 2839–2849.

Borges, L.D., Moura, N.N., Costa, A.A. et al. (2013). Investigation of biodiesel production by HUSY and Ce/HUSY zeolites: influence of structural and acidity parameters. *Applied Catalysis A: General* 450: 114–119.

Bradford, M.M. (1976). A rapid and sensitive method for the quantitation of microgram quantities of protein utilizing the principle of protein-dye binding. *Analytical Biochemistry* 72: 248–254.

Brzozowski, A.M., Derewenda, Z.S., Dodson, E.J. et al. (1992). Structure and molecular model refinement of *Rhizomucor miehei* triacyglyceride lipase: a case study of the use of simulated annealing in partial model refinement. *Acta Crystallographica Section B: Structural Science* 48: 307–319.

Brzozowski, A.M., Savage, H., Verma, C.S. et al. (2000). Structural origins of the interfacial activation in *Thermomyces (Humicola) lanuginosus* lipase. *Biochemistry* 39: 15071–15082.

Case, D.A. (1994). Normal mode analysis of protein dynamics. *Current Opinion in Structural Biology* 4: 285–290.

Chew, T.L. and Bhatia, S. (2009). Effect of catalyst additives on the production of biofuels from palm oil cracking in a transport riser reactor. *Bioresource Technology* 100: 2540–2545.

Chouhan, A.P.S. and Sarma, A.K. (2011). Modern heterogeneous catalysts for biodiesel production: a comprehensive review. *Renewable and Sustainable Energy Reviews* 15: 4378–4399.

De Vasconcellos, A. (2015). Nanozeólitas como suportes sólidos para imobilização enzimática: Síntese, caracterização de complexos nanozeólitas/enzimas e sua aplicação como catalisadores heterogêneos para produção de biodiesel via rota etílica. Doctoral dissertation, São Paulo State University, São Paulo, Brazil.

De Vasconcellos, A., Bergamasco, J., Miller, A.H. et al. (2015). Potential new biocatalysts for biofuel production: the fungal lipases of *Thermomyces lanuginosus* and *Rhizomucor miehei* immobilized on zeolitic supports ion exchanged with transition metals. *Microporous and Mesoporous Materials* 214: 166–180.

De Vasconcellos, A., Miller, A.H., Aranda, D., and Nery, J.G. (2018). Biocatalysts based on nanozeolite-enzyme complexes: effects of alkoxysilane surface functionalization and biofuel production using microalgae lipids feedstock. *Colloids and Surfaces. B, Biointerfaces* 165: 150–157.

Derewenda, U., Brzozowski, A.M., Lawson, D.M., and Derewenda, Z.S. (1992). Catalysis at the interface: the anatomy of a conformational change in a triglyceride lipase. *Biochemistry* 31: 1532–1541.

Deshpande, S.R., Sunol, A.K., and Philippidis, G. (2017). Status and prospects of supercritical alcohol transesterification for biodiesel production. *WIREs Energy and Environment* 6: e252.

Fan, X. (2012). Enzymatic biodiesel production – the way of the future. *Lipid Technology* 24: 31–32.

Flanigen, E.M. (1973). A review and new perspectives in zeolite crystallization. *Advances in Chemistry* 121: 119–139.

Frising, T. and Leflaive, P. (2008). Extra framework cation distributions in X and Y faujasite zeolites: a review. *Microporous and Mesoporous Materials* 114: 27–63.

Guldhe, A., Singh, B., Mutanda, T. et al. (2015). Advances in synthesis of biodiesel via enzyme catalysis: novel and sustainable approaches. *Renewable and Sustainable Energy Reviews* 41: 1447–1464.

Hanefeld, U., Gardossi, L., and Magner, E. (2008). Understanding enzyme immobilization. *Chemical Society Reviews* 38: 453–468.

Hartmann, M. and Kostrov, X. (2013). Immobilization of enzymes on porous silicas – benefits and challenges. *Chemical Society Reviews* 42: 77–89.

Herrgård, S., Gibas, C.J., and Subramaniam, S. (2000). Role of an electrostatic network of residues in the enzymatic action of the *Rhizomucor miehei* lipase family. *Biochemistry* 39: 2921–2930.

Hoekmana, S.K., Brocha, A., Robbinsa, C. et al. (2012). Review of biodiesel composition, properties, and specifications. *Renewable & Sustainable Energy Reviews* 16: 143–169.

Jegannathan, K.R., Abang, S., Poncelet, D. et al. (2008). Production of biodiesel using immobilized lipase – a critical review. *Critical Reviews in Biotechnology* 28: 253–264.

Kannan, K. and Jasra, R.V. (2011). Improved catalytic hydrolysis of carboxy methyl cellulose using cellulase immobilized on functionalized meso cellular foam. *Journal of Porous Materials* 18: 409–416.

Leung, D.Y.C., Wu, X., and Leung, M.K.H. (2010). A review on biodiesel production using catalyzed transesterification. *Applied Energy* 87: 1083–1095.

Ma, F.R. and Hanna, M.A. (1999). Biodiesel production: a review. *Bioresource Technology* 70: 1–15.

Macario, A., Giordano, G., Setti, L. et al. (2007). Study of lipase immobilization on zeolitic support and transesterification reaction in a solvent free-system. *Biocatalysis and Biotransformation* 25: 328–335.

Marx, S. (2016). Glycerol-free biodiesel production through transesterification: a review. *Fuel Processing Technology* 151: 139–147.

Meher, L.C., Sagar, D.V., and Naik, S.N. (2006). Technical aspects of biodiesel production by transesterification – a review. *Renewable & Sustainable Energy Reviews* 10: 248–268.

Rocchia, W. and Neshich, G. (2007). Electrostatic potential calculation for biomolecules: creating a database of pre-calculated values reported on a per residue basis for all PDB protein structures. *Genetics and Molecular Research Journal* 6: 923–936.

Rodrigues, R.C., Ortiz, C., Berenguer-Murcia, A. et al. (2013). Modifying enzyme activity and selectivity by immobilization. *Chemical Society Reviews* 42: 6290–6307.

Russbueldt, B.M.E. and Hoelderich, W.F. (2010). New rare earth oxide catalysts for the transesterification of triglycerides with methanol resulting in biodiesel and pure glycerol. *Journal of Catalysis* 271: 290–304.

Shu, Q., Yang, B., Yuan, H. et al. (2007). Synthesis of biodiesel from soybean oil and methanol catalyzed by zeolite beta modified with La^{3+}. *Catalysis Communications* 8: 2159–2165.

Silva, C. and Oliveira, J.V. (2014). Biodiesel production through non-catalytic supercritical transesterification: current state and perspectives. *Brazilian Journal of Chemical Engineering* 31: 271–285.

Singh, A. and Gaurav, K. (2018). Advancement in catalysts for transesterification in the production of biodiesel: a review. *Journal of Biochemical Technology* 9 (1): 17–27.

Singh, S.P. and Singh, D. (2010). Biodiesel production through the use of different sources and characterization of oils and their esters as the substitute of diesel: a review. *Renewable and Sustainable Energy Reviews* 14 (1): 200–216.

Skjaerven, L., Hollup, S.M., and Reuter, N. (2009). Normal mode analysis for proteins. *Journal of Molecular Structure: THEOCHEM* 898: 42–48.

Szczesna-Antczak, M., Kubiak, A., Antczak, T., and Bielecki, S. (2009). Enzymatic biodiesel synthesis-key factors affecting the efficiency of the process. *Renewable Energy* 34: 1185–1194.

Tan, K.T. and Lee, K.T. (2011). A review on supercritical fluid (SCF) technology in sustainable biodiesel production: potential and challenges. *Renewable and Sustainable Energy Reviews* 15: 2452–2456.

Tan, T., Lu, J., Nie, K. et al. (2010). Biodiesel production with immobilized lipase: a review. *Biotechnology Advances* 28: 628–634.

Tosheva, L. and Valtchev, V.P. (2005). Nanozeolites: synthesis, crystallization mechanism, and applications. *Chemistry of Materials* 17: 2494–2513.

Vieira, S.S., Magriotis, Z.M., Santos, N.A.V. et al. (2013). Biodiesel production by free fatty acid esterification using lanthanum (La^{3+}) and HZSM-5 based catalysts. *Bioresource Technology* 133: 248–255.

Vyas, A.P., Verma, J.L., and Subrahmanyam, N. (2010). A review on FAME production processes. *Fuel* 89: 1–9.

Wong, K.Y., Jo-Han, N., Chong, C.T. et al. (2019). Biodiesel process intensification through catalytic enhancement and emerging reactor designs: a critical review. *Renewable and Sustainable Energy Reviews* 116: 109399.

Yagiz, F., Kazan, D., and Akin, A.N. (2007). Biodiesel production from waste oils by using lipase immobilized on hydrotalcite and zeolites. *Chemical Engineering Journal* 134: 262–267.

Yang, K.S., Sohn, J.-H., and Kim, H.K. (2009). Catalytic properties of a lipase from Photobacterium lipolyticum for biodiesel production containing a high methanol concentration. *Journal of Bioscience and Bioengineering* 107: 599–604.

Zhang, B., Weng, Y., Xu, H., and Mao, Z. (2012). Enzyme immobilization for biodiesel production. *Applied Microbiology and Biotechnology* 93: 61–70.

12

Lipase-Immobilized Magnetic Nanoparticles: Promising Nanobiocatalysts for Biodiesel Production

Tooba Touqeer[1], Muhammad Waseem Mumtaz[1] and Hamid Mukhtar[2]

[1]*Department of Chemistry, University of Gujrat, Gujrat, Pakistan*
[2]*Institute of Industrial Biotechnology, Government College University, Lahore-54000, Pakistan*

12.1 Introduction

Finding alternative energy sources is the need of hour, as the depletion of nonrenewable energy reserves has been accelerated with modernization, and emissions from the burning of fossil fuels are adversely disrupting natural phenomena. These environmental and economic issues have spurred the production of green, low-priced, renewable energy sources. Biofuels have emerged as substitutes for petroleum due to their renewable nature and biodegradability, along with many other factors (Ali et al. 2018). Biodiesel is considered a domestic fuel because the feedstocks used for its production are most likely to be the locally available oils. Thus, it can provide a country with energy security.

Biodiesel is a green fuel, produced from various edible, non-edible, and waste feedstock oils using a number of different procedures (e.g. micro-emulsification, pyrolysis). Transesterification is the most frequently employed method, utilizing different catalysts, each with its own pros and cons. Lipase is the most desirable, due to its environmental advantages, absence of side-reactions, and easy post-treatment. Further, lipase can convert free fatty acids (FFAs) present in waste and non-edible oils into alkyl-esters/biodiesel, which chemical catalysis cannot do. High cost is the main impediment to enzymatic transesterification at an industrial level. Moreover, lipase denatures easily in the presence of short-chain alcohols, which are required for transesterification of oils.

Immobilizing the enzyme on different supports can make the process cost-effective and enhance the catalytic stability of the enzymes. Immobilized enzyme can be regenerated and used several times, further reducing the cost of the process. Enzymes are immobilized on macro- as well as micro-supports for applications in different fields. Cross-linking the enzyme aggregates (CLEAs) is an effective and inexpensive immobilization strategy, but CLEAs are mechanically less stable. Enzyme immobilization on inorganic supports (i.e. beads of calcium alginate) provides physical stability and reusability, but thus enzymes

Nano- and Biocatalysts for Biodiesel Production, First Edition. Edited by Avinash P. Ingle.

immobilized have high mass transfer restrictions, showing lower catalytic activity. In order to recover activity and make the enzymes reusable, nanomaterials represent potent immobilization agents. They show superior Brownian movement and low mass-transfer restrictions, so the activity of enzymes immobilized on them is not compromised (Zhong et al. 2020). Metallic/metallic oxide nanoparticles with magnetic properties and low toxicity have been widely used for enzyme immobilization, because enzymes thus immobilized can be recovered very easily using an external magnet. Oxides of iron (i.e. magnetite and maghemite) are mostly used as magnetic nanosupports for enzyme immobilization because of their high magnetizing power and biocompatibility (Touqeer et al. 2020).

Studies show that naked nanoparticles are susceptible to oxidation and may have less interaction with enzymes. To overcome these issues, nanoparticle surfaces are modified with different functionalities. Surface-modified nanomaterials make more stable complexes with enzymes as compared to naked ones.

This chapter is aimed at discussing the feasibility and advantages of magnetic nanoparticles as lipase immobilization supports for the synthesis of biodiesel. The preparation and characterization of nanoparticles are also discussed. Finally, it gives an overview of catalyst development for biodiesel production.

12.2 Transesterification for Biodiesel Production

Transesterification is the most potent method of biodiesel production (Balat and Balat 2010). Different strategies have been developed by researchers in order to obtain higher yields, higher purities, and faster reaction rates. Generally, transesterification is classified into two basic methods: (i) with catalyst and (ii) without catalyst. The catalytic method is frequently used because the presence of the catalyst increases the solubility of oil in alcohol and elevates the reaction rate (Karmakar et al. 2010). Catalysts used in transesterification may be either homogeneous or heterogeneous.

12.2.1 Homogenous Catalysts

Homogenous catalysts may be acid or alkaline. Alkaline catalysts are preferred because of their low cost, fast reaction rate, and high product purity, but when using feedstock with a high FFA content, they do not work efficiently (Fukuda et al. 2001). Vicente et al. (2004) compared the catalytic activities of different alkaline catalysts, including sodium/potassium hydroxides and their methoxides, and obtained about 100% biodiesel using the methoxide catalysts. Ghadge and Raheman (2005) produced biodiesel through a two-steps process: acid esterification of high-FFA Mahua oil followed by transesterification using KOH as catalyst. An additional acid esterification step can help increase the biodiesel yield by reducing the FFA content of the feedstock; however, reaction time and cost are also increased. Uzun et al. (2012) optimized the transesterification reaction and got a maximum yield of 96% with NaOH catalyst, under optimum conditions of 7.5 : 1 methanol to oil ratio, 50 °C reaction temperature, and 30 minutes reaction time. Carlini et al. (2014) produced biodiesel from waste cooking oil (WCO) using acid and basic catalysts, with reaction yields of 96% and 94.3%, respectively.

12.2.2 Heterogeneous Catalysts

Heterogeneous catalysts include enzymes and alkaline earth metals, potassium zirconias, titanium silicates, and sulfated amorphous zirconia. The advantage of using heterogeneous catalysts is their reusability over several cycles. Kim et al. (2004) developed a $Na/NaOH/\gamma\text{-}Al_2O_3$ alkaline heterogeneous catalyst, which they used for the production of biodiesel. Guo et al. (2010) cretaed a calcinated sodium silicate for biodiesel production and found that it sustained its activity for five cycles. Carrero et al. (2011) used hierarchical beta zeolites for biodiesel production from *Nannochloropsis gaditana*-produced lipids. Omar and Amin (2011) studied different metal zirconias for biodiesel production, and found Sr/ZrO_2 was the most potent, although it still gave a low yield as compared to other conventional catalysts. Patel et al. (2013) investigated the influence of different reaction parameters on biodiesel production using sulfated zirconia as a heterogeneous catalyst.

12.2.3 Enzymatic Catalysts

Enzymes show high tolerance to FFAs. Thus, while using waste oils as feedstock, enzyme catalysis is highly preferable (Chen et al. 2006). The enzyme used for transesterification can be a lipase from any source. Lipases are a set of water-soluble enzymes that function as interfaces between nonpolar and polar liquids, mediating the hydrolysis of the ester bond. This distinctive heterogeneous reaction is possible as a result of (i) the 3D structure of lipase molecules, which consist of three domains: a contact domain for distinction of the substrate surface, a hydrophobic domain chargeable to extract one substrate molecule, and a functional domain containing the catalytic sites; and (ii) the transition in conformation from closed to open in triglyceride phase. The catalytic activity of lipase is due to the attack on the carbonyl group of the ester by the nucleophiles present in the molecule. The main hurdle to its use is the cost of the enzyme, which is very high for commercial biodiesel production (Guncheva and Zhiryakova 2011).

Many types of lipases have been used in biodiesel production, and several strategies have been adopted by researchers in an effort to make them cost-effective and reusable. Each has its own pros and cons. Iso et al. (2001) immobilized *Pseudomonas fluorescens* lipase on kaolinite particles. Noureddini et al. (2005) entrapped *Pseudomonas cepacia* lipase in a chemically inert support made of sol-gel. Dizge et al. (2009) created an economic and novel biocatalyst by immobilization of *Thermomyces lanuginosus* lipase on microporous polymer to produce biodiesel from palm oil, sunflower oil, and WCO. Li and Yan (2010) used immobilized *P. cepacia* lipase and reported a 96.22% biodiesel yield. Li (2011) used lipase immobilized on silica gel for transesterification of biodiesel and optimized the reaction parameters using response surface methodology (RSM). Mukhtar et al. (2016) synthesized lipase from *Rhizopus oryzae* and used it for the production of biodiesel.

Figure 12.1 shows methods of transesterification based on different catalysts.

12.3 Advantages of Using Magnetic Nanobiocatalysts

Enzymes can be immobilized on various surfaces, including beads of calcium alginate, ceramics, and inorganic matrixes, in order to make them thermally stable and reusable,

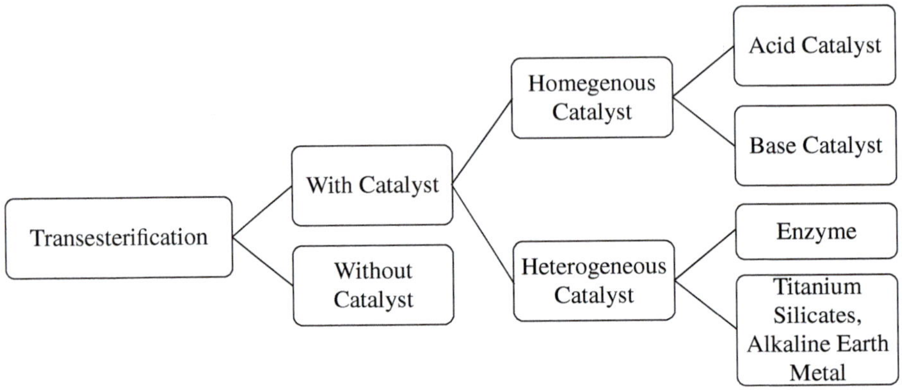

Figure 12.1 Methods of transesterification based on different catalysts.

but this will compromise their activity due to conformational restrictions and reductions in mass transfer. Immobilization of enzymes on magnetic nanoparticles has emerged as a potent method of biodiesel production because of the high specific surface area of nanoparticles, which provides high enzyme activity. Magnetic nanoparticles are also preferred because they allow easy recovery of the catalyst (Xie and Ma 2009). Nanoparticles of metals such as iron, nickel, manganese, cobalt, and chromium, as well as their oxides, possess considerable magnetic properties (Rossi et al. 2013). Among them, iron oxides are most commonly used for enzyme immobilization because of their biocompatibility, low toxicity, and strong magnetic moment. The advantages of using nanoparticles as enzyme supports are discussed in this section.

12.3.1 High Enzyme Loading and Surface Area to Volume Ratio

Nanomaterial shows worthwhile potential as compared with bulk materials as far as enzyme immobilization is concerned. Their high surface area to volume ratio and resulting high enzyme binding and biocatalytic activity make them more appropriate for enzyme immobilization than other materials (Hwang and Gu 2013). Enzyme loading depends upon the physical properties of the nanosupport and enzyme; for example, if some nanoparticles have a density of 5.18 g/cm^3 and an enzyme has a weight of 33 000 Da, mathematically about four enzyme molecules will bind to the surface of one nanoparticle (Shaw et al. 2006). Klein et al. (2012) used nano- and microsized chitosan particles to determine how enzyme loading is affected by particle size. Nanoparticles showed protein loading of 204 mg/g: much higher than that of macrosized particles, which were 40 mg/g of dry support only. Subsequently, the nanoparticles showed an enzyme activity of 18 330 units, while the macroparticles had an activity of 210 units. The study also indicated that, for enzyme immobilization, surface area to volume ratio is a better parameter as compared to surface area only because macroparticles show higher Brunauer–Emmett–Teller (BET) surface area (87 m^2/g) than nanoparticles (29 m^2/g) due to their mesoporosity. However, macroparticles also showed lower enzyme loading.

12.3.2 Low Mass Transfer Restriction and High Brownian Movement

The activity of enzyme immobilized on nanoparticles is greater than that of enzyme immobilized on macrosized support because of the reduced mass transfer restriction of nanostructures (Kim et al. 2008). Furthermore, the catalytic activity is significantly affected by the mobility of the catalyst and greater Brownian movement is revealed by monodispersed nanoparticles because of very small size, so the enzyme activity is significantly higher when nanomaterials are used as carriers as compared to bulk materials (Wang 2006).

12.3.3 Effortless Recovery and Reusability

Nanoparticles are potent enzyme carriers, but the recovery of nonmagnetic nanoparticles is difficult as it requires centrifugation or membrane separation. Superparamagnetic nanoparticles allow effortless recovery simply by using an external magnet. The recovered nanobiocatalysts can be reused after washing. For example, lipase immobilized on magnetic carriers can be reused for 10–15 cycles of transesterification (Osuna et al. 2015; Shi et al. 2016).

12.3.4 Stability

Enzyme immobilized on nanosurfaces exhibits adaptability toward harsh reaction conditions. Immobilized enzymes are thermally stable as compared to free enzymes. Similarly, immobilized enzymes are stable toward small pH and temperature fluctuations, perhaps due to bond formation between the enzyme and the surface (Dumri and Hung 2014; Touqeer et al. 2020).

12.4 Synthesis of Nanobiocatalysts

12.4.1 Preparation and Functionalization of Nanostructures

Nanomaterials are prepared by different techniques, which are evolving day by day. Basically, nanomaterial synthesis is divided into physical and chemical methods:

Physical methods: (i) Evaporation techniques; (ii) the lithographic process; (iii) use of hot and cold plasma; (iv) condensation of inert gas; (v) pulsed laser ablation; and (vi) spray pyrolysis.

Chemical methods: (i) Lyotropic template; (ii) sol-gel methods; (iii) electroless and electrochemical depositions; (iv) deposition of chemical vapors; and (vi) hydro/solvothermal methods (Rajput 2015; Rane et al. 2018).

The most widely used chemical methods for nanoparticles synthesis are sol-gel, co-precipitation, and hydro/solvothermal methods. Hydro- and solvothermal methods provide the edge of controlling the size and shape of nanoparticles simply by varying the time and temperature of reaction. In addition, nanostructures stable up to their melting point are produced by the solvothermal method (Chen et al. 2010). The reaction takes place at relatively low temperatures.

The solvothermal method for the synthesis of nanomaterials is defined as a precursor transformation in an enclosed solvent system at a temperature greater than the solvent boiling point. If the solvent used is water, the process is called the hydrothermal method. Choice of solvent and time and temperature of reaction are the crucial factors that affect the solvothermal process (Demazeau 2008). Metal oxides are mostly used for enzyme immobilization, and these are preferably produced by this method (Titirici et al. 2006; Chiu and Yeh 2007; Pang et al. 2019).

Metal oxide nanoparticles are chiefly used for catalysis because of their low price, low toxicity, and high chemical stability under various conditions. However, predicting their interaction with the ligand is difficult because of the complex surface chemistry. Moreover, the interaction of nanoparticles with different components and their agglomeration in different solvents depend upon their surface properties. Other limitations of using nanocatalysts include the formation of aggregates at certain pH levels and temperatures, toxicity, and maintenance of monodispersity. Therefore, in order to achieve better chemical, colloidal, and mechanical stability and less enzyme–particle interaction, nanoparticles can be modified with an organic moiety by various methods, such as adsorption of suitable surfactant and coating of a polymer on the surface (Kockmann et al. 2018). Polymers not only reduce the toxicity of nanoparticles by lowering their surface energy to make them biocompatible but also provide the functionality to the particle surface for the chemical attachment of the enzyme (Pavlidis et al. 2010). Commonly used methods for the grafting of polymers are as follows:

Grafting from: In this technique, the desired monomers are added in nanoparticle suspensions and self-assemble on the nanoparticle surface. Growth of the polymer chain also occurs at the surface. The polymerization process can be mediated by external mediators, or monomers can self-polymerize (Prucker and Rühe 1998).

Grafting to: In this strategy, a polymer chain with active functional groups reacts with nanoparticles and becomes attached to the surface (Francis, et al. 2014).

Organic, inorganic, natural, and synthetic polymers are used for the functionalization of nanoparticles. Xie and Ma (2009) used (3-aminopropyl) triethoxysilane to coat Fe_3O_4 nanoparticles in order to immobilize lipase upon them. Zhang et al. (2011) coated nanoparticles with amphiphilic polymers in order to change their phase interactions. Ren et al. (2011) investigated the feasibility of polydopamine coating on magnetic nanoparticles in order to immobilize lipase on them. Yang et al. (2017) coated magnetic iron oxide nanoparticles with polyethyleneimine, polyvinyl alcohol, and polyacrylic acid in order to form stable colloids. Lyu et al. (2019) immobilized *Bacillus subtilis* lipase on magnetic nanoparticles embedded in SiO_2.

12.4.2 Immobilizing Enzymes on Nanomaterials

There are a number of different techniques that can be applied to immobilize enzymes on a nanosupport. The most common ones are discussed in this section.

12.4.2.1 Adsorption Immobilization

In adsorption immobilization method, enzymes are adhered to nanosurfaces by means of weak physical interactions such as van der Waals forces and dipole–dipole interactions between their active sites. This method provides higher enzymatic activity, but the enzyme–matrix complex formed is not very stable (Hwang and Gu 2013; Eş et al. 2015). While these interactions are weak, however, they are large in number, providing adequate enzyme loading. On the other hand, enzyme loading is affected by the pH and ionic strength of the medium and by charges on the nanoparticles (Miletić et al. 2010).

Jiang et al. (2009) made nanoparticle-supported ionic liquids by covalent interaction between ionic liquids and silica-coated magnetic nanoparticles. Lipase was attached to the support by ionic adsorption. The resulting immobilized lipase showed 118% activity as compared to the free lipase. Ponvel et al. (2009) immobilized porcine pancreas lipase by adsorption on magnetic nanoparticles modified with alkyl benzene sulfonate. The nanobiocatalyst produced was reusable and easy to recover. Wang et al. (2012) immobilized *Candida rugosa* lipase by hydrophobic interactions on the surfaces of Fe_3O_4 nanoparticles modulated with alkyl silane; the lipase showed enhanced stability after immobilization and the nanobiocatalyst sustained 65% of its initial activity after seven cycles. Shi et al. (2016) tested different solvents for adsorption of lipase on $Fe_3O_4@SiO_2$ nanoparticles. Toluene was proved to be the most efficient, providing 97% immobilization and 124% relative activity. The nanobiocatalyst was reusable for 10 cycles with residual activity of 67%. Adeogun (2017) studied the thermodynamics of enzyme adsorption on $ZnFe_2O_4$ nanoparticles, finding that $ZnFe_2O_4$ is an efficient nanosupport for adsorption of lipase, protease, and amylase.

12.4.2.2 Covalent Immobilization

In this technique, the enzyme is attached to the nanosupport through covalent linkage. This provides a very stable enzyme–support complex, so leaching is well prevented. However, activity may or may not be affected by conformational restriction (Ahmad and Sardar 2015). Sometimes, nanoparticles are functionalized by certain groups to afford active sites for attachment of enzymes. Another possible modification is addition of a spacer before the enzyme to reduce the steric hindrance of the support and of other enzyme molecules. Covalent immobilization of enzyme on nanoparticles provides greater stability and reusability via strong interactions between the surface and enzymes.

Hu et al. (2009) modified the surfaces of Fe_3O_4 nanoparticles with 3-aminopropyltriethoxysilane and covalently immobilized *Serratia marcescens* lipase using glutaraldehyde as a coupling agent. The resulting nanobiocatalyst showed improved catalytic activity. Ren et al. (2011) used an easy coating method to immobilize lipase on magnetic nanoparticles, employing polydopamine as a surface modifier. They obtained 73.9% immobilization yield, and the bound lipase sustained 70% activity for 21 cycles. Dumri and Hung Anh (2014) used the Ren et al. (2011) method to immobilize porcine pancreatic lipase on silver nanoparticles. They obtained 95% transesterification yield using lipase bound to polydopamine coating. Ranjbakhsh et al. (2012) covalently immobilized the lipase form of porcine pancreases on silica-coated nanoparticles activated with 3-aminopropyltriethoxysilane and compared storage, temperature, and pH stability with free lipase, finding better stability and enhanced activity for the immobilized one. Zhu et al. (2014) prepared a novel nanobiocatalyst by covalently attaching porcine pancreas lipase on silica-coated Fe_3O_4 nanoparticles given amino functionality by treatment with 3-aminopropyltriethoxysilane and then tagged with carboxylic group using succinic anhydride. Bound enzyme exhibited an activity of 6.40 U/mg, as compared with 3.16 U/mg for free lipase. Osuna et al. (2015) made chitosan-coated magnetic nanoparticles and activated them with glutaraldehyde and glycidol, giving activities of 309.5 and 266.2 Unit, respectively. The synthesized nanobiocatalyst sustained 80% of initial activity even after 15 cycles. Malar et al. (2019) synthesized magnetite nanoparticles using the sol-gel method and incorporated $-NH_2$ groups on their surfaces using 3-aminopropyltriethoxysilane activated by glutaraldehyde, which helped immobilize lipase. They reported 92.3% enzyme loading and excellent thermal, pH, and storage stability of the bound enzyme.

Entrapment immobilization involves entrapment of enzymes in porous fibers and gels using nanoparticles, which are produced by reverse micelles or the sol-gel method. A single enzyme unit surrounded by a network of organic/inorganic porous nanocomposites produced by this method shows improved activity along with all other advantages of immobilization (Kim and Grate 2003).

12.5 Techniques for the Characterization of Nanobiocatalysts

Magnetic nanobiocatalysts are mostly characterized by X-ray diffraction (XRD), vibrating sample magnetometry (VSM), scanning electron microscopy (SEM), energy-dispersive X-ray spectroscopy (EDX), transmission electron microscopy (TEM), or Fourier transform infrared microscopy (FT-IR). The size and morphology of the nanomaterials are determined by SEM and TEM. TEM tells more about the internal structure of the sample but gives only 2D images, while SEM provides a 3D image of the surface.

The immobilization of enzymes on nanomaterials can be confirmed by a change in the size or shape of the nanosupport. XRD gives information about crystal phase and size. Xie and Ma (2010) have reported that there is no change in crystal phase after immobilization of enzyme on nanoparticles. The magnetic properties of the carrier nanoparticles and nanobiocatalysts are evaluated with the help of a vibrating sample magnetometer.

EDX is used for chemical characterization and elemental determination. EDX of polymer-coated and enzyme-immobilized nanoparticles gives information about the attachment of coatings and enzymes on nanoparticle surfaces (Park and Kang 2005; Ren et al. 2011).

FT-IR provides information about the functional groups present on nanomaterial surfaces. Polymer coatings on the surfaces of nanomaterials and immobilization of lipase are affirmed by the presence of specific absorption peaks in the FT-IR of nanomaterials that are also seen in the FT-IR of free enzymes and coatings (Ranjbakhsh et al. 2012; Thangaraj et al. 2016; Mehrasbi et al. 2017).

Immobilization efficiency or enzyme loading can be estimated by measuring the starting protein amount and protein content after immobilization through the Bradford method. Protein content can also be determined by the Kjeldahl, biuret, and Lowry methods. The activity of nanobiocatalysts is estimated by UV-visible spectrophotometry and various enzyme activity assay methods, such as hydrolytic activity assay and transesterification activity assay.

12.6 Transesterification Using Magnetic Nanobiocatalysts

Nano-immobilized lipase, being catalytically more active, gives mostly greater biodiesel yield compared to free lipase and provides the advantages of reusability, easy recovery, and stability in a wider range of pHs and temperatures, as well as having a greater storage life. Transesterification using nanobiocatalysts is carried out in shaking incubators and packed-bed reactors. In batch processes, the main concerns are reusability and sustainability, while in continuous processes, they are prevention of enzyme leakage and maintenance of stability. Magnetic nanoparticles are widely used in medical sciences due to their low toxicity and easy separation from the reaction medium. Fe_3O_4 (magnetite) is mostly used as a magnetic support; maghemite is also used for this purpose. Bare magnetite nanoparticles are less interactive with enzymes and more susceptible to oxidation. For this reason, nanoparticle surfaces are modified with different functional groups, which helps protect the nanosupport from oxidation and increases its enzyme carrying capacity (Zhong et al. 2020).

Natural polymers such as chitosan and polydopamine are used as surface modifiers of nanoparticles. The natural polymers provide the advantages of biocompatibility, biodegradability, and better support/enzyme adhesion due to the presence of abundant functionalities on the polymer surface. *C. rugosa* lipase immobilized on polydopamine-coated magnetite nanoparticles was prepared by Ren et al. (2011). It had a specific activity of 8.78 U, which was 73.9% of the free lipase activity of 11.88 U. Andrade et al. (2016) used free and immobilized lipase (Fe_3O_4@polydopamine@*P.cepacia*/Lipase) for the transesterification of soybean oil in biodiesel production. A yield of 93% was obtained using immobilized lipase at 37 °C with 12 hours of reaction time, while free lipase gave an 86% yield under the same conditions. The synthesized nanobiocatalyst was effective for three cycles. Chitosan-modified nanoparticles are prepared either by a single-step *in situ* process or by a two-step process in which they are first prepared and then modified with chitosan (Wu et al. 2009; Baghban et al. 2017).

Amino silanes are easily polymerized on their nanosurfaces to form amino-functionalized core shells, which can afford lipase immobilization using glutaraldehyde as a coupling agent. Xie and Ma (2009) investigated the effects of temperature and pH on free and immobilized lipase in biodiesel production. Nanobiocatalyst was produced using (3-aminopropyl)triethoxysilane-modified Fe_3O_4 as a lipase support and glutaraldehyde as a coupling reagent. Immobilized lipase was proved to be more thermally stable and to have greater adaptability toward higher pH values. Nanobiocatalyst was reusable for four cycles without a decrease in activity. In a related study, Fe_3O_4 core was coated with silica shell then treated with dimethyl octadecyl[3-(trimethoxysilyl) propyl] ammonium chloride to produce alkyl factionalized magnetic nanoparticles. Lipase from *Burkholderia* sp. was immobilized on synthesized nanoparticles. A 90% FAME yield was obtained within 30 hours using the 11% immobilized lipase. The nanobiocatalysts were recycled 10 times, with a significant decrease in activity (Tran et al. 2012).

Encapsulation of nanoparticles before attachment of amino functionality provides better stability and abundant silanol groups on the surface (Karimi 2016). Mehrasbi et al. (2017) have described the effects of immobilization on the enzyme activity of lipase. Lipase was immobilized on (3-glycidoxypropyl) trimethoxylsilane-functionalized $Fe_3O_4@SiO_2$ core-shell nanoparticles. Free lipase showed 6.5 U/mg activity, while immobilized lipase (*Candida antarctica* lipase immobilized on core-shell magnetic nanoparticles) showed 6.3 U/mg. This decrease in activity may be due to denaturation of the lipase during immobilization, but it retained its 100% initial activity for six cycles.

Nanoparticles can be encapsulated in graphene oxide (GO), as it has many functional groups on its surface (i.e. epoxides, hydroxyl, and carboxyl groups). The amino groups present in lipase can form covalent linkages with carboxyl groups on GO, forming a stable magnetic nanobiocatalyst (Xie and Huang 2018).

In order to improve the catalytic activity of magnetic nanobiocatalysts, nanoparticles of other metals such as gold can be grown on the surfaces of relatively large Fe_3O_4 nanoparticles in the presence of olice acid. Residual oleic acid groups provide interfacial activation of lipase, while the high surface polarity of Fe_3O_4/gold nanoparticles helps the lipase to achieve optimal orientation. Thus, the synthesized nanobiocatalyst has superior activity (Sarno and Iuliano 2019).

Cross-linked enzyme aggregates can also be immobilized on the surfaces of magnetic nanoparticles in order to overcome the low stability and high mass transfer restriction of the CLEAs and make them easily recoverable. Badoei-dalfard et al. (2019) immobilized CLEAs of Km12 lipases on the surfaces of amino-functionalized magnetic nanoparticles. The resulting biocatalyst preserved its activity for up to six enzyme cycles and retained 60% more activity than free enzyme on incubation at 4 °C for 24 days. Various studies conducted on biodiesel production from magnetic nanobiocatalysts are summarized in Table 12.1.

Nonmagnetic nanoparticles have also been used for the immobilization of lipase. They were employed in the synthesis of biofuel and gave comparable results in terms of yield. However, their recovery was more difficult than that of magnetic nanoparticles. Dumri and Hung (2014) reported 25.2 U/mg hydrolyzing activity of lipase immobilized on silver nanoparticles, while free lipase showed 23.9 U/mg activity. Immobilized lipase was also revealed to be thermally more stable than free lipase, with the transesterification yield decreased only 27% even after seven cycles. Iso et al. (2001) reported a noticeable increase in activity on lipase when it was immobilized on porous kaolinite particles. Moreover, an increase in thermal stability was also reported, and the catalyst was reusable without much decrease in activity.

12.7 Factors Affecting Enzymatic Transesterification

Parameters having pronounced effects on enzymatic transesterification include the type of alcohol used, solvent, reaction temperature, water content, alcohol to oil molar ratio, and source o flipase.

12.7.1 Type of Alcohol Used

Mostly methanol and ethanol are used in transesterification due to their low cost and easy availability. However, these alcohols cause noticeable inactivation of lipase enzyme as compared to long-chain alcohols. Nelson et al. (1996) have reported that *Rhizomucor miehei*

Table 12.1 Lipases immobilized on different magnetic nanosupports for biodiesel production.

Nanosupport	Lipase type	Feedstock	Biodiesel yield (%)	References
Hydrophobic carrier with Fe_3O_4	*Bacillus subtilis*	Waste cooking oil	90	Ying and Chen (2007)
Fe_3O_4 NPs	*Pseudomonas capacia*	Soybean oil	88	Wang et al. (2011)
Fe_3O_4-CHO	*Thermomyces lanuginosus*	Waste grease	99	Ngo et al. (2013)
Fe_3O_4–NH_2	*Candida antarctica*	Vegetable oils	90	Lopez et al. (2014)
Polyamidoamine-magnetic CNT	*Rhizomucor miehei*	Waste cooking oil	94	Fan et al. (2016)
Polydopamine – Fe_3O_4	*Pseudomonas cepacia*	Soybean oil	90	Andrade et al. (2016)
Fe_3O_4-SiO_2-NH_2	*Burkholderia cepacia*	Waste cooking oil	91	Karimi (2016)
Fe_3O_4 NPs encapsulated in GO	*Candida rugosa*	Soybean oil	92.8	Xie and Huang (2018)
Snowman-like Fe_3O_4/Au	*Thermomyces lanuginosus*	Tomato seed oil	98.5	Sarno and Iuliano (2019)
Polydopamine Fe_3O_4	*Aspergillus terreus* AH-F_2	Waste cooking oil	92	Touqeer et al. (2020)
Fe_3O_4-AP-GA	*Rhizopus oryzae*	Bio – oil from *Chlorella vulgaris* micro algae	69.8	Nematian et al. (2020)
Magnetic rice straw	*Thermomyces lanuginosus*	Soybean oil	94.3	Otari et al. (2020)
Fe_3O_4 NPs-TA	*Thermomyces lanuginosus*	Dairy waste scum oil	90	Sarno and Iuliano (2020)

CNT = carbon nanotubes, GO = graphene oxides, AP = 3-aminopropyl triethylenesilane, GA = glutaraldehyde, TA = tartaric acid.

lipase has a maximum biodiesel yield with primary alcohols while *C. Antarctica* gives the highest yield with secondary ones. Many strategies have been developed to reduce the inactivation of lipase by methanol, including stepwise addition. Lee et al. (2008) reported that stepwise addition of methanol gave 98.92% biodiesel yield where adding methanol in batch gave only 65%. Addition of solvent and use of acyl acceptors help avoid lipase inhabitation.

12.7.2 Solvent

Organic solvents such as hexane, heptane, cyclohexane, and pet-ether are mostly used for transesterification. The advantage of using a solvent system is the elevation of alcohol solubility, because insolubility of alcohol causes enzyme inactivation (Kumari et al. 2009).

Moreover, addition of solvent decreases the viscosity of the reaction mixture. Consequently, the mass transfer improves, along with the reaction rate (Fjerbaek et al. 2009). However, difficult recovery after completion of the reaction, flammability, and the toxic nature of some solvents make this process less appropriate for industrial-scale application (Soumanou and Bornscheuer 2003).

12.7.3 Reaction Temperature

Lipases have a large thermal stability range between 20 and 70 °C. Mostly, lipase has an optimum temperature for transesterification in the range of 30–60 °C. Immobilization of enzyme increases its optimum temperature (Fjerbaek et al. 2009), but the optimum temperature also depends upon other reaction parameters, such as type of solvent and feedstock to alcohol ratio (Antczak et al. 2009).

12.7.4 Water Content

Reactions catalyzed by lipase need a small amount of water to maintain their globular structure and enhance their catalytic activity. However, water content beyond the optimum will reduce the FAME yield, because hydrolysis will start to compete with methanolysis at higher water percentages. Another important factor is the type of lipase used, as *P. fluorescens*, *C. rugosa*, and *P. cepacia* do not show any activity without water (Akoh et al. 2007). On the other hand, *C. Antarctica* shows a negative effect on yield in the presence of water (Fjerbaek et al. 2009). Most studies reveal that the presence of a minimum amount of water increases the rate of reaction for lipase-catalyzed transesterification (Kaieda et al. 2001; Lu et al. 2009; Yu et al. 2013; Li and Dong 2016).

12.7.5 Alcohol to Oil Molar Ratio

Alcohol to oil molar ratio has a significant impact on the reaction rate and yield of transesterification. As transesterification is a reversible reaction, an excess of alcohol is needed to shift it toward the right side. Moreover, excess of alcohol obviates the need for solvent to reduce viscosity. However, excess alcohol – especially short-chain alcohol – reduces the enzyme activity as it forms droplets around the lipase and does not make a homogenous mixture with oil. Li and Dong (2016) have reported a maximum biodiesel yield under a 4.7 : 1 methanol to oil ratio with immobilized *Pichia guilliermondii* lipase. Thangaraj et al. (2016) used a 3 : 1 methanol to oil ratio to get a maximum yield using Lipase NS81006 attached on magnetic nanoparticles. Arumugam and Ponnusami (2017) achieved a maximum biodiesel yield using a 1 : 9 methanol to oil ratio.

12.7.6 Source of Lipase

Lipases can be obtained from different origins, such as fungi, yeasts, bacteria, plants, and animals. Lipases obtained from different sources have varying physical and catalytic properties. Thus, the rate of transesterification depends upon the type of lipase and the structure of the substrate (Zhong et al. 2020).

12.8 Conclusions

Biodiesel is gaining worldwide attention and being considered as a replacement transportation fuel because of the environmental and economic advantages associated with its use. Production of biodiesel from waste and non-edible oils is more promising than that from edible oils because it doesn't cause any food versus fuel conflict. Transesterification using different catalysts is the most widely used method of biodiesel production. Alkaline catalysts are commercially adopted to this end, but the production of a lot of hazardous waste and the need for purification of byproducts have increased the importance of enzymatic transesterification for feedstocks with high FFA contents. Hindering the commercial use of enzymatic transesterification is the high cost of enzymes. Thus, immobilization of lipase on magnetic nanomaterials has proved the most suitable strategy for producing robust, reusable, stable, and catalytically active biocatalysts as compared with other macro-sized supports. Optimization of reaction parameters and the washing of nanobiocatalyst before reuse can increase longevity and make the process cost-effective. Simple methods of nanomaterial synthesis, toxicity issues around some nanomaterials, and scale-up of biodiesel production using nanobiocatalysts all require further exploration in order to utilize the full potential of nanoscience in biofuel production.

References

Adeogun, A.I., Kareeem, S.O., Adebayo, O.S., and Balogun, S.A. (2017). Comparative adsorption of amylase, protease and lipase on $ZnFe_2O_4$: kinetics, isothermal and thermodynamics studies. *3 Biotech* 7 (3): 198.

Ahmad, R. and Sardar, M. (2015). Enzyme immobilization: an overview on nanoparticles as immobilization matrix. *Biochemistry and Analytical Biochemistry* 4 (2): 1–15.

Akoh, C., Chang, S., Lee, G., and Shaw, J. (2007). Enzymatic approach to biodiesel production. *Journal of Agricultural and Food Chemistry* 55 (22): 8995–9005.

Ali, M., Naqvi, B., and Watson, I.A. (2018). Possibility of converting indigenous *Salvadora persica* L. seed oil into biodiesel in Pakistan. *International Journal of Green Energy* 15 (7): 427–435.

Andrade, M.F.C., Parussulo, A.L.A., Netto, C.G.C.M. et al. (2016). Lipase immobilized on polydopamine-coated magnetite nanoparticles for biodiesel production from soybean oil. *Biofuel Research Journal* 3 (2): 403–409.

Antczak, M.S., Kubiak, A., Antczak, T., and Bielecki, S. (2009). Enzymatic biodiesel synthesis – key factors affecting efficiency of the process. *Renewable Energy* 34 (5): 1185–1194.

Arumugam, A. and Ponnusami, V. (2017). Production of biodiesel by enzymatic transesterification of waste sardine oil and evaluation of its engine performance. *Heliyon* 3 (12): e00486.

Badoei-dalfard, A., Malekabadi, S., Karami, Z., and Sargazi, G. (2019). Magnetic cross-linked enzyme aggregates of Km12 lipase: a stable nanobiocatalyst for biodiesel synthesis from waste cooking oil. *Renewable Energy* 141: 874–882.

Baghban, A., Heidarizadeh, M., Doustkhah, E. et al. (2017). Covalently bonded pancreatic lipase onto the dithiocarbamate/chitosan-based magnetite: stepwise fabrication of Fe_3O_4@CS/NHCS-lip as a novel and promising nanobiocatalyst. *International Journal of Biological Macromolecules* 103: 1194–1200.

Balat, M. and Balat, H. (2010). Progress in biodiesel processing. *Applied Energy* 87 (6): 1815–1835.

Carlini, M., Castellucci, S., and Cocchi, S. (2014). A pilot-scale study of waste vegetable oil transesterification with alkaline and acidic catalysts. *Energy Procedia* 45: 198–206.

Carrero, A., Vicente, G., Rodríguez, R. et al. (2011). Hierarchical zeolites as catalysts for biodiesel production from nannochloropsis microalga oil. *Catalysis Today* 167 (1): 148–153.

Chen, G., Ying, M., and Li, W. (2006). Enzymatic conversion of waste cooking oils into alternative fuel – biodiesel. *Applied Biochemistry and Biotechnology* 132 (3): 911–921.

Chen, H., Yu, S.M., Shin, D.W., and Yoo, J.B. (2010). Solvothermal synthesis and characterization of chalcopyrite $CuInSe_2$ nanoparticles. *Nanoscale Research Letters* 5 (1): 217.

Chiu, H.C. and Yeh, C.S. (2007). Hydrothermal synthesis of SnO_2 nanoparticles and their gas-sensing of alcohol. *The Journal of Physical Chemistry C* 111 (20): 7256–7259.

Demazeau, G. (2008). Solvothermal processes: new trends in materials chemistry. Paper presented at the Journal of Physics: Conference Series.

Dizge, N., Aydiner, C., Imer, D.Y. et al. (2009). Biodiesel production from sunflower, soybean, and waste cooking oils by transesterification using lipase immobilized onto a novel microporous polymer. *Bioresource Technology* 100 (6): 1983–1991.

Dumri, K. and Hung Anh, D. (2014). Immobilization of lipase on silver nanoparticles via adhesive polydopamine for biodiesel production. *Enzyme Research* 2014: 389739.

Eş, I., Vieira, J.D.G., and Amaral, A.C. (2015). Principles, techniques, and applications of biocatalyst immobilization for industrial application. *Applied Microbiology and Biotechnology* 99 (5): 2065–2082.

Fan, Y., Wu, G., Su, F. et al. (2016). Lipase oriented-immobilized on dendrimer-coated magnetic multi-walled carbon nanotubes toward catalysing biodiesel production from waste vegetable oil. *Fuel* 178: 172–178.

Fjerbaek, L., Christensen, K.V., and Norddahl, B. (2009). A review of the current state of biodiesel production using enzymatic transesterification. *Biotechnology and Bioengineering* 102 (5): 1298–1315.

Francis, R., Joy, N., Aparna, E.P., and Vijayan, R. (2014). Polymer grafted inorganic nanoparticles, preparation, properties, and applications: a review. *Polymer Reviews* 54 (2): 268–347.

Fukuda, H., Kondo, A., and Noda, H. (2001). Biodiesel fuel production by transesterification of oils. *Journal of Bioscience and Bioengineering* 92 (5): 405–416.

Ghadge, S.V. and Raheman, H. (2005). Biodiesel production from mahua (*Madhuca indica*) oil having high free fatty acids. *Biomass and Bioenergy* 28 (6): 601–605.

Guncheva, M. and Zhiryakova, D. (2011). Catalytic properties and potential applications of *Bacillus* lipases. *Journal of Molecular Catalysis B: Enzymatic* 68 (1): 1–21.

Guo, F., Peng, Z.G., Dai, J.Y., and Xiu, Z.L. (2010). Calcined sodium silicate as solid base catalyst for biodiesel production. *Fuel Processing Technology* 91 (3): 322–328.

Hu, B., Pan, J., Yu, H.L. et al. (2009). Immobilization of *Serratia marcescens* lipase onto amino-functionalized magnetic nanoparticles for repeated use in enzymatic synthesis of diltiazem intermediate. *Process Biochemistry* 44 (9): 1019–1024.

Hwang, E.T. and Gu, M.B. (2013). Enzyme stabilization by nano/microsized hybrid materials. *Engineering in Life Sciences* 13 (1): 49–61.

Iso, M., Chen, B., Eguchi, M. et al. (2001). Production of biodiesel fuel from triglycerides and alcohol using immobilized lipase. *Journal of Molecular Catalysis B: Enzymatic* 16 (1): 53–58.

Jiang, Y., Guo, C., Xia, H. et al. (2009). Magnetic nanoparticles supported ionic liquids for lipase immobilization: enzyme activity in catalyzing esterification. *Journal of Molecular Catalysis B: Enzymatic* 58 (1–4): 103–109.

Kaieda, M., Samukawa, T., Kondo, A., and Fukuda, H. (2001). Effect of methanol and water contents on production of biodiesel fuel from plant oil catalyzed by various lipases in a solvent-free system. *Journal of Bioscience and Bioengineering* 91 (1): 12–15.

Karimi, M. (2016). Immobilization of lipase onto mesoporous magnetic nanoparticles for enzymatic synthesis of biodiesel. *Biocatalysis and Agricultural Biotechnology* 8: 182–188.

Karmakar, A., Karmakar, S., and Mukherjee, S. (2010). Properties of various plants and animals feedstocks for biodiesel production. *Bioresource Technology* 101 (19): 7201–7210.

Kim, J. and Grate, J.W. (2003). Single-enzyme nanoparticles armoured by a nanometer-scale organic/inorganic network. *Nano Letters* 3: 1219–1222.

Kim, H.J., Kang, B.S., Kim, M.J. et al. (2004). Transesterification of vegetable oil to biodiesel using heterogeneous base catalyst. *Catalysis Today* 93: 315–320.

Kim, J., Grate, J.W., and Wang, P. (2008). Nanobiocatalysis and its potential applications. *Trends in Biotechnology* 26 (11): 639–646.

Klein, M.P., Nunes, M.R., Rodrigues, R.C. et al. (2012). Effect of the support size on the properties of b-galactosidase immobilized on chitosan: advantages and disadvantages of macro and nanoparticles. *Biomacromolecules* 13 (8): 2456–2464.

Kockmann, A., Porsiel, J.C., Saadat, R., and Garnweitner, G. (2018). Impact of nanoparticle surface modification on the mechanical properties of polystyrene-based nanocomposites. *RSC Advances* 8 (20): 11109–11118.

Kumari, A., Mahapatra, P., Garlapati, V.K., and Banerjee, R. (2009). Enzymatic transesterification of Jatropha oil. *Biotechnology for Biofuels* 2 (1): 1.

Lee, J.H., Lee, D.H., Lim, J.S. et al. (2008). Optimization of the process for biodiesel production using a mixture of immobilized *Rhizopus oryzae* and *Candida rugosa* lipases. *Journal of Microbiology and Biotechnology* 18 (12): 1927–1931.

Li, Y.X. (2011). Optimization of biodiesel production from rice bran oil via immobilized lipase catalysis. *African Journal of Biotechnology* 10: 72.

Li, Y.X. and Dong, B.X. (2016). Optimization of lipase-catalyzed transesterification of cotton seed oil for biodiesel production using response surface methodology. *Brazilian Archives of Biology and Technology* 59.

Li, Q. and Yan, Y. (2010). Production of biodiesel catalyzed by immobilized *Pseudomonas cepacia* lipase from *Sapium sebiferum* oil in micro-aqueous phase. *Applied Energy* 87 (10): 3148–3154.

Lopez, C., Cruz-Izquierdo, A., Pico, E.A. et al. (2014). Magnetic biocatalysts and their uses to obtain biodiesel and biosurfactants. *Frontiers in Chemistry* 2: 72.

Lu, J., Chen, Y., Wang, F., and Tan, T. (2009). Effect of water on methanolysis of glycerol trioleate catalyzed by immobilized lipase *Candida* sp. in organic solvent system. *Journal of Molecular Catalysis B: Enzymatic* 56 (3): 122–125.

Lyu, J., Li, Z., Men, J. et al. (2019). Covalent immobilization of *Bacillus subtilis* lipase a on Fe_3O_4 nanoparticles by aldehyde tag: an ideal immobilization with minimal chemical modification. *Process Biochemistry* 81: 63–69.

Malar, C.G., Seenuvasan, M., and Kumar, K.S. (2019). Basic study on lipase-immobilized magnetic nanoparticles. *Nanotechnology for Environmental Engineering* 4 (1): 2.

Mehrasbi, M.R., Mohammadi, J., Peyda, M., and Mohammadi, M. (2017). Covalent immobilization of *Candida antarctica* lipase on core-shell magnetic nanoparticles for production of biodiesel from waste cooking oil. *Renewable Energy* 101: 593–602.

Miletić, N., Abetz, V., Ebert, K., and Loos, K. (2010). Immobilization of *Candida antarctica* lipase B on polystyrene nanoparticles. *Macromolecular Rapid Communications* 31 (1): 71–74.

Mukhtar, H., Khursheed, S., Ikram ul, H. et al. (2016). Optimization of lipase biosynthesis from *Rhizopus oryzae* for biodiesel production using multiple oils. *Chemical Engineering and Technology* 39 (9): 1707–1715.

Nelson, L.A., Foglia, T.A., and Marmer, W.N. (1996). Lipase catalyzed production of biodiesel. *Journal of the American Oil Chemists' Society* 73 (9): 1191–1195.

Nematian, T., Salehi, Z., and Shakeri, A. (2020). Conversion of bio-oil extracted from *Chlorella vulgaris* micro algae to biodiesel via modified superparamagnetic nano-biocatalyst. *Renewable Energy* 146: 1796–1804.

Ngo, T.P., Li, A., Tiew, K.W., and Li, Z. (2013). Efficient transformation of grease to biodiesel using highly active and easily recyclable magnetic nanobiocatalyst aggregates. *Bioresource Technology* 145: 233–239.

Noureddini, H., Gao, X., and Philkana, R. (2005). Immobilized *Pseudomonas cepacia* lipase for biodiesel fuel production from soybean oil. *Bioresource Technology* 96 (7): 769–777.

Omar, W.N.N.W. and Amin, N.A.S. (2011). Biodiesel production from waste cooking oil over alkaline modified zirconia catalyst. *Fuel Processing Technology* 92 (12): 2397–2405.

Osuna, Y., Sandoval, J., Saade, H. et al. (2015). Immobilization of *Aspergillus niger* lipase on chitosan-coated magnetic nanoparticles using two covalent-binding methods. *Bioprocess and Biosystems Engineering* 38 (8): 1437–1445.

Otari, S.V., Patel, S.K., Kalia, V.C., and Lee, J.K. (2020). One-step hydrothermal synthesis of magnetic rice straw for effective lipase immobilization and its application in esterification reaction. *Bioresource Technology* 302: 122887.

Pang, J.H., Liu, Y., Li, J., and Yang, X.J. (2019). Solvothermal synthesis of nano-CeO_2 aggregates and its application as a high-efficient arsenic adsorbent. *Rare Metals* 38 (1): 73–80.

Park, O.K. and Kang, Y.S. (2005). Preparation and characterization of silica-coated TiO_2 nanoparticles. *Colloids and Surfaces A: Physicochemical and Engineering Aspects* 257: 261–265.

Patel, A., Brahmkhatri, V., and Singh, N. (2013). Biodiesel production by esterification of free fatty acid over sulfated zirconia. *Renewable Energy* 51: 227–233.

Pavlidis, I., Tsoufis, T., Enotiadis, A. et al. (2010). Functionalized multi-wall carbon nanotubes for lipase immobilization. *Advanced Engineering Materials* 12 (5): B179–B183.

Ponvel, K.M., Lee, D.G., Woo, E.J. et al. (2009). Immobilization of lipase on surface modified magnetic nanoparticles using alkyl benzenesulfonate. *Korean Journal of Chemical Engineering* 26 (1): 127–130.

Prucker, O. and Rühe, J. (1998). Synthesis of poly (styrene) monolayers attached to high surface area silica gels through self-assembled monolayers of azo initiators. *Macromolecules* 31 (3): 592–601.

Rajput, N. (2015). Methods of preparation of nanoparticles – a review. *International Journal of Advances in Engineering and Technology* 7 (6): 1806.

Rane, A.V., Kanny, K., Abitha, V., and Thomas, S. (2018). Methods for synthesis of nanoparticles and fabrication of nanocomposites. In: *Synthesis of Inorganic Nanomaterials* (eds. S.M. Bhagyaraj, O.S. Oluwafemi, N. Kalarikkal and S. Thomas), 121–139. Amsterdam: Elsevier.

Ranjbakhsh, E., Bordbar, A., Abbasi, M. et al. (2012). Enhancement of stability and catalytic activity of immobilized lipase on silica-coated modified magnetite nanoparticles. *Chemical Engineering Journal* 179: 272–276.

Ren, Y., Rivera, J.G., He, L. et al. (2011). Facile, high efficiency immobilization of lipase enzyme on magnetic iron oxide nanoparticles via a biomimetic coating. *BMC Biotechnology* 11 (1): 63–71.

Rossi, L.M., Costa, N.J.S., Silva, F.P., and Gonçalves, R.V. (2013). Magnetic nanocatalysts: supported metal nanoparticles for catalytic applications. *Nanotechnology Reviews* 2: 597–614.

Sarno, M. and Iuliano, M. (2019). Highly active and stable Fe_3O_4/Au nanoparticles supporting lipase catalyst for biodiesel production from waste tomato. *Applied Surface Science* 474: 135–146.

Sarno, M. and Iuliano, M. (2020). Biodiesel production from dairy waste scum by using an efficient nano-biocatalyst. *Chemical Engineering Transactions* 79: 181–186.

Shaw, S.Y., Chen, Y.J., Ou, J.J., and Ho, L. (2006). Preparation and characterization of *Pseudomonas putida* esterase immobilized on magnetic nanoparticles. *Enzyme and Microbial Technology* 39: 1089–1095.

Shi, Y., Liu, W., Tao, Q.L. et al. (2016). Immobilization of lipase by adsorption onto magnetic nanoparticles in organic solvents. *Journal of Nanoscience and Nanotechnology* 16 (1): 601–607.

Soumanou, M.M. and Bornscheuer, U.T. (2003). Improvement in lipase-catalyzed synthesis of fatty acid methyl esters from sunflower oil. *Enzyme and Microbial Technology* 33 (1): 97–103.

Thangaraj, B., Jia, Z., Dai, L. et al. (2016). Lipase NS81006 immobilized on Fe_3O_4 magnetic nanoparticles for biodiesel production. *Ovidius University Annals of Chemistry* 27 (1): 13–21.

Titirici, M.M., Antonietti, M., and Thomas, A. (2006). A generalized synthesis of metal oxide hollow spheres using a hydrothermal approach. *Chemistry of Materials* 18 (16): 3808–3812.

Touqeer, T., Mumtaz, M.W., Mukhtar, H. et al. (2020). Fe_3O_4-PDA-lipase as surface functionalized nano biocatalyst for the production of biodiesel using waste cooking oil as feedstock: characterization and process optimization. *Energies* 13 (1): 177.

Tran, D.T., Chen, C.L., and Chang, J.S. (2012). Immobilization of *Burkholderia* sp. lipase on a ferric silica nanocomposite for biodiesel production. *Journal of Biotechnology* 158 (3): 112–119.

Uzun, B.B., Kılıç, M., Özbay, N. et al. (2012). Biodiesel production from waste frying oils: optimization of reaction parameters and determination of fuel properties. *Energy* 44 (1): 347–351.

Vicente, G., Martınez, M., and Aracil, J. (2004). Integrated biodiesel production: a comparison of different homogeneous catalysts systems. *Bioresource Technology* 92 (3): 297–305.

Wang, P. (2006). Nanoscale biocatalyst systems. *Current Opinion in Biotechnology* 17 (6): 574–579.

Wang, X., Liu, X., Zhao, C. et al. (2011). Biodiesel production in packed-bed reactors using lipase–nanoparticle biocomposite. *Bioresource Technology* 102 (10): 6352–6355.

Wang, J., Meng, G., Tao, K. et al. (2012). Immobilization of lipases on alkyl silane modified magnetic nanoparticles: effect of alkyl chain length on enzyme activity. *PLoS One* 7 (8): e43478.

Wu, Y., Wang, Y., Luo, G., and Dai, Y. (2009). In situ preparation of magnetic Fe_3O_4-chitosan nanoparticles for lipase immobilization by cross-linking and oxidation in aqueous solution. *Bioresource Technology* 100 (14): 3459–3464.

Xie, W. and Huang, M. (2018). Immobilization of *Candida rugosa* lipase onto graphene oxide Fe_3O_4 nanocomposite: characterization and application for biodiesel production. *Energy Conversion and Management* 159: 42–53.

Xie, W. and Ma, N. (2009). Immobilized lipase on Fe_3O_4 nanoparticles as biocatalyst for biodiesel production. *Energy and Fuels* 23 (3): 1347–1353.

Xie, W. and Ma, N. (2010). Enzymatic transesterification of soybean oil by using immobilized lipase on magnetic nanoparticles. *Biomass and Bioenergy* 34 (6): 890–896.

Yang, J., Fan, L., Xu, Y., and Xia, J. (2017). Iron oxide nanoparticles with different polymer coatings for photothermal therapy. *Journal of Nanoparticle Research* 19 (10): 333.

Ying, M. and Chen, G. (2007). Study on the production of biodiesel by magnetic cell biocatalyst based on lipase-producing *Bacillus subtilis*. *Applied Biochemistry and Biotechnology* 137–140 (1–12): 793–803.

Yu, C.Y., Huang, L.Y., Kuan, I.C., and Lee, S.L. (2013). Optimized production of biodiesel from waste cooking oil by lipase immobilized on magnetic nanoparticles. *International Journal of Molecular Sciences* 14 (12): 24074–24086.

Zhang, F., Lees, E., Amin, F. et al. (2011). Polymer-coated nanoparticles: a universal tool for bio-labelling experiments. *Small* 7 (22): 3113–3127.

Zhong, L., Feng, Y., Wang, G. et al. (2020). Production and use of immobilized lipases in/on nanomaterials: a review from the waste to biodiesel production. *International Journal of Biological Macromolecules* 152: 207–222.

Zhu, Y.T., Ren, X.Y., Liu, Y.M. et al. (2014). Covalent immobilization of porcine pancreatic lipase on carboxyl-activated magnetic nanoparticles: characterization and application for enzymatic inhibition assays. *Materials Science and Engineering: C* 38: 278–285.

13

Technoeconomic Analysis of Biodiesel Production Using Different Feedstocks

Shemelis Nigatu Gebremariam

Hawassa University, Wondo Genet College of Forestry and Natural Resources, Shashemene, Ethiopia

13.1 Introduction

Population expansion and improvement of living standards in most developing countries are escalating the world energy demand. According to the World Energy Outlook 2018 report, global demand is expected to increase by more than a quarter by the end of 2040 (IAE 2018). This report emphasizes that most of the increase will come from developing countries, where population expansion is the big issue, along with rising income and the addition of about 1.7 billion people to urban areas (with a significant change in living standards) (IAE 2018). Globally, the use of conventional energy resources since the Industrial Revolution is believed to be the main source of greenhouse gases (GHGs) and subsequent global warming (Dreier et al. 2018). In addition, the world's conventional fossil fuel reserves are declining, creating energy insecurity in various energy-consuming sectors (Mohr et al. 2015). All these phenomena have brought the world to a turning point regarding the choice of energy resources for sustainable development. In this context, a number of alternative and renewable energy resources are being considered as candidates to replace the conventional fossil fuels, either directly or with some modifications to end-use devices. Biofuels, solar, wind, hydropower, and hydrogen are the most promising of such resources, proven to have minimal or zero environmental impact due to their exploitation, conversion, and utilization (Dincer 2000).

Biodiesel is one of the more promising of the biofuels. It possesses a number of environmental and fuel benefits over its counterparts, including a particularly low emission of GHGs and particulates across the whole of its life cycle, from cultivation of the feedstock, to conversion of feedstock to fuel, to final utilization (Corsini et al. 2015). It leads to zero emissions of aromatic compounds and other substances that might be harmful to the environment (Corsini et al. 2015). It is biodegradable and nontoxic in nature. As a fuel, it has some important advantages over conventional fossil diesel, such as high lubricity, which reduces engine wear, and high oxygen content, which encourages complete combustion, potentially further minimizing emissions.

A diesel engine can use plain vegetable oil directly as fuel. However, such application will have fatal consequences on the different parts of the engine due to gum formation through

Nano- and Biocatalysts for Biodiesel Production, First Edition. Edited by Avinash P. Ingle.

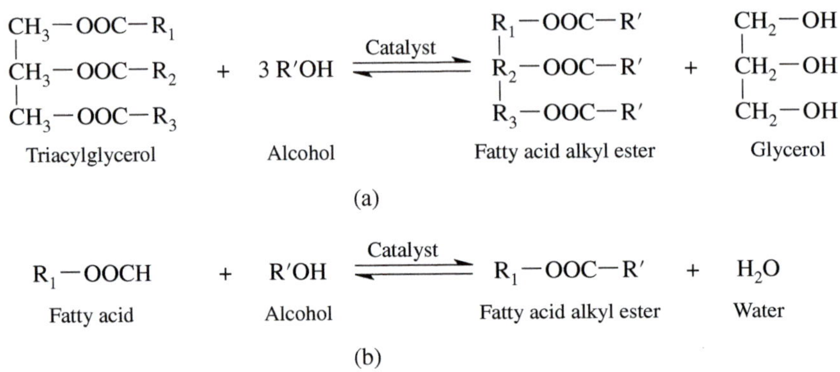

Figure 13.1 Schematics indicating (a) esterification reaction and (b) transesterification reactions for biodiesel production.

oxidation and polymerization during combustion as well as storage through time (Sidibe et al. 2020). In addition, the high viscosity of the oil results in incomplete combustion and thus the total amount of energy released is very much reduced (Hazar et al. 2019). The acid composition and free fatty acid (FFA) content of a vegetable oil can result in other sever problems, such as carbon deposits and lubricating oil thickening (Sidibe et al. 2020; Agrawal et al. 2019). Accordingly, there are four possible methods recommended to lower the viscosity or to improve the vital fuel properties of vegetable oil for direct application in a conventional diesel engine in an efficient manner: diluting/blending the oil (with alcohols, diesel fuel, or solvents), micro-emulsification, pyrolysis of the oil, and alcoholysis reaction. The latter is considered the most reliable, because the biodiesel produced resembles conventional fossil diesel in most of its fuel properties (Nabi et al. 2019).

Biodiesel is a mono alkyl ester of long-chain fatty acids produced either through transesterification when 1 mol of triglyceride reacts with 3 mol of organic alcohol (Figure 13.1a) or through esterification when 1 mol of FFA reacts with 1 mol of organic alcohol (Figure 13.1b) (Sandoval et al. 2017). Both transesterification and esterification can be catalyzed or noncatalyzed, with the latter occurring under the critical condition of the alcohol involved. There are a number of alcohols that can be applied in the process of biodiesel production, with methanol and ethanol being the most widely used. Other options are the short-chain alcohols propanol, butanol, isopropanol, tert-butanol, branched alcohols, and octanol, but these are very expensive and thus less used at industrial scale (Musa 2016).

A catalyzed transesterification reaction may use a number of catalyst technologies, both heterogeneous and homogeneous. However, the conventional method of transesterification for industrial-scale biodiesel production makes use of homogeneous base catalysts such as sodium hydroxide (NaOH) and potassium hydroxide (KOH) (Cong et al. 2020; Fadhil and Ali 2013; Abdullah et al. 2017), NaOH being the more often used as it is relatively cheaper (Gupta et al. 2020). The other catalyst technologies are homogeneous acid (Zhou et al. 2019) and heterogeneous acid (Tang et al. 2020; Deeba et al. 2020) catalysts, heterogeneous base catalysts (Putra et al. 2020; Chua et al. 2020), enzyme (bio)catalysts (Zhang et al. 2020; He et al. 2018), and ionic liquid catalysts (Xie and Wang 2020; Yan et al. 2020). There are also some more recent, less studied technologies that have proven to be efficient for the production of fuel-quality biodiesel from low-quality oil/fat feedstocks, such as membrane technology (Shuit and Tan 2019; Saidi and Moradi 2020; Noriega et al. 2018), reactive distillation technology (Petchsoongsakul et al. 2017; Boon-anuwat et al. 2015), and nanocatalyst technologies (Degirmenbasi et al. 2015; Banković–Ilić et al. 2017).

The major cost categories in the process of biodiesel production include the cost of raw materials, equipment cost, labor cost, and utility cost. Since most of the production cost (more than 80%) is associated with the cost of feedstock, there is a strong belief that the total cost of biodiesel production could be significantly reduced, without compromising the fuel quality, if we could only find an appropriate combination of efficient production technology and lowest-cost feedstock. In order to find such an effective and affordable option, a systematic investigation of the technoeconomic performances of the different possible production technologies is required. Accordingly, a number of investigations have been performed involving diverse feedstock types and production technologies. This chapter reviews the different approaches to the evaluation of technoeconomic performance in biodiesel production processes. In addition, it examines available production technologies and feedstock types.

13.2 Biodiesel Production Technologies

Diluting or blending, micro-emulsion, pyrolysis, and alcoholysis are the four possible methods enabling the use of oil or fat as fuel in diesel engines. Alcoholysis provides a fuel called biodiesel, which resembles fossil diesel in most of its fuel properties (Avhad and Marchetti 2015; Nabi et al. 2019)

The term "alcoholysis" covers both transesterification and esterification reactions. In a stoichiometric transesterification reaction, 1 mol of triglyceride reacts with 3 mol of alcohol to produce 3 mol of fatty acid alkyl esters (biodiesel) and 1 mol of glyceride. This reaction has three consecutive steps, each of them producing 1 mol of fatty acid alkyl ester along with intermediate products like di- and monoglycerides (Avhad and Marchetti 2015). The entire process of transesterification is a reversible reaction, which always requires more moles of alcohol than the stoichiometric amount to favor the forward reaction (Ambat et al. 2018). The other alcoholysis reaction, esterification, takes place when 1 mol of FFA reacts with 1 mol of alcohol to produce 1 mol of fatty acid alkyl ester and 1 mol of water. This reaction is typically important when the feedstock for biodiesel production has a considerable amount of FFAs, as in the case of waste cooking oil (WCO), animal fat, and non-edible plant oil (Sahar et al. 2018).

Transesterification can be either catalyzed or noncatalyzed. The noncatalyzed transesterification reaction makes use of alcohols under their critical conditions. Such conditions are created when a gas or liquid is under high pressure and temperature above its critical points, where there is no liquid or vapor phase and a single supercritical fluid prevails. This phenomenon helps overcome the reaction difficulties associated with poor immiscibility between the reactants (alcohol and oil). Supercritical transesterification can make use of low-quality feedstocks (those with high FFA and moisture contents) for biodiesel production, creating an opportunity for a considerable reduction of production costs arising from raw materials. The other advantages are associated with its very short reaction time and high yield of biodiesel per unit of feedstock. In addition, with regard to process design, no washing stage is required (unlike with most catalyzed processes), and it is very easy to design it as a continuous process (Gebremariam and Marchetti 2017). However, there are some disadvantages that might compromise the economic competitiveness of this process option for fuel-quality biodiesel production, including the requirements for high pressure and temperature (resulting in higher cost of utilities) and a high amount of alcohol. Together, these disadvantages can escalate operating costs (Sakdasri et al. 2017).

The catalyzed transesterification reaction can make use of a variety of catalysts, giving the opportunity to select among versatile combinations of feedstock and catalyst types for production of biodiesel fuel at minimum cost. Among them, acid catalysts, in both heterogeneous and homogeneous forms, can efficiently catalyze both esterification and transesterification reactions for biodiesel production from low-value and cheap feedstock (Ambat et al. 2018). This enables the use of any type of feedstock (including those with higher FFA contents). The other advantages of using acid catalysts for biodiesel production include the higher percentage of yield attained and the lower amount of energy input required compared to other catalyst categories (Gebremariam and Marchetti 2017). In addition, heterogeneous acid catalysts have the advantage of easy recovery and reuse for a number of cycles, creating the possibility of reducing catalyst purchasing costs as well as the cost of product separation. However, a slow rate of reaction and equipment corrosion are the main drawbacks generally associated with acid catalysts (Avhad and Marchetti 2015). Further, the use of homogeneous acid catalysts can create problems during separation of the product, as it requires the use of large amounts of water, possibly escalating the cost of production (Ambat et al. 2018).

The alkali or base catalysts are another option for catalyzing the transesterification reaction in biodiesel production. In general, base-catalyzed processes are known for their less energy-intensive and faster reaction conditions (Gebremariam and Marchetti 2017). The homogeneous base catalysts, such as NaOH and KOH, are well known for their industrial-level application in biodiesel fuel production using feedstocks with very low FFA contents. This makes the option economically noncompetitive, because such feedstocks are very expensive. However, their heterogeneous counterparts do create the possibility of easily recovering and reuseing the catalyst for a considerable number of cycles, reducing the cost for catalyst purchase. Among the most promising heterogeneous base catalysts are the alkaline earth metal oxides, which can give a fatty acid methyl ester (FAME) yield up to 98% (Colombo et al. 2017). Calcium oxide (CaO) is one such catalyst, capable of catalyzing feedstocks with considerable FFA contents, from which fuel quality biodiesel can be produced (Velickovic et al. 2016; Boey et al. 2012). The main difficulty in using CaO catalyst for biodiesel production is associated with its recycling over a number of cycles, whereby the catalyst surface absorbs and reacts with carbon dioxide and water vapor from the atmosphere to form an unreactive surface layer of calcium carbonate and calcium hydroxide, respectively (Soares et al. 2013). However, a number of solutions have been suggested to alleviate the problem, such as moistening the CaO catalyst with biodiesel, methanol, or oil (Acosta et al. 2020), which cover its surface and occupy the pores, significantly reducing the adsorption of carbon dioxide and water vapor and thus preserving the catalyst for multiple reuses.

Enzymes are the other most effective catalyst type for biodiesel production through transesterification of low-quality and cheap feedstock (Hama and Kondo 2013). Besides their ability to catalyze feedstock with high FFA contents, reusability over a number of cycles, easy recovery of the biodiesel product at high purity, and low reaction temperature requirements are the major advantages of using enzyme catalysts for biodiesel production (Canet et al. 2016; Hamd 2016). The most common drawbacks associated are low biodiesel yield, expensiveness of the enzyme, and the possible inactivation of the enzyme by alcohol (usually methanol) and glycerol, especially when it has been recycled several times (Gog et al. 2012; Cerveró et al. 2014).

13.3 Feedstock Types for Biodiesel Production

Feedstock for biodiesel production can be found from various sources, most commonly plant oil (edible and non-edible), animal fat, microbial oil, and algae. The chemical composition of these feedstock types is variable. Accordingly, there is variation in the quality of biodiesel product produced, both good and bad. This implies that selection of feedstock is a vital step in the production of fuel-quality biodiesel, affecting not only quality but also cost, composition, and yield (Singh et al. 2020). Regional availability and cost of feedstock production are very important factors affecting the selection of feedstock type for feasible production of biodiesel fuel. For example, given regional availability and production costs, canola oil is usually used as a feedstock for biodiesel in Canada, whereas in Brazil and the United States soybean is widely applied (Singh et al. 2020). Coconut and palm are common in Indonesia and Malaysia, while rapeseed is used in Italy, Germany, Finland, and the United Kingdom (Singh et al. 2020). In most parts of the world, the use of edible oil as a feedstock for biodiesel production has a considerable effect on the food chain, making the production very expensive (Anwar et al. 2019). In addition to its least-cost effect, the use of non-edible plant oil for biodiesel production brings benefits associated with its biodegradability, low sulfur content, absence of any effect on the food chain, low aromatic content, and abundance (Singh et al. 2020). As far as the cost of feedstock is concerned, there are a number of other alternatives with significantly reduced costs, including WCO, animal tallow, fish oil, micro-algae, and sludge. The problem with these type of least-cost feedstock alternatives is their higher content of FFA and moisture, which makes them inconvenient to use in the conventional base-catalyzed transesterification process.

The diverse feedstock alternatives for biodiesel production can be categorized as first-, second-, and third-generation feedstocks. First-generation feedstocks include edible vegetable oils such as soybean, sunflower, palm, rapeseed, coconut, corn, mustard, and olive. The use of these feedstock types for biodiesel production has the advantage of providing an easy and short conversion procedure. However, their major limitations are related to the risk of creating food supply limitations, being environment-specific for their adaptability/effective cultivation, the high cost of production, and their requirement for a large cultivation area (Singh et al. 2020).

Second-generation feedstocks include those oil feedstocks from non-edible plants such as jatropha, jojoba, karanja, castor, neem, and rubber tree seed. In general, compared to the first-generation feedstock types, they have such advantages as lower production cost and the ability to adapt to diverse environmental conditions (Singh et al. 2020). The availability of a number of alternative feedstock types with diverse properties under the second-generation category creates an opportunity to lower the overall cost of biodiesel production and make the biodiesel produced economically competitive with conventional fossil diesel. However, most have high FFA contents and thus require more conversion procedures if they are to be used in base-catalyzed processes; otherwise, a different production route other than the conventional base-catalyzed transesterification must be used, such as an acid- or enzyme-catalyzed procedure (Yaşar 2020). Some authors include animal fat and waste oil under the second-generation feedstock category (Singh et al. 2020), while others list them as third-generation (Yaşar 2020).

The third-generation feedstock category mainly comprises micro-algae. The main advantages associated with such feedstocks are their fast growth rate and high productivity, absence of effect on the food supply, absence of conflict over land for cultivation, and low greenhouse effect (Faried et al. 2017). In addition, since micro-algae are rich in protein and carbohydrate, the byproducts from the oil extraction process can be used for other valuable applications, such as animal feed (protein) and bioethanol/biogas production (carbohydrate) (Raheem et al. 2018). This will definitely help to maximize the overall economic output of the process. Additional advantages are related to the way the feedstock is produced; for example, the production scheme can be coupled with wastewater treatment, whereby the growing algae consume contaminates in the wastewater such as nitrates and phosphate, purifying the water (Arenas et al. 2017). Moreover, since the efficient growth of micro-algae requires concentrated carbon dioxide, the cultivation system can also be coupled with specific industries producing carbon dioxide in large amounts, such as cement industries. However, the industrial-scale application of this promising feedstock type is not well known, mainly due to the sophistication of the whole process, which increases the cost of production (Raheem et al. 2018).

13.4 Technical Performance Evaluation of Biodiesel Production

The technical performance of a biodiesel production process is usually evaluated in terms of the quality and quantity of the biodiesel product. Alternatively, the amount of biodiesel produced per amount of feedstock consumed is often used (Gebremariam and Marchetti 2019). It is also possible to consider the relative purity and quantity of valuable byproducts in order to compare the performances of different process options. The quality of biodiesel product, as a technical performance evaluation parameter, has various dimensions and is mostly dependent on the type of feedstock used.

Many alternative feedstocks are available for the production of biodiesel fuel. However, the question of fuel quality must always be considered as far as substitution of fossil diesel with biodiesel is concerned. Fuel quality is determined by the most important biodiesel properties, such as flash point, cold flow properties, cetane number, density, viscosity, and oxidation stability; most of which are directly affected by the properties of the feedstock, such as saturation level, carbon chain length, type and content of FFA, type of triglyceride, and moisture content (Fazal et al. 2019; Singh et al. 2020).

13.4.1 Fuel Properties of Biodiesel

The efficiency of fetching the maximum possible energy from biodiesel fuel is directly or indirectly affected by some important fuel properties, which are usually used to ascribe the quality of biodiesel as a fuel. This section looks at the main fuel properties, how they can be measured or quantified, and how they are affected by the type and quality of the feedstock used.

13.4.1.1 Flash Point

The flash point of a chemical substance refers to the lowest temperature at which there could be enough evaporation of the fluid to form a combustible concentration of gas (Fu 2019). It can be taken as an indication of how easily a fuel can catch fire, making it an essential safety-related property for fuel processing, transportation, and storage (Fu 2019). Apparently, the flash point of a fuel is directly correlated to its vapor pressure, such that the higher the vapor pressure of the fuel, the lower is its flash point. Thus, the flash point value of a fuel provides a measurement of both its flammability and its volatility (Fu 2019). A high flash point indicates delayed combustion, which leads to carbon deposition and large production of gaseous pollutants such as NOx (Kim et al. 2018). In general, since biodiesel is less volatile than conventional diesel, it has a higher flash point (i.e. it is more safe to handle biodiesel than fossil diesel); this is due to its high content of fatty acids with long carbon chains (Kim et al. 2018). In this respect, there is an expected difference in measurements/values of flash points among biodiesel fuels produced from different feedstock types, mainly due to the characteristics of the fatty acid chain in the feedstock used. Accordingly, feedstocks with short-chain fatty acids (C12 and less) such as coconut oil provide biodiesel with a convenient flash point in the range of 95–100 °C (Yaşar 2020). In contrast, used cooking oil and animal fat, which have high contents of long-chain fatty acids (C16, C18, and C20), provide biodiesel with low volatility and thus with a high flash point (Yaşar 2020). Based on American Society for Testing and Methods (ASTM) standard ASTM-D6751, a quality biodiesel fuel should have a flash point with a minimum value of 93 °C, while the European Committee for Standardization puts it at 101 °C, according to EN 14214 (Yesilyurt et al. 2020). Table 13.1 shows the flash point values of biodiesel fuels from various feedstocks.

Concerning measurements, the ASTM standards include two main types of testing methods for the flash points of liquid fuels in general: closed cup (e.g. ASTM D 56, Standard Test Method for Flash Point by Tag Closed Cup Tester; ASTM D 93, Standard Test Methods for Flash Point by Pensky-Martens Closed Cup Tester; ASTM D 3278, Standard Test Methods for Flash Point of Liquids by Small Scale Closed Cup Apparatus; ASTM D 3828, Standard Test Methods for Flash Point by Small Scale Closed Cup Tester) (Santos et al. 2020) and open cup (ASTM D 92, Standard Test Method for Flash and Fire Points by Cleveland Open Cup Tester, and D1310) (Santos et al. 2020).

13.4.2 Cold Flow Properties

The cold flow properties of biodiesel fuel are usually expressed in terms of cloud point, pour point, and cold filter plugging point (CFPP), all of which have a direct effect on fuel pumping and filtration capabilities (USDA 2016).

Table 13.1 Flash point values of biodiesel fuels from different feedstock types (Yaşar 2020).

Feedstock type	Flash point (°C)
Algae	140
Soybean	170
Sunflower	178
Corn	172
Cottonseed	175
Canola	137
Olive	177
Safflower	180
Hazelnut	171
Rapeseed	172

13.4.2.1 Cloud Point

The cloud point is the minimum temperature at which biodiesel starts to form visible solid crystals as it solidifies. Irrespective of the type of feedstock used, biodiesel has a higher cloud point than its counterpart, fossil diesel (Sia et al. 2020). This is mainly due to the higher content of saturated fatty acids in biodiesel, which may increase the cloud point (Abe et al. 2016). Comparing cloud point values among biodiesels from different feedstock types, the difference is mainly due to the presence or absence of long-chain saturated fatty acids in the feedstocks: the higher the content of long-chain saturated fatty acids, the higher the cloud point (Dunn et al. 2015). For example, biodiesel from tallow and palm oils has a high cloud point, whereas biodiesel from feedstocks with a lower saturated fatty acid content such as castor, sesame, and soybean has a lower cloud point, in relative terms (Kim et al. 2018). Measurement of the cloud point of biodiesel fuel is usually carried out through ASTM standards D6751, D5771, D5772, D5773, and D2500 (Sierra-Cantor and Guerrero-Fajardo 2017). According to ASTM-D6751, the standard value of the cloud point for biodiesel fuel is in the range of −3 to −12 °C (Yesilyurt et al. 2020).

13.4.2.2 Pour Point

In its simplest definition, the pour point of a biodiesel fuel refers to the temperature at which it stops flowing when it is cooled further under conditions specified in ASTM D 97 (Bazooyar et al. 2015). In general, biodiesel fuel produced from any type of feedstock has a higher pour point than that of fossil diesel (similar to the case with cloud point) (Bazooyar et al. 2015). In addition, biodiesel fuel has a much narrower range of temperatures between the cloud point and the pour point than fossil diesel fuel (Agarwal et al. 2010). Pour point can be measured according to several ASTM standards, such as D97, D5949, D6751, D5950, D5985, D6749, and D6982 (Monirul et al. 2015). Based on ASTM-D6751, the standard pour point of a quality biodiesel fuel is in the range of −15 to −16 °C (Yesilyurt et al. 2020).

Table 13.2 Cold filter plugging point (CFPP) values of biodiesel fuels from different feedstock types (Yaşar 2020).

Feedstock type	CFPP (°C)
Algae	−14
Soybean	−6
Sunflower	−4
Corn	−13
Cottonseed	−12
Canola	−13
Olive	−12
Safflower	−7
Hazelnut	−12
Rapeseed	−11

13.4.2.3 Cold Filter Plugging Point (CFPP)

The CFPP refers to the highest temperature at which a given volume of fuel fails to pass through a standardized filtration device in a specified time (Monirul et al. 2015). The CFPP can be a critical property associated with the lowest temperature at which biodiesel fuel can freely flow through filters in a diesel engine system. According to ASTM-D6751, the maximum value of CFPP for fuel-quality biodiesel should be +5 °C (Yesilyurt et al. 2020). Table 13.2 shows some CFPP values for biodiesel fuels from common feedstock types.

13.4.3 Cetane Number

Cetane number refers to the time delay between when a biodiesel fuel is injected into an engine's combustion chamber and when the combustion reaction begins (Giakoumis and Sarakatsanis 2018). A high cetane number reduces the ignition delay, meaning a fuel with a high cetane number will have a short time between ignition and combustion; accordingly, a higher cetane number indicates a better burning of the fuel within an engine. Nevertheless, both too low and too high cetane numbers can cause problems to the engine (Yaşar 2020). Too low a number means a very long delay between ignition and combustion, which is not desirable at all, but a very high number can cause pre-combustion of the fuel adjacent to the injection line, which may result in nozzle clogging and increased release of pollutants during combustion (Canakci and Sanli 2008). Moreover, particulate matter (PM) and carbon monoxide (CO) emissions may be lower if a biodiesel with a high cetane number is used, but NOx emission is expected to increase due to the higher oxygen content (Mofijur et al. 2013).

Irrespective of the feedstock type used for production, the cetane number of a biodiesel is usually higher than that of fossil diesel (Canakci and Sanli 2008). This is perhaps due to the existence of higher amounts of oxygen in the structure of biodiesel in comparison with fossil diesel (Bazooyar et al. 2015). The type of feedstock used for biodiesel production

Table 13.3 Cetane numbers of biodiesel fuels from different feedstock types (Yaşar 2020).

Feedstock type	Cetane number (dimensionless)
Algae	59
Soybean	51
Sunflower	50
Corn	56
Cottonseed	56
Canola	55
Olive	58
Safflower	53
Hazelnut	57
Rapeseed	54

directly determines the value of the cetane number of the fuel produced. In general terms, a feedstock with more long-chain fatty acids and with high contents of saturated fatty acids provides a higher cetane number (Pala-En et al. 2013). For example, biodiesel produced from WCO has a higher cetane number than that produced from soybean oil (Zaharin et al. 2017). The cetane number of a biodiesel fuel can also be related to its viscosity (Knothe 2005), where the higher the viscosity, the higher the cetane number. For example, if grease and animal fat, both of which have high viscosity, are used as feedstocks, the biodiesel produced will have a high cetane number (Canakci and Sanli 2008). The standard minimum value of cetane number for fuel-quality biodiesel is 47 according to ASTM-D6751 or 51 according to EN 14214 (Yesilyurt et al. 2020). Table 13.3 indicates the cetane numbers for biodiesel fuels produced from different common feedstock types.

13.4.4 Density

Higher biodiesel density is associated with very slow injection and inefficient atomization, which result in ineffective combustion of the fuel (Ejim et al. 2007). It might also cause greater emission of pollutants such as particulate matter (PM), nitrogen oxides (NOx), and carbon monoxide (CO) (Kim et al. 2018). When compared with fossil diesel, biodiesel has a higher density (Ramírez-Verduzco et al. 2012). The main feedstock properties affecting the density of biodiesel are the level of unsaturation (number of double bonds in carbon chains) and the hydroxyl ester contents in fatty acids (Öner and Altun 2009). Accordingly, biodiesel from jatropha and castor oils, which have high number of double bonds (as they possess more oleic, linoleic, and palmitoleic acids), has higher density than that from cooking vegetable oil, in general terms (Batista et al. 2018). ASTM-D6751 indicates that the standard density of fuel-quality biodiesel (at 15 °C) should be 880 kg/m^3, whereas EN 14214 puts it in the range of 860–900 kg/m^3 (Yesilyurt et al. 2020).

Table 13.4 Viscosities of biodiesel fuels from different feedstock types (Yaşar 2020).

Feedstock type	Viscosity (mm^2/s)
Algae	4.55
Soybean	4.37
Sunflower	4.30
Corn	4.42
Cottonseed	4.33
Canola	4.40
Olive	4.53
Safflower	4.60
Hazelnut	4.45
Rapeseed	4.48

13.4.5 Viscosity

Viscosity can directly affect the most important processes vital for efficient combustion of a fuel, such as injection and atomization (Boggavarapu and Ravikrishna 2013). For instance, high viscosity of biodiesel fuel is known to cause less efficient combustion due to the formation of large droplets during the injection process, as well as poor mixing with air (Ong et al. 2013, 2014). As in the case of most fuel properties, the viscosity of a biodiesel is directly affected by the type of feedstock used in its production. Feedstock with high contents of saturated fatty acids and long carbon chains will provide biodiesel with high viscosity (Wan et al. 2015). Thus, biodiesel from animal fat and tallow has higher viscosity than that from vegetable oils (Karmakar et al. 2010). ASTM-D6751 indicates that the standard viscosity of fuel quality biodiesel (at 40 °C) should be within the range of 1.9–6 mm^2/s, whereas EN 14214 puts it in the range of 3.5–5 mm^2/s (Yesilyurt et al. 2020). Table 13.4 shows the viscosities of biodiesel fuels (at 40 °C) from various feedstock alternatives.

13.4.6 Oxidation Stability

Oxidation stability is associated with the capability of a fuel to remain unchanged during extended storage. It is one of the most important fuel properties as far as biodiesel is concerned, as the extent of oxidation on storage and transport governs how badly the other properties are compromised (Pullen and Saeed 2012). Generally, the extent of oxidation of a biodiesel fuel is determined by the time of storage or transport and the degree of unsaturation of its fatty acid contents (Atabani et al. 2012). Oxidation of fatty acids in the biodiesel can change its fuel quality through the formation of gum, which may significantly affect its most important properties such as viscosity, cetane number, and density, leading eventually to inefficient combustion (Pullen and Saeed 2012; Jakeria et al. 2014). In addition, if oxidized biodiesel is combusted, it can cause severe carbon deposition in the combustion

Table 13.5 Oxidation stabilities of biodiesel fuels from different feedstock types (Yaşar 2020).

Feedstock type	Oxidation stability (h)
Algae	2.3
Soybean	1.5
Sunflower	0.9
Corn	1.3
Cottonseed	1.88
Canola	1.9
Olive	2.3
Safflower	2.4
Hazelnut	2.1
Rapeseed	2.2

chamber and produce increased NOx as a pollutant (Kalam and Masjuki 2008). Concerning the unsaturation of fatty acids, the higher the degree of unsaturation, the more likely the tendency of a biodiesel to oxidize (Atabani et al. 2012). Accordingly, an oil feedstock like soybean oil, with a high content of such fatty acids (linoleic and linolenic acids), could provide a biodiesel fuel with poor oxidation stability (Karmakar et al. 2010). In comparison, coconut oil and animal fat could provide a biodiesel with very high oxidation stability due to their high content of saturated fatty acids (Nakpong and Wootthikanokkhan 2010). ASTM-D6751 puts the minimum oxidation stability (at 110 °C) of a fuel-quality biodiesel at three hours, whereas EN 14214 puts it at eight hours (Yesilyurt et al. 2020). Table 13.5 shows the oxidation stabilities of biodiesel fuels from different common feedstock alternatives.

13.4.7 Biodiesel Quality Standards

There is no one universally accepted biodiesel quality specification globally. Rather, a number of standards are utilized around the world to specify the required quality of biodiesels on the market. Such diversification is due to a number of factors, including the types of diesel engines most common in a given region or country, the emissions regulations governing those engines, the development stage and climatic properties of the producing or using region or country, and the purpose and motivation for the use of the biodiesel (Barabás and Todoruţ 2011). Table 13.6 describes the most common diesel quality standards around the world.

13.5 Economic Performance Evaluation of the Biodiesel Production Process

In a nutshell, the economic performance of a given biodiesel production business is largely affected by the type of feedstock used and the technology applied to convert it to biodiesel.

Table 13.6 Most important biodiesel quality standards around the world (Barabás and Todoruţ 2011).

Country/region	Specification	Description
EU	EN 14213	Heating Fuels – Fatty Acid Methyl Esters (FAME) – Requirements and Test Methods
EU	EN 14214	Automotive Fuels – Fatty Acid Methyl Esters (FAME) for Diesel Engines – Requirements and Test Methods
USA	ASTM D 6751	Standard Specification for Biodiesel Fuel Blend Stock (B100) for Middle Distillate Fuels
Australia		Fuel Standard (Biodiesel) Determination 2003
Brazil	ANP 42	Brazilian Biodiesel Standard (Agência Nacional do Petróleo)
India	IS15607	Biodiesel (B 100) Blend Stock for Diesel Fuel – Specification
Japan	JASO M360	Automotive Fuel – Fatty Acid Methyl Ester (FAME) as Blend Stock
South Africa	SANS 1935	Automotive Biodiesel Fuel

The cheaper the feedstock, the lower the total production cost, significantly affecting the economic feasibility of the whole process. However, most of the cheaper feedstock types for biodiesel production, such as used cooking oil, animal fat, and non-edible plant oil, contain considerable amounts of FFA and moisture, potentially jeopardizing the quality of the fuel and the economics of the whole process if the relatively cheapest conventional production technology (homogeneous catalyzed transesterification) be applied. Even though there are a number of available technologies for the production of fuel-quality biodiesel from such low-cost feedstock alternatives, some are not economically feasible, for a number of reasons (e.g. use of sophisticated and expensive equipment, high cost of utilities due to operation under high pressure and temperature, expense of catalysts involved, requirement for a huge amount of raw materials such as alcohols and catalysts).

A sustainable business in biodiesel fuel production needs to carry out a detailed analysis of the relative economic performances of the available alternative technologies coupled with the available low-cost feedstock types. In this way, the relative economic performances of such production technologies can be evaluated in terms of total investment cost, unit production cost, internal rate of return, payback time, net present value, and gross margin of the business project. Various studies have been carried out to evaluate the economic performances of biodiesel production technologies in different market scenarios.

The total cost of biodiesel production depends on such factors as the type of production technology chosen, the production scale (plant size), and the type and market price of the raw materials used (Gebremariam and Marchetti 2018). The total cost can be categorized into fixed capital investment cost and working capital investment cost (operating cost) (Peters et al. 1991). Fixed capital investment cost includes capital costs required for the purchase of installed process equipment, including all facilities and auxiliaries. Working capital investment cost includes the cost of raw materials, utility costs, labor-dependent costs, facility-dependent costs, and other similar variable expenses required to produce a biodiesel fuel at a given rate (Gebremariam and Marchetti 2018). There are a number of approaches to the estimation of fixed and working capital costs for a given biodiesel production technology. This sections describes the most important ones in each case.

13.5.1 Fixed Capital Investment Cost

For any kind of chemical processing industry, fixed capital investment cost estimation typically takes five different forms (Turton et al. 2009; Hastak 2015). The difference between these forms is associated with the level of detail required according to the purpose of the cost estimation; because of this, they offer different accuracy levels. The forms are order-of-magnitude estimates (class 5), study estimates (class 4), preliminary estimates (class 3), definitive estimates (class 2), and detailed estimates (class 1). Class 5 estimation (order-of-magnitude estimate) is usually applied for preliminary feasibility analysis of more than two alternative processes in order to compare their respective total capital costs. It is done by extrapolating the values from similar plant costs, and the accuracy might be up to 30% (Shabani 2015). Class 4 estimation (study estimate) is also applied for preliminary analysis of the required capital investment cost for a given production process. It involves considering the costs of the major equipment involved, and again its accuracy can reach up to 30% (Shabani 2015). Class 3 estimation (preliminary estimate) and class 2 estimation (definitive estimate) are used to further perform accurate estimation of the capital costs of a process, which are usually initially provided using class 4 and class 5 estimates. Class 3 estimation is typically done for budget authorization and its accuracy can reach up to 20%, whereas class 2 estimation is done based on basic engineering and quotes from suppliers and contractors and its accuracy can be about 10%. The most accurate cost estimation for chemical processing industries is carried out using class 1 estimation (detailed estimate). It is typically done based on detailed engineering specifications and the accuracy can be up to 5% (Hastak 2015).

Fixed capital cost comprises both direct and indirect cost categories. Direct costs include equipment purchasing costs, equipment installation costs, piping and insulation costs, instrument and control costs, electrical equipment costs, building and site preparation costs, service facilities costs, and the cost of land. Indirect costs include engineering and supervision costs, construction expenses, contractors' fees, and contingency and startup expenses. Table 13.7 lists cost estimation classes and their respective primary and secondary characteristics.

The selection of a form of estimation is dependent on the level of progress of a project (Shabani 2015). For early-stage design of the process plant, estimation methods with low accuracy can be applied because such methods can be used to roughly estimate the capital cost for various alternative processes in order to compare between them and sort out which is most feasible. However, at the detailed-design specification level, more accurate estimation methods should be applied in order to estimate the approximate amount of investment required to start the process.

Thus, to select the biodiesel process with the best economic viability, an economic performance evaluation must be carried out among a number of available alternative technologies. Class 5 and class 4 estimates are commonly used for this as they are easy and effective (Apostolakou et al. 2009; Jegannathan et al. 2011). The class 5 estimate is less accurate and is

Table 13.7 Cost estimation classes and their respective primary and secondary characteristics (Shabani 2015).

	Primary characteristics	Secondary characteristics			
Estimate class	**Level of project definition (expressed as % of complete definition)**	**End usage Typical purpose of estimate**	**Methodology Typical estimating method**	**Expressed accuracy range Typical variation in low and high ranges**	**Preparation effort Typical degree of effort relative to least cost index of 1**
Class 5	1–2%	Concept screening	Capacity factored, parametric models, judgment, or analogy	L: −20% to −50% H: +30% to +30%	1
Class 4	1–15%	Study of feasibility	Equipment factored or parametric models	L: −15% to −30% H: +20% to +50%	2–4
Class 3	10–40%	Budget authorization or control	Semi-detailed unit cost with assembly-level line items	L: −10% to −20% H: +10% to +30%	3–10
Class 2	30–70%	Control estimate or bid/tender	Detailed unit cost with forced detailed take-off	L: −5% to −20% H: +5% to +20%	4–20
Class 1	50–100%	Control estimate or bid/tender	Detailed unit cost with take-off	L: −3% to −10% H: +3% to +5%	5–100

usually applied for project concept screening. It is the simplest method of estimating capital cost based on only plant capacity and plant cost (Hastak 2015). It is related to the six-tenth power method or capacity powered estimation (Hastak 2015), whereby the capital cost of the new plant is estimated based on the already known capital cost of another plant with a different capacity, as indicated in Eq. (13.1) (Shabani 2015). In this method, the ratio of Cap B to Cap A should be in the range of 0–2 (Hastak 2015).

$$\frac{\$B}{\$A} = \left(\frac{Cap\ B}{Cap\ \mathrm{A}}\right)^{e} \tag{13.1}$$

where $\$A$ refers known cost of the previous plant with capacity A (Cap A); $\$B$ is new estimated cost for the new plant with capacity B (Cap B); and the exponent e is usually taken as 0.6, although it might be different at higher capacities.

In general, the six-tenth power method demands the two process plants (the previous plant with known cost and the new plant) to have a similar process (e.g. continuous or batch) and a similar product, and the capital cost it estimates is considered the cost of the battery limit (Hastak 2015). The other most important point in using the six-tenth method is that the plant capacities and costs should relate to the same time period as the calculation does not consider the time value of the money involved.

Class 4 estimates can make use of some preliminary design data, such as capacity, individual equipment cost, and equipment size, so that the estimation is done to analyze feasibility among different plant alternatives. This method is related to equipment-factored estimation and parametric models, where the design model is progressed between 1% and 15% (Shabani 2015). A biodiesel production project screened through this method can be directly authorized/approved for further estimation using class 3 estimation. Here, the cost of equipment is multiplied by factors to estimate the total plant cost (TPC) (Shabani 2015). Eq. (13.1) can be used for this method by replacing the equipment capacity with parameters such as surface area for heat exchangers and flow rate for pipes and by using the given e values for some common equipment taken from the literature (Lemmens 2016). The cost of the equipment can be found from manufacturers, vendors, and previous similar projects or can be estimated using such methods as the Peters and Thamaures method, which makes use of the latest Chemical Engineering Plant Cost Index (CEPCI) (Chemical Engineering n.d.). This helps to get the latest estimated cost of equipment based on the known cost from previous years. The CEPCI designates the possible time value of the money involved, which may change through time due to inflation or deflation.

The equipment-factored estimation method makes use of the known cost of the main equipment to calculate the TPC. A typical example is the Lang factor method, as indicated in Eq. (13.2) (Shabani 2015):

$$(TPC) = (TEC)\,X\,(E.F) \tag{13.2}$$

where TPC is the total plant cost; TEC is the total equipment cost; and E, F are factors related to the process types, given as 3.1 for solid, 3.63 for solid and fluid, and 4.74 for fluid processes (Shabani 2015). Van Kasteren and Nisworo (2007) used a similar approach to estimate the capital cost for a biodiesel production operation using WCO as feedstock in a supercritical methanol transesterification process. The equipment cost was determined using data taken from the Dutch Association of Cost Engineers (DACE) price book, corrected according to the CEPCI (Van Kasteren and Nisworo 2007). The fixed capital or inside battery limit cost of the plant processing unit, which includes mixers, reactors, heat exchangers, pumps, and pipes, was calculated by multiplying the purchased cost of the equipment by a factor of 5 (Van Kasteren and Nisworo 2007), which gave a result more or less equal to the Lang factor for fluid processes, given as 4.74 (Lemmens 2016). Another method that makes use of equipment-factored estimation is the Hand method (Hastak 2015), which uses multiple factors for the equipment involved rather than only a single factor for the plant, as used in the Lang factor. This makes the Hand method relatively more accurate (Lemmens 2016).

In addition, there are a number of techniques to calculate the total capital investment cost for a given biodiesel production process. However, in each of them, the cost of equipment is taken as an input to estimate the total capital cost of the project. This implies that the precision of the total capital cost estimation is largely dependent on the estimated or acquired cost of equipment. In this respect, there are a number of methods for obtaining the estimated cost of equipment for a biodiesel production process plant. However, before doing capital cost estimation for any type of biodiesel production process, some vital activities must be performed, including designing the complete process flow, selecting the equipment types and determining their respective sizes, determining the type of material to be used for construction of the equipment, and, eventually, performing an energy and material balance for the process (Peters et al. 1991).

13.5.2 Working Capital (Operating) Cost

The working capital cost of a biodiesel production project includes the costs of raw materials, labor, utilities, maintenance, tax, and depreciation, among others. Raw materials for biodiesel production are oil or fat feedstock, alcohol, catalyst, co-solvent, and washing water. Since the cost of feedstock represents the largest share of the total biodiesel production cost, care has to be taken in considering the latest feedstock costs when undertaking an economic performance evaluation of the process. The amount of feedstock to be used is directly determined by the required production capacity. In the meantime, the exact amounts of the other raw materials, especially the alcohol and catalyst, are determined by consideration of their relative amounts with respect to the total amount of feedstock required under optimum reaction conditions in order to achieve the maximum possible conversion. For example, if a biodiesel production process with a given feedstock capacity has an optimum reaction condition at 7% catalyst with respect to oil mass and 9 : 1 molar ratio of alcohol to oil, the amounts of catalyst and alcohol required can easily be determined based on the total mass and the molar mass of the oil used, respectively. The costs of raw materials can easily be acquired from manufacturers, relevant vendors, existing similar projects, and one's own documents.

The type and quantity of the utilities involved in a biodiesel production process are largely determined by the type of process technology selected, the type, size, and efficiency of the equipment involved, and the capacity of the whole production process. Steam, cooling water, and electricity arc the common utilities usually applied in biodiesel production processes. Similarly, the number and qualification of the labor force are dependent on the whole production capacity, the level of sophistication in the production processes, the type of process employed, and the frequency of production. The cost of utilities can be found from relevant service providers or from existing projects using utilities of the same type. Labor cost is usually given in terms of daily or hourly rate and can be determined based on the cost rate for similar labor forces in existing projects involving the same or similar labor qualifications or perhaps from relevant administrative documents indicating the possible wage rates for professionals. The common labor force required in the process of biodiesel production includes the reactor operator, assistant operator, technician for repair and maintenance, and process supervisor.

Once all the costs involved in the process of biodiesel production are estimated, the economic performance of a given biodiesel production technology can be analyzed using the latest commercial software, such as Aspen Plus (Aspentech, Bedford, MA, USA) (Aspentech 2018) and Super Pro (Intelligen Inc., Scotch Plains, NJ, USA) (Intelligen 1996). These kinds of software can help develop new processes or improve existing ones through modeling and simulation. They also make it very easy to evaluate the technoeconomic performances of various possible scenarios that might be created by changing the cost of equipment and raw materials, production capacities, labor costs, utility costs, type of materials, and so on.

13.6 Conclusions

The ever-increasing world energy demand, coupled with negative environmental effects due to production and utilization of fossil fuels, urges the world to tend to renewable energy resources such as biodiesel. However, the widespread use of biodiesel in place of fossil diesel is hindered by its relatively higher cost of production. Fortunately, the availability of numerous technological options and feedstock alternatives provides enormous opportunities to identify technically efficient and economically viable alternatives for the production of fuel-quality biodiesel.

The technical performance among different alternatives can be determined based on the quality and quantity of the biodiesel produced as well as the biodiesel yield relative to the amount of feedstock used. The fuel quality of biodiesel is usually expressed in terms of its flash point, cold flow properties, cetane number, oxidation stability, density, and viscosity. There is no one universally acceptable fuel quality standard for biodiesel; rather, a number of standards are utilized around the world to specify the required quality of biodiesel on the market. Such diversification is due to a number of factors differing from region to region, including the types of diesel engines commonly in use, the emissions regulations related to those engines, the development stage and climatic properties of the producing or using region, and the purpose and motivation behind the use of the biodiesel.

Evaluation of the economic performance of a biodiesel production technology/process demands a thorough design of the unit procedures required for quality fuel production and a determination of the equipment size and the type of material to be used in its construction. The next step is to conduct a material and energy balance throughout the process, based on which the required quantity of raw materials and the expected amount and quality of product can be determined. Energy and material balances can easily be carried out using process simulation software like Super Pro and Aspen Plus.

The fixed capital cost of a biodiesel production business can be determined following one of five estimation methods common to the chemical processing industry, depending on the required level of accuracy, as well as by using Lung factor and Hand methods, among others. It is essential first to determine the latest market values of the equipment to be used, based on which other components of the fixed capital cost (e.g., installation, auxiliary, piping, maintenance, etc.) can easily be determined through the equipment-factor method. In addition to using price data from similar projects and vendors, the latest equipment costs can be determined using the Peter and Thamaures methods, which consider the latest cost values based on the CEPCI values for a given month. The variable cost of biodiesel production includes the cost of raw materials, cost of utilities, and cost of labor, which can be determined directly based on the market situation in a given area. The cost of feedstock represents the highest share of total biodiesel production costs and largely determines the economic feasibility of the business. At the same time, the properties of the feedstock (FFA type and content, length of carbon chain, type of triglyceride) determine the quality of the biodiesel produced. This implies that a technoeconomically efficient biodiesel production business requires a technology that can effectively convert cheaper feedstock to fuel-quality biodiesel.

References

Abdullah, R.N.R.S., Ariyani, D., and Nata, I.F. (2017). Conversion of palm oil sludge to biodiesel using alum and KOH as catalysts. *Sustainable Environment Research* 27: 291–295.

Abe, M., Hirata, S., Komatsu, H. et al. (2016). Thermodynamic selection of effective additives to improve the cloud point of biodiesel fuels. *Fuel* 171: 94–100.

Acosta, P.I., Campedelli, R.R., Correa, E.L. et al. (2020). Efficient production of biodiesel by using a highly active calcium oxide prepared in presence of pectin as heterogeneous catalyst. *Fuel* 271: 117651.

Agarwal, M., Singh, K., and Chaurasia, S.P. (2010). Prediction of biodiesel properties from fatty acid composition using linear regression and ANN techniques. *Indian Chemical Engineer* 52: 347–361.

Agrawal, B.N., Sinha, S., Kuzmin, A.V., and Pinchuk, V.A. (2019). Effect of vegetable oil share on combustion characteristics and thermal efficiency of diesel engine fueled with different blends. *Thermal Science and Engineering Progress* 14: 100404.

Ambat, I., Srivastava, V., and Sillanpää, M. (2018). Recent advancement in biodiesel production methodologies using various feedstock: a review. *Renewable and Sustainable Energy Reviews* 90: 356–369.

Anwar, M., Rasul, M.G., and Ashwath, N. (2019). The efficacy of multiple-criteria design matrix for biodiesel feedstock selection. *Energy Conversion and Management* 198: 111790.

Apostolakou, A.A., Kookos, I.K., Marazioti, C., and Angelopoulos, K.C. (2009). Techno-economic analysis of a biodiesel production process from vegetable oils. *Fuel Processing Technology* 90: 1023–1031.

Arenas, E.G., Rodriguez Palacio, M.C., Juantorena, A.U. et al. (2017). Microalgae as a potential source for biodiesel production: techniques, methods, and other challenges. *International Journal of Energy Research* 41: 761–789.

Aspentech. (2018). Aspen Plus®. https://www.aspentech.com/products/engineering/aspen-plus (accessed 19 Decebmer 2020).

Atabani, A.E., Silitonga, A.S., Badruddin, I.A. et al. (2012). A comprehensive review on biodiesel as an alternative energy resource and its characteristics. *Renewable and Sustainable Energy Reviews* 16: 2070–2093.

Avhad, M.R. and Marchetti, J.M. (2015). A review on recent advancement in catalytic materials for biodiesel production. *Renewable and Sustainable Energy Reviews* 50: 696–718.

Banković-Ilić, I.B., Miladinović, M.R., Stamenković, O.S., and Veljković, V.B. (2017). Application of nano CaO–based catalysts in biodiesel synthesis. *Renewable and Sustainable Energy Reviews* 72: 746–760.

Barabás, I. and Todoruţ, I.-A. (2011). *Biodiesel Quality, Standards and Properties*. Cluj-Napoca: Technical University of Cluj-Napoca.

Batista, F.R.M., Lucchesi, K.W., Carareto, N.D.D. et al. (2018). Properties of microalgae oil from the species chlorella protothecoides and its ethylic biodiesel. *Brazilian Journal of Chemical Engineering* 35: 1383–1394.

Bazooyar, B., Ghorbani, A., and Shariati, A. (2015). Physical properties of methyl esters made from alkali-based transesterification and conventional diesel fuel. *Energy Sources, Part A: Recovery, Utilization, and Environmental Effects* 37: 468–476.

Boey, P.-L., Ganesan, S., Maniam, G.P., and Khairuddean, M. (2012). Catalysts derived from waste sources in the production of biodiesel using waste cooking oil. *Catalysis Today* 190: 117–121.

Boggavarapu, P. and Ravikrishna, R.V. (2013). A review on atomization and sprays of biofuels for IC engine applications. *International Journal of Spray and Combustion Dynamics* 5: 85–121.

Boon-anuwat, N.-n., Kiatkittipong, W., Aiouache, F., and Assabumrungrat, S. (2015). Process design of continuous biodiesel production by reactive distillation: comparison between homogeneous and heterogeneous catalysts. *Chemical Engineering and Processing: Process Intensification* 92: 33–44.

Canakci, M. and Sanli, H. (2008). Biodiesel production from various feedstocks and their effects on the fuel properties. *Journal of Industrial Microbiology and Biotechnology* 35: 431–441.

Canet, A., Bonet-Ragel, K., Dolors Benaiges, M., and Valero, F. (2016). Lipase-catalysed transesterification: viewpoint of the mechanism and influence of free fatty acids. *Biomass and Bioenergy* 85: 94–99.

Cerveró, J.M., Álvarez, J.R., and Luque, S. (2014). Novozym 435-catalyzed synthesis of fatty acid ethyl esters from soybean oil for biodiesel production. *Biomass and Bioenergy* 61: 131–137.

Chemical Engineering. (n.d.). The chemical engineering plant cost index. http://www.chemengonline.com/pci-home (accessed 19 December 2020).

Chua, S.Y., Periasamy, L.A.P., Goh, C.M.H. et al. (2020). Biodiesel synthesis using natural solid catalyst derived from biomass waste – a review. *Journal of Industrial and Engineering Chemistry* 81: 41–60.

Colombo, K., Ender, L., and Barros, A.A.C. (2017). The study of biodiesel production using CaO as a heterogeneous catalytic reaction. *Egyptian Journal of Petroleum* 26: 341–349.

Cong, W.-J., Wang, Y.-T., Hu, L. et al. (2020). Direct production of biodiesel from waste oils with a strong solid base from alkalized industrial clay ash. *Applied Energy* 264: 114735.

Corsini, A., Fanfarillo, G., Rispoli, F., and Venturini, P. (2015). Pollutant emissions in common-rail diesel engines in extraurban cycle: rapeseed oils vs diesel fuel. *Energy Procedia* 82: 141–148.

Deeba, F., Kumar, B., Arora, N. et al. (2020). Novel bio-based solid acid catalyst derived from waste yeast residue for biodiesel production. *Renewable Energy* 159: 127–139.

Degirmenbasi, N., Coskun, S., Boz, N., and Kalyon, D.M. (2015). Biodiesel synthesis from canola oil via heterogeneous catalysis using functionalized CaO nanoparticles. *Fuel* 153: 620–627.

Dincer, I. (2000). Renewable energy and sustainable development a crucial review. *Renewable and Sustainable Energy Reviews* 4: 157–175.

Dreier, D., Silveira, S., Khatiwada, D. et al. (2018). Well-to-wheel analysis of fossil energy use and greenhouse gas emissions for conventional, hybrid-electric and plug-in hybrid-electric city buses in the BRT system in Curitiba, Brazil. *Transportation Research Part D: Transport and Environment* 58: 122–138.

Dunn, R.O., Ngo, H.L., and Haas, M.J. (2015). Branched-chain fatty acid methyl esters as cold flow improvers for biodiesel. *Journal of the American Oil Chemists' Society* 92: 853–869.

Ejim, C.E., Fleck, B.A., and Amirfazli, A. (2007). Analytical study for atomization of biodiesels and their blends in a typical injector: surface tension and viscosity effects. *Fuel* 86: 1534–1544.

Fadhil, A.B. and Ali, L.H. (2013). Alkaline-catalyzed transesterification of Silurus triostegus Heckel fish oil: optimization of transesterification parameters. *Renewable Energy* 60: 481–488.

Faried, M., Samer, M., Abdelsalam, E. et al. (2017). Biodiesel production from microalgae: processes, technologies and recent advancements. *Renewable and Sustainable Energy Reviews* 79: 893–913.

Fazal, M.A., Rubaiee, S., and Al-Zahrani, A. (2019). Overview of the interactions between automotive materials and biodiesel obtained from different feedstocks. *Fuel Processing Technology* 196: 106178.

Fu, J. (2019). Flash points measurements and prediction of biofuels and biofuel blends with aromatic fluids. *Fuel* 241: 892–900.

Gebremariam, S.N. and Marchetti, J.M. (2017). Biodiesel production technologies: review. *AIMS Energy* 5 (3): 425–457.

Gebremariam, S.N. and Marchetti, J.M. (2018). Economics of biodiesel production: review. *Energy Conversion and Management* 168: 74–84.

Gebremariam, S.N. and Marchetti, J.M. (2019). Techno-economic performance of a bio-refinery for the production of fuel-grade biofuel using a green catalyst. *Biofuels, Bioproducts and Biorefining* 13: 936–949.

Giakoumis, E.G. and Sarakatsanis, C.K. (2018). Estimation of biodiesel cetane number, density, kinematic viscosity and heating values from its fatty acid weight composition. *Fuel* 222: 574–585.

Gog, A., Roman, M., Toşa, M. et al. (2012). Biodiesel production using enzymatic transesterification: current state and perspectives. *Renewable Energy* 39: 10–16.

Gupta, J., Agarwal, M., and Dalai, A.K. (2020). An overview on the recent advancements of sustainable heterogeneous catalysts and prominent continuous reactor for biodiesel production. *Journal of Industrial and Engineering Chemistry* 88: 58–77.

Hama, S. and Kondo, A. (2013). Enzymatic biodiesel production: an overview of potential feedstocks and process development. *Bioresource Technology* 135: 386–395.

Hamd, H.T. (2016). Kinetic processes simulation for production of the biodiesel using enzyme as catalyst. *Journal of Natural Science Research* 6 (3).

Hastak, M. (2015). *Skills and Knowledge of Cost Engineering*. Morgantown, WV: AACE International.

Hazar, H., Sevinc, H., and Sap, H. (2019). Performance and emission properties of preheated and blended fennel vegetable oil in a coated diesel engine. *Fuel* 254: 115677.

He, Y., Wu, T., Wang, X. et al. (2018). Cost-effective biodiesel production from wet microalgal biomass by a novel two-step enzymatic process. *Bioresource Technology* 268: 583–591.

IAE. (2018). World Energy Outlook 2018, excutive summary. International Energy Agency. https://www.iea.org/weo2018 (accessed 19 December 2020).

Intelligen. (1996). SuperPro Designer®. https://www.intelligen.com/products/superpro-overview/ (accessed 19 December 2020).

Jakeria, M.R., Fazal, M.A., and Haseeb, A.S.M.A. (2014). Influence of different factors on the stability of biodiesel: a review. *Renewable and Sustainable Energy Reviews* 30: 154–163.

Jegannathan, K.R., Eng-Seng, C., and Ravindra, P. (2011). Economic assessment of biodiesel production: comparison of alkali and biocatalyst processes. *Renewable and Sustainable Energy Reviews* 15: 745–751.

Kalam, M.A. and Masjuki, H.H. (2008). Testing palm biodiesel and NPAA additives to control NOx and CO while improving efficiency in diesel engines. *Biomass and Bioenergy* 32: 1116–1122.

Karmakar, A., Karmakar, S., and Mukherjee, S. (2010). Properties of various plants and animals feedstocks for biodiesel production. *Bioresource Technology* 101: 7201–7210.

Kim, D.-S., Hanifzadeh, M., and Kumar, A. (2018). Trend of biodiesel feedstock and its impact on biodiesel emission characteristics. *Environmental Progress & Sustainable Energy* 37: 7–19.

Knothe, G. (2005). Dependence of biodiesel fuel properties on the structure of fatty acid alkyl esters. *Fuel Processing Technology* 86: 1059–1070.

Lemmens, S. (2016). Cost engineering techniques and their applicability for cost estimation of organic Rankine cycle systems. *Energies* 9: 485.

Mofijur, M., Atabani, A.E., Masjuki, H.H. et al. (2013). A study on the effects of promising edible and non-edible biodiesel feedstocks on engine performance and emissions production: a comparative evaluation. *Renewable and Sustainable Energy Reviews* 23: 391–404.

Mohr, S.H., Wang, J., Ellem, G. et al. (2015). Projection of world fossil fuels by country. *Fuel* 141: 120–135.

Monirul, I.M., Masjuki, H.H., Kalam, M.A. et al. (2015). A comprehensive review on biodiesel cold flow properties and oxidation stability along with their improvement processes. *RSC Advances* 5: 86631–86655.

Musa, I.A. (2016). The effects of alcohol to oil molar ratios and the type of alcohol on biodiesel production using transesterification process. *Egyptian Journal of Petroleum* 25: 21–31.

Nabi, M.N., Rasul, M.G., Anwar, M., and Mullins, B.J. (2019). Energy, exergy, performance, emission and combustion characteristics of diesel engine using new series of non-edible biodiesels. *Renewable Energy* 140: 647–657.

Nakpong, P. and Wootthikanokkhan, S. (2010). High free fatty acid coconut oil as a potential feedstock for biodiesel production in Thailand. *Renewable Energy* 35: 1682–1687.

Noriega, M.A., Narváez, P.C., and Habert, A.C. (2018). Simulation and validation of biodiesel production in liquid-liquid film reactors integrated with PES hollow fibers membranes. *Fuel* 227: 367–378.

Öner, C. and Altun, Ş. (2009). Biodiesel production from inedible animal tallow and an experimental investigation of its use as alternative fuel in a direct injection diesel engine. *Applied Energy* 86: 2114–2120.

Ong, H.C., Silitonga, A.S., Masjuki, H.H. et al. (2013). Production and comparative fuel properties of biodiesel from non-edible oils: *Jatropha curcas*, *Sterculia foetida* and *Ceiba pentandra*. *Energy Conversion and Management* 73: 245–255.

Ong, H.C., Masjuki, H.H., Mahlia, T.M.I. et al. (2014). Optimization of biodiesel production and engine performance from high free fatty acid *Calophyllum inophyllum* oil in CI diesel engine. *Energy Conversion and Management* 81: 30–40.

Pala-En, N., Sattler, M., Dennis, B.H. et al. (2013). Measurement of emissions from a passenger truck fueled with biodiesel from different feedstocks. *Journal of Environmental Protection* 04: 74–82.

Petchsoongsakul, N., Ngaosuwan, K., Kiatkittipong, W. et al. (2017). Process design of biodiesel production: hybridization of ester- and transesterification in a single reactive distillation. *Energy Conversion and Management* 153: 493–503.

Peters, M.S., Timmerhaus, K.D., West, R.E. et al. (1991). *Plant Design and Economics for Chemical Engineers*. Singapore: McGraw-Hill.

Pullen, J. and Saeed, K. (2012). An overview of biodiesel oxidation stability. *Renewable and Sustainable Energy Reviews* 16: 5924–5950.

Putra, M.D., Nata, I.F., and Irawan, C. (2020). Biodiesel production from waste cooking oil using heterogeneous catalyst: biodiesel product data and its characterization. *Data in Brief* 28: 104879.

Raheem, A., Prinsen, P., Vuppaladadiyam, A.K. et al. (2018). A review on sustainable microalgae based biofuel and bioenergy production: recent developments. *Journal of Cleaner Production* 181: 42–59.

Ramírez-Verduzco, L.F., Rodríguez-Rodríguez, J.E., and del Rayo Jaramillo-Jacob, A. (2012). Predicting cetane number, kinematic viscosity, density and higher heating value of biodiesel from its fatty acid methyl ester composition. *Fuel* 91: 102–111.

Sahar, S.S., Iqbal, J., Ullah, I. et al. (2018). Biodiesel production from waste cooking oil: an efficient technique to convert waste into biodiesel. *Sustainable Cities and Society* 41: 220–226.

Saidi, M. and Moradi, P. (2020). Conversion of biodiesel synthesis waste to hydrogen in membrane reactor: theoretical study of glycerol steam reforming. *International Journal of Hydrogen Energy* 45: 8715–8726.

Sakdasri, W., Sawangkeaw, R., and Ngamprasertsith, S. (2017). An entirely renewable biofuel production from used palm oil with supercritical ethanol at low molar ratio. *Brazilian Journal of Chemical Engineering* 34: 1023–1034.

Sandoval, G., Casas-Godoy, L., Bonet-Ragel, K. et al. (2017). Enzyme-catalyzed production of biodiesel as alternative to chemical-catalyzed processes: advantages and constraints. *Current Biochemical Engineering* 4: 109–141.

Santos, S.M., Nascimento, D.C., Costa, M.C. et al. (2020). Flash point prediction: reviewing empirical models for hydrocarbons, petroleum fraction, biodiesel, and blends. *Fuel* 263: 116375.

Shabani, M.R. (2015). Suitable method for capital cost estimation in chemical process industries. *Cost Engineering* 48 (3).

Shuit, S.H. and Tan, S.H. (2019). Esterification of palm fatty acid distillate with methanol via single-step pervaporation membrane reactor: a novel biodiesel production method. *Energy Conversion and Management* 201: 112110.

Sia, C.B., Kansedo, J., Tan, Y.H., and Lee, K.T. (2020). Evaluation on biodiesel cold flow properties, oxidative stability and enhancement strategies: a review. *Biocatalysis and Agricultural Biotechnology* 24: 101514.

Sidibe, S., Blin, J., Daho, T. et al. (2020). Comparative study of three ways of using *Jatropha curcas* vegetable oil in a direct injection diesel engine. *Scientific African* 7: e00290.

Sierra-Cantor, J.F. and Guerrero-Fajardo, C.A. (2017). Methods for improving the cold flow properties of biodiesel with high saturated fatty acids content: a review. *Renewable and Sustainable Energy Reviews* 72: 774–790.

Singh, D., Sharma, D., Soni, S.L. et al. (2020). A review on feedstocks, production processes, and yield for different generations of biodiesel. *Fuel* 262: 116553.

Soares, D., Paula, A., Puna, J. et al. (2013). Effect of the oil acidity on the methanolysis performances of lime catalyst biodiesel from waste frying oils (WFO). *Fuel Processing Technology* 116: 94–100.

Tang, Z.-E., Lim, S., Pang, Y.-L. et al. (2020). Utilisation of biomass wastes based activated carbon supported heterogeneous acid catalyst for biodiesel production. *Renewable Energy* 158: 91–102.

Turton, R.C.B., Whiting, W.B., and Shaeiwitz, J.A. (2009). *Analysis Synthesis and Design of Chemical Processes*, 3e. Upper Saddle River, NJ: Prentice Hall.

USDA. (2016). EU Biofuels Annual 2016. https://gain.fas.usda.gov/Recent%20GAIN%20Publications/Biofuels%20Annual_The%20Hague_EU-28_6-29-2016.pdf (accessed 19 December 2020).

Van Kasteren, J.M.N. and Nisworo, A.P. (2007). A process model to estimate the cost of industrial scale biodiesel production from waste cooking oil by supercritical transesterification. *Resources, Conservation and Recycling* 50: 442–458.

Velickovic, A., Avramovic, J., Stamenkovic, O., and Veljkovic, V. (2016). Kinetics of the sunflower oil ethanolysis using CaO as catalyst. *Chemical Industry and Chemical Engineering Quarterly* 22: 409–418.

Wan, G., Maawa, W.N., Mamat, R. et al. (2015). Effects of biodiesel from different feedstocks on engine performance and emissions: a review. *Renewable and Sustainable Energy Reviews* 51: 585–602.

Xie, W. and Wang, H. (2020). Synthesis of heterogenized polyoxometalate-based ionic liquids with Brönsted-Lewis acid sites: a magnetically recyclable catalyst for biodiesel production from low-quality oils. *Journal of Industrial and Engineering Chemistry* 87: 162–172.

Yan, J., Zhao, Y., Li, K. et al. (2020). Efficient production of biodiesel from ionic liquid catalyzed esterification using ultrasonic-microwave combined intensification. *Chemical Engineering and Processing: Process Intensification* 149: 107870.

Yaşar, F. (2020). Comparison of fuel properties of biodiesel fuels produced from different oils to determine the most suitable feedstock type. *Fuel* 264: 116817.

Yesilyurt, M.K., Cesur, C., Aslan, V., and Yilbasi, Z. (2020). The production of biodiesel from safflower (*Carthamus tinctorius* L.) oil as a potential feedstock and its usage in compression ignition engine: a comprehensive review. *Renewable and Sustainable Energy Reviews* 119: 109574.

Zaharin, M.S.M., Abdullah, N.R., Najafi, G. et al. (2017). Effects of physicochemical properties of biodiesel fuel blends with alcohol on diesel engine performance and exhaust emissions: a review. *Renewable and Sustainable Energy Reviews* 79: 475–493.

Zhang, H., Liu, T., Zhu, Y. et al. (2020). Lipases immobilized on the modified polyporous magnetic cellulose support as an efficient and recyclable catalyst for biodiesel production from Yellow horn seed oil. *Renewable Energy* 145: 1246–1254.

Zhou, Z., Zhang, X., Yang, F., and Zhang, S. (2019). Polymeric carbon material from waste sulfuric acid of alkylation and its application in biodiesel production. *Journal of Cleaner Production* 215: 13–21.

Index

Nano- and Biocatalysts for Biodiesel Production, First Edition. Edited by Avinash P. Ingle.